Engineering Materials and Processes

Series Editor

Professor Brian Derby, Professor of Materials Science
Manchester Materials Science Centre, Grosvenor Street, Manchester, M1 7HS, UK

Other titles published in this series

Prakash M. Dixit • Uday S. Dixit

Modeling of Metal Forming and Machining Processes

by Finite Element and Soft Computing Methods

 Springer

Professor Prakash M. Dixit
Department of Mechanical Engineering
Indian Institute of Technology Kanpur
Kanpur 208016
India

Professor Uday S. Dixit
Department of Mechanical Engineering
Indian Institute of Technology Guwahati
Guwahati 781039
India

ISBN 978-1-84800-188-6   e-ISBN 978-1-84800-189-3

DOI 10.1007/978-1-84800-189-3

Engineering Materials and Processes ISSN 1619-0181

British Library Cataloguing in Publication Data
Dixit, Prakash Mahadeo
  Modeling of metal forming and machining processes : by
  finite element and soft computing methods. - (Engineering
  materials and processes)
  1. Metal-work - Mathematical models 2. Machining -
  Mathematical models 3. Deformations (Mechanics) -
  Mathematical models 4. Elastoplasticity 5. Finite
  element method 6. Soft computing
  I. Title II. Dixit, Uday S.
  671.3'015118
ISBN-13: 9781848001886

Library of Congress Control Number: 2008926767

Cover design: eStudio Calamar S.L., Girona, Spain

Printed on acid-free paper

9 8 7 6 5 4 3 2 1

springer.com

Dedicated to the loving memory of my parents:
Parvati M. Dixit and Mahadeo V. Dixit
-Prakash M. Dixit

Dedicated to my teachers
-Uday S. Dixit

Preface

The use of computational techniques is increasing day by day in the manufacturing sector. Process modeling and optimization with the help of computers can reduce expensive and time consuming experiments for manufacturing good quality products. Metal forming and machining are two prominent manufacturing processes. Both of these processes involve large deformation of elasto-plastic materials due to applied loads. In metal forming, the material is plastically deformed without causing fracture. On the other hand, in machining, the material is deformed till fracture, in order to remove material in the form of chips. To understand the physics of metal forming and machining processes, one needs to understand the kinematics of large deformation (dependence of deformation and its rate on displacement) as well as the constitutive behavior of elasto-plastic materials (dependence of internal forces on deformation and its rate). Once the physics is understood, these phenomena have to be converted to mathematical relations in the form of differential equations. The interaction of the work-piece with the tools/dies and other surroundings also needs to be expressed in a mathematical form (known as the boundary and initial conditions).

In this book, the first four chapters essentially discuss the physics of metal forming and machining processes. The physical behavior of the work-piece during the processes is modeled in the form of differential equations and boundary and initial conditions. One of the well-known mathematical techniques to solve differential equations and boundary and initial conditions is the finite element method. Chapters 5–7 describe the finite element formulations of metal forming processes using Eulerian and updated Lagrangian approaches and that of machining process using an Eulerian approach. Instead of physics-based modeling, the metal forming and machining processes can also be modeled by another approach using only empirical data and soft computing techniques. Chapter 8 introduces some soft computing techniques like neural networks, fuzzy set theory and genetic algorithms. Chapter 9 discusses the application of the soft computing techniques to metal forming and machining processes. Chapter 10 deals with optimization of metal forming and machining processes. Chapter 11 concludes the book. We feel that the physics-based finite element modeling and soft computing-based modeling are complementary to each other. However, readers interested only in finite element modeling may go through Chapters 1–7. Similarly, the readers interested only in soft computing based modeling may read only Chapter 1 and Chapters 8–10.

This book is essentially for graduate students and researchers in the field of computational manufacturing. Some background in the areas of solid mechanics, finite element method and soft computing is desirable. For the benefit of readers, a brief review of these subjects is provided in Chapters 2, 5 and 8. The book can also be used as a textbook for a three-semester graduate level course (which can also be taken by senior undergraduate students) on modeling of metal forming and machining processes: the first course on theory of plasticity covering the first four chapters, the second course on finite element modeling of metal forming and machining processes covering Chapters 5–7 and the third course on soft computing modeling of metal forming and machining processes covering Chapters 8–10.

The major objective of this book is to stimulate the interest of readers in the area of computational manufacturing. We expect the book to be used as a source of direction rather than information. In order to provide an optimized treatment of the subject, we had to make quite a few simplifying assumptions. Although we have taken the utmost care to avoid errors, we would welcome details of errors and/or suggestions (preferably by e-mail) for improving future editions of the book. A number of books and papers have been consulted while preparing the draft of the book. A list of references has been provided at the end of each chapter. There may be some important works that may have been unintentionally omitted. We request the readers to bring any omissions to our notice.

The authors of this book have a long association with each other, since the second author (USD) came as a graduate student to the Indian Institute of Technology (IIT) Kanpur in 1991. The two authors have worked together in the area of finite element modeling for several years. In 1998, the second author shifted to IIT Guwahati, about 1500 km from Kanpur, as a faculty member. However, e-mails and inexpensive telephonic communications compensated for the geographical distance. Both authors have been teaching the courses on Solid Mechanics, Plasticity, Metal Forming and Machining at IIT Kanpur and IIT Guwahati for the past several years. Further, they have supervised several masters and doctoral students in the area of finite element and soft computing applications to metal forming and machining processes and other plasticity problems. Interaction with these students (whom they taught as well as supervised) has certainly helped in preparing the draft of this book. The authors thank all these students. The complete list is long. However, the first author (PMD) cannot avoid mentioning the names of the following: his past Ph.D. students—Sankar Dhar, N. Venkata Reddy (now colleague at IIT Kanpur), Uday S. Dixit (the second author of the book), his present Ph.D. students—Ravindra K. Saxena, Anupam Agrawal and Sachin S. Gautam and his past M.Tech. student—S.N. Vardhan. The first author (PMD) would also like to thank his family for providing the moral support while writing the book: his wife Rekha, his daughter Rashmi, his and his wife's brothers and sisters-in-law, his and his wife's sisters and brothers-in-law and the son-in-law.

We thank C. Venu Madhav and Sharad Tiwari, past M.Tech. students of the Department of Mechanical Engineering at IIT Kanpur, for preparing some figures in the first draft of the book. We acknowledge the help of Mr. P.P. Gudur,

Research Scholar in the Department of Mechanical Engineering at IIT Guwahati, for his help in drawing the figures and formatting the chapters. We acknowledge the cooperation offered by the staff of Springer Verlag during the planning and production of the book. Our special thanks are to Mr. Anthony Doyle, Senior Editor Engineering at Springer London for his help. We also appreciate the help of Mr. Simon Rees of Springer London in going through the proofs and offering valuable suggestions.

Prakash M. Dixit
pmd@iitk.ac.in

Uday S. Dixit
uday@iitg.ernet.in

Contents

1

Metal Forming and Machining Processes

1.1 Introduction

Two prominent methods of converting raw material into a product have been metal forming and machining. Metal forming involves changing the shape of the material by permanent plastic deformation. After converting non-porous metal into product form by metal forming processes, the mass as well as the volume remains unchanged. However, in the case of metal forming of porous metal, volume does not remain unchanged. The advantages of metal forming processes include no wastage of the raw material, better mechanical properties of the product and faster production rate. Machining is the process of removing the material in the form of chips by means of a wedge shaped tool. In ductile materials, a significant amount of plastic deformation occurs before the material fractures. In brittle materials, very little plastic deformation takes place. Hence, the mechanics of machining is quite different for ductile and brittle materials. In machining, the work-piece is subjected to shear, bending and compression by the tool. Combined loading effects as well as heat generation due to plastic deformation and friction influences the chip formation. Removal of metal in the form of chips causes wastage of the material, but machining can achieve good surface finish and dimensional accuracy. As a result of machining, the material properties are altered only at the surface or just below it. Even complex shapes can be produced with economy, thanks to computer numerically controlled (CNC) machines.

The need to manufacture high precision items and to machine difficult-to-cut materials led to the development of the newer machining processes. These are called non-traditional or non-conventional machining processes, notwithstanding that the definition of conventional and traditional changes with time. Unlike conventional machining processes, non-conventional machining processes are not based on the removing the metal in the form of chips using a wedge shaped tool. There are a variety of ways by which the material may be removed in non-conventional machining processes. Some of them are abrasion by abrasive particles, impact of water, thermal action, chemical action *etc*.

This book focuses on modeling of metal forming and machining processes. Hence, in this chapter, we will introduce some of the basic metal forming and machining processes. The primary aim is to introduce the process and performance parameters of interest in various processes. The need to model these processes and the difficulties in the modeling will be introduced. This chapter will be useful for understanding the terminology one frequently encounters in the modeling of metal forming and machining processes.

1.2 Metal Forming

Metal forming is the process of plastically deforming the raw material into product form. It is broadly classified into two classes—bulk metal forming and sheet metal forming. In the bulk metal forming processes, usually the work-piece has a high volume to surface area ratio. Examples of such processes are rolling, wire drawing, extrusion, forging *etc*. In the sheet metal forming processes, usually the work-piece sheet has a low volume to surface area ratio. The sheets usually have a thickness less than 6 mm. In sheet metal working, the change in thickness during plastic deformation is not desirable. Examples of sheet metal forming processes are deep drawing, stretch forming, bending, spinning *etc*. In the following subsection we describe various bulk metal forming processes.

1.2.1 Bulk Metal Forming

Bulk metal forming processes are characterized by high volume to surface area ratio. These processes can be carried out in hot or cold conditions. If the temperature during processing is more than the recrystallization temperature of the metal, the process is called hot working. Recrystallization temperature is the temperature above which new equiaxed and strain-free grains are formed replacing the old grains. For most metals, recrystallization homologous temperature ranges approximately between 0.3 and 0.5, where homologous temperature is the absolute working temperature (in Kelvin) divided by the absolute melting temperature of the work-piece. However, in hot working, the homologus temperature is generally more than 0.6. If the temperature is less than the recrystallization temperature, the process is called cold working. In between cold working and hot working falls the warm working. In warm working, the heating of the work-piece reduces the flow stress; however the temperature is not high enough to cause recrystallization. The relative ease with which metal can be shaped through plastic deformation is called workability. It is dependent on strain, strain rate, temperature and inherent flow characteristics of the material. Some typical bulk metal forming processes are described here.

1.2.1.1 Forging
Forging is the process of plastically deforming metal by pressing or hammering. It is perhaps the oldest metal forming process. Forging may be performed in cold, warm or hot state of the metal. There are mainly two types of forging processes: open die forging and closed die forging. Open die forging is carried out between flat dies or dies of simple shape. In this process, on certain surfaces, material flows in an unconstrained manner. One example of open die forging operation is the

upsetting of a cylindrical work-piece between two flat dies, as shown in Figure 1.1. In this, the work-piece is kept on a fixed platen and the top surface is pressed by a moving platen. Due to friction between the work-piece and platens, the material faces a restraint in its flow at the top and bottom surfaces, whilst the middle portion flows freely. Because of this, the work-piece adopts a barreled shape. The amount of bulging may be used as an indirect measurement of friction at the tool-job interface [1]. Open die forging is often employed to pre-form material for subsequent metal forming processes.

In closed die forging, also called impression die forging, the work-piece is deformed between two die halves, which carry the impression of the desired final shape. The hammering or pressing causes the metal to flow so as to fill completely the die cavity. Excess metal is squeezed out around the periphery of the die cavity to form a flash. In the die design, the design of a proper flash gap is very crucial. Besides providing an outlet for excess metal, it helps in proper filling of the die cavity. The flash is trimmed off after the forging operation is complete. This causes a significant amount (of the order of 20%) of wastage of material. Figure 1.2 illustrates a closed die forging process schematically. In Figure 1.2b, the dies have reached the final position and the deformation is complete. Note the formation of a thin ribbon of excess metal called flash. This flash can be removed with a trimming die. In flashless forging, the flash is not produced. However, the design of a flashless process is difficult. In this process, the work-piece should be of proper size. Also, the design of work-piece and die is very important.

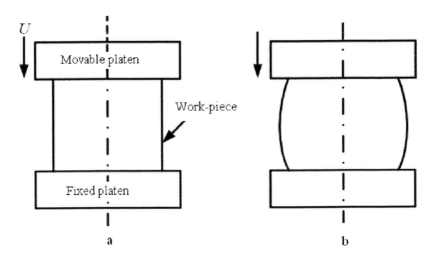

Figure 1.1. Open die forging process. **a** Before deformation. **b** After deformation

The strain rates in a forging operation can be of the order of 10^{-3} to 10^{2} per second. The coefficient of friction ranges typically from 0.05 to 0.15 for cold forging and 0.1 to 0.5 for hot forging. However, Coulomb's law may not be applicable. At various locations at the die-work interface, the sticking may take place, *i.e.*, the work-piece may not move relative to tool. Lubrication plays a very important role in forging. Besides affecting friction and die wear, it also acts as a thermal barrier between die and job. For hot forging, graphite, molybdenum disulfide and glass are

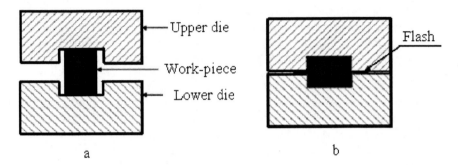

Figure 1.2. Closed die forging process. **a** Before deformation. **b** After deformation

used as lubricants. For cold forging, mineral oils, synthetic oils or soap solutions can be used.

Besides determining the loads to carry out forging operation, the determination of stress, strain, strain rate and temperature distribution in the material during processing as well as the determination of residual stresses after processing is very important for producing a good quality product. Typical forging defects are flash cracking extending up to the main product during trimming, surface cracking due to tensile stress generated and ductile fracture. Ductile fracture occurs due to micro-void nucleation, growth and coalescence into micro-cracks. There are many criteria for describing the process of void-nucleation. These are based on critical stress, critical strain or critical strain energy in and around an inclusion. After nucleation, the void grows to a characteristic volume and shape depending on the material properties and process conditions. Stress triaxiality, *i.e.*, the ratio of hydrostatic stress to equivalent stress, greatly influences void growth. When voids start coalescing, fracture occurs. Thus, the knowledge of stress-strain distribution along with microscopic details of the material is important for predicting the onset of ductile fracture. Free surface folding, also called laps, is another defect occurring in the forging process due to inappropriate design of die or initial billet. To predict this defect, metal flow behavior has to be understood.

Three-dimensional analysis of the forging process can bring out the detailed information about the process. However, sometimes the process may be modeled as a plane strain or an axisymmetric process. In a plane strain process, deformation is essentially two-dimensional. A simple slab method analysis shows that while compressing the plate in plane strain, the forging pressure decreases from the center towards the end [2]. This type of pressure distribution is called friction-hill. In the absence of friction, there will be no friction-hill and the pressure will be uniform. In the presence of friction, the resistance to deformation increases as the ratio of contact length to thickness increases [2]. Thus, in closed-die forging, the deformation resistance of the flash is very high because of its high length to thickness ratio. Therefore, the pressure in the die becomes high enough to ensure complete filling of the die cavity.

Axisymmetric problems are problems with radial symmetry, such as compression of solid or hollow disks. As in the case of plane strain, in the case of compression of a solid disk, the friction-hill behavior is present with a pressure peak at the center. The role of friction increases for large radius to thickness ratio. In forging problems, the excessive pressure may create sticking friction, where there is no relative motion between the work-piece and the dies. Actually, the frictional behavior of the surface is very complex. The coefficient of friction may vary along the surface. Study of friction, lubrication and wear forms the science of the tribology. In metal forming, the friction at the die and work-piece interface is usually found by a ring compression test, which is basically the open die forging of a hollow disk. The test uses a ring, usually with height not exceeding one-third of the outer diameter and inner diameter about 50% of the outer diameter. The friction is determined by measuring the percentage of change in the inside diameter when the disk is compressed. For low friction, the internal diameter increases as the disk is compressed. For high friction, the internal diameter of the ring reduces as the disk is compressed. Calibration curves are available to show the percentage change in inside diameter with the percentage reduction in height for different coefficients of friction. With the help of these curves, one can find the coefficient of friction by conducting a ring compression test. This test was originated by Kunogi [3] and improved by Male and Cockcroft [4]. Avitzur [5] and Hawkyard and Johnson [6] carried out theoretical study of ring-compression. The method needs only the measurement of dimensions and does not require the measurement of compression load. It is generally assumed that calibration curves do not depend on material properties. However, recently Sofuoglu *et al.* [7] carried out a series of ring compression tests on different materials to investigate the effects of material properties, strain rate sensitivity and barreling on the behavior of friction calibration curves. They also carried out simulations using an elastic-plastic finite element code. The results of the experiments and finite element analysis indicated that material properties, strain rate and barreling do influence the friction calibration curves. Thus, although the ring compression test is an effective method for determining friction, its accuracy will be enhanced when material behavior and test conditions are taken into account. An accurate modeling of the ring compression process will help in the reduction of experiments for generating friction calibration curves.

In three-dimensional open die forging of a block of rectangular cross-section, dimensions change along the thickness and length as well as width. When the thickness is reduced by compression, increase in length is called "elongation" and increase in width is called "spread". If a solid work-piece of initial thickness t_i, width w_i and length l_i changes to final thickness t_f, width w_f and length l_f, then a coefficient of spread can be defined as [8]

$$S = \frac{\ln(w_f / w_i)}{\ln(t_i / t_f)}. \tag{1.1}$$

Due to volume constancy,

$$l_i w_i t_i = l_f w_f t_f .$$
(1.2)

Taking logarithms of both sides and transferring the terms of the right side to left side, we get

$$\ln(l_f / l_i) + \ln(w_f / w_i) + \ln(t_f / t_i) = 0 .$$
(1.3)

Expressing the middle term in the above equation in terms of the other two terms and substituting it in Equation 1.1, the coefficient of spread can also be expressed as

$$S = \frac{-\ln(t_f / t_i) - \ln(l_f / l_i)}{\ln(t_i / t_f)} = 1 - \frac{\ln(l_f / l_i)}{\ln(t_i / t_f)} .$$
(1.4)

The term $(1-S)$ is called coefficient of elongation in the literature.

Apart from its application in producing parts by metal forming, open die forging is often used for finding material behavior. Maximum strain in a tensile test is much smaller than that encountered in a metal working application. Therefore, compression of a short cylinder between anvils is used for measuring the flow stress variation with reduction. In the test, the friction is minimized by using smooth and hardened compression platens along with a good lubricant. The ends of the specimen are grooved in order to retain the lubricant. Sometimes the test is carried out in increments, so that the lubricants can be replaced at intervals. Another way to eliminate the effect of friction in the flow stress vs reduction curve is to plot the load-deformation curves for a number of different specimens with different diameter to height ratios and extrapolating the results to the case when diameter to height ratio becomes zero [2]. This is because when the diameter to height ratio is zero, friction is not having any influence on the process.

1.2.1.2 Rolling

Rolling is a process of metal forming in which raw material is shaped by passing it between two counter-rotating cylinders. The process can be used for reducing the thickness of slab, plate or sheet. It can also be used to produce products of different cross-sections. Both hot and cold rolling can be performed on a mill with one stand or several stands, the latter being called a tandem mill. Figure 1.3 shows a tandem mill with three stands. In this figure, each stand has two work-rolls, which are in contact with the material being rolled. On the other side, the work rolls are in contact with the backup rolls. Backup rolls are used to minimize the deflection of the work rolls.

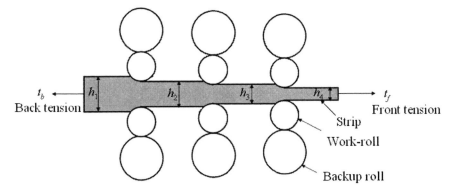

Figure 1.3. A four-high, three-stand tandem cold-rolling mill

Thick ingots are hot rolled into blooms, billets or slabs. A bloom has a square or rectangular cross-section with a thickness greater than 150 mm and width less than twice the thickness. A further reduction by hot-rolling results in a billet. The billet has a square or circular cross-section. A slab refers to a hot rolled ingot having width greater than twice the thickness. Slabs can be further rolled into plate, sheet and strip. Generally, plate has a thickness greater than 6 mm, whilst the thickness of sheet and strip is less than 6 mm. The difference between sheet and strip is that the former has a greater width to thickness ratio. Plates, sheets and strips are cold rolled to thinner sheets. This process is called cold flat rolling.

In cold flat rolling, the roll radius is usually more than 50 times the initial strip thickness. If the width of the strip is at least five times the length of the arc of contact, the material in the elastic zone prevents the lateral spread and deformation takes place effectively under plane strain conditions. As the strip enters the roll gap, it is first deformed elastically and is pulled by the frictional forces between roll and strip. Its thickness keeps on reducing and speed keeps on increasing. After certain distance of travel in the roll gap, a point comes where the speed of strip becomes equal to the peripheral speed of the roll. This point is the point of "no slip" and is called the neutral point. After that the strip speed keeps on increasing but the direction of frictional forces changes. Once the strip comes out of the roll, it is unloaded elastically. Some residual stresses get set up in the product.

During the rolling process, only a small fraction of the roll periphery is in contact with the strip. This produces significant amounts of contact stresses in the roll and causes the roll to flatten. For relatively softer strips and large draft (difference of inlet and exit thicknesses), the deformation in the roll is small and the deformed surface is circular with increased radius of curvature. Hitchcock [9] has provided a formula to compute the radius of curvature of the deformed surface. It is based on the assumption that the roll pressure distribution is elliptical. According to this formula, the ratio of the radius of the deformed arc of contact R' to roll radius R is given by

$$\frac{R'}{R} = \left(1 + \frac{F_r}{C\delta}\right), \tag{1.5}$$

where F_r is the roll force per unit width of the rolled strip and $\delta = h_1 - h_2$ is the draft. The constant C depends on the material of the rolls and is given by

$$C = \frac{\pi E_r}{16\left(1 - v_r^2\right)}, \tag{1.6}$$

where E_r is the Young's modulus of elasticity and v_r is the Poisson's ratio of the material of the roll. For getting less roll deflection, the material of the roll should have high E_r. However, this is not the only consideration in choosing a roll material. The work roll material should have good thermal, mechanical and tribological properties. Common roll materials are alloy steel and high chromium cast iron. The roll force itself is dependent on the radius of deformed arc of contact; it increases with increasing roll flattening. Thus, the deformed roll radius has to be found iteratively using Equation 1.5. As is clear from Equation 1.5, roll flattening increases with reduction in the value of draft.

Besides getting flattened, rolls also get deflected due to bending and shear. Rolls supported on bearings may be regarded as beams. If the strip is passed between the rolls symmetrically between two bearings, the maximum roll deflection will be encountered in the middle of two bearings. Therefore, after exit from the roll gap, the strip will be thicker in the middle with gradually decreasing thickness towards the end. Thus, the strip will adopt a convex profile. To compensate for this effect, rolls are ground so that their diameters at the center are slightly more than the edges and keep on decreasing towards edges. Thus, the roll adopts a convex profile, which is called camber. With proper camber during the rolling process, the rolls get deflected in a manner such that a uniform gap exists between the rolls across their widths and a flat sheet is produced. However, roll deflection is dependent on the process parameters. Thus, rolls with a particular camber are suitable for operation at particular process parameters. For precise control of flatness at various process parameters, the best method is to apply the load through backup rolls to counteract the roll pressure.

Besides flattening and deflection of rolls, temperature also influences the work roll profile. The part of the roll in contact with the work-piece is hotter than the part outside the width of the work-piece. This causes differential radial expansion across the length of the roll and the roll profile gets distorted. The difference in roll diameter across the length of the roll due to temperature is called thermal camber. It can be controlled by providing a proper arrangement for cooling of the rolls. For this purpose, modeling of heat transfer phenomena plays an important role.

Apart from the above roll deformation, additional roll displacement takes place due to elastic deformation of the structure under heavy loads acting on it. The combined effect causes the actual roll gap to be increased. The ratio of roll force to increase in roll gap is called mill modulus. Its value can be determined experimentally. Once the mill modulus is known, the roll gap can be set properly to obtain accurate final thickness. The following example illustrates the procedure of obtaining the mill modulus and using it to obtain the actual roll gap. In the example, the mill modulus of a laboratory mill is calculated [10].

Example 1.1: Aluminum strips of width 38 mm and different thickness values were cold rolled in a laboratory rolling mill. First, material testing in a universal testing machine was carried out and the following equation for the flow stress was obtained:

$$\sigma = 67\left(1 + \frac{\varepsilon}{0.0016}\right)^{0.121}.$$

The data obtained in the rolling mill is given in Table 1.1. The work-roll diameter is 200 mm.

Table 1.1. Data obtained in the laboratory rolling mill

Experiment number	Thickness of strip (mm)	Roll gap setting (mm)	Size of strip after rolling (mm)
1	3	1.7	2.2
2	3	1.4	2.0
3	2.3	1.4	1.8
4	3	1.95	2.38
5	3	2.10	2.5
6	3	2.25	2.58
7	3	2.40	2.7

From this data, determine the mill modulus of the rolling mill and develop an equation to set the roll gap for obtaining a prescribed thickness after rolling.

Solution: The mill modulus of a rolling mill can be found from the relation

$$G_a = G_0 + \frac{F_r w}{M}, \tag{1.7}$$

where G_a is the actual roll gap in mm during the rolling (equal to final strip thickness neglecting elastic recovery), G_0 is the roll gap setting in mm, done before the rolling of the strip, w is the width of the strip in mm, F_r is the roll force per unit width in MN/mm and M is the mill modulus in MN/mm. The total load $F_r w$ may be calculated from a rigid-plastic finite element method (FEM) code. As there are some inherent assumptions in the calculation of the load, it is better to replace $F_r w$ by $a + bF$, where F is the total load in MN computed by FEM code. Thus, Equation 1.7 may be written as

$$G_a = G_0 + \frac{(a + bF)}{M}, \tag{1.8}$$

or

$$G_a - G_0 = A + BF ,$$ (1.9)

where A is defined as a/M and B is defined as b/M. This equation provides a linear relationship between the $(G_a - G_0)$ and F. In the range of Coulomb coefficient of friction between 0.1 to 0.2 (usual range for dry rolling), the friction coefficient 0.16 provided the best linear relation. For this coefficient of friction, the computed forces are shown in Table 1.2.

Table 1.2. Forces computed by plane strain FEM code

Experiment number	F (MN)
1	0.05984
2	0.07277
3	0.04815
4	0.04901
5	0.04669
6	0.04182
7	0.03270

Calculating the constants A and B with the data of first two experiments, we get, A=0.0372 and B=7.73395. With these, the gap is calculated for the other experiments and it is given in Table 1.3. It is seen that these values are very close to the exit thicknesses obtained.

Table 1.3. Calculated actual roll gap

Experiment number	G_a (mm)
3	1.81
4	2.37
5	2.50
6	2.61
7	2.69

It is to be noted that, although this method provides an accurate estimate of the exit thickness, the actual roll forces need not be same as shown in Table 1.2. Mainly, the variation of friction across the roll-strip contact and exact values are not known. However, if we assume that the values of the roll force calculated by FEM are sufficiently accurate, *i.e.*, $a = 0$ and $b = 1$, then $B = 1/M$, which provides $M = 0.129$ MN/m. This may be treated as an approximate estimate of mill modulus.

One interesting feature of the rolling process is that in this process, the presence of friction is a must, notwithstanding that it also causes wastage of energy by generating heat. If the strip has to be drawn by the rolls themselves, the minimum coefficient of friction should be

$$\mu_{min} = \sqrt{\frac{h_1 r}{R}} \, , \tag{1.10}$$

where r is the fractional reduction given by

$$r = \frac{h_1 - h_2}{h_1} . \tag{1.11}$$

Once the rolling process has sustained, the minimum coefficient of friction is given by [11]

$$\mu_{min} \approx \frac{1}{2}\sqrt{\frac{h_1 r}{2R}} . \tag{1.12}$$

This equation has been derived by considering the equilibrium of forces acting on the strip and taking the roll diameter as twice the actual roll diameter to take into account the roll flattening effect. If the coefficient of friction is less than the value given by Equation 1.12, the rolls will skid over the strip surface and the strip will not be drawn.

The common defects in the sheet rolling are alligatoring, central burst, tearing and buckling of the strip. Alligatoring is also called split end defect. Split end defect initiates as a crack, forming along the central plane of the deformed material. As the rolling proceeds, the two halves of the material separate from each other and split end defect (alligatoring) occurs. The central burst defect is caused by internal void formation. This defect is promoted by small roll radius, large initial thickness of the sheet, small percentage reduction and front/back tension [12]. More or less the same conditions promote alligatoring [13]. If the front and back tensions are more than about one-third of the yield stress, the strip may tear [14]. On the other hand, if the tension is less than the maximum longitudinal compressive residual stress, the strip may buckle.

The rolling process has many variants. Sheet or strip rolling is the basic process. Foil rolling is the process of rolling of sheets less than about 0.2 mm. The thickness of the rolled foil may be as low as 0.01 mm. Under these circumstances, there is no definite neutral point. Instead a zone exists in which there is no slip between the strip and the rolls. The thickness of the strip in this no–slip zone is effectively uniform. Roll deformation in such cases cannot be calculated by Hitchcock's formula.

Temper rolling or skin-pass rolling is the process of giving very light reductions (0.5–4%) to sheet. The process is used to improve surface finish, impart a degree of hardness, improve flatness and eliminate stretcher strains (Lueder's bands). A brief description of Lueder's bands is provided in Section 1.2.2.1. Another type of rolling process is asymmetric rolling, which is often undesirable and can occur if the friction conditions are different at the upper and lower rolls. However, sometimes it is employed for obtaining the desired curvature in the sheet. Rod rolling is used to produce rods of different diameters either from a round or a rectangular cross-section. Pack rolling is a process in which two or more layers of metal are rolled together for improving the productivity. For producing various structural shapes, like channel, T and L sections, shape rolling is used.

1.2.1.3 Wire Drawing

Wire drawing is a process of pulling wires through tapered dies resulting in the reduction of its cross-section and increase in its length. The diameter of a wire may range from 0.025 mm to 15 mm, with research going on for producing the wires of even smaller diameter. A similar process is rod drawing in which, instead of wires, rods are pulled through dies for reducing the cross-sectional area. The drawing operation is accomplished with or without a back tension applied to the wire at its entrance. The back tension increases the requirement of drawing force, but lowers the die pressure. The normal force per unit area exerted by the wire on the die is called the die pressure. Reduction in die pressure increases the die-life. The back tension also tries to keep the input work-piece straight. Drawing is normally a cold working process in which good tolerances and surface finish can be obtained. Usually, the die angles vary from 4° to 30°. The reduction in wire drawing is defined as

$$r = \frac{A_i - A_f}{A_i},$$

(1.13)

where A_i is the initial cross-sectional area and A_f is the final cross-section area. Larger reduction may be obtained by passing the wire through a series of dies, the maximum reduction in one pass being limited to about 0.45. Sometimes intermediate annealing may also be required between two passes of wire drawing. For most wire drawing operations, the drawing speed (the speed of the wire at the exit of the die) ranges from 10 m/min to 3000 m/min.

A very important component of wire drawing process is the die. Dies are made of hardened steel, tungsten carbide or diamond. The die pressure is a major factor in deciding the selection of die material. Generally, a die has the shape of a truncated cone followed by a cylindrical zone. Sufficient relief is provided at entry and exit. Figure 1.4 shows the schematic diagram of a wire drawing process through a conical die. In the figure, α is the semi-die angle, E_z is the entry zone, L_c is the conical zone, L_b is the straight bearing length and D_r is the die relief. Dies other than simple conical dies are also very common.

Lubrication is employed in the process to reduce the wear of the die, drawing force and interface temperature. Wet and dry methods are two methods for applying the lubricants. In wet drawing, the entire drawing zone is submerged in a liquid bath. In dry drawing, the soaps placed ahead of the die are employed as lubricants. The average coefficient of friction ranges from 0.01 to 0.10 in a practical wire drawing operation. Though wire drawing is a cold working process, sometimes the temperature rise in the process is very significant, requiring the incorporation of thermal effects in the process modeling.

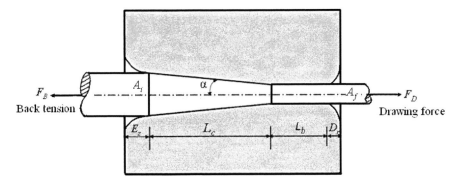

Figure 1.4. A wire drawing process

For a successful design of drawing equipment, sufficiently accurate estimation of drawing force, maximum die pressure and interfacial stresses *etc.* is required. For improving the quality of product, the knowledge of stress-strain distribution in the material during the process is very useful. Common defects in the wire drawing process are central bursting, surface cracking and residual stresses. Central bursting defects occur at large cone angles and small reduction, when the plastic deformation is mostly confined away from the central portion. In such a case, the fracture may take place at the center portion leading to chevron cracking. Many times chevron cracks may remain undetected and may cause failure of the wire during its service period. Surface cracking may occur due to improper selection of process parameters and insufficient lubrication. Compressive surface residual stresses are beneficial from the point of view of improving fatigue life; however tensile surface residual stresses are undesirable. Sometimes residual stresses lead to stress-corrosion cracking and warping of the wire/rod during the subsequent machining process.

A variant of the wire drawing process is tube drawing. In this process, the thickness and/or diameter of the tube are reduced by pulling through a die. In the process, conventionally called sinking, the tube is drawn through a die without supporting the inside surface of the tube. As the inside surface of the tube is not supported in tube sinking, wall thickness and internal surface becomes uneven. For controlling the thickness, the plug or mandrel can be used as shown in Figure 1.5.

A relatively new process in wire/rod/tube drawing is dieless drawing [15, 16]. In this process the cross-section of wire, rod or tube is reduced by pulling it without a die. The zone in which the reduction in cross-section takes place is located between a heating and a cooling zone. The drawing force is created by making use of two pairs of rollers rotating at different speeds. The slow rotating pair is located on the entry side and the faster one towards exit side. The main advantage of the process is that friction and the need for lubricants are eliminated.

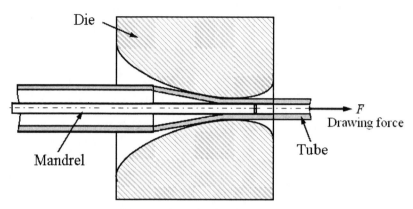

Figure 1.5. A tube drawing process

1.2.1.4 Extrusion

Extrusion is the process in which the work-piece (billet) is placed in a chamber and is forced through an opening by means of a moving ram. Figure 1.6 shows a direct extrusion process through a square die, in which the work-piece and ram move in the same direction. There is a lot of friction force generated at the work-piece and container interface. To reduce the friction, vegetable and petroleum oils are generally used as lubricant. Extrusion can be performed in the hot or cold state of the material. In hot extrusion, a dummy block is kept between the ram and the material. The purpose of the dummy block is to avoid sticking of the ram with the work material. The material flow in the process is shown by arrows. In the square die, a portion of the material near the entry to the wall does not move. This is called the dead metal zone.

In indirect extrusion, shown in Figure 1.7, instead of the movement of the work-piece, the die moves. Unlike direct extrusion, in this process ram and extruded product move in different directions. This causes less friction compared to direct extrusion. There is no friction between the container walls and the work-piece. Friction is present at the interface of die and work-piece, which is a small region. In the hydrostatic extrusion, shown in Figure 1.8, the chamber is filled with hydraulic fluid. There is no friction between the work-piece and container. The impact extrusion (Figure 1.9) is the process of producing hollow products, such as toothpaste tubes. As the name implies, in this process the ram makes an impact on the work-piece.

The extrusion ratio R is the ratio of initial cross-sectional area of the billet, A_i to the final cross-section area after extrusion, A_f. The extrusion ratio can also be expressed in terms of the fractional reduction as follows:

$$R = \frac{A_i}{A_f} = \frac{1}{1 - \dfrac{(A_i - A_f)}{A_i}} = \frac{1}{1-r}. \qquad (1.14)$$

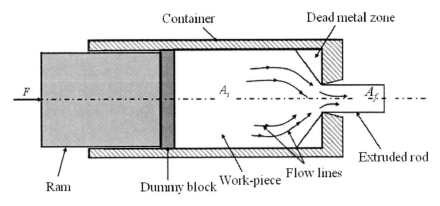

Figure 1.6. Direct extrusion process

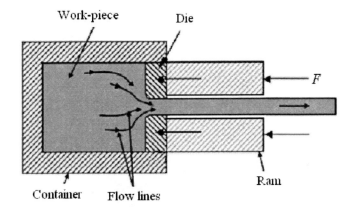

Figure 1.7. Indirect extrusion process

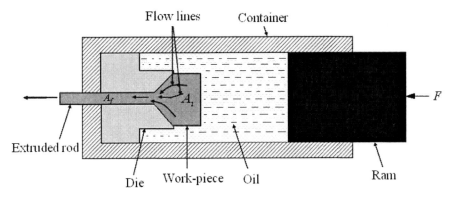

Figure 1.8. Hydrostatic extrusion process

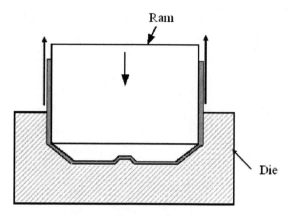

Figure 1.9. Impact extrusion process

The extrusion ratio reaches about 40:1 for hot extrusion of steel and may be as high as 400:1 for aluminum. The extrusion ratio provides a better physical feel compared to fractional reduction, as is clear from the following example:

Example 1.2: Find the extrusion ratios corresponding to the fractional reduction of 0.9 and 0.95.

Solution: For $r = 0.9$, using Equation 1.14,

$$R = \frac{1}{1-0.9} = 10 .$$

Similarly, for $r = 0.95$,

$$R = \frac{1}{1-0.95} = 20 .$$

We see that even though there is only a slight change in the fractional reduction from 0.9 to 0.95, the extrusion ratio doubled. If the final blank area is same, in the first case the initial area is 10 times, while in the second case it is 20 times.

Similar to other bulk metal forming processes, in this process a number of defects might also occur. Prominent defects are surface cracking, pipe and central burst. Surface cracking occurs due to longitudinal tensile stresses. This defect is promoted by high temperature, high speed and friction. Surface defects occurring due to high speed is known as speed cracking. These defects may occur due to hot shortness. Hot shortness is due to local melting of some of the constituents along the grain boundaries. This causes weakening of that region and produces a crack during hot extrusion. A type of surface defect known as bamboo cracking may occur due to high friction. An extrusion defect (also known as extrusion defect, pipe, tailpipe, and fishtailing defect) occurs due to the typical flow pattern in extrusion [17]. The particular flow pattern tends to draw surface oxide and impurities towards the center of the billet. This defect can be reduced by modifying the flow pattern. Central burst or chevron cracking can occur at low extrusion ratios. This defect is promoted whenever the material encounters more resistance to flow at the center rather than at the surface. In this case, high frictional resistance at the die-work-piece interface produces a sound product while center burst occurs when friction is low.

Many shapes that are not possible by rolling can be produced by extrusion. Since the deformation is compressive, the amount of reduction is limited only by the capacity of the equipment. Tolerances of ±0.05 mm/mm can easily be obtained.

1.2.2 Sheet Metal Forming Processes

Sheets are produced by a rolling operation. They have high ratio of surface area to thickness. Unlike bulk metal forming, in sheet metal forming, thickness is not generally reduced. It does, however, change due to Poisson's effect. Too much decrease in the thickness may lead to necking.

The number of sheet metal forming processes is quite large. It is not possible to cover every process in this chapter. Here, we first describe the deep drawing process, which is one of the most popular and widely investigated sheet metal forming process. The description will highlight many modeling issues. Then we describe sheet/tube bending processes. Afterwards, we describe punching and blanking. They are considered sheet metal forming processes, although they are unique in the sense that material removal takes place in these processes. Finally, we describe a number of other processes very briefly.

1.2.2.1 Deep Drawing

In the deep drawing process, a flat sheet metal blank is formed into a cylindrical or box-shaped part by means of a punch, which presses the blank into a die cavity (Figure 1.10). The blank is held in place with a blank-holder or a hold-down ring with a certain force. When the punch moves down, the portion beneath the blank holder is subjected to radial tensile stresses. Radial tensile stresses lead to compressive hoop stresses in that portion. Thus, the portion beneath the blank holder elongates in the radial direction and compresses in the hoop direction. Compression in the hoop direction may cause wrinkling of the flange during drawing. To avoid this, the blank-holder should apply sufficient amount of holding pressure. However, if the blank holding is excessive, the fracture of the sheet may occur at the end of the punch stroke. Thus, the optimization of blank holding pressure by the finite element method is becoming popular [18]. The cup wall is subjected to tensile stresses, which may cause thinning of the walls.

If precise control of the thickness is desired, the clearance between the punch and die should be less than the thickness. This causes some reduction in thickness. This process is known as ironing. Because of the volume constancy, the length of the cup produced by ironing will be more than that produced by deep drawing with large clearance.

The workability in the deep drawing process is assessed by limiting drawing ratio. The limiting drawing ratio is the maximum ratio of blank diameter to punch diameter that can be drawn without failure. Failure occurs by thinning of the cup wall. It can be shown that the theoretical limiting drawing ratio is 2.718 [17]. The limiting drawing ratio increases with normal anisotropy.

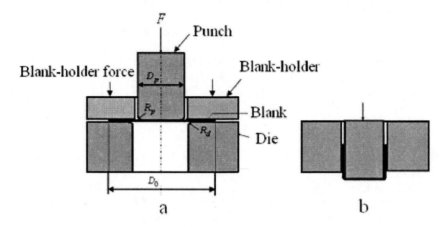

Figure 1.10. Deep drawing process. **a** Before deformation. **b** After deformation

Anisotropy plays an important role in the performance of deep drawing processes. The anisotropy is of two types. In normal anisotropy the properties differ in the thickness direction. In planar anisotropy, the properties vary with the orientation in the plane of the sheet. Whereas deep drawability of sheets increases with normal anisotropy, planar anisotropy leads to the formation of ears in cup drawing [19]. Ears cause the wavy edge of a drawn cup. A sketch of earing defect is shown in Figure 1.11.

Figure 1.11. An earing defect showing the formation of four ears

In sheet metal working, yield point elongation is undesirable. Annealed mild steel exhibits this type of behavior, in which the region between upper and lower yield point is called yield point elongation. Because of this, material stretches in certain regions, with no yielding at the other regions. After the material has been strained enough to cross the yield point elongation, the entire specimen gets deformed uniformly. A sheet with yield point elongation produces Lueder's band or stretcher strains. These bands are elongated depressions on the surface of the sheet and are also called worms. To avoid this problem, sheet is subjected to temper rolling.

Another common defect in deep drawn product is orange peeling. This defect occurs when the grain size is large and individual grains deform independent of each other. Because of this, surface roughness increases.

1.2.2.2 Bending
Bending is the process by which a straight length is transformed to a curved length. Bending of sheets is called sheet bending, whereas bending of tubes is called tube bending. Figure 1.12 shows the bent sheet. In the figure, α is the bend angle, R is the bend radius, a is the bend allowance and L is the length of the bend. The length of the bend is the width of the sheet perpendicular to the plane of bending. Figure 1.13 shows two of the various possible methods of bending: air bending and roll bending. During bending the inner fibers are subjected to compressive strain and the outer fibers are subjected to tensile strain. In between, there are fibers, which have zero strain. The fibers of zero strain in the plane of bending are called the neutral axis. Its location is more towards the inner radius. The radial position of the neutral axis depends on the bend radius and bend angle. Knowing the correct location of the bend radius, one can find the bend allowance.

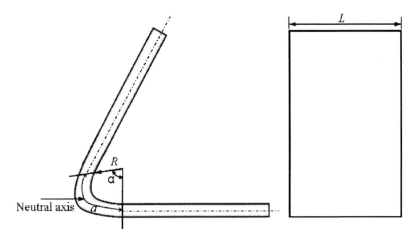

Figure 1.12. A bent sheet

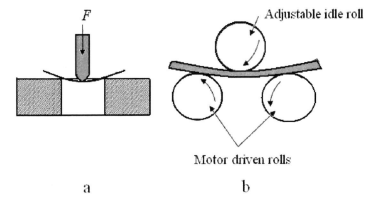

Figure 1.13. Bending processes. **a** Air bending. **b** Roll bending

As the bending radius to thickness ratio decreases, the tensile stress at the outer fibers increases. The limiting bend radius below which the cracking starts is called the minimum bend radius. It is usually expressed in terms of the thickness. For example, a bend radius of $4t$ means that the smallest radius to which the sheet can be bent without cracking is four times its thickness. The bendability of a sheet can be improved by heating, applying hydrostatic pressure or by applying compressive forces in the plane of bending. In bending, proper allowance must be given to the elastic recovery, which follows plastic deformation. This recovery is called springback.

Tube bending is similar to sheet bending, although there are procedural differences. One of the oldest methods of tube bending is to pack the tube with sand particles and then bend around a fixed block or die. This avoids inward buckling of the tube. In conventional tube bending, different dies are required for bending to different radii. In die-less bending of U-bent tube, bending to different radii can be accomplished without a die [20]. The machine consists of feeding rolls, a bending roller and a rotating arm with a clamp which pulls the tube. The bending radius after springback is controlled by the position of the bending lever and the velocity ratio of tube feeding to the rotating arm.

1.2.2.3 Punching and Blanking

Both punching and blanking are shearing processes. In these processes, a portion of the sheet is removed from the rest of the part by shear. In punching, the objective is to make a hole; therefore, the material which is removed from the sheet is scrap. In blanking, the portion which is removed from the sheet is the desired product. Strictly speaking, the shearing process is not a metal forming process, but is a metal removing process. However, it is different from a dominant class of metal removing processes, i.e., machining, in the sense that the material is not removed in the form of chips.

Figure 1.14 illustrates a shearing process. In analytical models of the shearing process, the process is often assumed to be pure shearing, which does not represent the real situation. During the start of the process, the sheet is pushed into the die and the blank material is deformed elastically. The material between the punch and the die can be considered as a plate subjected to load. Thus, in the blank material, the stresses due to combined bending and shear are generated. The blank material is also subjected to thinning. Near the edges of the punch and die, the stresses concentrate in the blank material. Damage initiation followed by its growth takes place from the edges of the punch and die. With proper clearance, the cracks that initiate at the two edges will propagate through the metal and meet near the center of the thickness to provide a clean fracture surface. Insufficient clearance produces a ragged fracture and requires more energy to shear. Too large a clearance produces burrs. In the shearing of hard and brittle material, the clearance should be kept small, as there is much less plastic deformation. In the shearing of ductile material, the clearance should be kept greater to provide enough zone for plastic deformation.

Two recent models of the shearing process are the pure shear model of Atkin [21] and the tension model of Zhou and Wierzbicki [22]. The former assumes that the pure shear model is responsible for fracture, while the latter considers the tension to be responsible for it. Klingenberg and Singh [23] improved the shear model by allowing for additional stretching due to bending of the fibers surrounding the sheared edge of the hole. The main points of interest in the shearing process are the design of die and punch, the clearance between die and punch and the force required.

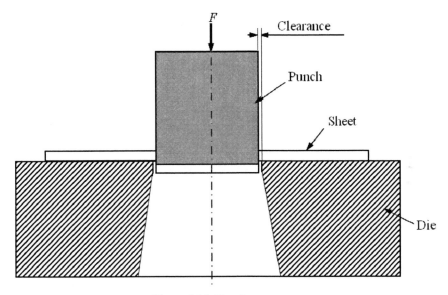

Figure 1.14. Shearing process

If the perimeter of the hole and the thickness of the sheet are large, a huge amount of force acts during the shearing operation. This requires more power and rigidity from the machine. To reduce the force on the punch, the blade is tapered, so that the entire punch does not come in contact with the sheet metal all at once. This taper is called shear. With this, the sheet can be cut with less force but more cutting stroke.

One of the recent blanking processes is fine blanking, which produces close dimensional tolerance and burr-free smooth edges. A rigid machine and tools are employed for this process. The punch penetrates the full thickness of the material. Sideways flow of the material is prevented by an impingement ring having a V-projection. Besides preventing the flow of the material, the V-projection provide a compressive environment near the shear zone.

1.2.2.4 Some Other Sheet Metal Working Processes
There are a number of other popular sheet metal forming techniques. Flat strips can be roll formed into complex sections. The process involves progressive bending of metal strips as they pass through a series of rolls. By this process, a flat sheet can be converted to tubular or other cross-sections.

Stretch forming illustrated in Figure 1.15 is used for producing large sheet metal parts in limited quantities. In this process, a sheet of metal is gripped by two sets of jaws that stretch it and wrap it around a form block. In this process, most of the deformation is induced by the tensile stretching and the forces on the form block are lesser than in bending. There is a very little springback and the work-piece conforms very closely to the shape of the form blank.

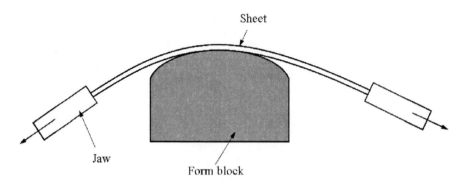

Figure 1.15. Stretch forming process

Spinning (Figure 1.16) is a process of forming axisymmetric sheet metal products. The process can be performed on a lathe machine. A form block (mandrel) is attached with the lathe spindle and rotated. The work-piece (blank) is clamped against it by tailstock. The blank is progressively formed against the form block with a tool.

In explosive forming, the sheet-metal blank is placed over a die cavity and an explosive charge is detonated in water at a certain standoff distance from the blank. The shock wave propagating from the explosion provides a force to deform the blank. Instead of explosives, electric discharge in the form of sparks can also be used to generate a shock wave in a fluid. This is the principle of electro-hydraulic forming.

Certain materials exhibit the phenomenon of superplasticity at high temperature and low strain rate combination. During the superplastic stage they can be stretched to up to about 200% reduction without breakage. Therefore, the superplastic sheets can be formed using the conventional sheet metal working techniques as well as polymer processing techniques with low force requirement. However, the tooling required for superplastic forming must be able to withstand high temperature. The strain rates in the process have to be slow; therefore a superplastic forming takes longer time. A superplastic material behaves like a non-Newtonian fluid and its formability is dependent on the strain rate sensitivity. The phenomenon of superplasticity has been observed in alloys of aluminum, titanium, ceramics and metallic composites having low grain size. Typical homologous temperature is more than 0.5, and strain rate less than 10^{-3} s^{-1}.

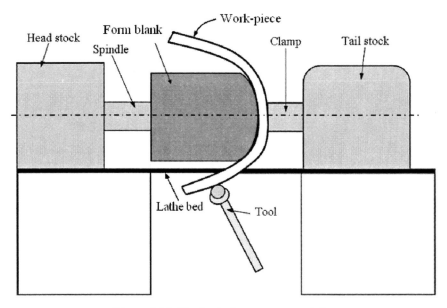

Figure 1.16. Metal spinning process on a lathe

1.3 Machining

Machining processes are far more difficult to model physically than metal forming processes. This is because of the distinct behavior of the machining processes. In almost all machining processes, a wedge shaped tool or a number of tools make contact with the work-piece and remove the material in the form of chips. The process of chip formation is not known properly. In most of the textbooks on machining, a single shear plane model is described. This is based on the assumption that material removal takes place through shear over a very narrow zone. As argued by Astakhov [24], this model suffers from a number of drawbacks *viz.* infinite strain rate, unrealistically high shear strain, unrealistic behavior of the work material, improper accounting for the resistance of the work material to cut, unrealistic representation of the tool–work-piece contact, inapplicability for cutting brittle work materials, incorrect velocity diagram, incorrect force diagram and inability to explain chip curling. Actually the machining process is so complex that no existing physics-based model seems to describe the process properly. Some of the difficulties in the modeling of the machining processes are:

- The machining takes place under the condition of high strain rate and temperature. The proper constitutive model of the material for these conditions should be known.
- The friction at tool-chip interface is very high and simple Coulomb's model cannot be used. Estimation of the friction as well as the proper model for describing it poses a challenge.

- Whereas metal forming processes involve plastic deformation in all the processes, cutting mechanics varies a lot from material to material. There are different methods of fracture for ductile and brittle materials, and they are also dependent on the process conditions.
- In machining processes, the estimation of surface roughness and tool wear is of paramount importance. They are highly sensitive to a number of factors and are highly statistical in nature. Therefore, prediction of these quantities just by physical modeling is a big challenge.

There are a number of machining processes but in all of them there is a relative motion between the tool and work-piece, which is called cutting speed. In addition to main cutting motion, the tool is traversed perpendicular to cutting motion, which is called feed motion. If the cutting edge is perpendicular to the cutting velocity, the process is called orthogonal cutting. If the cutting edge is not perpendicular to the cutting velocity, the process is called oblique cutting. Most machining processes are oblique cutting processes. However, because of their simplicity orthogonal machining processes have been studied a lot and findings have been extended to oblique cutting.

In this section we shall describe the salient points related to the modeling of the machining process under two headings—turning and milling. These are two widely used machining processes. Turning is used mainly for producing axisymmetric components, whilst milling is mainly used for producing flat surfaces or prismatic shapes. The discussion here is mainly intended to provide a background for modeling of machining processes by finite element methods and soft computing techniques. For technological details of the processes, the reader can refer to a number of textbooks on this subject [17, 25–27].

1.3.1 Turning

Turning is a process of removing excess material from the work-piece to produce an axisymmetric surface, in which the work-piece (job) rotates in a spindle and the tool moves in a plane perpendicular to the surface velocity of the job at the tool-job contact point. Turning operations are performed on a machine tool called a lathe. Modern computer numerically controlled (CNC) lathes with a tool magazine for mounting a number of tools are called turning centers. In straight turning, the tool moves parallel to the job axis to machine the rotating job for producing a cylindrical surface. In taper turning, a conical surface is produced. With the exception of taper turning by a form tool, in this process the tool moves simultaneously along the axis of the job and a radial direction to produce a conical surface. In taper turning by a form tool, a tool having the form of the taper is fed in a radial direction. The method is limited to turning short lengths of taper only. Form turning is used to produce axisymmetric surfaces of various types. It may be accomplished by using a form tool, by tracing a template or by providing simultaneous motion to the tool along longitudinal and radial directions, the latter being more common in CNC turning centers.

The process parameters of interest in turning are cutting speed, feed and depth of cut. The cutting speed can be defined as the relative surface speed of the work-piece with respect to the tool, which is responsible for material removal. Thus, in turning it is the surface speed of the job. If D is the outer diameter of the job in mm and N is the revolution per minute (RPM) of the spindle, then the cutting speed v in m/minute is given by

$$v = \frac{\pi DN}{1000}.$$

(1.15)

Usually the cutting speed is expressed in m/min and not in m/s. It is important to know commonly used units of various variables in metal cutting and be careful in converting them to other units, particularly when empirical relations are involved. The cutting speed at the tool nose is slightly less than this conventionally accepted definition of cutting speed. The relative motion of the tool with respect to the job in a direction perpendicular to the direction of cutting speed for the purpose of reaching unmachined surface is called feed, f. In straight turning, it is the distance moved in one revolution along the longitudinal axis of the work-piece and is expressed in mm/revolution. The tool travel speed in mm/min is the feed multiplied by the spindle RPM. The depth of cut (d) is the penetration of the tool into the job beneath the job surface. In turning, it is the radial distance from the unmachined surface of the job to the tool tip. In one pass of the tool across the longitudinal axis of the job, the job diameter will be reduced by twice the depth of cut.

The main objectives in the optimization of a turning process are minimization of the cost of production, maximization of the production rate, maximization of the profit rate, minimization of the surface roughness of the machined surface, and minimization of the dimensional deviation. The material removal rate (MRR) in mm^3/min in a turning process is given by

MRR= $1000\,f v\, d.$

(1.16)

The above equation shows that to increase the material removal rate, one has to increase feed, cutting speed and depth of cut. However, increasing these quantities beyond a limit has an adverse effect on the tool life, surface finish and dimensional deviation, thus affecting productivity and quality. Hence, deciding the proper values of the process parameters in a given context forms an interesting optimization problem.

The cutting tool has great influence on the performance parameters. In turning, usually a single point cutting tool (Figure 1.17) is employed. Figure 1.17 shows the tool angles according to American Standard Association (ASA). According to this system, the seven elements that comprise the signature of a single point cutting tool are stated in the following order: back rake angle, side rake angle, end relief angle, side relief angle, end cutting edge angle, side cutting edge angle and nose radius. The angles are expressed in degrees and nose radius in mm. Relations have been developed, which transform tool angles from one system to the other [27]. The tool signature has a great influence on the machining performance.

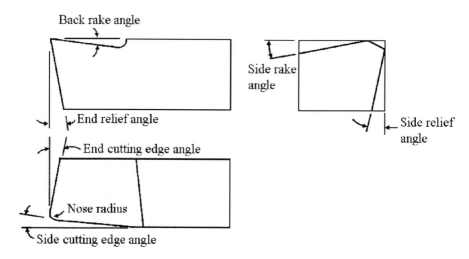

Figure 1.17. Tool angles of a single point cutting tool

Figure 1.18 shows a schematic representation of the straight turning process. In general, it is an oblique cutting process. The portions of the tool drawn in solid and dotted lines represent the positions of the tool before and after one revolution of the job, respectively. It is seen that the tool side cutting edge digs into the job perpendicular to the edge by an amount t, which may be called uncut chip thickness, as this is the thickness of the material that forms the chip. If ψ is the side cutting edge angle, the uncut chip thickness can be expressed as

$$t = f \cos\psi .\tag{1.17}$$

Similarly, the width of the uncut chip is

$$w = d / \cos\psi .\tag{1.18}$$

Thus, it is seen that the width of the uncut chip is related to depth of cut and the uncut chip thickness is related to feed. (Undergraduate students often make the mistake of relating uncut chip thickness to depth of cut.) If the chip thickness is t_c, the cutting ratio, r, is defined as

$$r = \frac{t}{t_c} .\tag{1.19}$$

This ratio is always less than unity and its reciprocal is called chip compression ratio.

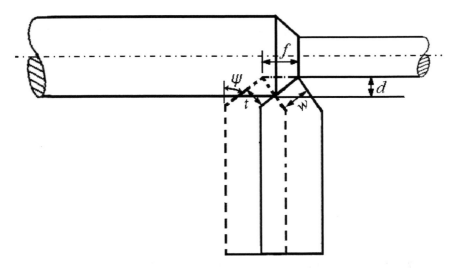

Figure 1.18. A straight turning process

Cutting tool material affects the performance of the process to a great extent. A good tool material must have toughness (ability to withstand shocks) as well as hot hardness (ability to retain its hardness at high temperature). The common tool materials are high-speed steel (HSS), cemented carbide, coated carbide, cermets, ceramics and polycrystalline materials. In high-speed steel, the major constituents are iron, carbon, tungsten, chromium, vanadium and cobalt. The common 18-4-1 high-speed steel means a high-speed steel with 18% tungsten, 4% chromium and 1% vanadium. The addition of cobalt with an amount of 5–10%, increases tool life at high speed. Such a high speed is called super high speed. Cemented carbide is produced by a powder metallurgy process and consists of hard particles in a binder metal. Earlier a tool with 94% tungsten carbide and 6% cobalt was very common. Now there are many grades of carbide material with tungsten carbide, titanium carbide, tantalum carbide and niobium carbide in cobalt binder. A cemented carbide tool can perform turning at a speed of 2–3 times the speed used by HSS tools. However, their toughness is lower. Coating the cemented carbide tools with 6–12-μm coating of titanium carbide, titanium nitride, titanium carbonitride, aluminum oxide and aluminum oxynitride further improves its wear resistance. Alumina-based ceramic cutting tools are attractive alternatives to carbide tools for the machining of steels and cast irons because of their higher hot hardness. Two main types of these ceramics are pure/white oxide ceramic and mixed/black oxide ceramic. White oxide ceramic containing Al_2O_3 with sintered additives and without a metallic binder phase is relatively brittle. Its toughness can be improved by embedding fine zirconia (ZrO_2) particles by an amount of 3–5% into the aluminum oxide matrix. Such a ceramic is called a dispersion ceramic. White oxide ceramic is used in rough machining of gray cast iron, nodular cast iron and chilled cast iron. The black oxide ceramic contains, besides aluminum oxide, titanium oxide and/or titanium carbonitride in the order of about 30% by weight. It is generally used for machining of hard materials and finish machining of cast iron. Polycrystalline diamonds are the hardest among modern tool materials. Two main types are polycrystalline diamond (PCD) and cubic boron nitride (CBN). CBN has been a popular tool material in hard turning, *i.e.*, the machining of material having hardness between 45–70 HRC. One interesting phenomenon in the hard turning of ferrous materials is the formation of about 10 μm deep white layer, which generally consists of a hard phase and leads to the surface becoming brittle [28].

The cutting fluids or coolants play an important role in the turning process. They are used to reduce friction, wear and temperature. Common cutting fluids are soap solution and mineral oils. Soap solution is used in the machining of soft steel, kerosene in the turning of aluminum and mineral lard oil in the turning of brass. Cast iron is machined dry. In recent years, dry turning or turning with minimum lubrication is gaining popularity due to environmental hazards posed by coolants.

1.3.2 Milling

Milling is the process of producing flat surfaces and prismatic shapes. Unlike turning, it employs a multipoint cutting tool called a cutter and having many teeth. The main cutting motion is imparted to the cutter and linear feed motion to the work-piece. Two popular types of milling machines are horizontal and vertical milling machines. In horizontal milling machine the spindle is horizontal and in the vertical milling machine it is vertical. Two main classifications of milling processes are peripheral milling and face milling. In peripheral milling the cutting edges are primarily on the periphery of the cutter. There are various types of peripheral milling processes. Slab milling is used to produce flat surfaces, slot milling is used to produce slots and form milling is used to produce a form. Peripheral milling is performed on horizontal milling machines. It can be further divided into two types—up (conventional) milling and down (climb) milling. In up milling, job feed motion is opposite to cutter motion, whilst in down milling the cutter motion and job motion are in the same direction (Figure 1.19). In up milling, the chip varies in thickness from a minimum at tooth entrance to a maximum at tooth exit, whilst the reverse is the case in down milling.

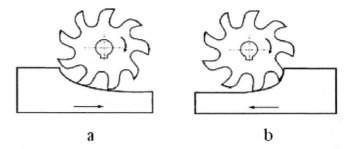

a   b

Figure 1.19. Peripheral milling. **a** Up milling. **b** Down milling

In down milling, the net force is directed downward and its vertical component keeps pressing the job against the table. Thus, it requires a simplified fixture. It requires less power in feed motion as the horizontal component of the net cutter force assists in feed motion. The cutting operation is smoother as the tooth starts from maximum thickness and uncut chip thickness keeps reducing till the tooth leaves the machined surface at zero uncut chip thickness. Due to smoother operation, the process has a lesser tendency to chatter. In up milling, when the cutter starts from zero thickness, it rubs with the work-piece for sometime before biting into the material. This rubbing action reduces tool life. This disadvantage is not present in down milling, although if the surface is sandy or scaly, down milling may reduce the tool life than up milling in which the cut starts from the machined neat surface. In spite of all the advantages of the down milling, it is not used in a machine without a backlash eliminator in the lead screw-nut drive of the table. Without this, the table has a tendency to encounter jerky motion whenever there is a fluctuation in the cutting force.

Most of the discussion pertaining to turning is also applicable to milling. However, the milling process is more complex than turning as it is not a continuous cutting process. The number of teeth in a cutter is an important tool variable as well as the geometry of an individual tooth. The higher the number of teeth the smoother the operation; however, reducing the tooth width beyond a limit weakens the individual tooth and provides less space for the chip. Wider cutters in peripheral milling are helical instead of straight in order to reduce the cutting force and provide smoother cutting action, although at the expense of increased axial thrust.

Face milling is performed on vertical milling machines. As illustrated in Figure 1.20, in this process part of the material is subjected to up milling and another part is subjected to down milling. Besides the geometry of an individual tool, the number of teeth and the diameter of the cutter are important tool parameters. Some tool manufacturers recommend that the cutter's diameter to width of cut ratio should be approximately 3:2 for steel, 5:4 for cast iron and 5:3 for light alloys [29]. The surface generated by face milling has circular marks. End milling, a process used to machine slots, is a combination of peripheral and face milling.

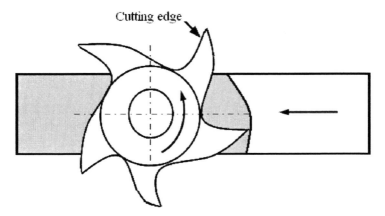

Figure 1.20. Plan view of the face milling process

1.3.3 Some Other Machining Processes

Turning and milling are two popular and typical machining processes. There are a number of other machining processes. Here a few of them are described very briefly. Drilling is a process of making a hole. The hole is generated by the edges of a cutting tool known as the drill. In drilling machines, the drill rotates, whereas if drilling is performed on a lathe, the work-piece rotates and the drill advances in the axial direction. The most common type of drill is the twist drill with two spiral flutes or grooves that run lengthwise around the body of the drill. The flutes help to remove the chips from inside the hole. The determination of the torque and thrust in drilling has been an interesting research topic. However, the surface finish prediction of machined holes was not paid much attention, perhaps because drilling is followed by reaming for generating a better surface finish or boring for enlarging the size of the hole.

In the broaching process, a long tapered tool called the broach is passed across the stationary work-piece. There are a number of teeth on the broach, each of successively greater height. Linear cutting speed is provided to the broach. The feed per tooth is the difference between the heights of the tooth doing cutting and the preceding tooth. The cutting speed in broaching lies between 1 m/min and 25 m/min. The feed per tooth is kept small (less than 0.1 mm/tooth), hence the length of the broach is high. Broaching is classified into internal broaching for machining holes (finishing the hole or making splines and internal gears) and external (surface) broaching for machining an external surface. The broach may be pushed or pulled over the job. The pull type is more common as it places the broach in tension and avoids buckling of the broach. The typical load in broaching ranges from about 50 kN to 250 kN.

Grinding is an abrasive finishing process. A number of grains participate in removing the chips, each removing only a small portion. Silicon carbide and alumina are two abrasive materials. The rake angles of individual abrasive grains may range from $-60°$ to $+45°$. Sometimes the shape of the grains is spherical. Grains that participate in machining are called active grains. The size and distance from the center of different active grains are different. During grinding, the grains keep breaking and forming new cutting edges. Thus, grinding is a self-sharpening process. A grinding wheel is called hard or soft depending on the bond strength of the grinding wheel and not on the abrasive grains. For grinding of the hard material a soft wheel is used and *vice versa*. In grinding, cutting speeds are very high (1500–2000 m/min) and feed per abrasive grain is very small. Hence, due to the size effect, power requirement is high. Cylindrical objects are ground on cylindrical grinding machine, in which the job rotates at a slower speed and the grinding wheel at a higher speed. In surface grinding the job moves linearly and grinding wheel rotates. Prediction of surface roughness in the grinding process is an interesting topic. Another parameter of interest is wear ratio, the ratio of volume of material removed and volume of wheel wear.

1.4 Summary

In this chapter, a number of metal forming and machining processes have been introduced in order to provide a background for understanding the chapters on the modeling of these processes. Metal forming processes may be divided into bulk metal forming and sheet metal forming processes. Forging, rolling, extrusion and wire drawing are the bulk metal forming processes described in this chapter. The important process and performance parameters of these processes have been highlighted. The parameters in other metal forming processes are also similar. For sheet metal forming processes we have provided a relatively detailed discussion on deep drawing, bending and shearing processes, because of their popularity and distinct nature. Some other sheet metal forming processes have been discussed briefly and illustrated with the help of figures. In machining, the turning process has been described in detail followed by milling. Many concepts discussed for these two processes are common to almost all machining processes. We have also discussed drilling, broaching and grinding. A number of references have been provided, to which the reader can refer for more details.

1.5 References

[1]   Ebrahimi, R. and Najafizadeh, A. (2004), A new method for evaluation of friction in bulk metal forming, Journal of Materials Processing Technology, Vol. 152, pp. 136–143.

[2]   Dieter, G.E. (1988), Mechanical Metallurgy, McGraw-Hill, London.

[3]   Kunogi, M. (1956), A new method of cold extrusion, Journal of Science Research Institute Tokyo, Vol. 50, pp. 215–246.

[4]   Male, A.T., and Cockcroft, M.G. (1964–65), A method for the determination of the coefficient of friction of metals under condition of bulk plastic deformation, Journal of the Institute of Metals, Vol. 93, pp. 38–46.

[5]   Avitzur, B. (1964), Forging of Hollow Disks, Israel Journal of Technology, Vol. 9, pp. 295–304.

[6]   Hawkyard, J.B., and Johnson, W. (1967), An Analysis of the changes in geometry of short hollow cylinder during axial compression, International Journal of Mechanical Sciences, Vol. 9, pp. 163–182.

[7]   Sofuoglu, H., Gedikli, H. and Rasty, J. (2001), Determination of friction coefficient by employing the ring compression test, Vol. 123, Transaction of the ASME, Journal of Engineering Materials and Technology, pp. 338–348.

[8]   Tomlinson, A. and Stringer, J.D. (1959), Spread and elongation in flat tool forging, The Journal the Iron and Steel Institute, London, Vol. 193, pp. 157–162.

[9]   Hitchcock, J.H. (1935), Elastic deformation of roll during cold rolling, Report of Special Research Committee on Roll Neck Bearings, New York, pp. 33–41.

[10]   Sarma, D.K. (2001), Design and fabrication of a cold rolling mill using a fuzzy set based methodology, M.Tech. Thesis, IIT Guwahati.

[11]   Roberts, W.L. (1978), Cold Rolling of Steel, Marcel Dekker, New York.

[12]   Avitzur, B., Van Tyne, C.J. and Turczyn, S. (1988), The prevention of central bursts during rolling, Transaction of ASME, Journal of Engineering for Industry, Vol. 110, pp. 173–178.

[13] Zhu, Y.D. and Avitzur, B. (1988), Criteria for the prevention of central bursts during rolling, Transaction of ASME, Journal of Engineering for Industry, Vol. 110, pp. 162–172.

[14] Bryant, G.F., Halliday, J.M. and Spooner, P.D. (1973), Optimal scheduling of a tandem cold-rolling mill, Automation of Tandem Mills, Bryant, G.F. ed., The Iron and Steel Institute, London.

[15] Alexander, J. M. and Turner, T. W. (1974), A preliminary investigation of the die-less drawing of titanium and some steels, Proceedings of Machine Tool Design and Research, Vol. 15, pp. 525–537.

[16] Fortunier, R., Sassoulas, H. and Montheillet, F. (1997), A thermo-mechanical analysis of stabilityin dieless wire drawing, International Journal of Mechanical Sciences, Vol. 39, pp. 615–627.

[17] Kalpakjian, S. and Schmid, S.R. (2003), Manufacturing Processes for Engineering Materials, Pearson Education, Singapore.

[18] Chengzhi, S., Guanlong , C., Zhongqin, L. (2005), Determining the optimum variable blank-holder forces using adaptive response surface methodology (ARSM), International Journal of Advanced Manufacturing Technology, Vol. 26, pp. 23–29.

[19] Van Houutte, P. (1992), Anisotropic plasticity, in Hartley, P., Pillingar, I. And Sturgess, C. (eds.), Numerical Modelling of Material Deformation Process: Research, Development and Applications, Springer-Verlag, London.

[20] Kuboki T., Ohsaka, S., Ono T. and Furugen, M. (2004), Die-less bending method for precision u-bent tube and optimization of the bending conditions, Proceedings of the 7th Japan-India Joint Seminar, 16–21 February 2004, Tokyo.

[21] Atkin, A.G. (1980), On cropping and related processes, International Journal of Mechanical Sciences, Vol. 22, pp. 215–231.

[22] Zhou, Q. and Wierzbicki, T. (1996), A tension model of blanking and tearing of ductile metal plates, International Journal of Mechanical Sciences, Vol. 38, pp. 303–324.

[23] Klingenberg, W. and Singh, U.P. (2005), Comparison of two analytical models of blanking and proposal of a new model, International Journal of Machine Tools & Manufacture, Vol. 45, pp. 519–527.

[24] Astakhov, V.P. (2005), On the inadequacy of the single-shear plane model of chip formation, International Journal of Mechanical Sciences, Vol. 47, pp. 1649–1672.

[25] Hajra Choudhary, S.K. and Hajra Choudhary A.K. (1986), Elements of Workshop Technology, Vol. II, Media Promoters, Mumbai.

[26] Shaw, M.C. (2005), Metal Cutting Principles, 2nd ed., Oxford University Press, Oxford.

[27] Bhattacharyya, A. (1984), Metal Cutting: Theory and Practice, Central Book Publishers, Kolkata.

[28] Bosheh, S.S. and Mativenga, P.P. (2006), White layer formation in hard turning of H13 tool steel at high cutting speeds using CBN tooling, International Journal of Machine Tools & Manufacture, Vol. 46, pp. 225–233.

[29] Sadasivan, T.A. and Sarthy, D. (1999), Cutting Tools for Productive Machining, Widia (India) Ltd., Bangalore.

2

Review of Stress, Linear Strain and Elastic Stress-Strain Relations

2.1 Introduction

In metal forming and machining processes, the work-piece is subjected to external forces in order to achieve a certain desired shape. Under the action of these forces, the work-piece undergoes *displacements* and deformation and develops internal forces. A measure of deformation is defined as *strain*. The intensity of internal forces is called *stress*. The displacements, strains and stresses in a deformable body are interlinked. Additionally, they all depend on the geometry and material of the work-piece, external forces and supports. Therefore, to estimate the external forces required for achieving the desired shape, one needs to determine the displacements, strains and stresses in the work-piece. This involves solving the following set of governing equations: (i) strain-displacement relations, (ii) stress-strain relations and (iii) equations of motion.

In this chapter, we develop the governing equations for the case of *small deformation of linearly elastic materials*. While developing these equations, we disregard the molecular structure of the material and assume the body to be a *continuum*. This enables us to define the displacements, strains and stresses at every point of the body.

We begin our discussion on governing equations with the concept of stress at a point. Then we carry out the analysis of stress at a point to develop the ideas of stress invariants, principal stresses, maximum shear stress, octahedral stresses and the hydrostatic and deviatoric parts of stress. These ideas will be used in the next chapter to develop the theory of plasticity. Next we discuss the conditions which the principle of balance of linear momentum places on the derivatives of the stress components. These conditions lead to *equations of motion*. The concept of *linear strain*, which is a measure of small deformation, is discussed next. For linear strain, the *strain-displacement relations* are linear. The linear strain measure is not directly useful in the analysis of plastic deformation, but it does provide a qualitative understanding of deformation in solid bodies. We can draw upon it to develop a measure for large deformation which is to be used in the theory of plasticity. The analysis of linear strain at a point, similar to the analysis of stress at a point, is also carried out to develop the ideas of strain invariants, principal strains, maximum shear, volumetric strain and the hydrostatic and deviatoric parts of strain. Finally, the *stress-strain relations for small deformation of linearly elastic materials* are developed. Even though these relations are not directly useful for analyzing plastic behavior, their development provides a methodology of expressing qualitative material behavior into quantitative form. This will be useful for developing the plastic stress-strain relations in the next chapter.

Since the stress and strain at a point are *tensor* quantities, a simple definition of tensors involving transformation of components with respect to two Cartesian coordinate systems is provided. Essential elements of tensor algebra and calculus needed to develop the governing equations are discussed. For more elaborate definitions of tensor and for more details of tensor algebra and calculus, the reader is advised to refer to other books. There are quite a few well-written books on these topics like those by Jaunzemis [1], Malvern [2], Fung [3], Sokolnikoff [4] *etc*.

Both tensor and vector quantities are denoted by bold-face letters. Whether the quantity is a tensor or a vector can be understood from the context. Some tensor quantities, like the displacement gradient tensor, involve the use of symbol like the capital Greek letter delta. Most tensors used in the book are of second order. However, for brevity, the adjective "second order" is dropped. Thus, the word tensor without any qualifier means second order tensor. Higher order tensors are referred to by their order. For example, the tensor relating stress and strain tensors in the stress-strain relations is of fourth order and is referred to as such. The governing equations and some intermediate equations are expressed in *tensor notation*. This is done to emphasize the fact that these equations have a form which is independent of the coordinate system. However, while doing calculations, one needs a form of these equations which depends on the coordinate system being used. *Index notation* and the associated *summation convention* are useful for writing the component form of these equations in a condensed fashion. Since the reader is not expected to be familiar with the index notation and summation convention, both are discussed at length right from the beginning. Sometimes, for calculation purposes, an *array notation* is useful for writing the component form of these equations. This involves knowledge of matrix algebra. It is expected that the reader will have sufficient background in matrix algebra and the associated array notation. Wherever possible, the equations are expressed in all three notations: tensor, index and array notations. The calculations are carried out either in index notation or in array notation depending on the convenience of the situation.

The organization of this chapter is as follows. In Section 2.2, we introduce index notation and summation convention. The idea of stress at a point is developed in Section 2.3. Further, the analysis of stress at a point is also carried out. Equations of motion involving the derivatives of stress components are also presented in this section. The concept of linear strain tensor and associated strain-displacement relations are developed in Section 2.4. Additionally, analysis of the linear strain tensor and compatibility conditions for the strain components are also discussed in Section 2.4. Section 2.5 is devoted to the development of stress-strain relations for small deformation of linearly elastic materials. Finally, the whole chapter is summarized in Section 2.6. Worked out examples are provided at the end of Sections 2.2–2.5 to elaborate the concepts discussed in that section.

2.2 Index Notation and Summation Convention

In the modeling of manufacturing processes, we encounter physical quantities in the form of scalars, vectors and tensors. Definition of a tensor is provided in Section 2.3. (In this book, a tensor means the tensor of order two unless stated otherwise). In three-dimensional space, a vector has three components and tensor has nine components. The *index notation* can be employed to represent these components as well as expressions and equations involving scalars, vectors and tensors. In the index notation, the coordinate axes (x,y,z) are labeled as $(1,2,3)$. Thus, to represent a velocity vector (v_x, v_y, v_z), we use the notation v_i, where it is implied that the index i takes the values 1, 2 and 3 in three-dimensional space. In two-dimensional space, it will take the values 1 and 2. Similarly, the notation I_{ij} with the indices i and j is used to represent the following nine components of an inertia tensor: $I_{xx}, I_{xy}, I_{xz}, I_{yx}, I_{yy}, I_{yz}, I_{zx}, I_{zy}, I_{zz}$.

Einstein's *summation convention* is employed for writing the sum of various terms in a condensed form. In this convention, if an index occurs twice in a term, then the term represents the sum of all the terms involving all possible values of the index. For example, $a_i b_i$ means $a_1 b_1 + a_2 b_2 + a_3 b_3$ in three-dimensional space. Similarly, I_{ii} means $I_{11} + I_{22} + I_{33}$. The repeated index is called a *dummy index*, while the non-repeated index is called a *free index*. Thus, in the term $c_{ij} b_j$, i is a free index and j is a dummy index. Any symbol can be used for a dummy index. Therefore, the expression $c_{ij} b_j$ can also be written as $c_{ik} b_k$. When there are two dummy indices, it means the sum over both. Thus in three-dimensions, it will contain nine terms. As an example, the term $p_{ij} q_{ij}$ means $p_{11} q_{11} + p_{12} q_{12} + p_{13} q_{13} + p_{21} q_{21} + p_{22} q_{22} + p_{23} q_{23} + p_{31} q_{31} + p_{32} q_{32} + p_{33} q_{33}$. If an index is repeated more than twice, then it is an *invalid* expression. An expression or equation containing no free index represents a scalar expression or scalar equation. Similarly, an expression or equation containing one free index denotes a vector expression or equation. An expression or equation containing two free indices represents a tensor expression or equation. As an example, the term I_{ii} represents a scalar, the term $c_{ij} b_j$ containing the free index i represents a vector while the term $p_{ij} q_{jk}$ containing the free indices i and k represents a tensor. Similarly, the equation

$$a_i b_i = d \tag{2.1}$$

represents a scalar equation. Further, the equations

$$\sigma_{ij}n_j = t_i, \quad \text{(free index } i \text{, dummy index } j \text{)}, \tag{2.2}$$

$$p_{ij} = q_{ik}r_{kj}, \quad \text{(free indices } i \text{ and } j \text{, dummy index } k \text{)} \tag{2.3}$$

denote vector and tensor equations respectively. In an equation, all the terms should have the same number of free indices. Further, the notation for free indices should be the same in all the terms. Thus, the equations

$$I_{ii} = a_j, \quad \text{(no free index on left side)} \tag{2.4}$$

and

$$p_{ij} = q_{kl}, \text{(the two free indices have different notation on two sides)} \tag{2.5}$$

are *invalid* expressions.

Example 2.1: Expand the following expression:

$$t_i = \sigma_{ij}n_j. \tag{2.6}$$

Solution: This is a vector equation as there is only one free index, namely i, on each side of the equation. Dummy index j on the left side indicates that it is a sum of three terms. Expanding this sum, the equation becomes

$$t_i = \sigma_{i1}n_1 + \sigma_{i2}n_2 + \sigma_{i3}n_3. \tag{2.7}$$

Now, since i is a free index and takes the values 1, 2 and 3, the above vector equation actually represents the following three scalar equations:

$$\begin{aligned}
t_1 &= \sigma_{11}n_1 + \sigma_{12}n_2 + \sigma_{13}n_3, \\
t_2 &= \sigma_{21}n_1 + \sigma_{22}n_2 + \sigma_{23}n_3, \\
t_3 &= \sigma_{31}n_1 + \sigma_{32}n_2 + \sigma_{33}n_3.
\end{aligned} \tag{2.8}$$

Example 2.2: Write in index notation the following expression:

$$\begin{aligned}
\sigma_n &= \sigma_{11}n_1^2 + \sigma_{22}n_2^2 + \sigma_{33}n_3^2 + (\sigma_{12} + \sigma_{21})n_1n_2 + (\sigma_{23} + \sigma_{32})n_2n_3 \\
&\quad + (\sigma_{31} + \sigma_{13})n_3n_1.
\end{aligned} \tag{2.9}$$

Solution: Note that there are nine terms. Therefore, the index notation must involve two dummy indices. In order to write the above equation in terms of the dummy indices, we rearrange the right side as follows:

$$\sigma_n = (\sigma_{11}\, n_1\, n_1 + \sigma_{12}\, n_1\, n_2 + \sigma_{13}\, n_1\, n_3) + (\sigma_{21}\, n_2\, n_1 + \sigma_{22}\, n_2\, n_2 + \sigma_{23}\, n_2\, n_3)$$
$$+ (\sigma_{31}\, n_3\, n_1 + \sigma_{32}\, n_3\, n_2 + \sigma_{33}\, n_3\, n_3). \tag{2.10}$$

Note that in each parenthesis, there is a sum over the second index of σ and the index of second n. This sum can be expressed using a dummy index which we denote by j. Then, the above expression becomes

$$\sigma_n = \sigma_{1j} n_1 n_j + \sigma_{2j} n_2 n_j + \sigma_{3j} n_3 n_j. \tag{2.11}$$

Now, there is a sum over the first index of σ and the index of first n. We express this sum using another dummy index which we denote by i. Thus, the final expression in terms of the index notation can be written as

$$\sigma_n = \sigma_{ij} n_i n_j. \tag{2.12}$$

Note that, as stated earlier, the symbols for the dummy indices can be other than i and j.

Two symbols often used to simplify and shorten expressions in index notation are *Kronecker's delta* and *permutation symbol*. The Kronecker's delta is defined by

$$\delta_{ij} = 1 \quad \text{if} \quad i = j,$$
$$= 0 \quad \text{if} \quad i \neq j. \tag{2.13}$$

The permutation symbol is defined by

$$\in_{ijk} = \; 0 \quad \text{if two or more indices are equal,}$$
$$= +1 \quad \text{if } (i,j,k) \text{ are even permutations of } (1,2,3),$$
$$= -1 \quad \text{if } (i,j,k) \text{ are odd permutations of } (1,2,3). \tag{2.14}$$

The symbols δ and $\in$ satisfy the following identities:

$$a_i \delta_{ij} = a_j, \quad A_{ij} \delta_{jk} = A_{ik}, \quad \delta_{ij} \delta_{jk} = \delta_{ik}, \tag{2.15}$$
$$\in_{ijk} \in_{pqr} = \delta_{ip} \left(\delta_{jq} \delta_{kr} - \delta_{jr} \delta_{kq} \right) + \delta_{iq} \left(\delta_{jr} \delta_{kp} - \delta_{jp} \delta_{kr} \right) + \delta_{ir} \left(\delta_{jp} \delta_{kq} - \delta_{jq} \delta_{kp} \right). \tag{2.16}$$

Example 2.3: Expand the following expressions:

(a) $c = a_i b_j \delta_{ij}.$ \hfill (2.17)

(b) $d = \in_{ijk} \hat{i}_i a_j b_k.$ \hfill (2.18)

Solution: (a) This is a scalar equation involving two dummy indices i and j. Thus, it involves a sum of nine terms. First expanding the sum over i, we get the following three terms on the left side of Equation 2.17:

$$c = a_1 b_j \delta_{1j} + a_2 b_j \delta_{2j} + a_3 b_j \delta_{3j}. \tag{2.19}$$

Now, while expanding the sum over j in each of the three terms, we use Equation 2.13 to substitute the values of δ. Since the value of δ is zero when its two indices are different, we get only one non-zero term in each expansion over j. Thus, the final expanded expression becomes

$$c = a_1 b_1 + a_2 b_2 + a_3 b_3. \tag{2.20}$$

Note that the expression on the right side of Equation 2.20 is the expansion of $a_i b_i$. Thus, we get an identity

$$a_i b_j \delta_{ij} = a_i b_i. \tag{2.21}$$

(b) This is a vector equation involving three dummy indices. Therefore, it is a sum of 27 terms. However, the value of the permutation symbol $\in$ is zero when two of its indices are equal. Therefore, 21 terms are zero. The expansion with the remaining six non-zero terms is

$$d = \in_{123} \hat{i}_1 a_2 b_3 + \in_{132} \hat{i}_1 a_3 b_2 + \in_{231} \hat{i}_2 a_3 b_1 + \in_{213} \hat{i}_2 a_1 b_3$$
$$+ \in_{312} \hat{i}_3 a_1 b_2 + \in_{321} \hat{i}_3 a_2 b_1. \tag{2.22}$$

Now, we use Equation 2.14 to substitute the values of the permutation symbol. Then we get

$$d = \hat{i}_1 (a_2 b_3 - a_3 b_2) + \hat{i}_2 (a_3 b_1 - a_1 b_3) + \hat{i}_3 (a_1 b_2 - a_2 b_1). \tag{2.23}$$

Note that the expression on the right side is the cross product of the vectors a and b. Thus, we can write

$$a \times b = \in_{ijk} \hat{i}_i a_j b_k. \tag{2.24}$$

Example 2.4: Determinant of a matrix $[A]$ is defined by

$$\det[A] = \frac{1}{6} \in_{lmn} \in_{xyz} A_{lx} A_{my} A_{nz}. \tag{2.25}$$

There are the following constraints on the components of $[A]$.

(i) The matrix $[A]$ is symmetric, that is, its non-diagonal components satisfy the relation:

$$A_{ij} = A_{ji}. \tag{2.26}$$

(ii) Further, the sum of the diagonal components is zero:

$$A_{kk} = 0. \tag{2.27}$$

Using the above constraints, show that the expression for the determinant (Equation 2.25) reduces to

$$\det[A] = \frac{1}{3} A_{lm} A_{mn} A_{nl}. \tag{2.28}$$

Solution: Using the identity at Equation 2.16, the determinant of $[A]$ (Equation 2.25) can be expressed in terms of δ :

$$\det[A] = \frac{1}{6}[\delta_{lx}(\delta_{my}\delta_{nz} - \delta_{mz}\delta_{ny}) + \delta_{ly}(\delta_{mz}\delta_{nx} - \delta_{mx}\delta_{nz}) \\ + \delta_{lz}(\delta_{mx}\delta_{ny} - \delta_{my}\delta_{nx})]A_{lx}A_{my}A_{nz}. \tag{2.29}$$

The above expression can be modified using the identity at Equation 2.15 in each of the six terms:

$$\det[A] = \frac{1}{6}(A_{ll}A_{mm}A_{nn} - A_{ll}A_{mn}A_{nm} + A_{ln}A_{ml}A_{nm} - A_{lm}A_{ml}A_{nn} \\ + A_{lm}A_{mn}A_{nl} - A_{ln}A_{mm}A_{nl}). \tag{2.30}$$

Further modification in the second, fourth and sixth terms arises because of the symmetry of $[A]$ (Equation 2.26):

$$\det[A] = \frac{1}{6}(A_{ll}A_{mm}A_{nn} - A_{ll}A_{mn}^2 + A_{ln}A_{ml}A_{nm} - A_{lm}^2 A_{nn} \\ + A_{lm}A_{mn}A_{nl} - A_{ln}^2 A_{mm}). \tag{2.31}$$

Next we use the constraint on the diagonal terms (Equation 2.27) to simplify the above equation. Note that the index k in Equation 2.27 is a dummy index, and thus can be replaced by any other index. Therefore, the first, second, fourth and sixth terms become zero. Then, Equation 2.31 becomes

$$\det[A] = \frac{1}{6}(A_{ln}A_{ml}A_{nm} + A_{lm}A_{mn}A_{nl}). \tag{2.32}$$

Next we modify the first term using the symmetry of $[A]$:

$$\det[A] = \frac{1}{6}(A_{nl}A_{lm}A_{mn} + A_{lm}A_{mn}A_{nl}). \tag{2.33}$$

Finally, shuffling the order in the first term, we find that both the terms are identical. Combining the two terms, we get the desired expression:

$$\det[A] = \frac{1}{6}(A_{lm}A_{mn}A_{nl} + A_{lm}A_{mn}A_{nl}) = \frac{1}{3}A_{lm}A_{mn}A_{nl}. \tag{2.34}$$

Example 2.5: Express the derivative of A_{ij} with respect to A_{pq} in index notation.

Solution: Note that the derivate of A_{ij} with respect to A_{pq} is 1 only if both the indices p and q are exactly equal to i and j. If any one index is different, then the derivative is zero. For example, choose $i = 2$ and $j = 3$. Then, if both $p = 2$ and $q = 3$, then the derivative of A_{23} with respect to A_{23} is one. However, the derivative of A_{23} with respect to A_{p3} for $p = 1,3$ or with respect to A_{2q} for $q = 1, 2$ is zero. Thus, we get

$$\frac{\partial A_{ij}}{\partial A_{pq}} = \delta_{ip}\delta_{jq}. \tag{2.35}$$

The first partial derivative of a component with respect to x_j is indicated by a *comma* followed by j. For example, $u_{i,j}$ means $\partial u_i / \partial x_j$, which in turn represents nine expressions, because both i and j vary from 1 to 3.

Example 2.6: Expand the following expression:

$$\sigma_{ij,j} = 0. \tag{2.36}$$

Solution: This is a vector equation as there is one free index, namely i. Dummy index j represents a sum over three terms. Further, the comma before j indicates differentiation with respect to x_j. Hence, after expanding the sum over j, the above vector equation takes the form

$$\frac{\partial \sigma_{i1}}{\partial x_1} + \frac{\partial \sigma_{i2}}{\partial x_2} + \frac{\partial \sigma_{i3}}{\partial x_3} = 0. \tag{2.37}$$

Since i is a free index and takes the values 1, 2 and 3, the above vector equation represents the following three scalar equations:

$$\frac{\partial \sigma_{11}}{\partial x_1} + \frac{\partial \sigma_{12}}{\partial x_2} + \frac{\partial \sigma_{13}}{\partial x_3} = 0,$$

$$\frac{\partial \sigma_{21}}{\partial x_1} + \frac{\partial \sigma_{22}}{\partial x_2} + \frac{\partial \sigma_{23}}{\partial x_3} = 0, \tag{2.38}$$

$$\frac{\partial \sigma_{31}}{\partial x_1} + \frac{\partial \sigma_{32}}{\partial x_2} + \frac{\partial \sigma_{33}}{\partial x_3} = 0.$$

2.3 Stress

As stated in the introduction, the stresses in a body vary from point to point. In this section, we first discuss the concept of stress at a point. Then we carry out the analysis of stress at a point to develop the ideas of stress invariants, principal stresses, maximum shear stress, octahedral stresses and the hydrostatic and deviatoric parts of stress. Finally, we discuss the equations of motion which involve the derivatives of stress components. These equations arise as a consequence of the balance of linear momentum.

2.3.1 Stress at a Point

In this subsection, we first define the stress vector at a point. Then the ideas of stress tensor and its relation with stress vector are developed. Definition of a tensor (or a second order tensor to be precise) is provided involving the transformation of components with a change in Cartesian coordinate system.

2.3.1.1 Stress Vector
Stress is a measure of the intensity of internal forces generated in a body. In general, stresses in a body vary from point to point. To understand the concept of stress at a point, consider a body subjected to external forces and supported in a suitable fashion, as shown in Figure 2.1. Note that, as soon as the forces are applied, the body gets deformed and sometimes displaced if the supports do not restrain the rigid body motion of the body. Thus, Figure 2.1 shows the deformed configuration. In fact, throughout this section, the configuration considered will be the deformed configuration. First, we define the stress vector (on a plane) at point P of the body. For this, pass a plane (called cutting plane) through point P having a unit normal $\hat{n}$. On each half of the body there are distributed internal forces acting on the cutting plane and exerted by the other half. On the left half, consider a small

area ΔA around point P of the cutting plane. Let ΔF be the resultant of the distributed internal forces (acting on ΔA) exerted by the right half. Then the *stress vector* (or traction) at point P (on the plane with normal $\hat{n}$) is defined as

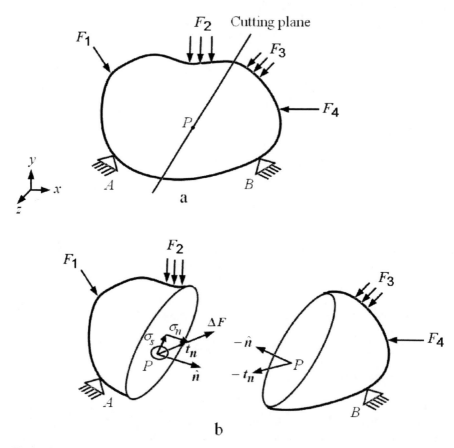

Figure 2.1. Stress vector at a point on a plane. **a** Cutting plane passing through point P of the deformed configuration. **b** Stress vector t_n, normal stress component σ_n and shear stress component σ_s acting at point P on the cutting plane

$$t_n = \lim_{\Delta A \to 0} \frac{\Delta F}{\Delta A}. \tag{2.39}$$

The component of t_n normal to the plane is called the *normal stress* component. It is denoted by σ_n and given by

$$\sigma_n = (t_n)_i \, n_i. \tag{2.40}$$

The component of t_n along the plane is called the *shear stress* component. It is denoted by σ_s and given by

$$|\sigma_s| = \left[|t_n|^2 - (\sigma_n)^2\right]^{1/2}. \tag{2.41}$$

Note that, on the right half, the normal to the cutting plane will be $-\hat{n}$ and the stress vector at P will be $-t_n$ as per the Newton's third law.

2.3.1.2 State of Stress at a Point, Stress Tensor

One can pass an infinite number of planes through point P to obtain an infinite number of stress vectors at point P. The set of stress vectors acting on every plane passing through a point describes the *state of stress* at that point.

It can be shown that a stress vector on any arbitrary plane can be uniquely represented in terms of the stress vectors on *three* mutually orthogonal planes. To show this, we consider x, y and z planes as the three planes, having normal vectors along the three Cartesian directions x, y and z respectively. Let the stress vectors on x, y and z planes be denoted by t_x, t_y and t_z respectively. Further, we denote their components along x, y and z directions as follows:

$$t_x = \sigma_{xx}\hat{i} + \sigma_{xy}\hat{j} + \sigma_{xz}\hat{k}, \tag{2.42}$$

$$t_y = \sigma_{yx}\hat{i} + \sigma_{yy}\hat{j} + \sigma_{yz}\hat{k}, \tag{2.43}$$

$$t_z = \sigma_{zx}\hat{i} + \sigma_{zy}\hat{j} + \sigma_{zz}\hat{k}, \tag{2.44}$$

where ($\hat{i}, \hat{j}, \hat{k}$) are the unit vectors along (x, y, z) axes. The stress vectors and their components are shown in Figure 2.2. To derive the above result, we consider a small element at point P whose shape is that of a tetrahedron. The three sides of the tetrahedron are chosen perpendicular to x, y and z axes and the slant face is chosen normal to vector $\hat{n}$. Then, equilibrium of the tetrahedron in the limit as its size goes to zero leads to the following result [1–5]:

$$t_n = t_x n_x + t_y n_y + t_z n_z, \tag{2.45}$$

where n_x, n_y and n_z are the components of the normal vector $\hat{n}$. This result is true for every stress vector at point P no matter what the orientation of the normal vector $\hat{n}$ is. Further, this result remains valid even if the body forces are not zero or the body is accelerating.

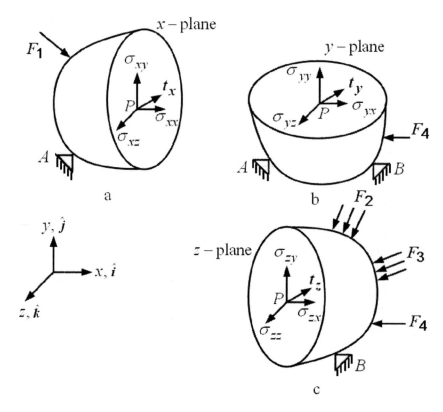

Figure 2.2. Stress vectors and their components on x, y and z planes. **a** Stress vector and its components on x plane. **b** Stress vector and its components on y plane. **c** Stress vector and its components on z plane

Let the components of the stress vector t_n be

$$t_n = (t_n)_x \, \hat{i} + (t_n)_y \, \hat{j} + (t_n)_z \, \hat{k} \, . \tag{2.46}$$

Substituting Equations 2.42–2.44 and 2.46, we get the component form of Equation 2.45 as follows:

$$\begin{Bmatrix} (t_n)_x \\ (t_n)_y \\ (t_n)_z \end{Bmatrix} = \begin{bmatrix} \sigma_{xx} & \sigma_{yx} & \sigma_{zx} \\ \sigma_{xy} & \sigma_{yy} & \sigma_{zy} \\ \sigma_{xz} & \sigma_{yz} & \sigma_{zz} \end{bmatrix} \begin{Bmatrix} n_x \\ n_y \\ n_z \end{Bmatrix} \, . \tag{2.47}$$

In array notation, this can be written as

$$\{t_n\} = [\sigma]^T \{n\} \, , \tag{2.48}$$

where the matrix $[\sigma]$ is

$$[\sigma] = \begin{bmatrix} \sigma_{xx} & \sigma_{xy} & \sigma_{xz} \\ \sigma_{yx} & \sigma_{yy} & \sigma_{yz} \\ \sigma_{zx} & \sigma_{zy} & \sigma_{zz} \end{bmatrix}. \tag{2.49}$$

In index notation, it can be expressed as

$$(t_n)_i = \sigma_{ij}^{\mathrm{T}} \, n_j. \tag{2.50}$$

Equation 2.47 or 2.48 or 2.50 is called the Cauchy's relation. Equations 2.45 and 2.47 indicate that the stress at a point can be completely described by means of just three stress vectors t_x, t_y and t_z acting on mutually orthogonal planes or by their nine components: $\sigma_{xx}, \sigma_{xy}, \sigma_{xz}, \sigma_{yx}, \sigma_{yy}, \sigma_{yz}, \sigma_{zx}, \sigma_{zy}$ and σ_{zz}. Thus, the stress at a point is conceptually different to a *scalar* which has only one component or a *vector* which has three components (in three dimensions). In the next paragraph, we shall discuss a characteristic of the stress at a point which will indicate that it is a *tensor* (*of order two*).

2.3.1.3 Transformation Relations

Note that we can represent the stress vector t_n (at a point) as a combination of the stress vectors on any three mutually orthogonal planes. These planes can be x', y' and z' (Figure 2.3) instead of x, y and z. Then, following the earlier procedure, the stress vector t_n in the component form can be written as

$$\begin{Bmatrix} (t_n)_{x'} \\ (t_n)_{y'} \\ (t_n)_{z'} \end{Bmatrix} = \begin{bmatrix} \sigma_{x'x'} & \sigma_{y'x'} & \sigma_{z'x'} \\ \sigma_{x'y'} & \sigma_{y'y'} & \sigma_{z'y'} \\ \sigma_{x'z'} & \sigma_{y'z'} & \sigma_{z'z'} \end{bmatrix} \begin{Bmatrix} n_{x'} \\ n_{y'} \\ n_{z'} \end{Bmatrix}, \tag{2.51}$$

or

$$\{t_n'\} = [\sigma']^{\mathrm{T}} \{n'\}. \tag{2.52}$$

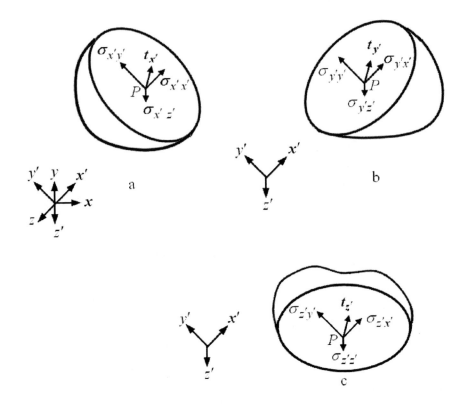

Figure 2.3. Stress vectors and their components on x', y' and z' planes. (Forces acting on the body and supports not shown). **a** Stress vector and its components on x' plane. **b** Stress vector and its components on y' plane. **c** Stress vector and its components on z' plane

Obviously, the components of the matrices $[\sigma]$ and $[\sigma']$ must be related as the stress vector t_n (at point P) has a unique magnitude and direction. To get this relation, we consider equilibrium of three tetrahedra (at point P) whose three faces are perpendicular to x, y and z directions. The fourth face is normal to x' direction for the first tetrahedron, normal to y' direction for the second tetrahedron and normal to z' direction for the third tetrahedron. Three equilibrium equations for each of the three tetrahedra lead to the following result:

$$
\begin{bmatrix}
\sigma_{x'x'} & \sigma_{x'y'} & \sigma_{x'z'} \\
\sigma_{y'x'} & \sigma_{y'y'} & \sigma_{y'z'} \\
\sigma_{z'x'} & \sigma_{z'y'} & \sigma_{z'z'}
\end{bmatrix}
=
\begin{bmatrix}
\ell_1 & m_1 & n_1 \\
\ell_2 & m_2 & n_2 \\
\ell_3 & m_3 & n_3
\end{bmatrix}
\begin{bmatrix}
\sigma_{xx} & \sigma_{xy} & \sigma_{xz} \\
\sigma_{yx} & \sigma_{yy} & \sigma_{yz} \\
\sigma_{zx} & \sigma_{zy} & \sigma_{zz}
\end{bmatrix}
\begin{bmatrix}
\ell_1 & \ell_2 & \ell_3 \\
m_1 & m_2 & m_3 \\
n_1 & n_2 & n_3
\end{bmatrix}.
$$

$$(2.53)$$

Here, if ($\hat{i}', \hat{j}', \hat{k}'$) are the unit vectors along (x', y', z') axes, then (ℓ_1, m_1, n_1) denote the direction cosines of $\hat{i}'$ with respect to (x, y, z) axes. Similarly, (ℓ_2, m_2, n_2) denote the direction cosines of $\hat{j}'$ with respect to (x, y, z) axes and (ℓ_3, m_3, n_3) denote the direction cosines of $\hat{k}'$ with respect to (x, y, z) axes. Define the matrix $[Q]$ as

$$[Q] = \begin{bmatrix} \ell_1 & m_1 & n_1 \\ \ell_2 & m_2 & n_2 \\ \ell_3 & m_3 & n_3 \end{bmatrix}. \tag{2.54}$$

Then, the relation at Equation 2.53 can be written as

$$[\sigma'] = [Q][\sigma][Q]^\mathrm{T}, \tag{2.55}$$

or, in index notation, it can be expressed as

$$\sigma'_{ij} = Q_{ik} \sigma_{kl} Q^\mathrm{T}_{lj}. \tag{2.56}$$

The result of Equation 2.53 or 2.55 or 2.56 remains valid even if the body forces are not zero or the body is accelerating.

Any quantity whose components with respect to two Cartesian coordinate systems transform according to the relation at Equation 2.53 or 2.55 or 2.56 is called a *tensor* (or *tensor of second order*). Thus, the stress at a point is a tensor, known as *stress tensor*. We denote it by the symbol σ. It is related to the stress vector on plane with normal $\hat{n}$ by the relation at Equation 2.47 or 2.48 or 2.50. In tensor notation, this relation can be written as

$$t_n = \sigma^\mathrm{T} \hat{n}. \tag{2.57}$$

The relation at Equation 2.53 or 2.55 or 2.56 is called as the *tensor transformation relation*. The stress tensor σ is called the Cauchy stress tensor. In the next chapter we shall discuss other types of stress tensors.

There is a difference between a tensor and its matrix. A tensor represents a physical quantity which has an existence independent of the coordinate system being used. On the other hand, a matrix of a tensor contains its components with respect to some coordinate system. If the coordinate system is changed, the components change giving a different matrix. Matrices with respect to two different coordinate systems are related by the tensor transformation relation.

Let (a_x, a_y, a_z) be the components of a vector a with respect to the coordinate system (x, y, z). Further, denote the components of a with respect to the coordinate system (x', y', z') as (a'_x, a'_y, a'_z). Then these two sets of components are related by the following transformation law:

$$\begin{Bmatrix} a'_x \\ a'_y \\ a'_z \end{Bmatrix} = \begin{bmatrix} \ell_1 & m_1 & n_1 \\ \ell_2 & m_2 & n_2 \\ \ell_3 & m_3 & n_3 \end{bmatrix} \begin{Bmatrix} a_x \\ a_y \\ a_z \end{Bmatrix},$$ (2.58)

or

$$\{a\}' = [Q]\{a\},$$ (2.59)

or, in index notation

$$a'_i = Q_{ij} \, a_j.$$ (2.60)

The relation at Equation 2.58 or 2.59 or 2.60 is called as the *vector transformation relation*. The matrix $[Q]$, which appears in vector and tensor transformation relations, is called the *transformation matrix*. It can easily be verified that $[Q]$ is an *orthogonal matrix*, that is, it satisfies the relation

$$Q_{ik}Q_{kj}^\mathrm{T} = Q_{ik}^\mathrm{T}Q_{kj} = \delta_{ij}.$$ (2.61)

There are two types of orthogonal matrices. The first type represents the rotation of the coordinate axes and its determinant is +1. The second type represents the reflection of the coordinate axes and its determinant is -1. It can be shown that the matrix $[Q]^\mathrm{T}$ represents the rotation of the (x, y, z) coordinate axes to (x', y', z') axes and therefore it is known as the *rotation matrix*. Its determinant is +1.

2.3.1.4 Stress Components

A tensor component is always represented by two subscript indices. In the case of a component of the stress tensor, the meaning of the indices is as follows. The first index describes the direction of the normal to the plane on which the stress component acts while the second index represents the direction of the stress component itself. Thus, σ_{xy} indicates a stress component acting in the y-direction on the x-plane. When both the indices are same, it means the stress component is along the normal to the plane on which it acts. It is called the normal stress component. Thus, σ_{xx}, σ_{yy} and σ_{zz} are the *normal stress* components. When the two indices are different, it means the direction of the component is within the plane. Such a component is called the *shear stress* component. Thus, σ_{ij} where

$i \ne j$ are the shear stress components. We adopt the following *sign convention* for the stress components. We first define positive and negative planes. A plane i is considered positive if the outward normal to it points in the positive i direction, otherwise it is considered as negative. A stress component is considered positive if it acts in a positive direction on a positive plane or in negative direction on a negative plane. Otherwise, it is considered as negative. Figure 2.4 illustrates positive and negative normal and shear stress components.

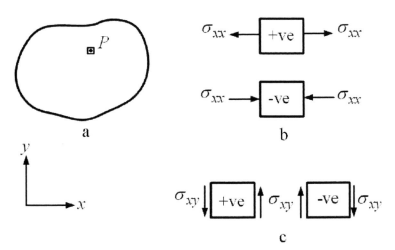

Figure 2.4. Sign convention for normal and shear stress components. **a** Small element at point 'P' in the deformed configuration. Forces on the body and supports are not shown. **b** Positive and negative 'σ_{xx}'. **c** Positive and negative 'σ_{xy}'

2.3.1.5 Symmetry of Stress Tensor

By considering the moment equilibrium of a small element (of parallelepiped shape) at point P in the limit as the size of the element tends to zero, it can be shown that [2]

$$\sigma_{ij} = \sigma_{ji}. \tag{2.62}$$

Thus, the stress tensor is *symmetric*. Now, the Cauchy relation (Equation 2.48 or 2.50) may be written as

$$\{t_n\} = [\sigma]\{n\} \tag{2.63}$$

or

$$(t_n)_i = \sigma_{ij}n_j. \tag{2.64}$$

In tensor notation, it can be expressed as

$$t_n = \sigma \hat{n} .$$ (2.65)

The result of Equation 2.62 is valid even if the body forces are not zero or the body is accelerating.

Example 2.7: Components of the stress tensor σ at point P of the beam of Figure 2.5, with respect to (x, y, z) coordinate system, are given as

$$[\sigma] = \begin{bmatrix} 35 & -25 & 0 \\ -25 & -15 & 0 \\ 0 & 0 & 0 \end{bmatrix} \text{ (MPa)}.$$ (2.66)

(a) Find the stress vector t_n on the plane whose normal is given by

$$\hat{n} = (1/\sqrt{3})(\hat{i} + \hat{j} + \hat{k}) .$$ (2.67)

Find the normal (σ_n) and shear (σ_s) components of t_n.
(b) Find the components of σ with respect to the rotated coordinate system (x', y', z'). The unit vectors $(\hat{i}', \hat{j}', \hat{k}')$ along the (x', y', z') axes are given as

$$\hat{i}' = 0.6\hat{i} + 0.8\hat{k},$$
$$\hat{j}' = \hat{j},$$ (2.68)
$$\hat{k}' = -0.8\hat{i} + 0.6\hat{k} .$$

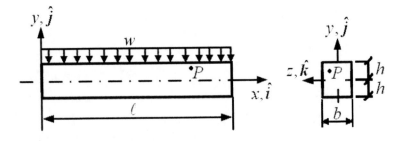

Figure 2.5. A cantilever beam subjected to uniformly distributed load on top surface

Solution: (a) We use the Cauchy's relation in array form to evaluate the stress vector t_n. As per Equation 2.46, we denote its components with respect to (x, y, z) coordinate system by $(t_n)_x$, $(t_n)_y$ and $(t_n)_z$. Further, the given components of the unit normal vector $\hat{n}$ are

$$n_x = n_y = n_z = 1/\sqrt{3}. \tag{2.69}$$

Writing the components of t_n and $\hat{n}$ in the array form and using Equation 2.63, we get

$$\begin{Bmatrix} (t_n)_x \\ (t_n)_y \\ (t_n)_z \end{Bmatrix} = \begin{bmatrix} 35 & -25 & 0 \\ -25 & -15 & 0 \\ 0 & 0 & 0 \end{bmatrix} \begin{Bmatrix} \dfrac{1}{\sqrt{3}} \\ \dfrac{1}{\sqrt{3}} \\ \dfrac{1}{\sqrt{3}} \end{Bmatrix} = \begin{Bmatrix} \dfrac{10}{\sqrt{3}} \\ -\dfrac{40}{\sqrt{3}} \\ 0 \end{Bmatrix}. \tag{2.70}$$

Thus, the stress vector is

$$t_n = \frac{10}{\sqrt{3}}\hat{i} - \frac{40}{\sqrt{3}}\hat{j} \text{ (MPa).} \tag{2.71}$$

Then, using Equation 2.40, we get the normal component of the stress vector:

$$\sigma_n = (t_n)_i n_i = \left(\frac{10}{\sqrt{3}}\right)\left(\frac{1}{\sqrt{3}}\right) + \left(-\frac{40}{\sqrt{3}}\right)\left(\frac{1}{\sqrt{3}}\right) + (0)\left(\frac{1}{\sqrt{3}}\right) = -10 \text{ (MPa).} \tag{2.72}$$

Further, using Equation 2.41, we get the magnitude of the shear component of the stress vector:

$$|\sigma_s| = \left[|t_n|^2 - (\sigma_n)^2\right]^{1/2} = \left[\left(\frac{10}{\sqrt{3}}\right)^2 + \left(\frac{-40}{\sqrt{3}}\right)^2 - (-10)^2\right]^{1/2} = \frac{10\sqrt{14}}{\sqrt{3}} \text{ (MPa).} \tag{2.73}$$

(b) To find the components of σ with respect to (x', y', z') coordinate system, we first evaluate the transformation matrix $[Q]$. We get the direction cosines of the unit vectors $(\hat{i}', \hat{j}', \hat{k}')$ from Equation 2.68. Substituting them in Equation 2.54, we get the following expression for $[Q]$:

$$[Q] = \begin{bmatrix} 0.6 & 0 & 0.8 \\ 0 & 1 & 0 \\ -0.8 & 0 & 0.6 \end{bmatrix}. \tag{2.74}$$

Using the tensor transformation relation (Equation 2.55), we get the following matrix of the components of the stress tensor with respect to (x', y', z') coordinate system:

$$[\sigma]' = [Q][\sigma][Q]^{\mathrm{T}} = \begin{bmatrix} 0.6 & 0 & 0.8 \\ 0 & 1 & 0 \\ -0.8 & 0 & 0.6 \end{bmatrix} \begin{bmatrix} 35 & -25 & 0 \\ -25 & -15 & 0 \\ 0 & 0 & 0 \end{bmatrix} \begin{bmatrix} 0.6 & 0 & -0.8 \\ 0 & 1 & 0 \\ 0.8 & 0 & 0.6 \end{bmatrix},$$

$$= \begin{bmatrix} 12.6 & -15 & -16.8 \\ -15 & -15 & 20 \\ -16.8 & 20 & 22.4 \end{bmatrix} \text{(MPa)}.$$

$$\tag{2.75}$$

Equation 2.75 shows that the matrix of σ is symmetric with respect to the coordinate system (x', y', z') as well.

Note that the stress components σ_{xz}, σ_{yz} and σ_{zz} are zero at point P of the beam (Equation 2.66). Such a state is called as the *state of plane stress* (*at a point*) *in* $x - y$ *plane*. When these stress components are zero at every point of the body and if, additionally, the remaining stress components σ_{xx}, σ_{yy} and σ_{xy} are independent of z coordinate, it is called as the *state of plane stress* (*in a body*) *in* $x - y$ *plane*. It can be shown that the state of stress in the beam of Figure 2.5 is of this type.

2.3.2 Analysis of Stress at a Point

As stated earlier, in this subsection we carry out the analysis of stress at a point to discuss the concepts of principal stresses and principal directions, principal invariants, maximum shear stress, octahedral stresses and the hydrostatic and deviatoric parts of stress.

2.3.2.1 Principal Stresses, Principal Planes and Principal Directions
There exist at least three mutually perpendicular planes (in the deformed configuration) such that there are no shear stress components on them, *i.e.*, the stress vector is normal to these planes. These planes are known as the *principal planes* and normals to these planes are called the *principal directions* (of stress). The values of the normal stress components are called the *principal stresses*. We denote the principal stresses as σ_1, σ_2 and σ_3. The unit vectors along the principal directions are normally denoted as $\hat{e}_1$, $\hat{e}_2$ and $\hat{e}_3$. We arrange the principal stresses as

$$\sigma_1 \geq \sigma_2 \geq \sigma_3. \tag{2.76}$$

The senses of the unit vectors are so chosen that they always form a right-sided system. Thus,

$$\hat{e}_1.\hat{e}_2 \times \hat{e}_3 = +1. \tag{2.77}$$

Since the stress vector on a principal plane i (*i.e.*, the plane perpendicular to the principal direction i) has only the normal component equal to σ_i, the components of the stress tensor, with respect to the principal directions as the coordinate system, become

$$[\sigma]^P = \begin{bmatrix} \sigma_1 & 0 & 0 \\ 0 & \sigma_2 & 0 \\ 0 & 0 & \sigma_3 \end{bmatrix}. \tag{2.78}$$

It can easily be verified that, at a point, *maximum* value of the *normal stress* component (σ_n) with respect to the orientation of the normal vector $\hat{n}$ is σ_1. Further, the *minimum* value is σ_3.

It can be shown that the principal stresses are the *eigenvalues* or *principal values* and the unit vectors along the principal directions are the normalized *eigenvectors* of the stress tensor [2–4]. Before we write the equation which the eigen values and eigenvectors of a tensor satisfy, we define a *unit tensor*. It is denoted by the symbol **1**. A unit tensor is defined as a tensor whose components with respect to every coordinate system are given by the following array:

$$\begin{bmatrix} 1 & 0 & 0 \\ 0 & 1 & 0 \\ 0 & 0 & 1 \end{bmatrix}. \tag{2.79}$$

Thus, in index notation, the components of the unit tensor are represented as δ_{ij}. If λ is an eigenvalue of the stress tensor σ and if x is the corresponding eigenvector, λ and x satisfy the following equation:

$$\left([\sigma] - \lambda [1] \right) \{x\} = \{0\}. \tag{2.80}$$

Here, the arrays $[\sigma]$, $[1]$ and $\{x\}$ contain the components of respectively σ, **1** and x with respect to the given coordinate system (x, y, z). For x to be an eigen vector of σ, Equation 2.80 should have a non-trivial solution. For this to happen, the determinant of the coefficient matrix $\left([\sigma] - \lambda [1] \right)$ must be zero. This condition leads to the following cubic equation in λ :

$$\lambda^3 - I_\sigma \lambda^2 - II_\sigma \lambda - III_\sigma = 0, \tag{2.81}$$

where

$$I_\sigma = \sigma_{ii}, \tag{2.82}$$

$$II_\sigma = \frac{1}{2}(\sigma_{ij}\sigma_{ij} - \sigma_{ii}\sigma_{jj}), \tag{2.83}$$

$$III_\sigma = \in_{ijk} \sigma_{1i}\sigma_{2j}\sigma_{3k}. \tag{2.84}$$

Thus, the principal stresses σ_i are found as the roots of the above equation. Once σ_i are determined, the unit vectors $\hat{e}_i$ along the principal directions are found from the following equation:

$$\left([\sigma] - \sigma_i [1] \right)\{e_i\} = \{0\}, \quad \text{(no sum over } i \text{).} \tag{2.85}$$

where the array $\{e_i\}$ contains the components of $\hat{e}_i$ with respect to the given coordinate system (x, y, z).

2.3.2.2 Principal Invariants

Trace of tensor σ (denoted by $tr\sigma$) is a scalar function of σ which is defined as

$$tr\sigma = \sigma_{ii}. \tag{2.86}$$

Thus, using Equation 2.82, we get

$$I_\sigma = tr\sigma. \tag{2.87}$$

Note that, in Equation 2.86, the scalar function $tr\sigma$ is evaluated using the components of σ with respect to the given coordinate system (x, y, z). Let σ'_{ij} be the components of σ in a rotated coordinate system (x', y', z'). If we use the rotated coordinate system to evaluate the scalar function $tr\sigma$, then it would be

$$tr\sigma = \sigma'_{ii}. \tag{2.88}$$

Using the tensor transformation relation (Equation 2.56), and the orthogonality of $[Q]$ (Equation 2.61), it can be shown that

$$\sigma'_{ii} = \sigma_{ii}. \tag{2.89}$$

Thus, Equations 2.86, 2.88 and 2.89 show that the value of $tr\sigma$ is independent of the coordinate system. A scalar function of a tensor whose value is independent of the coordinate system is called an *invariant* of the tensor. Thus, I_σ is an invariant of the tensor σ. Similarly, it can be shown that II_σ and III_σ are also the invariants of the tensor σ. Using the definition of the trace and the symmetry of σ, it can be shown that

$$II_\sigma = \frac{1}{2}\left\{ tr(\sigma^2) - (tr\sigma)^2 \right\}. \tag{2.90}$$

Further, it can be shown that

$$III_\sigma = \det \sigma, \tag{2.91}$$

where $\det \sigma$ means the *determinant* of the matrix of σ in any coordinate system.

Every other invariant of σ can be expressed in terms of these three invariants [1]. Therefore, I_σ, II_σ and III_σ are called as the *principal invariants* of the tensor σ.

2.3.2.3 Maximum Shear Stress

It can be shown that, at a point, *maximum* value of the *shear stress* component with respect to the orientation of the normal vector $\hat{n}$ is [4]

$$\left| \sigma_s \right|_{max} = \frac{(\sigma_1 - \sigma_3)}{2}. \tag{2.92}$$

Further, the normals to the planes on which $\left| \sigma_s \right|_{max}$ acts are given by [4]

$$\hat{n} = \pm\frac{1}{\sqrt{2}}(\hat{e}_1 \pm \hat{e}_3). \tag{2.93}$$

This result will be useful when we discuss yield criteria later.

2.3.2.4 Octahedral Stresses

A plane whose normal is equally inclined to the three principal directions is called *octahedral plane*. Let $\hat{n}$ be the unit normal to an octahedral plane. Further, let n_i be its components with respect to the principal directions $\hat{e}_i$. Since n_i are equal in magnitude and

$$n_i n_i = 1 \tag{2.94}$$

we get

$$n_i = \pm \frac{1}{\sqrt{3}}.$$

(2.95)

From Equation 2.95, we get eight different normal vectors: $(1/\sqrt{3}, 1/\sqrt{3}, 1/\sqrt{3})$, $(1/\sqrt{3}, 1/\sqrt{3}, -1/\sqrt{3}), \ldots \ldots, (-1/\sqrt{3}, -1/\sqrt{3}, -1/\sqrt{3})$. Thus there are *eight* octahedral planes.

Let t_n be the stress vector on an octahedral plane. Further, let $(t_n)_i$ be its components with respect to the principal directions $\hat{e}_i$. Substituting the components of σ and $\hat{n}$ with respect to the principal directions (Equations 2.78 and 2.95) in Equation 2.64, we get the following expression for $(t_n)_i$:

$$(t_n)_i = \pm \frac{1}{\sqrt{3}} \sigma_i.$$

(2.96)

Substituting Equations 2.95 and 2.96 for n_i and $(t_n)_i$ in Equation 2.40, we get the following expression for the normal stress component (denoted by σ_{oct}) on the octahedral planes:

$$\sigma_{oct} = \frac{1}{3}(\sigma_1 + \sigma_2 + \sigma_3) = \frac{I_\sigma}{3}.$$

(2.97)

Similarly, substituting Equations 2.96 and 2.97 for $(t_n)_i$ and σ_{oct} in Equation 2.41, we get the following expression for the magnitude of the shear stress component on the octahedral planes (denoted by $|\tau_{oct}|$):

$$|\tau_{oct}| = \left[\frac{1}{3}\left(\sigma_1^2 + \sigma_2^2 + \sigma_3^2 \right) - \frac{1}{9}\left(\sigma_1 + \sigma_2 + \sigma_3 \right)^2 \right]^{1/2} = \left[\frac{2}{9}\left(I_\sigma^2 + 3II_\sigma \right) \right]^{1/2}.$$

(2.98)

Note that when the stress tensor at a point has only the deviatoric part, then the octahedral planes are free of the normal stress component. The expression for the shear stress on the octahedral planes will be useful when we discuss the yield criteria in Chapter 3.

2.3.2.5 Decomposition into the Hydrostatic and Deviatoric Parts

Every tensor can be decomposed into a sum of a scalar multiple of a unit tensor **1** and a traceless tensor. Thus, for the stress tensor σ, we can write

$$\sigma = \left(\frac{1}{3}tr\sigma\right)\mathbf{1} + \sigma', \qquad (tr\sigma' = 0). \tag{2.99}$$

In index notation, this can be written as

$$\sigma_{ij} = \left(\frac{1}{3}\sigma_{kk}\right)\delta_{ij} + \sigma'_{ij}, \; (\sigma'_{kk} = 0). \tag{2.100}$$

Note that, since σ is a symmetric tensor, σ' is also a symmetric tensor. The unit tensor $\mathbf{1}$ is of course symmetric. The stress vector corresponding to the first part is always normal to the plane and has the same magnitude on every plane, namely $(1/3)tr\sigma$. Thus, this part of the stress tensor is similar to the state of stress in water at rest, except that whereas $(1/3)tr\sigma$ may be tensile or compressive, the state of stress in water is always compressive. Therefore, this part of the stress tensor is known as the *hydrostatic part* of σ. The second part is called the *deviatoric part* of σ and represents a pure shear state.

In *isotropic materials*, the deformation caused by the hydrostatic part consists of only a change in volume (or size) but no change in shape. On the other hand, the deformation caused by the deviatoric part consists of no change in volume but only the change in shape. We shall see in Chapter 3 that, in an isotropic ductile material, yielding is caused only by the deviatoric part of the stress tensor.

2.3.2.6 Principal Invariants of the Deviatoric Part
The principal invariants of σ' are denoted by J_1, J_2 and J_3. Like the principal invariants of σ (Equations 2.82–2.84, 2.87, 2.90, 2.91), they are defined as

$$J_1 = tr\sigma' = \sigma'_{ii}, \tag{2.101}$$

$$J_2 = \frac{1}{2}\left\{tr(\sigma'^2) - (tr\sigma')^2\right\} = \frac{1}{2}(\sigma'_{ij}\sigma'_{ij} - \sigma'_{ii}\sigma'_{jj}), \tag{2.102}$$

$$J_3 = \det\sigma' = \epsilon_{ijk}\,\sigma'_{1i}\sigma'_{2j}\sigma'_{3k}. \tag{2.103}$$

Since $tr\sigma' = 0$ (Equation 2.99), J_1 has the value zero. Further, J_2 also gets simplified. Thus,

$$J_1 = 0, \tag{2.104}$$

$$J_2 = \frac{1}{2}tr\left(\sigma'^2\right) = \frac{1}{2}\sigma'_{ij}\sigma'_{ij}. \tag{2.105}$$

The expressions for these invariants will be useful while discussing the yield criteria of isotropic materials in Chapter 3.

Example 2.8: Components of the stress tensor σ at a point, with respect to the (x, y, z) coordinate system, are given as

$$[\sigma] = \begin{bmatrix} 18 & 24 & 0 \\ 24 & 32 & 0 \\ 0 & 0 & -20 \end{bmatrix} \text{ (MPa)}. \tag{2.106}$$

(a) Find the principal invariants of σ .

(b) Find the principal stresses σ_i and the unit vectors $\hat{e}_i$ along the principal directions. Arrange σ_i such that $\sigma_1 \geq \sigma_2 \geq \sigma_3$. Further, choose the senses of $\hat{e}_i$ such that $\hat{e}_1 \cdot \hat{e}_2 \times \hat{e}_3 = +1$.

(c) Find the maximum shear stress $|\sigma_s|_{max}$ and the normals to the planes on which $|\sigma_s|_{max}$ acts. Express the normals in terms of the unit vectors $(\hat{i}, \hat{j}, \hat{k})$.

(d) Find the octahedral normal (σ_{oct}) and shear (τ_{oct}) stresses .

(e) Find the hydrostatic and deviatoric parts of σ .

Solution: (a) Substituting the values of σ_{ij} from Equation 2.106 and the values of permutation symbol $\in_{ijk}$ from Equation 2.14, we get

$$I_\sigma = \sigma_{ii} = 18 + 32 - 20 = 30 \text{ (MPa)}; \tag{2.107}$$

$$II_\sigma = \frac{1}{2}(\sigma_{ij}\sigma_{ij} - \sigma_{ii}\sigma_{jj}),$$

$$= \frac{1}{2}\left[(18)^2 + 2(24)^2 + (32)^2 + (-20)^2 + 4(0)^2 - (30)(30)\right] = 1000\left(\text{MPa}\right)^2 ; \tag{2.108}$$

$$III_\sigma = \in_{ijk} \sigma_{1i}\sigma_{2j}\sigma_{3k},$$
$$= \sigma_{11}(\sigma_{22}\sigma_{33} - \sigma_{23}\sigma_{32}) + \sigma_{12}(\sigma_{23}\sigma_{31} - \sigma_{21}\sigma_{33}) + \sigma_{13}(\sigma_{21}\sigma_{32} - \sigma_{22}\sigma_{31}),$$
$$= 18[32 \times (-20) - 0 \times 0] + 24[0 \times 0 - 24 \times (-20)] + 0[24 \times 0 - 32 \times 0],$$
$$= 0 \left(\text{MPa}\right)^3 . \tag{2.109}$$

(b) Substituting the values of I_σ , II_σ and III_σ from part (a), the cubic equation for λ (Equation 2.81) becomes

$$\lambda^3 - 30\lambda^2 - 1000\lambda - 0 = 0 . \tag{2.110}$$

The roots of this equation are $\lambda = 0, -20, 50$. Arranging them in decreasing order, we get the following values of the principal stresses:

$$\sigma_1 = 50 \text{ MPa}, \quad \sigma_2 = 0 \text{ MPa}, \quad \sigma_3 = -20 \text{ MPa}. \tag{2.111}$$

To find the unit vectors $\hat{e}_i$ along the principal directions, we use Equation 2.85. Let the unit vector along the first principal direction be

$$\hat{e}_1 = e_{1x}\hat{i} + e_{1y}\hat{j} + e_{1z}\hat{k}. \tag{2.112}$$

Then for $i = 1$, Equation 2.85 becomes

$$\begin{bmatrix} \sigma_{11} - \sigma_1 & \sigma_{12} & \sigma_{13} \\ \sigma_{21} & \sigma_{22} - \sigma_1 & \sigma_{23} \\ \sigma_{31} & \sigma_{32} & \sigma_{33} - \sigma_1 \end{bmatrix} \begin{Bmatrix} e_{1x} \\ e_{1y} \\ e_{1z} \end{Bmatrix} = \begin{Bmatrix} 0 \\ 0 \\ 0 \end{Bmatrix}. \tag{2.113}$$

Substituting $\sigma_1 = 50$ and the values of σ_{ij} from Equation 2.106 and expanding the above equation, we get

$$\begin{aligned} (18-50)e_{1x} + 24e_{1y} + 0e_{1z} &= 0, \\ 24e_{1x} + (32-50)e_{1y} + 0e_{1z} &= 0, \\ 0e_{1x} + 0e_{1y} + (-20-50)e_{1z} &= 0. \end{aligned} \tag{2.114}$$

From the third equation we obtain $e_{1z} = 0$. Note that the first two equations are linearly dependent. Each of them gives $e_{1y} = (4/3)e_{1x}$. Since $\hat{e}_1$ is a unit vector, we have

$$e_{1x}^2 + e_{1y}^2 + e_{1z}^2 = 1. \tag{2.115}$$

Substituting $e_{1z} = 0$ and $e_{1y} = (4/3)e_{1x}$ in the above equation, we obtain $e_{1x} = \pm(3/5)$. Choosing the positive sign, we get the following expression for the unit vector along the first principal direction:

$$\hat{e}_1 = \frac{3}{5}\hat{i} + \frac{4}{5}\hat{j}. \tag{2.116}$$

Similarly, we get the following expressions for the unit vectors along the other two principal directions:

$$\hat{e}_2 = \frac{4}{5}\hat{i} - \frac{3}{5}\hat{j}, \tag{2.117}$$

$$\hat{e}_3 = -\hat{k}. \tag{2.118}$$

Note that whereas the sense of the second unit vector has been chosen to be arbitrary, that of the third one has been selected so as to satisfy the condition $\hat{e}_1 \cdot \hat{e}_2 \times \hat{e}_3 = +1$.

(c) Maximum shear stress is given by Equation 2.92. Substituting the values of σ_1 and σ_3 from part (b) in this equation, we get

$$\left|\sigma_s\right|_{max} = \frac{\sigma_1 - \sigma_3}{2} = \frac{50-(-20)}{2} = 35 \ (\text{MPa}). \tag{2.119a}$$

The normals $\hat{n}$ to the planes on which $\left|\sigma_s\right|_{max}$ acts are given by Equation 2.93. Substituting the expressions for $\hat{e}_1$ and $\hat{e}_3$ from part (b) in this equation, we obtain the following expressions for $\hat{n}$:

$$\hat{n} = \pm\frac{1}{\sqrt{2}}(\hat{e}_1 \pm \hat{e}_3) = \pm\frac{1}{\sqrt{2}}\left(\frac{3}{5}\hat{i} + \frac{4}{5}\hat{j} \mp \hat{k}\right). \tag{2.119b}$$

(d) Octahedral normal (σ_{oct}) and shear (τ_{oct}) stresses are calculated using Equations 2.97 and 2.98. Substituting the values of I_σ and II_σ from part (a), we get

$$\sigma_{oct} = \frac{I_\sigma}{3} = \frac{30}{3} = 10 \ (\text{MPa}), \tag{2.120a}$$

$$\left|\tau_{oct}\right| = \left[\frac{2}{9}(I_\sigma^2 + 3II_\sigma)\right]^{1/2} = \left\{\frac{2}{9}[(30)^2 + 3(1000)]\right\}^{1/2} = \frac{10\sqrt{78}}{3}(\text{MPa}). \tag{2.120b}$$

(e) As *per* Equation 2.100, components of the hydrostatic part are given by $[(1/3)\sigma_{kk}]\delta_{ij}$. Since $\sigma_{kk} = 30$ from part (a), the matrix of the hydrostatic part of σ becomes

$$\begin{bmatrix} 10 & 0 & 0 \\ 0 & 10 & 0 \\ 0 & 0 & 10 \end{bmatrix} (\text{MPa}). \tag{2.121}$$

Using $\sigma_{kk} = 30$ and Equation 2.100, components of the deviatoric part can be expressed as

$$\sigma'_{ij} = \sigma_{ij} - 10\delta_{ij}. \tag{2.122a}$$

Using the values of σ_{ij} from Equation 2.106, we get the following expression for the matrix of the deviatoric part:

$$[\sigma'] = \begin{bmatrix} 18 & 24 & 0 \\ 24 & 32 & 0 \\ 0 & 0 & -20 \end{bmatrix} - \begin{bmatrix} 10 & 0 & 0 \\ 0 & 10 & 0 \\ 0 & 0 & 10 \end{bmatrix} = \begin{bmatrix} 8 & 24 & 0 \\ 24 & 22 & 0 \\ 0 & 0 & -30 \end{bmatrix} (\text{MPa}). \tag{2.122b}$$

In the state of stress given by Equation 2.106, σ_{zz} is not zero. Therefore, it is not a state of plane stress (at a point) in $x - y$ plane. However, since the principal stress σ_2 is zero, it is a state of plane stress (at a point) in the plane perpendicular to $\hat{e}_2$.

2.3.3 Equations of Motion

Let

$$a = a_x \hat{i} + a_y \hat{j} + a_z \hat{k} \tag{2.123}$$

be the *acceleration vector* at a point of the deformed configuration. The acceleration vector is related to the time derivatives of the displacement vector and velocity vector. But, that relation will be discussed later. Further, let

$$b = b_x \hat{i} + b_y \hat{j} + b_z \hat{k} \tag{2.124}$$

be the *body force vector per unit mass* acting on the body. We shall denote the density in the deformed configuration by the symbol ρ. Note that, in general, a, b and ρ vary from point to point. Thus, they are functions of the coordinates (x, y, z).

Now, we apply the principle of balance of linear momentum in x, y and z directions to a small element (of parallelepiped shape) at a point of the deformed configuration. In the limit as the size of the element tends to zero, it leads to the following three *equations of motion*:

$$\rho\, a_x = \rho\, b_x + \frac{\partial \sigma_{xx}}{\partial x} + \frac{\partial \sigma_{yx}}{\partial y} + \frac{\partial \sigma_{zx}}{\partial z},$$

$$\rho a_y = \rho b_y + \frac{\partial \sigma_{xy}}{\partial x} + \frac{\partial \sigma_{yy}}{\partial y} + \frac{\partial \sigma_{zy}}{\partial z}, \qquad (2.125)$$

$$\rho\, a_z = \rho\, b_z + \frac{\partial \sigma_{xz}}{\partial x} + \frac{\partial \sigma_{yz}}{\partial y} + \frac{\partial \sigma_{zz}}{\partial z}.$$

When the acceleration vector is zero, we get the *equilibrium equations*.

As stated in the introduction, there are three sets of equations which govern the displacements, strains and stresses in a body. Equations 2.125 represent the *first set of governing equations*. The other two sets will be discussed in the remaining sections.

Divergence of the stress tensor σ is denoted by $\nabla \cdot \sigma$. It is a vector and defined by

$$\nabla \cdot \sigma = \left(\frac{\partial \sigma_{xx}}{\partial x} + \frac{\partial \sigma_{xy}}{\partial y} + \frac{\partial \sigma_{xz}}{\partial z} \right) \hat{i} + \left(\frac{\partial \sigma_{yx}}{\partial x} + \frac{\partial \sigma_{yy}}{\partial y} + \frac{\partial \sigma_{yz}}{\partial z} \right) \hat{j}$$
$$+ \left(\frac{\partial \sigma_{zx}}{\partial x} + \frac{\partial \sigma_{zy}}{\partial y} + \frac{\partial \sigma_{zz}}{\partial z} \right) \hat{k}. \qquad (2.126)$$

In index notation, the component i of $\nabla \cdot \sigma$ can be written as

$$(\nabla \cdot \sigma)_i = \sigma_{ij,j}. \qquad (2.127)$$

Using the definition of $\nabla \cdot \sigma$, the equations of motion (Equation 2.125) become

$$\rho a = \rho b + \nabla \cdot \sigma^{\mathrm{T}}. \qquad (2.128)$$

In index notation, they can be expressed as

$$\rho\, a_i = \rho\, b_i + \sigma_{ji,j}. \qquad (2.129)$$

But, since σ is a symmetric tensor, the above equations can be written as

$$\rho a = \rho b + \nabla \cdot \sigma, \qquad (2.130)$$

or

$$\rho\, a_i = \rho\, b_i + \sigma_{ij,j}. \qquad (2.131)$$

Example 2.9: For the beam of Figure 2.5, expressions of the stress components with respect to (x, y, z) coordinate system are

$$\sigma_{xx} = \frac{3wb}{6I_{zz}}(\ell - x)^2 y,$$

$$\sigma_{yy} = -\frac{wb}{6I_{zz}}\left(3h^2 y - y^3 + 2h^3\right),$$

$$\sigma_{xy} = -\frac{3wb}{6I_{zz}}(l - x)\left(h^2 - y^2\right),$$

$$\sigma_{xz} = \sigma_{yz} = \sigma_{zz} = 0.$$

(2.132)

Here, w is the uniform stress acting on the top surface of the beam in negative y direction and b, ℓ and h are the geometric dimensions of the beam (Figure 2.5). Further, I_{zz} is the moment of inertia of the cross-section of the beam about z-axis. Assuming the body force vector b to be zero, verify that the above stress expressions satisfy the equations of motion (Equation 2.125).

Solution: Since the beam is in equilibrium, the acceleration vector is zero. Therefore,

$$a_x = a_y = a_z = 0.$$

(2.133)

Further, since the body force vector is given as zero,

$$b_x = b_y = b_z = 0.$$

(2.134)

Then, the equations of motion (Equation 2.125) reduce to

$$\frac{\partial \sigma_{xx}}{\partial x} + \frac{\partial \sigma_{yx}}{\partial y} + \frac{\partial \sigma_{zx}}{\partial z} = 0,$$

$$\frac{\partial \sigma_{xy}}{\partial x} + \frac{\partial \sigma_{yy}}{\partial y} + \frac{\partial \sigma_{zy}}{\partial z} = 0,$$

$$\frac{\partial \sigma_{xz}}{\partial x} + \frac{\partial \sigma_{yz}}{\partial y} + \frac{\partial \sigma_{zz}}{\partial z} = 0.$$

(2.135)

They are known as the equilibrium equations since the acceleration vector is zero. Differentiating the expressions for σ_{ij} (Equation 2.132) and substituting the derivatives in the first two equilibrium equations, we get

$$\frac{\partial \sigma_{xx}}{\partial x} + \frac{\partial \sigma_{yx}}{\partial y} + \frac{\partial \sigma_{zx}}{\partial z} = \frac{3wb}{6I_{zz}}[-2(\ell - x)y - (\ell - x)(-2y)] + 0 = 0, \quad (2.136a)$$

$$\frac{\partial \sigma_{xy}}{\partial x} + \frac{\partial \sigma_{yy}}{\partial y} + \frac{\partial \sigma_{zy}}{\partial z} = \frac{wb}{6I_{zz}}\left[(-3)(-1)\left(h^2 - y^2\right) - \left(3h^2 - 3y^2\right)\right] + 0 = 0.$$

$$(2.136b)$$

Because σ_{xz}, σ_{yz} and σ_{zz} are zero (Equation 2.132), the third equilibrium equation is identically satisfied.

2.4 Deformation

While discussing stresses in a body, we considered only the deformed configuration. However, for describing the deformation of a body, one must consider both the initial (undeformed) and the deformed configurations of the body. Those are shown in Figure 2.6. However, the forces acting on the deformed configuration and the supports are not shown as they are not necessary to discuss the deformation. For the sake of clarity, overlapping of the undeformed and deformed configurations is avoided by assuming the translation of the body to be very large as shown in the figure. Deformation in a body varies from point to point. Deformation at a point has two aspects. When the body is deformed, a small line element P_0Q_0 at a point undergoes a change in its initial length (Figure 2.6). In general, this happens for the line elements in all directions at that point. Similarly, a pair of line elements P_0R_0 and P_0S_0 undergo a change in their initial angle (Figure 2.6). Again, generally, this happens for every pair of line elements at that point. Strain at a point is a measure of the deformation at that point. Thus, strain at a point consists of the following two infinite sets:

- A measure of change in linear dimension in every direction at that point
- A measure of change in angular dimension for every pair of directions at that point

One can choose various measures to define the strain at a point. For example, one can choose either the change in length per unit length or the change in square length per unit square length or the logarithm of the ratio of new length to the initial length as measures of the change in linear dimension. Further, one can choose the change in angle, the sine function of the change in angle *etc.* as the measures of the change in angle. For specifying the measure of change in angle, usually, the initial angle is chosen to be $\pi/2$ radians.

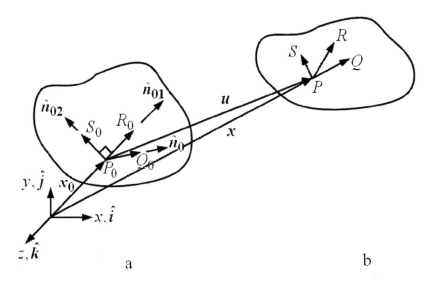

Figure 2.6. Deformation at a point. The length P_0Q_0 changes to PQ in the deformed configuration. The angle $S_0P_0R_0$ changes to SPR in the deformed configuration. **a** Undeformed configuration. **b** Deformed configuration

Deformation at a point is related to the displacement of the neighborhood of that point. The neighborhood of a point is defined as a set of points in the close vicinity of that point. The displacement consists of three parts: (i) displacement due to translation of the neighborhood of that point, (ii) displacement due to rotation of the neighborhood of that point and (iii) displacement due to deformation of the neighborhood of that point. If we consider only the relative displacement of a point with respect to the center of its neighborhood, then it contains the displacement only due to rotation and deformation of the neighborhood. We start our discussion on linear strain tensor at a point with displacement gradient tensor which is a measure of the relative displacement.

2.4.1 Linear Strain Tensor

In this section we first define the displacement gradient tensor at a point. Then we decompose it into symmetric and antisymmetric parts. It is shown that the symmetric part can completely describe the deformation at a point when the deformation is small. It is called the linear strain tensor. The antisymmetric part represents the rotation when the rotation is small.

2.4.1.1 Displacement Gradient Tensor
Let

$$\boldsymbol{u} = u_x\hat{\boldsymbol{i}} + u_y\hat{\boldsymbol{j}} + u_z\hat{\boldsymbol{k}} \qquad (2.137)$$

be the displacement vector at point P_0 whose position vector in the *initial* configuration is given by

$$x_0 = x_0\hat{i} + y_0\hat{j} + z_0\hat{k} \tag{2.138}$$

(Figure 2.6). Consider the following array:

$$[\nabla_0 u] = \begin{bmatrix} \dfrac{\partial u_x}{\partial x_0} & \dfrac{\partial u_x}{\partial y_0} & \dfrac{\partial u_x}{\partial z_0} \\[2mm] \dfrac{\partial u_y}{\partial x_0} & \dfrac{\partial u_y}{\partial y_0} & \dfrac{\partial u_y}{\partial z_0} \\[2mm] \dfrac{\partial u_z}{\partial x_0} & \dfrac{\partial u_z}{\partial y_0} & \dfrac{\partial u_z}{\partial z_0} \end{bmatrix}. \tag{2.139}$$

The subscript zero is used with the symbol ∇ to emphasize the fact that the derivatives are to be taken with respect to the coordinates in the *initial* configuration. Consider a rotated coordinate system (x', y', z') with unit vectors $(\hat{i}', \hat{j}', \hat{k}')$ along them (not shown in Figure 2.6). Further, let the components of the displacement vector u and the position vector x_0 with respect to the rotated coordinates be

$$u' = u_x'\hat{i}' + u_y'\hat{j}' + u_z'\hat{k}', \tag{2.140}$$

$$x_0' = x_0'\hat{i}' + y_0'\hat{j}' + z_0'\hat{k}'. \tag{2.141}$$

In (x', y', z') coordinate system, the array of the displacement derivatives can be written as

$$[\nabla_0 u]' = \begin{bmatrix} \dfrac{\partial u_x'}{\partial x_0'} & \dfrac{\partial u_x'}{\partial y_0'} & \dfrac{\partial u_x'}{\partial z_0'} \\[2mm] \dfrac{\partial u_y'}{\partial x_0'} & \dfrac{\partial u_y'}{\partial y_0'} & \dfrac{\partial u_y'}{\partial z_0'} \\[2mm] \dfrac{\partial u_z'}{\partial x_0'} & \dfrac{\partial u_z'}{\partial y_0'} & \dfrac{\partial u_z'}{\partial z_0'} \end{bmatrix}. \tag{2.142}$$

Using the vector transformation relation (Equation 2.58) for the components of u and x_0, and the chain rule for the derivatives, it can be shown that

$$[\nabla_0 u]' = [Q][\nabla_0 u][Q]^{\mathrm{T}}, \tag{2.143}$$

where the matrix $[Q]$ (Equation 2.54) represents the transformation from (x, y, z) coordinate system to (x', y', z') system. Thus, the components of the array $[\nabla_0 u]$ are the components of a tensor. It is denoted by $\nabla_0 u$ and is called the *displacement gradient tensor* at the point.

2.4.1.2 Linear Strain Tensor

Every tensor can be decomposed into a sum of symmetric and antisymmetric parts. Thus,

$$\nabla_0 u = \frac{1}{2}\left(\nabla_0 u + (\nabla_0 u)^{\mathrm{T}}\right) + \frac{1}{2}\left(\nabla_0 u - (\nabla_0 u)^{\mathrm{T}}\right). \tag{2.144}$$

Here, the first part is the symmetric part of the tensor $\nabla_0 u$ while the second part is the antisymmetric part. At a point, define tensor ε as the symmetric part of $\nabla_0 u$:

$$\varepsilon = \frac{1}{2}\left(\nabla_0 u + (\nabla_0 u)^{\mathrm{T}}\right). \tag{2.145}$$

In matrix notation, this can be written as

$$[\varepsilon] = \frac{1}{2}\left([\nabla_0 u] + [\nabla_0 u]^{\mathrm{T}}\right), \tag{2.146}$$

while in index notation, it can be expressed as

$$\varepsilon_{ij} = \frac{1}{2}\left(u_{i,j} + u_{j,i}\right), \tag{2.147}$$

where it is understood that the comma indicates the derivatives with respect to the coordinates in the *initial* configuration.

Assume that the components of the tensor $\nabla_0 u$ are small compared to 1 everywhere in the body. In many aerospace, civil and mechanical engineering applications, the components of $\nabla_0 u$ are of the order of 10^{-4} to 10^{-6}. Therefore, this assumption is not very restrictive. Let ε_n denote the *unit extension* along the direction $\hat{n}_0$ at point P_0 of the initial configuration (Figure 2.6), *i.e.*, the change in length per unit length at P_0 along the direction $\hat{n}_0$. Further, let γ_{n1n2} denote the *shear* associated with the directions $\hat{n}_{01}$ and $\hat{n}_{02}$ at point P_0 of the initial configuration (Figure 2.6), *i.e.*, the change in angle between the two perpendicular directions $\hat{n}_{01}$ and $\hat{n}_{02}$ at P_0. We denote the arrays of the components of $\hat{n}_0$, $\hat{n}_{01}$ and $\hat{n}_{02}$ with respect to (x, y, z) coordinates by $\{n_0\}$, $\{n_{01}\}$ and $\{n_{02}\}$. Then, under the above assumption, it can be shown that [5]

$$\varepsilon_n = \{n_0\}^{\mathrm{T}} [\varepsilon] \{n_0\}, \tag{2.148}$$

$$\gamma_{n1n2} = 2\{n_{01}\}^{\mathrm{T}} [\varepsilon] \{n_{02}\}. \tag{2.149}$$

Therefore, under the above assumption, if the tensor ε is given at a point, we can find the change in length per unit length along any direction at that point. Further, we can find the change in angle between any pair of perpendicular directions at that point. Thus, under the above assumption, the tensor ε can completely describe the deformation at a point and, therefore, can be used as a strain tensor. It is known as *linear or infinitesimal strain tensor*. Note that the assumption of the components of the tensor $\nabla_0 u$ being small implies that the components of the tensor ε are also small. Therefore, this assumption is called as the *small deformation assumption*. Thus, ε can be used as a strain tensor, only when the deformation is small. The plastic deformation is often not small. Therefore, to describe the plastic deformation, we shall have to look for some other tensor. Such tensors are discussed in Chapter 3.

Note that, by definition (Equation 2.145), the tensor ε is symmetric. Therefore, its components with respect to (x, y, z) coordinate system can be expressed as

$$[\varepsilon] = \begin{bmatrix} \varepsilon_{xx} & \varepsilon_{xy} & \varepsilon_{zx} \\ \varepsilon_{xy} & \varepsilon_{yy} & \varepsilon_{yz} \\ \varepsilon_{zx} & \varepsilon_{yz} & \varepsilon_{zz} \end{bmatrix}. \tag{2.150}$$

Substituting Equations 2.150 and 2.139 into Equation 2.146, we get the following expressions for the strain components:

$$\varepsilon_{xx} = \frac{\partial u_x}{\partial x_0}, \; \varepsilon_{yy} = \frac{\partial u_y}{\partial y_0}, \; \varepsilon_{zz} = \frac{\partial u_z}{\partial z_0}.$$

$$\varepsilon_{xy} = \frac{1}{2} \left(\frac{\partial u_x}{\partial y_0} + \frac{\partial u_y}{\partial x_0} \right),$$

$$\varepsilon_{yz} = \frac{1}{2} \left(\frac{\partial u_y}{\partial z_0} + \frac{\partial u_z}{\partial y_0} \right), \tag{2.151}$$

$$\varepsilon_{zx} = \frac{1}{2} \left(\frac{\partial u_z}{\partial x_0} + \frac{\partial u_x}{\partial z_0} \right).$$

These are known as the *strain-displacement relations*. The tensor, array and index forms of these equations are given by Equations 2.145–2.147. Note that the strain-displacement relations are *linear* when the deformation is small. For plastic deformation, the strain-displacement relations may be non-linear. They are discussed in Chapter 3.

As stated in the introduction, there are three sets of equations which govern the displacements, strains and stresses in a body. This is the *second set of governing equations* when the deformation is small.

By substituting $\hat{n}_0 = \hat{i}$ in Equation 2.148, we find that the component ε_{xx} represents the unit extension (*i.e.*, the change in length per unit length) along the direction which was initially along x-axis. Similarly, the components ε_{yy} and ε_{zz} denote the unit extensions along the directions which were respectively along y and z axes in the initial configuration. These three components, which represent the *deformation in linear dimension* along *three mutually perpendicular directions*, are called *normal strain components*. By substituting $\hat{n}_{01} = \hat{i}$ and $\hat{n}_{02} = \hat{j}$ in Equation 2.149, we find that the component ε_{xy} represents *half* the shear (*i.e.*, half the change in angle) associated with the directions which were along x and y axes in the initial configuration. Similarly, the component ε_{yz} denotes half the shear associated with the directions which were initially along y and z axes. Further, the component ε_{zx} represents half the change in angle between the directions which were originally along z and x axes. These three components, which represent the *deformation in angular dimension* associated with the same *three mutually perpendicular directions*, are called *shear strain components*. The sign convention for the strain components is as follows. A *normal strain* component is considered *positive* if there is *elongation* in that direction and *negative* if there is *compression*. A *shear strain* component is considered *positive* if the *angle decreases* and *negative* if the *angle increases*. Note that the sign convention for the shear strain components is different than what you might expect. However, it is chosen to ensure that a positive shear stress would cause a positive shear strain and *vice versa*.

2.4.1.3 Infinitesimal Rotation Tensor

At a point, define tensor ω as the antisymmetric part of the displacement gradient tensor $\nabla_0 u$:

$$\omega = \frac{1}{2}\left(\nabla_0 u - (\nabla_0 u)^T\right). \tag{2.152}$$

In matrix notation, this can be written as

$$[\omega] = \frac{1}{2}\left([\nabla_0 u] - [\nabla_0 u]^T\right), \tag{2.153}$$

whilst in index notation, it can be expressed as

$$\omega_{i,j} = \frac{1}{2}(u_{i,j} - u_{j,i}), \tag{2.154}$$

where it is understood that the comma indicates the derivatives with respect to the coordinates in the *initial* configuration. It can be shown that when components of the tensor $\nabla_0 u$ are small compared to 1, the tensor ω represents rotation of a neighborhood of the point. Note that when the components of $\nabla_0 u$ are small, the components of ω are also small. Thus, ω represents the rotation only when it is small. We call ω the *infinitesimal rotation tensor*.

The diagonal components of ω, namely ω_{xx}, ω_{yy} and ω_{zz} are zero. The expressions for the *non-diagonal components* of ω are as follows:

$$\omega_{zy} = -\omega_{yz} = \frac{1}{2}\left(\frac{\partial u_z}{\partial y_0} - \frac{\partial u_y}{\partial z_0}\right),$$

$$\omega_{xz} = -\omega_{zx} = \frac{1}{2}\left(\frac{\partial u_x}{\partial z_0} - \frac{\partial u_z}{\partial x_0}\right), \tag{2.155}$$

$$\omega_{yx} = -\omega_{xy} = \frac{1}{2}\left(\frac{\partial u_y}{\partial x_0} - \frac{\partial u_x}{\partial y_0}\right).$$

The components ω_{zy}, ω_{xz} and ω_{yx} represent the angle of rotation respectively about x, y and z axes. They are considered *positive* if they are *counterclockwise* and *negative* if *clockwise*. Since an antisymmetric tensor has only three non-zero components, one can always associate a vector with it. The vector which can be associated with ω is given by

$$\omega_{zy}\hat{i} + \omega_{xz}\hat{j} + \omega_{yx}\hat{k} = \frac{1}{2}\left[\left(\frac{\partial u_z}{\partial y_0} - \frac{\partial u_y}{\partial z_0}\right)\hat{i} + \left(\frac{\partial u_x}{\partial z_0} - \frac{\partial u_z}{\partial x_0}\right)\hat{j} + \left(\frac{\partial u_y}{\partial x_0} - \frac{\partial u_x}{\partial y_0}\right)\hat{k}\right],$$

$$= \frac{1}{2}\,\epsilon_{ijk}\,\frac{\partial u_k}{\partial x_{0j}}\hat{i}_i,$$

$$= \frac{1}{2}\nabla_0 \times u. \tag{2.156}$$

This is consistent with the fact that only small rotation can be expressed as a vector.

Example 2.10: For the beam of Figure 2.7, components of the displacement vector u at a point (x_0, y_0, z_0), with respect to (x, y, z) coordinate system, are given as

$$u_x = A\left\{\frac{1}{2}a^2 y_0 + \left(\frac{1}{2}x_0^2 - \ell x_0\right)y_0 - \frac{1}{4}\left(y_0^3 + y_0 z_0^2\right)\right\},$$ (2.157a)

$$u_y = A\left\{\frac{1}{2}a^2 x_0 + \frac{1}{2}\ell x_0^2 - \frac{1}{6}x_0^3 + \frac{1}{4}(\ell - x_0)(y_0^2 - z_0^2)\right\},$$ (2.157b)

$$u_z = A\left\{\frac{1}{2}(\ell - x_0)y_0 z_0\right\},$$ (2.157c)

where

$$A = \frac{4F_y}{\pi E a^4}.$$ (2.157d)

Here, a, ℓ and F_y are as shown in Figure 2.7. Further, E is a material constant which is defined in Section 2.5.1

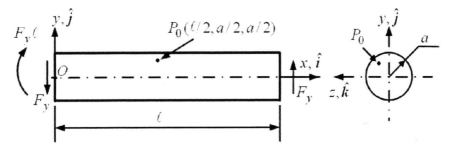

Figure 2.7. A beam of circular cross-section subjected to shear forces and bending moment. The point O is fixed against the translation and rotation. Further, since the deformation is small, the deformed and undeformed configurations almost overlap

(a) Find the components of the displacement gradient tensor $\nabla_0 u$.

(b) Find the components of the linear strain tensor ε and the infinitesimal rotation tensor ω .

(c) Evaluate the strain components at point P_0 (Figure 2.7) whose coordinates are $(x_0, y_0, z_0) = (\ell/2, a/2, a/2)$. Further, at P_0 , find the unit extension along the direction

$$\hat{n}_0 = (1/\sqrt{3})(\hat{i} + \hat{j} + \hat{k}) .$$ (2.158)

and the shear associated with the directions

$$\hat{n}_{01} = (1/5)(3\hat{i} - 4\hat{j}), \quad \hat{n}_{02} = (1/5)(4\hat{i} + 3\hat{j}).$$ (2.159)

(d) Evaluate the non-diagonal components of ω at P_0 .

Solution: (a) Differentiating Equations 2.157a–2.157c, we get the components of the displacement gradient tensor $\nabla_0 u$:

$$\frac{\partial u_x}{\partial x_0} = A[(x_0 - \ell)y_0],$$

$$\frac{\partial u_x}{\partial y_0} = A\left[\frac{1}{2}a^2 + \left(\frac{1}{2}x_0^2 - \ell x_0\right) - \frac{1}{4}\left(3y_0^2 + z_0^2\right)\right],$$

$$\frac{\partial u_x}{\partial z_0} = A\left[-\frac{1}{2}y_0 z_0\right],$$

$$\frac{\partial u_y}{\partial x_0} = A\left[\frac{1}{2}a^2 + \ell x_0 - \frac{1}{2}x_0^2 - \frac{1}{4}\left(y_0^2 - z_0^2\right)\right],$$

$$\frac{\partial u_y}{\partial y_0} = A\left[\frac{1}{2}(\ell - x_0)y_0\right],$$

$$\frac{\partial u_y}{\partial z_0} = A\left[-\frac{1}{2}(\ell - x_0)z_0\right],$$

$$\frac{\partial u_z}{\partial x_0} = A\left[-\frac{1}{2}y_0 z_0\right],$$

$$\frac{\partial u_z}{\partial y_0} = A\left[\frac{1}{2}(\ell - x_0)z_0\right],$$

$$\frac{\partial u_z}{\partial z_0} = A\left[\frac{1}{2}(\ell - x_0)y_0\right]. \tag{2.160}$$

(b) Substituting the expressions of the displacement derivatives of part (a) into the strain-displacement relations (Equation 2.151), we get the components of the linear strain tensor ε :

$$\varepsilon_{xx} = \frac{\partial u_x}{\partial x_0} = A(x_0 - \ell)y_0, \quad \varepsilon_{yy} = \frac{\partial u_y}{\partial y_0} = \frac{A}{2}(\ell - x_0)y_0, \quad \varepsilon_{zz} = \frac{\partial u_z}{\partial z_0} = \frac{A}{2}(\ell - x_0)y_0,$$

$$\varepsilon_{xy} = \frac{1}{2}\left(\frac{\partial u_x}{\partial y_0} + \frac{\partial u_y}{\partial x_0}\right) = \frac{A}{2}\left(a^2 - y_0^2\right), \quad \varepsilon_{yz} = \frac{1}{2}\left(\frac{\partial u_y}{\partial z_0} + \frac{\partial u_z}{\partial y_0}\right) = 0,$$

$$\varepsilon_{zx} = \frac{1}{2}\left(\frac{\partial u_z}{\partial x_0} + \frac{\partial u_x}{\partial z_0}\right) = -\frac{A}{2}y_0 z_0. \tag{2.161}$$

Again, substituting the expressions of the displacement derivatives of part (a) into the rotation-displacement relations (Equation 2.155), we get the non-diagonal components of the infinitesimal rotation tensor ω :

$$\omega_{zy} = -\omega_{yz} = \frac{1}{2}\left(\frac{\partial u_z}{\partial y_0} - \frac{\partial u_y}{\partial z_0}\right) = \frac{A}{2}(\ell - x_0)z_0,$$

$$\omega_{xz} = -\omega_{zx} = \frac{1}{2}\left(\frac{\partial u_x}{\partial z_0} - \frac{\partial u_z}{\partial x_0}\right) = 0, \qquad\qquad (2.162)$$

$$\omega_{yx} = -\omega_{xy} = \frac{1}{2}\left(\frac{\partial u_y}{\partial x_0} - \frac{\partial u_x}{\partial y_0}\right) = \frac{A}{2}\left[x_0(2\ell - x_0) + \frac{1}{2}\left(y_0^2 + z_0^2\right)\right].$$

The diagonal components of ω, namely ω_{xx}, ω_{yy} and ω_{zz}, are of course zero.

(c) We obtain values of the strain components at point P_0 by substituting $(x_0, y_0, z_0) = (\ell/2, a/2, a/2)$ in the expressions of the strain components of part (b). Then, the strain matrix at point P_0 becomes

$$[\varepsilon] = \begin{bmatrix} \varepsilon_{xx} & \varepsilon_{xy} & \varepsilon_{zx} \\ \varepsilon_{xy} & \varepsilon_{yy} & \varepsilon_{yz} \\ \varepsilon_{zx} & \varepsilon_{yz} & \varepsilon_{zz} \end{bmatrix} = \frac{Aa}{8} \begin{bmatrix} -2\ell & 3a & -a \\ 3a & \ell & 0 \\ -a & 0 & \ell \end{bmatrix}. \qquad (2.163)$$

To get the unit extension along the direction $\hat{n}_0$ at point P_0, we substitute the above equation along with the components of $\hat{n}_0$ (Equation 2.158) in Equation 2.148:

$$\varepsilon_n = \{n_0\}^{\mathrm{T}}[\varepsilon]\{n_0\} = \left\{\frac{1}{\sqrt{3}} \quad \frac{1}{\sqrt{3}} \quad \frac{1}{\sqrt{3}}\right\} \frac{Aa}{8} \begin{bmatrix} -2\ell & 3a & -a \\ 3a & \ell & 0 \\ -a & 0 & \ell \end{bmatrix} \begin{Bmatrix} \dfrac{1}{\sqrt{3}} \\ \dfrac{1}{\sqrt{3}} \\ \dfrac{1}{\sqrt{3}} \end{Bmatrix} = \frac{1}{6}Aa^2.$$

$$(2.164a)$$

To get the shear associated with the directions $\hat{n}_{01}$ and $\hat{n}_{02}$ at point P_0, we substitute Equation 2.163 along with the components of $\hat{n}_{01}$ and $\hat{n}_{02}$ (Equation 2.159) in Equation 2.149:

$$\gamma_{n1n2} = 2\{n_{01}\}^{T}[\varepsilon]\{n_{02}\} = 2\left\{\frac{3}{5} \quad -\frac{4}{5} \quad 0\right\}\frac{Aa}{8}\begin{bmatrix} -2\ell & 3a & -a \\ 3a & \ell & 0 \\ -a & 0 & \ell \end{bmatrix}\left\{\begin{matrix} \frac{4}{5} \\ \frac{3}{5} \\ 0 \end{matrix}\right\},$$

$$= -\frac{1}{100}Aa(36\ell + 21a).$$

(2.164b)

(d) We obtain values of the non-diagonal rotation components at point P_0, by substituting $(x_0, y_0, z_0) = (\ell/2, a/2, a/2)$ in the rotation-displacement equations of part (b). We get

$$\omega_{zy} = \frac{A}{2}(\ell - x_0)z_0 = \frac{1}{8}A\ell a,$$

$$\omega_{xz} = 0,$$

(2.165)

$$\omega_{yx} = \frac{A}{2}\left[x_0(2\ell - x_0) + \frac{1}{2}\left(y_0^2 + z_0^2\right)\right] = \frac{1}{8}A\left(3\ell^2 + a^2\right).$$

For the following values of geometric, force and material parameters:

$$\ell = 200\,\text{mm}, \quad a = 10\,\text{mm}, \quad F_y = 100\,\text{N}, \quad E = 2\times10^5\,\text{N}/\text{mm}^2$$

(2.166a)

we get

$$A = \frac{4F_y}{\pi Ea^4} = 6.34\times10^{-8}.$$

(2.166b)

Then we obtain

$$\varepsilon_n = \frac{1}{6}Aa^2 = 1.06\times10^{-6}, \quad \gamma_{n1n2} = -\frac{1}{100}Aa(36\ell + 21a) = -47.17\times10^{-6}\,\text{rad},$$

(2.167a)

$$\omega_{zy} = \frac{A\ell a}{8} = 1.59\times10^{-5}\,\text{rad}, \quad \omega_{yx} = \frac{1}{8}A\left(3\ell^2 + a^2\right) = 9.51\times10^{-4}\,\text{rad},$$

(2.167b)

Thus, for a typical situation, the deformation and rotation are quite small.

2.4.2 Analysis of Strain at a Point

As stated earlier, in this section we carry out the analysis of strain at a point to discuss the concepts of principal strains and principal directions, principal invariants, maximum shear, volumetric strain and the hydrostatic and deviatoric parts of strain.

2.4.2.1 Principal Strains, Principal Directions and Principal Invariants

There exist at least three mutually perpendicular directions (in the initial configuration) such that the shear (γ_{n1n2}) associated with these directions is zero. It means these directions also remain perpendicular in the deformed configuration. These directions are called as the *principal directions* (of strain). The unit extensions (ε_n) along these directions are called the *principal strains*. We denote the principal strains as ε_1, ε_2 and ε_3 and the unit vectors along the principal directions (of strain) as $\hat{e}_1$, $\hat{e}_2$ and $\hat{e}_3$. Recall that the same notation has been used for the unit vectors along the principal directions (of stress). However, whether we are referring to the principal directions of stress or strain will be clear from the context. Further, the principal directions of stress exist in the deformed configuration whereas the principal directions of strain exist in the initial configuration. We arrange the principal strains as

$$\varepsilon_1 \ge \varepsilon_2 \ge \varepsilon_3 . \tag{2.168}$$

The senses of the unit vectors along the principal directions are so chosen that they always form a right-sided system. Thus,

$$\hat{e}_1 \cdot \hat{e}_2 \times \hat{e}_3 = +1 . \tag{2.169}$$

Since the unit extension along a principal direction i $(i = 1,2,3)$ is ε_i and the shear associated with these principal directions is zero, the components of the linear strain tensor, with respect to the principal directions as the coordinate system, become

$$[\varepsilon]^P = \begin{bmatrix} \varepsilon_1 & 0 & 0 \\ 0 & \varepsilon_2 & 0 \\ 0 & 0 & \varepsilon_3 \end{bmatrix} . \tag{2.170}$$

It can easily be verified that, at a point, *maximum* value of the *unit extension* (ε_n) with respect to the orientation of the direction $\hat{n}_0$ is ε_1. Further, the *minimum* value is ε_3.

It can be shown that the principal strains are the eigenvalues or the principal values and the unit vectors along the principal directions are the eigenvectors of the linear strain tensor. The principal strains are determined as the roots of the following equation:

$$\lambda^3 - I_\varepsilon \lambda^2 - II_\varepsilon \lambda - III_\varepsilon = 0, \tag{2.171}$$

where

$$I_\varepsilon = \varepsilon_{ii}, \tag{2.172}$$

$$II_\varepsilon = \frac{1}{2}(\varepsilon_{ij}\varepsilon_{ij} - \varepsilon_{ii}\varepsilon_{jj}), \tag{2.173}$$

$$III_\varepsilon = \epsilon_{ijk}\,\varepsilon_{1i}\varepsilon_{2j}\varepsilon_{3k}. \tag{2.174}$$

Here, I_ε, II_ε and III_ε are the three *principal invariants* of the linear strain tensor. After finding the principal strains, the unit vectors $\hat{e}_i$ along the principal directions are found from an equation similar to Equation 2.85:

$$\left([\varepsilon] - \varepsilon_i[1]\right)\{e_i\} = \{0\} \quad (\text{no sum over } i). \tag{2.175}$$

2.4.2.2 Maximum Shear
It can be shown that, at a point, *maximum* value of the *shear* (γ_{n1n2}) with respect to the orientation of the directions $\hat{n}_{01}$ and $\hat{n}_{02}$ is

$$\left|\gamma_{n1n2}\right|_{\max} = \varepsilon_1 - \varepsilon_3. \tag{2.176}$$

Further, the directions associated with the maximum shear are given by

$$\hat{n}_{01} = \pm\frac{1}{\sqrt{2}}(\hat{e}_1 + \hat{e}_3), \; \hat{n}_{02} = \pm\frac{1}{\sqrt{2}}(\hat{e}_1 - \hat{e}_3). \tag{2.177}$$

2.4.2.3 Volumetric Strain, Decomposition into Hydrostatic and Deviatoric Parts
The change in volume per unit volume of a small element around a point is known as the *volumetric strain* and is denoted by ε_v. It can be shown that when the deformation is small (*i.e.*, when the components of the tensor $\nabla_0 u$ are small compared to 1), ε_v is given by [2,4]

$$\varepsilon_v = tr\varepsilon = \varepsilon_{ii}. \tag{2.178}$$

Similar to the decomposition of the stress tensor σ (Equation 2.99), the linear strain tensor ε also can be decomposed as

$$\varepsilon = \left(\frac{1}{3}tr\varepsilon\right)\mathbf{1} + \varepsilon', \quad (tr\varepsilon' = 0).$$ (2.179)

In index notation, this can be written as

$$\varepsilon_{ij} = \left(\frac{1}{3}\varepsilon_{kk}\right)\delta_{ij} + \varepsilon'_{ij}, \quad (\varepsilon'_{ii} = 0).$$ (2.180)

Note that, since ε is a symmetric tensor, ε' is also a symmetric tensor. The first part of Equations 2.179 and 2.180 is called the *hydrostatic* part of ε while the second part is called the *deviatoric* part of ε. Since, $tr\varepsilon$ is volumetric strain, the hydrostatic part of ε represents a deformation in which there is only change in volume (or size) but no change in shape. Such a deformation is known as *dilatation*. Since $tr\varepsilon'$ is zero, the deviatoric part of ε represents a deformation in which there is no change in volume but only change in shape. Such deformation is called *distortion*.

As stated earlier, in *isotropic materials*, the hydrostatic part of stress tensor causes only dilatation type of deformation while the deviatoric part causes only the distortion type of deformation. The yielding consists of only the distortion type of deformation. Therefore, as we shall see in Chapter 3, in isotropic ductile materials yielding is caused only by the deviatoric part of stress tensor.

Example 2.11: Components of the linear strain tensor ε at a point with respect to the (x, y, z) coordinate system are given by

$$[\varepsilon] = \begin{bmatrix} 0 & 2 & 2 \\ 2 & 0 & 2 \\ 2 & 2 & 0 \end{bmatrix} 10^{-4}.$$ (2.181)

(a) Find the principal invariants of ε.
(b) Find the principal strains ε_i and the unit vectors $\hat{e}_i$ along the principal directions. Arrange ε_i such that $\varepsilon_1 \geq \varepsilon_2 \geq \varepsilon_3$. Further, choose the senses of $\hat{e}_i$ such that $\hat{e}_1 \cdot \hat{e}_2 \times \hat{e}_3 = +1$.
(c) Find the maximum shear (γ_{n1n2}) and the directions $\hat{n}_{01}$ and $\hat{n}_{02}$ associated with maximum shear.
(d) Using the tensor transformation relation, find the components of ε with respect to the principal directions as the coordinate system.

Solution: (a) Substituting the values of ε_{ij} from Equation 2.181 and the values of $\in_{ijk}$ from Equation 2.14 in Equations 2.172–2.174, we get

$$I_\varepsilon = \varepsilon_{ii} = (0+0+0)\times10^{-4} = 0, \tag{2.182a}$$

$$II_\varepsilon = \frac{1}{2}(\varepsilon_{ij}\varepsilon_{ij} - \varepsilon_{ii}\varepsilon_{jj}) = \frac{1}{2}\left[3(0)^2 + 6(2)^2 - (0)(0)\right]\times10^{-8} = 12\times10^{-8}, \tag{2.182b}$$

$$III_\varepsilon = \in_{ijk} \varepsilon_{1i}\varepsilon_{2j}\varepsilon_{3k},$$
$$= \varepsilon_{11}(\varepsilon_{22}\varepsilon_{33} - \varepsilon_{23}\varepsilon_{32}) + \varepsilon_{12}(\varepsilon_{23}\varepsilon_{31} - \varepsilon_{21}\varepsilon_{33}) + \varepsilon_{13}(\varepsilon_{21}\varepsilon_{32} - \varepsilon_{22}\varepsilon_{31})$$
$$= \left[0(0\times0 - 2\times2) + 2(2\times2 - 2\times0) + 2(2\times2 - 0\times2)\right]\times10^{-12}$$
$$= 16\times10^{-12}. \tag{2.182c}$$

(b) Substituting the values of I_ε, II_ε and III_ε from part (a), the cubic equation for λ (Equation 2.171) becomes

$$\lambda^3 - 0\lambda^2 - 12\times10^{-8}\lambda - 16\times10^{-12} = 0. \tag{2.183}$$

The roots of this equation are $\lambda = 4\times10^{-4}, -2\times10^{-4}, -2\times10^{-4}$. Thus, we have a double eigenvalue. Arranging the roots in decreasing order, we get the following values of the principal strains:

$$\varepsilon_1 = 4\times10^{-4}, \ \varepsilon_2 = -2\times10^{-4}, \ \varepsilon_3 = -2\times10^{-4}. \tag{2.184}$$

To find the unit vectors $\hat{e}_i$ along the principal directions, we follow the procedure of Example 2.8(b). Thus, corresponding to the first eigenvalue $\varepsilon_1 = 4\times10^{-4}$, we get the following expression for the first unit vector:

$$\hat{e}_1 = \frac{1}{\sqrt{3}}(\hat{i} + \hat{j} + \hat{k}). \tag{2.185a}$$

While finding the eigenvector corresponding to the eigenvalue $\varepsilon_2 = -2\times10^{-4}$, it is observed that only one scalar equation out of the three equations (Equation 2.175) satisfied by the components of $\hat{e}_2$ is linearly independent. This means the eigenvector has no unique direction. In fact, it can be shown that every vector in the plane perpendicular to $\hat{e}_1$ is an eigenvector of $\varepsilon_2 = -2\times10^{-4}$. This happens because it is a double eigenvalue. Therefore, we choose any pair of orthonormal vectors (i.e., any two unit vectors perpendicular to each other) in the plane perpendicular to $\hat{e}_1$ as the vectors $\hat{e}_2$ and $\hat{e}_3$. We make the following choice:

$$\hat{e}_2 = \frac{1}{\sqrt{2}}(\hat{i} - \hat{j}), \qquad\qquad\qquad (2.185b)$$

$$\hat{e}_3 = \frac{1}{\sqrt{6}}(\hat{i} + \hat{j} - 2\hat{k}). \qquad\qquad\qquad (2.185c)$$

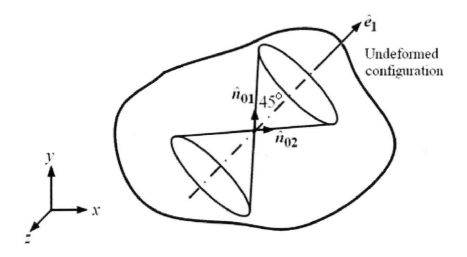

Figure 2.8. Conical surface on which the directions $\hat{n}_{01}$ and $\hat{n}_{02}$ associated with maximum shear lie when the second and third principal stresses are equal. The vector $\hat{e}_1$ represents the first principal direction

(c) Maximum shear is given by Equation 2.176. Substituting the values of ε_1 and ε_3 from part (b) in this equation, we get

$$\left| \gamma_{n1n2} \right|_{max} = \varepsilon_1 - \varepsilon_3 = [4 - (-2)] \times 10^{-4} = 6 \times 10^{-4}. \qquad\qquad (2.186)$$

The directions $\hat{n}_{01}$ and $\hat{n}_{02}$ associated with $\left| \gamma_{n1n2} \right|_{max}$ are given by Equation 2.177. So we can obtain them by substituting the expressions for $\hat{e}_1$ and $\hat{e}_3$ from part (b) into this equation. However, the vector $\hat{e}_3$ has no unique direction. As stated earlier, it can have any direction in the plane perpendicular to $\hat{e}_1$. Equation 2.185c is just one such direction. Therefore, the directions $\hat{n}_{01}$ and $\hat{n}_{02}$ associated with $\left| \gamma_{n1n2} \right|_{max}$ are also not unique. Equation 2.177 shows that whereas $\hat{n}_{01}$ (with + sign) makes an angle of $45°$ with both $\hat{e}_1$ and $\hat{e}_3$ directions, $\hat{n}_{02}$ (with + sign) makes an angle of $45°$ with $\hat{e}_1$ direction but $135°$ with $\hat{e}_3$ direction. Thus, the directions $\hat{n}_{01}$ and $\hat{n}_{02}$ lie on the surface of a cone whose axis is along $\hat{e}_1$ and semi-cone angle is 45° (Figure 2.8).

(d) To find the components of ε with respect to the principal directions, we first evaluate the transformation matrix $[Q]$. For that purpose, we substitute the direction cosines of the principal directions as given by Equations 2.185a–c into the expression (Equation 2.54) for $[Q]$. Thus, we get

$$[Q] = \begin{bmatrix} \dfrac{1}{\sqrt{3}} & \dfrac{1}{\sqrt{3}} & \dfrac{1}{\sqrt{3}} \\ \dfrac{1}{\sqrt{2}} & -\dfrac{1}{\sqrt{2}} & 0 \\ \dfrac{1}{\sqrt{6}} & \dfrac{1}{\sqrt{6}} & -\dfrac{2}{\sqrt{6}} \end{bmatrix}. \tag{2.187}$$

Using the tensor transformation relation (Equation 2.55), we get the following matrix of the components of the strain tensor with respect to the principal directions:

$$[\varepsilon]^P = [Q][\varepsilon][Q]^{\mathrm{T}},$$

$$= \begin{bmatrix} \dfrac{1}{\sqrt{3}} & \dfrac{1}{\sqrt{3}} & \dfrac{1}{\sqrt{3}} \\ \dfrac{1}{\sqrt{2}} & -\dfrac{1}{\sqrt{2}} & 0 \\ \dfrac{1}{\sqrt{6}} & \dfrac{1}{\sqrt{6}} & -\dfrac{2}{\sqrt{6}} \end{bmatrix} \begin{bmatrix} 0 & 2 & 2 \\ 2 & 0 & 2 \\ 2 & 2 & 0 \end{bmatrix} \times 10^{-4} \begin{bmatrix} \dfrac{1}{\sqrt{3}} & \dfrac{1}{\sqrt{2}} & \dfrac{1}{\sqrt{6}} \\ \dfrac{1}{\sqrt{3}} & -\dfrac{1}{\sqrt{2}} & \dfrac{1}{\sqrt{6}} \\ \dfrac{1}{\sqrt{3}} & 0 & -\dfrac{2}{\sqrt{6}} \end{bmatrix},$$

$$= \begin{bmatrix} 4 & 0 & 0 \\ 0 & -2 & 0 \\ 0 & 0 & -2 \end{bmatrix} \times 10^{-4}. \tag{2.188}$$

Note that even if we choose any other pair of orthonormal vectors (in the plane perpendicular to $\hat{e}_1$) as the second and third principal directions, then also the tensor transformation relations will lead to the same expression for $[\varepsilon]^P$.

Example 2.12: Components of the linear strain tensor ε at point O of the thin plate of Figure 2.9, with respect to (x, y, z) coordinate system, are given as

$$[\varepsilon] = \frac{\sigma_0}{E} \begin{bmatrix} 1-v & 0 & 0 \\ 0 & 1-v & 0 \\ 0 & 0 & -2v \end{bmatrix}.$$
(2.189)

Here, σ_0 is the maximum value of the parabolically varying tensile stresses acting on the edges of the plate (Figure 2.9). Further, E and v are the material constants which are defined in Section 2.5.1.

(a) Find the volumetric strain ε_v at point O.

(b) Find the hydrostatic and deviatoric parts of ε at point O.

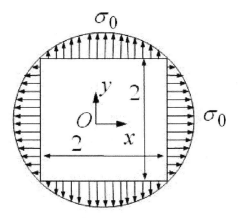

Figure 2.9. A thin square plate subjected to in-plane stresses. Since the deformation is small, the initial and deformed configurations almost overlap

Solution: (a) Using the values of ε_{ij} from Equation 2.189, we get

$$\varepsilon_{ii} = \frac{\sigma_0}{E}[(1-v)+(1-v)+(-2v)] = \frac{\sigma_0}{E} 2(1-2v).$$
(2.190)

Substituting this expression into Equation 2.178 for the volumetric strain, we get

$$\varepsilon_v = \varepsilon_{ii} = \frac{\sigma_0}{E} 2(1-2v).$$
(2.191)

(b) As per Equation 2.180, components of the hydrostatic part are given by $[(1/3)\varepsilon_{kk}]\delta_{ij}$. Since $\varepsilon_{kk} = \frac{\sigma_0}{E} 2(1-2v)$ from part (a), the matrix of the hydrostatic part of ε becomes

$$\frac{\sigma_0}{E}\begin{bmatrix} \frac{2}{3}(1-2v) & 0 & 0 \\ 0 & \frac{2}{3}(1-2v) & 0 \\ 0 & 0 & \frac{2}{3}(1-2v) \end{bmatrix}. \tag{2.192}$$

Using $\varepsilon_{kk} = \dfrac{\sigma_0}{E}2(1-2v)$ and Equation 2.180, components of the deviatoric part can be expressed as

$$\varepsilon'_{ij} = \varepsilon_{ij} - \frac{\sigma_0}{E}\frac{2}{3}(1-2v)\delta_{ij}. \tag{2.193}$$

Using the values of ε_{ij} from Equation 2.189, we get the following expression for the matrix of the deviatoric part:

$$[\varepsilon]' = \frac{\sigma_0}{E}\begin{bmatrix} 1-v & 0 & 0 \\ 0 & 1-v & 0 \\ 0 & 0 & -2v \end{bmatrix} - \frac{\sigma_0}{E}\begin{bmatrix} \frac{2}{3}(1-2v) & 0 & 0 \\ 0 & \frac{2}{3}(1-2v) & 0 \\ 0 & 0 & \frac{2}{3}(1-2v) \end{bmatrix},$$

$$= \frac{\sigma_0}{E}\begin{bmatrix} \frac{1+v}{3} & 0 & 0 \\ 0 & \frac{1+v}{3} & 0 \\ 0 & 0 & -\frac{2(1+v)}{3} \end{bmatrix}. \tag{2.194}$$

2.4.3 Compatibility Conditions

Suppose the linear strain tensor at a point is known as a function of initial coordinates (x_0, y_0, z_0) of the point and we wish to find the displacement vector *u* at that point by integrating the strain-displacement relations (Equation 2.151). Then, we have six scalar equations to solve but only three scalar unknowns to be determined. These unknowns are the components (u_x, u_y, u_z) of the displacement vector. Is it possible to get a single-valued solution in this case? The *necessary condition* to get the single-valued displacements, in this case, is that the strain components should satisfy the following constraints [2–4]:

$$E_1 \equiv \frac{\partial^2 \varepsilon_{xx}}{\partial y_0^2} + \frac{\partial^2 \varepsilon_{yy}}{\partial x_0^2} - 2\frac{\partial^2 \varepsilon_{xy}}{\partial x_0 \partial y_0} = 0,$$

$$E_2 \equiv \frac{\partial^2 \varepsilon_{yy}}{\partial z_0^2} + \frac{\partial^2 \varepsilon_{zz}}{\partial y_0^2} - 2\frac{\partial^2 \varepsilon_{yz}}{\partial y_0 \partial z_0} = 0,$$

$$E_3 \equiv \frac{\partial^2 \varepsilon_{zz}}{\partial x_0^2} + \frac{\partial^2 \varepsilon_{xx}}{\partial z_0^2} - 2\frac{\partial^2 \varepsilon_{zx}}{\partial z_0 \partial x_0} = 0, \qquad (2.195)$$

$$E_4 \equiv \frac{\partial^2 \varepsilon_{xx}}{\partial y_0 \partial z_0} - \frac{\partial}{\partial x_0}\left[-\frac{\partial \varepsilon_{yz}}{\partial x_0} + \frac{\partial \varepsilon_{zx}}{\partial y_0} + \frac{\partial \varepsilon_{xy}}{\partial z_0} \right] = 0,$$

$$E_5 \equiv \frac{\partial^2 \varepsilon_{yy}}{\partial z_0 \partial x_0} - \frac{\partial}{\partial y_0}\left[\frac{\partial \varepsilon_{yz}}{\partial x_0} - \frac{\partial \varepsilon_{zx}}{\partial y_0} + \frac{\partial \varepsilon_{xy}}{\partial z_0} \right] = 0,$$

$$E_6 \equiv \frac{\partial^2 \varepsilon_{zz}}{\partial x_0 \partial y_0} - \frac{\partial}{\partial z_0}\left[\frac{\partial \varepsilon_{yz}}{\partial x_0} + \frac{\partial \varepsilon_{zx}}{\partial y_0} - \frac{\partial \varepsilon_{xy}}{\partial z_0} \right] = 0.$$

These conditions are known as the strain compatibility conditions or integrability conditions.

While finding three unknowns from six equations, it would seem that only three constraints are needed. But we have six conditions. However, it can be shown that only three out of the six compatibility conditions are independent [2].

It can be shown that conditions (Equation 2.195) are also *sufficient* for getting the single-valued displacements, but only for *simply-connected regions* [2–4]. For multiply-connected regions, additional compatibility conditions are required. Further, when the conditions at Equation 2.195 are satisfied in a simply-connected region, only the *non-rigid part of the displacement vector* becomes single-valued. Uniqueness of the *rigid part of the displacement vector* depends on the displacement boundary conditions of the problem.

Example 2.13: Components of the linear strain tensor ε at a point (x_0, y_0, z_0), with respect to (x, y, z) coordinate system, are given as

$$\varepsilon_{xx} = a\left(x_0^2 + y_0^2 \right),$$

$$\varepsilon_{yy} = b\left(x_0^2 + y_0^2 \right),$$

$$\varepsilon_{xy} = cx_0 y_0, \qquad (2.196)$$

$$\varepsilon_{xz} = \varepsilon_{yz} = \varepsilon_{zz} = 0,$$

where a, b and c are constants. Check whether this state of strain is compatible.

Solution: Note that the strain components ε_{xz}, ε_{yz} and ε_{zz} are zero. Further, the components ε_{xx}, ε_{yy} and ε_{xy} are independent of z_0. Therefore, the last five compatibility conditions (Equation 2.195) are identically satisfied. Substituting the expressions (Equation 2.196) for ε_{xx}, ε_{yy} and ε_{xy} in the first compatibility condition we get

$$E_1 \equiv \frac{\partial^2 \varepsilon_{xx}}{\partial y_0^2} + \frac{\partial^2 \varepsilon_{yy}}{\partial x_0^2} - 2\frac{\partial^2 \varepsilon_{xy}}{\partial x_0 \partial y_0} = 2a + 2b - 2c. \tag{2.197}$$

Therefore, the given state of strain is compatible if

$$a + b = c. \tag{2.198}$$

Note that when the strain components ε_{xz}, ε_{yz} and ε_{zz} are zero at a point, the state of deformation is called as the *state of plane strain (at a point) in $x - y$ plane*. When these strain components are zero at every point of the body and if, additionally, the remaining strain components ε_{xx}, ε_{yy} and ε_{xy} are independent of z_0, it is known as the *state of plane strain (in a body) in $x - y$ plane*. It is seen that the state of strain described by Equation 2.196 is of this type.

2.5 Material Behavior

Relations which characterize various responses (like mechanical, thermal, electrical *etc.*) of a material are called the *constitutive equations*. It is these relations which differentiate one material from another. These relations are based on experimental observation.

In this section, we shall consider only mechanical response. It is possible that a mechanical response may be caused by non-mechanical stimuli like a change in temperature or an application of an electromagnetic field. But we shall consider only *purely mechanical response*, that is, a mechanical response caused by a mechanical stimulus. Constitutive equation for such a response is usually expressed as a relation between the applied forces and the resulting deformation. In order to eliminate effects of the shape and size of the body and the nature and point of application of the loading, normally the constitutive equation is formulated for a material particle rather than for the whole body. For a purely mechanical response, such an equation is expressed as a relation between the stress and a measure of deformation (strain) and/or a measure of rate of deformation (strain rate).

There are various types of mechanical responses. The basic responses are (i) elastic response, (ii) plastic response and (iii) viscous response. Sometimes the response consists of a combination of the basic responses. Further, a material may exhibit different types of responses over different ranges of deformation. For example, metals behave elastically at small deformation but exhibit plastic behavior at large deformation. As a result, it is quite difficult to express the complete mechanical behavior of a material over the entire range of deformation through just one single equation. Therefore, we simplify the constitutive equation by restricting ourselves to only *small deformation*. As stated earlier, metals behave elastically at small deformation. Therefore, in this section we shall develop constitutive equation for *elastic* behavior of metals at small deformation. In elastic response, the stress depends on the instantaneous value of strain. Further, this relation is one-to-one. It means if the external forces acting on the body are removed (*i.e.*, if the stress is reduced to the value zero), the strain will also attain the value zero, thereby bringing the body to the original undeformed configuration.

2.5.1 Elastic Stress-Strain Relations for Small Deformation

For small deformation, the linear strain tensor ε can be used as a measure of the deformation. Therefore, for small deformation, the constitutive equation becomes a relation between σ and ε.

2.5.1.1 One-Dimensional Experimental Observations

As stated earlier, constitutive equations are based on experimental observation. Therefore, let us first see what the experimental observation is about the relation between σ and ε. The simplest experiment is the *tension test*. In tension test, a rod of uniform cross-section is subjected to an (axial) tensile force F_x as shown in Figure 2.10. The geometry and loading are such that it is reasonable to assume that the state of stress is one-dimensional and homogeneous in the region away from the ends. That is, the only non-zero stress component is σ_{xx} and it is constant. Further, the state of strain also can be assumed to be homogeneous in the region away from the ends. But, the number of non-zero strain components is not one. Only the shear strain components can be assumed to be zero. Thus, there are three non-zero strain components, namely ε_{xx}, ε_{yy} and ε_{zz} and all are constant.

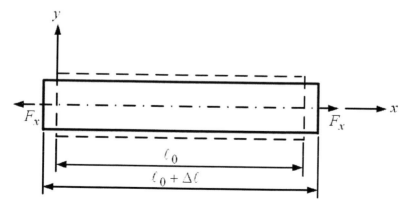

Figure 2.10. Rod subjected to axial tensile forces. The dashed lines indicate the undeformed configuration

For the rod of Figure 2.10, we define the following:

$$\sigma_0 = \frac{F_x}{A_0}, \qquad\qquad (2.199)$$

$$e = \frac{\Delta\ell}{\ell_0}, \qquad\qquad (2.200)$$

where A_0 is the initial area of the cross-section of the rod, ℓ_0 is the initial length of the rod and $\Delta\ell$ is the change in length corresponding to the (axial) tensile force F_x. Note that, when the deformation is small (*i.e.*, when the area A_0 does not change much), σ_0 is almost equal to σ_{xx} component of the stress tensor. However, when the change in area is large, σ_0 does not represent the *true stress*.

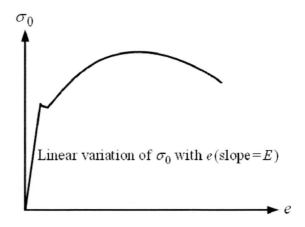

Figure 2.11. Variation of engineering stress with engineering strain for a ductile material in tension test

Therefore, we call σ_0 *engineering* or *nominal stress*. Again, when the deformation is small (*i.e.*, when the change in length $\Delta\ell$ is small), e is equal to $\partial u/\partial x$ and thus represents ε_{xx} component of the *linear* or *infinitesimal strain* tensor. But when the change in length is large, ε_{xx} or $\partial u/\partial x$ does not become equal to e. Therefore, we call e the *engineering strain*.

Figure 2.11 shows the variation of σ_0 with e up to fracture for a typical metal (mild steel). The figure shows that σ_0 varies linearly with e when the deformation is small. But, for small deformation, σ_0 is same as σ_{xx} and e is equal to ε_{xx}. Therefore, for small deformation, σ_{xx} varies linearly with ε_{xx}.

It should be noted that the stress-strain relations need not be linear for all elastic materials. For a material like rubber, which is elastic in nature, the stress-strain relations are non-linear.

2.5.1.2 Generalization to Three-Dimensional Case

One can generalize the one-dimensional experimental observation of Figure 2.11 (for small deformation) as follows. For small deformation, one can assume that each stress component depends linearly on all the components of the linear strain tensor. Thus,

$$\sigma_{xx} = C_{xxxx}\varepsilon_{xx} + C_{xxxy}\varepsilon_{xy} + C_{xxxz}\varepsilon_{xz} + C_{xxyx}\varepsilon_{yx} + \ldots\ldots\ldots + C_{xxzz}\varepsilon_{zz},$$

$$\sigma_{xy} = C_{xyxx}\varepsilon_{xx} + C_{xyxy}\varepsilon_{xy} + C_{xyxz}\varepsilon_{xz} + C_{xyyx}\varepsilon_{yx} + \ldots\ldots\ldots + C_{xyzz}\varepsilon_{zz},$$

$$\ldots\ldots\ldots\ldots\ldots\ldots\ldots\ldots\ldots\ldots\ldots\ldots\ldots\ldots\ldots\ldots\ldots\ldots\ldots\ldots\ldots\ldots\ldots\ldots\ldots\ldots\ldots\ldots,$$

$$\ldots\ldots\ldots\ldots\ldots\ldots\ldots\ldots\ldots\ldots\ldots\ldots\ldots\ldots\ldots\ldots\ldots\ldots\ldots\ldots\ldots\ldots\ldots\ldots\ldots\ldots\ldots\ldots,$$

$$\sigma_{zz} = C_{zzxx}\varepsilon_{xx} + C_{zzxy}\varepsilon_{xy} + C_{zzxz}\varepsilon_{xz} + C_{zzyx}\varepsilon_{yx} + \ldots\ldots\ldots + C_{zzzz}\varepsilon_{zz}.$$

$$(2.201)$$

The stress-strain relations given by Equation 2.201 have 81 material constants. These constants characterize the elastic response of the metal at small deformation. These constants need to be determined by experiments.

In index notation, Equation 2.201 can be written as

$$\sigma_{ij} = C_{ijkl}\varepsilon_{kl}. \tag{2.202}$$

Note that C_{ijkl} are the components of a *fourth order tensor* C which is called the *elasticity tensor*. In three dimensions, a fourth order tensor has $3^4 = 81$ components.

2.5.1.3 Restrictions on Elasticity Tensor C

One can reduce the number of constants in the stress-strain relations as follows. Since σ_{ij} and ε_{kl} are *symmetric* tensors, that is,

$$\sigma_{ij} = \sigma_{ji}, \ \varepsilon_{kl} = \varepsilon_{lk}, \tag{2.203}$$

the components C_{ijkl} must satisfy the following relations:

$$C_{ijkl} = C_{jikl}, C_{ijkl} = C_{ijlk}. \tag{2.204}$$

These relations imply that the tensor C has only 36 independent components.

Further simplification can be achieved by using the conservative nature of the internal forces generated by elastic response. For a certain class of elastic materials, work done by the internal forces, during deformation, is path-independent. As a result, the work of deformation (per unit volume) during an infinitesimal deformation can be expressed as an exact differential of a scalar quantity which has the dimensions of energy per unit volume (called the *strain energy density*). The work of deformation (per unit volume) during an infinitesimal deformation is $\sigma_{ij}d\varepsilon_{ij} = C_{ijkl}\varepsilon_{kl}d\varepsilon_{ij}$. For this expression to be an exact differential, the tensor C must be symmetric in the first two and the last two indices:

$$C_{ijkl} = C_{klij}. \tag{2.205}$$

Equations 2.204 and 2.205 imply that the tensor C has only 21 independent components.

For *isotropic* materials, the number of independent components of C can be reduced further. For isotropic material, the response of the material is the same in every direction. Mathematically it means the constants in the stress-strain relations remain invariant with a change in the coordinate system. Equation 2.202 represents the stress-strain relations in (x, y, z) coordination system. Let σ'_{ij} and ε'_{kl} represent respectively the stress and strain components in (x', y', z') coordinate system. Then, for isotropic materials, the stress-strain relations in (x', y', z') coordinate system can be written as

$$\sigma'_{ij} = C_{ijkl}\varepsilon'_{kl}. \tag{2.206}$$

Note that, since the material is isotropic, the constants C_{ijkl} appearing in the stress-strain relations are the same both in (x, y, z) and (x', y', z') coordinate systems. Note that, since σ is a second order tensor, its components σ'_{ij} and σ_{mn} with respect to two coordinate systems are related by the tensor transformation relation (Equation 2.56). Rewriting this relation with the change of indices, we get

$$\sigma'_{ij} = Q_{im}\sigma_{mn}Q^T_{nj}, \tag{2.207}$$

where the matrix $[Q]$ (Equation 2.54) represents the transformation from (x, y, z) coordinate system to (x', y', z') system. Since ε is also a second order tensor, its components ε'_{kl} and ε_{pq} are also related by a similar relation:

$$\varepsilon'_{kl} = Q_{kp}\varepsilon_{pq}Q^T_{ql}. \tag{2.208}$$

Substituting Equations 2.207 and 2.208 in Equation 2.206 and using the orthogonality of matrix $[Q]$ (Equation 2.61), we get

$$\sigma_{mn} = \left(Q_{mi}^{T} Q_{jn} C_{ijkl} Q_{kp} Q_{ql}^{T} \right) \varepsilon_{pq} . \tag{2.209}$$

In changed indices, Equation 2.202 can be rewritten as

$$\sigma_{mn} = C_{mnpq} \varepsilon_{pq} . \tag{2.210}$$

Comparing Equations 2.209 and 2.210, we get the following restriction on the components of C due to isotropy:

$$C_{mnpq} = Q_{im} Q_{jn} C_{ijkl} Q_{kp} Q_{lq} . \tag{2.211}$$

Equation 2.211 must hold for all rotations of a coordinate system, *i.e.,* for all orthogonal matrices whose determinant is $+1$.

Equations 2.204, 2.205 and 2.211 imply that the six components C_{1122}, C_{1133}, C_{2211}, C_{2233}, C_{3311} and C_{3322} are equal. Further, these equations imply that the twelve components C_{1212}, C_{1221}, C_{2112}, C_{2121}, C_{2323}, C_{2332}, C_{3223}, C_{3232}, C_{3131}, C_{3113}, C_{1331} and C_{1313} are also equal but their value is different to the value of the first set of components. Additionally, these equations imply that the three components C_{1111}, C_{2222} and C_{3333} are also equal and their value is related to the values of the first and second sets of components. If the value of the first set is λ and that of the second set is μ, then the value of the third set is $\lambda + 2\mu$. Thus, we have the following relations between the 21 components of the tensor C:

$$\begin{aligned}
C_{1122} &= C_{1133} = C_{2211} = C_{2233} = C_{3311} = C_{3322} = \lambda, \\
C_{1212} &= C_{1221} = C_{2112} = C_{2121} = C_{2323} = C_{2332} = C_{3223} = C_{3232} \\
&= C_{3131} = C_{3113} = C_{1331} = C_{1313} = \mu, \\
C_{1111} &= C_{2222} = C_{3333} = \lambda + 2\mu.
\end{aligned} \tag{2.212}$$

Finally, these equations imply that the remaining 60 components of the tensor C are zero. Thus, for isotropic materials, there are only two independent components of the tensor C [2,4].

2.5.1.4 Stress-Strain Relations for Isotropic Materials
Substituting the values of 21 components of C from Equation 2.212 in Equation 2.202 and setting the remaining components of C to zero, the stress-strain relations for isotropic materials become

$$\sigma_{ij} = \lambda \varepsilon_{kk} \delta_{ij} + 2\mu \varepsilon_{ij} . \tag{2.213}$$

In tensor notation, they can be expressed as

$$\sigma = \lambda(tr\varepsilon)\mathbf{1} + 2\mu\varepsilon .$$
(2.214)

Further, in component forms, they can be written as

$$\sigma_{xx} = \lambda(\varepsilon_{xx} + \varepsilon_{yy} + \varepsilon_{zz}) + 2\mu\varepsilon_{xx},$$
$$\sigma_{yy} = \lambda(\varepsilon_{xx} + \varepsilon_{yy} + \varepsilon_{zz}) + 2\mu\varepsilon_{yy},$$
$$\sigma_{zz} = \lambda(\varepsilon_{xx} + \varepsilon_{yy} + \varepsilon_{zz}) + 2\mu\varepsilon_{zz},$$
$$\sigma_{xy} = 2\mu\varepsilon_{xy},$$
$$\sigma_{yz} = 2\mu\varepsilon_{yz},$$
$$\sigma_{zx} = 2\mu\varepsilon_{zx}.$$
(2.215)

Expressions for the remaining three shear stress components are not needed because of the symmetry of the stress tensor. The constants λ and μ are known as the Lame's constants.

As stated in the introduction, there are three sets of equations which govern the displacements, strains and stresses in a body. This is the *third set of governing equations* when the deformation is small and the material is isotropic linearly elastic.

Sometimes, we need inverse relations. That is, we need expressions for the strain components in terms of the stress components. They can be obtained by inverting Equation 2.215. When we do that, we get the following relations:

$$\varepsilon_{xx} = \frac{1}{E}\left[-v(\sigma_{xx} + \sigma_{yy} + \sigma_{zz}) + (1+v)\sigma_{xx}\right],$$
$$\varepsilon_{yy} = \frac{1}{E}\left[-v(\sigma_{xx} + \sigma_{yy} + \sigma_{zz}) + (1+v)\sigma_{yy}\right],$$
$$\varepsilon_{zz} = \frac{1}{E}\left[-v(\sigma_{xx} + \sigma_{yy} + \sigma_{zz}) + (1+v)\sigma_{zz}\right],$$
$$\varepsilon_{xy} = \frac{(1+v)}{E}\sigma_{xy},$$
$$\varepsilon_{yz} = \frac{(1+v)}{E}\sigma_{yz},$$
$$\varepsilon_{zx} = \frac{(1+v)}{E}\sigma_{zx},$$
(2.216)

where

$$E = \frac{\mu(3\lambda + 2\mu)}{\lambda + \mu}, \quad v = \frac{\lambda}{2(\lambda + \mu)}.$$
(2.217)

In index notation, Equation 2.216 can be expressed as

$$\varepsilon_{ij} = \frac{1}{E}\left[-\nu\sigma_{kk}\delta_{ij} + (1+\nu)\sigma_{ij}\right],$$
(2.218)

and, in tensor notation, it can be written as

$$\varepsilon = \frac{1}{E}[-\nu(tr\sigma)\mathbf{1} + (1+\nu)\sigma].$$
(2.219)

It can be shown that the constant E is the slope of the straight portion of the one-dimensional stress-strain curve (Figure 2.11):

$$E = \frac{\sigma_{xx}}{\varepsilon_{xx}}.$$
(2.220)

It is called the *Young's modulus*. Further, the constant ν can be shown to be negative of the ratio of the transverse normal strain to the axial or longitudinal normal strain in tension test. Thus,

$$\nu = -\frac{\varepsilon_{yy}}{\varepsilon_{xx}} = -\frac{\varepsilon_{zz}}{\varepsilon_{xx}}.$$
(2.221)

It is known as the *Poisson's ratio*. Equations 2.215 or 2.216 are called the *generalized Hooke's law*.

Elimination of λ from both parts of Equation 2.217 gives the following expression for μ :

$$\mu = \frac{E}{2(1+\nu)}.$$
(2.222)

Similarly, elimination of μ from both parts of Equation 2.217 gives the following expression for λ :

$$\lambda = \frac{E\nu}{(1+\nu)(1-2\nu)}.$$
(2.223)

2.5.1.5 Alternate Form of Stress-Strain Relations for Isotropic Materials
If we substitute the decompositions of stress and strain tensors (Equations 2.100 and 2.180) in the stress-strain relations (Equation 2.213) and equate the hydrostatic and deviatoric parts on each side, we get the following relations:

$$\left(\frac{1}{3}\sigma_{kk}\right) = (3\lambda + 2\mu)\left(\frac{1}{3}\varepsilon_{kk}\right),$$ (2.224)

$$\sigma'_{ij} = 2\mu\varepsilon'_{ij}.$$ (2.225)

In tensor notation, they become

$$\left(\frac{1}{3}tr\boldsymbol{\sigma}\right) = (3\lambda + 2\mu)\left(\frac{1}{3}tr\boldsymbol{\varepsilon}\right).$$ (2.226)

$$\boldsymbol{\sigma}' = 2\mu\boldsymbol{\varepsilon}'.$$ (2.227)

This is the third form of the stress-strain relations. It relates the hydrostatic and deviatoric parts of stress and strain tensors separately. This is possible only in isotropic materials. Equation 2.226 is a scalar equation. Because of the symmetry of $\boldsymbol{\sigma}'$ and $\boldsymbol{\varepsilon}'$, the tensor equation (Equation 2.227) represents six scalar equations. So, it appears that this form of the stress-strain relations consists of seven scalar relations. However, it is not so. Because of the constraints $tr\boldsymbol{\sigma}' = 0$ (Equation 2.99) and $tr\boldsymbol{\varepsilon}' = 0$ (Equation 2.179), only five out of six equations from the set (Equation 2.227) are independent.

Equation 2.225 or 2.227 shows that, in isotropic materials, the elastic constant μ relates the deviatoric parts of stress and strain tensors. Therefore, it is called the *shear modulus*. These equations imply that, in isotropic materials, the change in shape (without change in volume) is caused only by the deviatoric part of stress tensor. It also means, in isotropic materials, the hydrostatic part of stress tensor causes only the change in volume (without change in shape).

Besides the four elastic constants λ, μ, E and v, there is one more elastic constant that is often used. It is called the *bulk modulus* and is denoted by K. It is defined as the ratio of the hydrostatic part of stress to the volumetric strain. In small deformation, the volumetric strain is given by $tr\varepsilon = \varepsilon_{ll}$ (Equation 2.178). Thus, for small deformation, K is defined as

$$K = \frac{(1/3)\sigma_{kk}}{\varepsilon_{ll}} = \frac{(1/3)tr\boldsymbol{\sigma}}{tr\boldsymbol{\varepsilon}}.$$ (2.228)

This shows that, when the deformation is small, the bulk modulus K relates the hydrostatic parts of stress and strain tensors. Combining Equations 2.224 and 2.228 we get the following expression for the bulk modulus in terms of λ and μ :

$$K = \frac{(3\lambda + 2\mu)}{3}.$$ (2.229)

By taking the trace of Equation 2.219 and using Equation 2.228 for K, we get

$$K = \frac{E}{3(1-2v)} . \tag{2.230}$$

Using the sign conventions for stress and strain components described in Sections 2.3.1.4 and 2.4.1.2, experimental observations in real materials show that the signs of E, v, μ, λ and K are all positive. Equation 2.230 shows that, for compressible materials (finite K), v has to be less that (1/2). For incompressible materials ($K \to \infty$), v must be (1/2).

Example 2.14: Using the stress-strain relations at Equation 2.215, find the expressions for the stress components corresponding to the strain expressions of Example 2.10 (Equation 2.161).

Solution: Using the strain expressions of Example 2.10 (Equation 2.161), we get

$$\varepsilon_{xx} + \varepsilon_{yy} + \varepsilon_{zz} = A(x_0 - \ell)y_0 \left[1 - \frac{1}{2} - \frac{1}{2} \right] = 0 . \tag{2.231}$$

Note that Equation 2.231 implies that the volumetric strain ε_v is zero. This is expected, since the material is incompressible. Further, it implies that the hydrostatic part of the strain tensor is zero. Thus, the whole strain tensor is identical to its deviatoric part.

Substituting the strain expressions of Example 2.10 (Equation 2.161) along with Equation 2.231 in the stress-strain relations at Equation 2.215, we get

$$\begin{aligned}
\sigma_{xx} &= \lambda(\varepsilon_{xx} + \varepsilon_{yy} + \varepsilon_{zz}) + 2\mu\varepsilon_{xx} = 0 + 2\mu A(x_0 - \ell)y_0, \\
\sigma_{yy} &= \lambda(\varepsilon_{xx} + \varepsilon_{yy} + \varepsilon_{zz}) + 2\mu\varepsilon_{yy} = 0 + \mu A(\ell - x_0)y_0, \\
\sigma_{zz} &= \lambda(\varepsilon_{xx} + \varepsilon_{yy} + \varepsilon_{zz}) + 2\mu\varepsilon_{zz} = 0 + \mu A(\ell - x_0)y_0, \\
\sigma_{xy} &= 2\mu\varepsilon_{xy} = \mu A\left(a^2 - y_0^2\right), \\
\sigma_{yz} &= 2\mu\varepsilon_{yz} = 0, \\
\sigma_{zx} &= 2\mu\varepsilon_{zx} = -\mu A y_0 z_0.
\end{aligned} \tag{2.232}$$

2.6 Summary

In this chapter, first the index notation and the associated summation convention which have been used throughout the book have been explained. Then the equations which govern the displacements, strains and stresses in a deformable body have been developed for the case of small deformation of linearly elastic materials. These equations have been developed in the following stages. First, the concept of stress at a point has been discussed. Since the stress at a point is a tensor (a second order tensor to be precise), a simple definition of tensor has been provided. The analysis of stress at a point has been carried out to provide a background material for developing the theory of plasticity in Chapter 3. The equations of motion which the stress components satisfy have also been discussed. Next, the linear strain tensor at a point, which is a measure of small deformation, has been developed. The associated strain-displacement relations have been presented. The linear strain tensor is not applicable to the analysis of plastic deformation. However, it does provide an insight into the deformation of solids which would be useful while developing a measure of plastic deformation in the next chapter. Analysis of the linear strain at a point has also been carried out similar to the analysis of stress at a point. Finally, the stress-strain relations, for the case of small deformation of linearly elastic solids, have been developed. These relations provide an introduction to the material behavior and therefore provide a useful foundation for developing the plastic stress-strain relations of Chapter 3.

2.7 References

[1] Jaunzemis, W. (1967), Continuum Mechanics, The Macmillan Company, New York.
[2] Malvern, L.E. (1969), Introduction to the Mechanics of a Continuous Medium, Prentice-Hall Inc., Englewood Cliffs, New Jersey.
[3] Fung, Y.C., (1965), Foundations of Solid Mechanics, Prentice-Hall Inc., Englewood Cliffs, New Jersey.
[4] Sokolnikoff, I.S. (1956), Mathematical Theory of Elasticity, McGraw-Hill Book Company Inc., New York.
[5] Timoshenko, S.P. and Goodier, J.N. (1982), Theory of Elasticity, McGraw-Hill Book Company, Singapore, International Edition.

3

Classical Theory of Plasticity

3.1 Introduction

In metal forming processes, the material is deformed plastically to obtain the desired shape. On the other hand, in machining processes, the desired shape is achieved by removing the material in the form of chips. In machining of ductile materials, a significant amount of plastic deformation takes place before the chips fracture. To estimate the external forces required for achieving the desired shape, it is necessary to determine the plastic deformation and the stresses developed due to this deformation.

In the last chapter, we developed three governing equations for determining the displacements, deformation (strains) and stresses for the case of small deformation of linearly elastic materials: (i) strain-displacement relations, (ii) stress-strain relations and (ii) equations of motion. The equations of motion remain the same for the case of plastic deformation. But, the first two equations need modification, as the plastic deformation involved in the metal forming and machining processes differs from the small deformation of linearly elastic materials in two respects. The first difference is that this plastic deformation is quite large. Therefore, we cannot use the measure of small deformation, namely linear strain tensor, developed in the last chapter. A new measure of deformation, applicable for large deformation, needs to be developed. This leads to a different set of strain-displacement relations. It needs to be emphasized that, if the elastic deformation is large as happens in rubber like materials, a measure of large deformation is required for the analysis of elastic deformation as well. A second difference is that the material behavior responsible for the plastic deformation differs significantly from the elastic behavior. When a body is deformed plastically, then it does not return to the original undeformed configuration after the external forces are removed. This behavior is described by a set of stress-strain relations which are not one-to-one like the elastic stress-strain relations. It also means, in plastic deformation, that the material behaviors in loading and unloading are different. In fact, in unloading the behavior is elastic. In this chapter, we plan to develop: (i) *measures of plastic deformation* and corresponding strain-displacement relations and (ii) *plastic stress-strain relations*.

Besides the measure of plastic deformation and the plastic stress-strain relations, we need four more things for the analysis of plastic deformation. These four things are as follows. The first is that, in metal forming and machining processes, the metal first behaves elastically for small deformation and then behaves plastically as the deformation grows. Therefore, we need a criterion which tells us when the elastic behavior ends and the plastic behavior begins. Such a criterion is called the yield criterion. It is usually represented as a scalar function of the stress components. So we need to develop the *(initial) yield criterion* of metals. Second, for achieving continued or subsequent plastic deformation beyond initial yielding, additional stress needs to be applied. It means, in subsequent yielding, the (initial) yield criterion keeps changing with the level of plastic deformation. This phenomenon is called *hardening*. To model the hardening behavior, we need to develop the *criterion for subsequent yielding*. Third, when a combination of the stress components decreases, the material again behaves elastically. This is called unloading phenomenon. To model this phenomenon, we need to develop the *unloading criterion*. Fourth, the constitutive equation for plastic behavior is usually expressed either in the rate form or in the increment form. The stress tensors which appear in these constitutive equations have to be objective, *i.e.*, they have to be invariant under a change of reference frame. Whereas the Cauchy stress tensor (introduced in the last chapter) is objective, its rate or increment is not objective. Therefore, we need to develop the *objective stress rate and objective incremental stress measures*. Constitutive equation for large elastic deformation is also sometimes expressed in the rate or incremental form. In that case, the stress measures appearing in these constitutive equations also have to be objective.

Organization of this chapter is as follows. First, we describe the one-dimensional experimental observations on plasticity based on tension test. This is done in Section 3.2. These observations provide a valuable insight into the phenomenon of plasticity. Further, they provide a useful basis for the development of three-dimensional yield criterion, hardening relations, unloading criterion *etc*. In Section 3.3, we discuss two (initial) *yield criteria* for *isotropic materials*: (i) *Mises yield criterion* and (ii) *Tresca yield criterion*. We also discuss their experimental validation. The Mises yield criterion is found to have better agreement with experimental predictions on yielding. Two common measures of plastic deformation, namely the *incremental linear strain tensor* and *strain rate tensor* are developed in Section 3.4. The first is valid only for *small incremental deformation*. A relation between the two measures is also discussed. The incremental linear strain tensor is useful in the analysis of processes like forging, deep drawing, and sheet bending *etc*. which are amenable to incremental formulation, called the updated Lagrangian formulation. The strain rate measure is employed in the analysis of rolling, drawing, extrusion *etc*. where the analysis is carried out by fixing a region in space (called the control volume) and observing the deformation of the material particles as they pass through the control volume. This formulation is known as the Eulerian formulation. In Section 3.5, hardening behavior is modeled by developing a criterion for subsequent yielding. While doing so, it is assumed that the *hardening is isotropic*. This assumption does not have much experimental support. However, in the absence of required experimental data, it is difficult to develop a better hardening model.

Section 3.6 is devoted to the development of *plastic stress-strain relations* for *isotropic materials*. The first approach for developing the plastic stress-strain relations is based on Drucker's postulate for stable plastic material. The second approach is based on the *postulate of plastic potential*. In this book, we follow the second approach because it is less mathematical. Starting from the postulate of plastic potential, we first discuss the associated flow rule and then develop the following two constitutive equations: (i) elastic-plastic incremental stress-strain relation for the updated Lagrangian formulation and (ii) elastic-plastic stress-strain rate relations for the Eulerian formulation. While developing these relations, it is assumed that the *elastic and plastic parts of the deformation are additive*. This is true for the incremental linear strain tensor only when the *incremental deformation is small*. For the strain rate tensor, it is true if *the rotation is small*. Unloading criterion is presented in Section 3.7. The concept of objective stress rate and objective incremental stress measures is discussed in Section 3.8. A commonly used objective stress rate measure, namely the *Jaumann stress rate tensor*, is also presented. The objective incremental stress tensor is taken to be the product of the Jaumann stress rate and the time increment. This measure is objective only when the *incremental rotation is small*. Section 3.9 describes the *Eulerian and updated Lagrangian formulations for the metal forming processes*. These formulations are illustrated through the examples of wire drawing and forging of cylindrical block respectively. The boundary and initial conditions for these two problems are also described. The *Eulerian formulation for* the simplest machining process, namely *orthogonal cutting*, is presented in Section 3.10. The discussion includes the boundary and initial conditions as well. The last section, namely Section 3.11, summarizes the chapter. Worked out examples are provided at the end of Sections 3.3, 3.4, 3.5 and 3.6.

All the above material falls under the domain of classical plasticity [1–4] as the discussion is confined to small incremental deformation and isotropic materials only. In the next chapter, we shall discuss the plasticity of finite incremental deformation and anisotropic behavior. The tensor, array and index notations used in this chapter have already been introduced in the previous chapter.

3.2 One-Dimensional Experimental Observations on Plasticity

Consider a rod of uniform cross-section subjected to an axial tensile force F_x as shown in Figure 2.10. Let A_0 be its initial area of cross-section, ℓ_0 its initial length and $\Delta\ell$ the change in length corresponding to F_x. The *engineering* (or *nominal*) stress σ_0 and the *engineering strain* e are defined by Equations 2.199 and 2.200. Further, the variation of σ_0 with e is plotted in Figure 2.11. Note that, after certain deformation, the value of F_x and along with that the value of σ_0 decreases. However, the *true stress* does not decrease with the deformation. Further, it is observed that the value of e at fracture, for most metals, is more than 0.5. Therefore, plastic deformation is usually quite large. It is more appropriate then to use a measure of deformation which can represent large deformation. One such measure is the *logarithmic strain*. The stress-strain diagram involving the true stress and logarithmic strain is more useful for studying the phenomenon of plasticity. Therefore, we construct such a diagram.

In the one-dimensional case, we denote the logarithmic strain by ε and define it as

$$\varepsilon = \ln \frac{\ell}{\ell_0}, \tag{3.1}$$

where ℓ is the current length (*i.e.*, the length in the deformed configuration). It is also called the *natural strain*. Since

$$\ell = \ell_0 + \Delta\ell, \tag{3.2}$$

using Equations 3.1, 3.2 and 2.200, we get the following relationship between ε and e:

$$\varepsilon = \ln(1 + e). \tag{3.3}$$

Note that, when the deformation is small (*i.e.*, when $e < 0.05$), ε is approximately equal to e. The expression for true stress σ is given by

$$\sigma = \frac{F_x}{A}, \tag{3.4}$$

where A is the current area of cross-section. It is observed that the volume remains constant during plastic deformation. This condition implies that A is related to A_0 by

$$A = \frac{\ell_0}{\ell} A_0. \tag{3.5}$$

Substituting Equations 3.5, 3.2, 2.199 and 2.200 in Equation 3.4, we get the following relationship between σ and σ_0:

$$\sigma = (1 + e)\sigma_0. \tag{3.6}$$

Note that, when the deformation is small (*i.e.*, when $e < 0.05$), σ is approximately equal to σ_0. Using Equations 3.3 and 3.6, we convert the variation of σ_0 with e (Figure 2.11) into the graph of σ vs ε which is shown in Figure 3.1. From this figure, we can make the following observations about the plasticity.

- Elastic region

We denote the value of stress at point Y (end of the straight portion of Figure 3.1) as σ_Y. If the rod is stressed up to any level less than σ_Y (say up to point A), then it attains the original undeformed configuration on unloading. Therefore, the straight portion OY corresponds to the elastic behavior.

- Yield stress

We observe that when the stress reaches the value σ_Y (Figure 3.1), the material yields, that is, it starts flowing suddenly, leading to large deformation. The value σ_Y is called the *yield stress*. It marks the transition from elastic to plastic behavior.

Thus, in one-dimensional state of stress, *initial yielding* occurs when the condition

$$\sigma - \sigma_Y = 0 \tag{3.7}$$

is satisfied. Generalization of this initial yield condition to three-dimensional state of stress is discussed in next section.

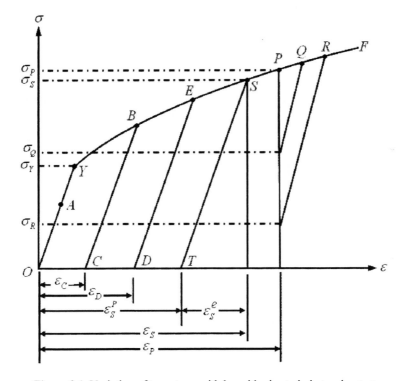

Figure 3.1. Variation of true stress with logarithmic strain in tension test

In some materials, in the neighborhood of σ_Y, the actual stress-strain curve [1,2] differs a little from the curve of Figure 3.1. For example, the end of elastic behavior does not coincide with the end of the straight portion of the curve. Further, there is a drop in the stress after initial yielding leading to upper and lower yield points. However, for ease of mathematical modeling, we neglect these finer aspects of yielding and assume the existence of a sharp yield point at the end of the straight portion. For materials like aluminum where there is a continuous change of slope at the end of the straight portion, the yield stress is defined as the stress corresponding to 0.2% permanent strain. The concept of permanent strain is defined later.

It is observed that the value of σ_Y is more if the tension test is conducted at higher rate of loading. Further, the value of σ_Y is less if the test is conducted at elevated temperature. Thus, σ_Y increases with strain rate but decreases with temperature.

- Plastic region

The curved portion of Figure 3.1 beyond point Y corresponds to the plastic behavior. Some of the characteristics of plastic behavior are as follows.

Imagine that the rod has been stressed beyond yielding up to point B. If we continue to increase the load, then the stress-strain curve will follow the path BF leading to fracture at point F. The portion YF is called the loading path. However, if we unload from point B to zero stress level, then the stress-strain curve will follow the straight path BC leaving a permanent strain ε_C in the rod. Thus, if the rod is stressed beyond the level σ_Y, then it does not attain the initial undeformed configuration on unloading. Instead, it acquires some permanent strain, also called the plastic strain. Now imagine that the rod has an initial plastic strain ε_D. If we load this rod, then the stress-strain curve will be a straight line from point D to point E and then it will follow the curved portion EF. It means the rod will behave elastically up to point E and will yield at the stress level corresponding to point E, which is greater than σ_Y. Thus, a rod which has some initial plastic strain yields at a higher stress level than the undeformed rod. This is called *subsequent or continued yielding*. Condition for this yielding is developed later.

The stress-strain relationship corresponding to plastic behavior is not *one-to-one*. To see this, assume that the rod has been strained up to the strain level ε_P and let us find the corresponding value of stress. It will be equal to σ_P if we are on the loading path. However, it will be equal to σ_Q if we stress the rod up to point Q and then unload it to the strain level ε_P. Further, it will be equal to σ_R if we stress the rod up to point R and then unload it to the strain level ε_P. Thus, the stress corresponding to the strain level ε_P is not unique but depends on the *history* of deformation. Further, there is one type of stress-strain relationship if we are on the loading path and a different one when we are on the unloading path. Generalization to three-dimensional plastic stress-strain relations is discussed in Section 3.6

To avoid the mathematical complexity in the analysis of plastic behavior, plastic stress-strain relations are sometimes simplified by approximating or idealizing the actual stress-strain behavior. With reference to the stress-strain curve of Figure 3.1, these idealizations can be stated as follows. In metals, elastic strain is very small compared to the plastic strain. First simplification arises by neglecting the elastic strain. Then, the stress-strain curve of Figure 3.1 starts from point Y and has only the plastic portion YF. Such a material is called *rigid-plastic*. Otherwise, the material is called *elastic-plastic*. In the second simplification, we assume that the portion YF is straight. Such a material is called *linearly hardening*. (The phenomenon of hardening is discussed in the next paragraph.) In the third simplification, the portion YF is assumed to be straight as well as parallel to the strain axis. Such a material is called *ideal or perfectly plastic*. Various combinations of these simplifications result in the following four idealizations: (i) rigid perfectly plastic material, (ii) rigid-plastic material with linear hardening, (iii) elastic perfectly plastic material and (iv) elastic-plastic material with linear hardening.

- Strain hardening

It is observed that, beyond point Y of Figure 3.1, stress increases with strain. It means, beyond initial yielding, the stress required to cause subsequent yielding or continued material flow increases with the strain. This phenomenon is called *strain hardening*. The yield stress in subsequent yielding depends on the plastic part of deformation. Therefore, to develop a mathematical expression for subsequent yielding, we need a graph which gives the variation of stress with the plastic part of strain. We construct such a graph from Figure 3.1 as follows. To find the plastic part of strain corresponding to σ_S (*i.e.*, the stress at point S of Figure 3.1), we unload from point S to the zero stress level (*i.e.*, to point T). Then OT is the plastic part of strain corresponding to σ_S. The remaining part is the elastic part of strain. In this way, we find the plastic part of strain corresponding to all values of stress greater than σ_Y. We denote the plastic part by ε^P and the elastic part by ε^e. The graph of σ vs ε^P is shown in Figure 3.2.

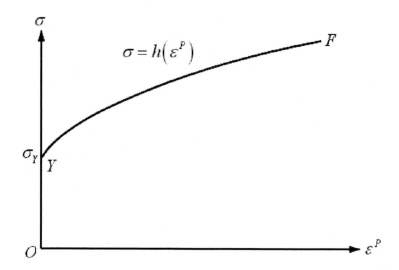

Figure 3.2. Variation of true stress with plastic part of logarithmic strain in tension test

The curve of Figure 3.2 can be represented mathematically as

$$\sigma - h(\varepsilon^P) = 0,\tag{3.8}$$

where the function h is called hardening function. This equation represents the criterion for *subsequent* or *continued yielding* for one-dimensional state of stress. When ε^P is zero, the value of h is equal to σ_Y. Thus, for initial yielding $(\varepsilon^P = 0)$, Equation 3.8 reduces to the criterion for initial yielding (Equation 3.7). Generalization of Equation 3.8 to the three-dimensional case is discussed in Section 3.5.

Several forms of function h have been proposed to fit the experimental stress-strain curves. Some commonly used forms are listed below [3, 4]. The original expressions for these functions are in terms of the total strain. Here, they have been appropriately modified to express them in terms of the plastic part of strain. Further, the symbols for the material constants also have been changed.

1. Ludwik's expression:

$$\sigma = \sigma_Y + K(\varepsilon^P)^n.\tag{3.9}$$

This expression does not give a good fit at large strains as the experimental stress-strain curves of most metals have a constant slope at large strain.

2. Swift's expression:

$$\sigma = \sigma_Y[1 + K\varepsilon^P]^n.\tag{3.10}$$

This expression gives a better fit of experimental stress-strain curves at large strains than the Ludwik's expression.

3. Voce's expression:

$$\sigma = \sigma_Y + K[1 - e^{-(n\varepsilon^P)}]. \tag{3.11}$$

This expression gives a good fit of experimental stress-strain curves at moderate values of strain.

In all the three expressions above, K and n are the material constants known as *hardening parameters*, which are to be determined by fitting the above equations with the experimental curves of true stress *vs* the plastic part of logarithmic strain (Figure 3.2). Note that, when ε^P is zero (*i.e.*, at initial yielding), σ reduces to σ_Y (*i.e.*, to the initial yield stress) in all the three equations above. When n is equal to 1, Equations 3.9 and 3.10 represent a *linear hardening* curve.

Here, we have not presented the mathematical expressions proposed by Prager and Ramberg and Osgood [3]. Prager's expression is essentially for non-hardening material. Expression due to Ramberg and Osgood represents a continuous transition from elastic to plastic behavior, and therefore not suitable for modeling subsequent yielding.

- Temperature softening

If we conduct the tension test at elevated temperature, we observe that, beyond initial yielding, the stress required to cause further material flow decreases with temperature rise. This phenomenon is called *temperature softening*. In this case, the function h of Equation 3.8 depends on temperature also. This effect needs to be included in the plastic stress-strain relations while analyzing hot forming processes or machining processes or if the temperature rise in cold processes is quite large.

- Viscoplasticity

If we conduct the tension test at higher rate of loading, we observe that, beyond initial yielding, the stress required to cause further material flow increases with strain rate or the rate of deformation. This phenomenon is called *viscoplasticity*. This increase in the stress is due to the *viscous* resistance of the material to further yielding. In this case, the function h of Equation 3.8 also depends on some measure of the rate of deformation. This effect needs to be incorporated in the plastic stress-strain relations while analyzing hot or high speed metal forming processes or machining processes as the material becomes viscoplastic at elevated temperature and high strain rate.

- Isochoric deformation

As stated earlier, it is observed that the volume remains constant during plastic deformation. Thus, the plastic deformation is *isochoric*. This imposes a constraint on plastic deformation.

- Large deformation

As stated earlier, the deformation in plastic region is quite large. As a result, we cannot use the linear or infinitesimal strain tensor ε as the measure of deformation. We have to look for some other measure of deformation to represent the plastic deformation. One such measure is the *logarithmic strain*, whose definition for the one-dimensional case has been given earlier. Measures used for describing plastic deformation are discussed in Section 3.4.

- Hysteresis

Suppose the rod, which has been stressed upto point B (Figure 3.3), is unloaded up to zero stress level. Then it will follow the straight path BC leaving a plastic strain ε_C in the rod. If we load it now, then the initial straight path CD, which the stress-strain curve follows, has a slightly different slope to the unloading path BC. This phenomenon is called *hysteresis* and the loop BCD is called the hysteresis loop.

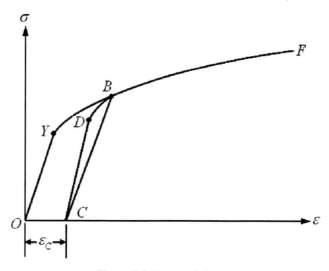

Figure 3.3. Hysteresis loop

In Figure 3.3, the hysteresis loop has been exaggerated. The actual hysteresis loop is quite small and thus its effect on the plastic stress-strain relationship can be neglected. Therefore, we assume that the slopes of both the straight line portions BC as well as CD are identical and are equal to the Young's modulus.

- Bauschinger effect

If the rod is subjected to axial compressive force instead of tensile force, then the numerical value of the *yield stress in compression* is found to be exactly equal to σ_Y. However, this numerical equality of yield stress in tension and compression does not hold in reversed loading after the yielding.

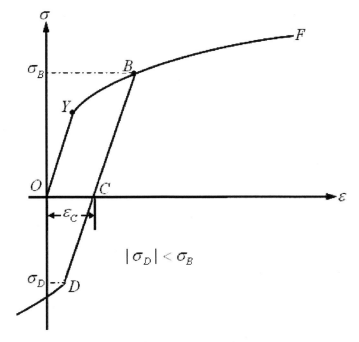

Figure 3.4. Bauschinger effect

Suppose the rod, which has been stressed up to point B (Figure 3.4), is unloaded to the zero stress level (*i.e.*, to point C) leaving a plastic strain ε_C in the rod. Next, it is loaded in *compression*. Then it will follow the path CD, where the new yield stress σ_D (in compression) is smaller in magnitude than the stress σ_B (the yield stress in tension corresponding to the initial strain of ε_C). This phenomenon is called the *Bauschinger effect*. This lowering of the yield stress in reversed loading is caused by the residual stresses (at the *microscopic scale*) left in the rod after unloading. The Bauschinger effect can be removed after mild annealing. In our analysis, we shall neglect the Bauschinger effect and assume that the yield stress in tension and compression are numerically equal. We shall discuss modeling of Bauschinger effect and *kinematic hardening* in next chapter.

- Necking or one-dimensional plastic instability

At a certain value of the force F_x, necking of the rod starts. Then, the deformation of the rod will not be as shown in Figure 2.10. This happens due to instability of the one-dimensional state of stress existing in the rod. Disturbance for the instability is provided by the nucleation of voids which starts at that value of F_x. The one-dimensional state of stress, after the disturbance, gives rise to a three-dimensional (or triaxial) state of stress which manifests itself in the form of necking. Necking initiates when the value of F_x starts decreasing, or when F_x attains the maximum value.

To determine the values of σ and ε at the onset of necking, we differentiate Equations 3.4, 3.5 and 3.1:

$$dF_x = \sigma dA + A d\sigma , \tag{3.12a}$$

$$A d\ell + \ell dA = 0 , \tag{3.12b}$$

$$d\varepsilon = \frac{d\ell}{\ell} . \tag{3.12c}$$

Eliminating dA and $d\ell$ from these three equations, we get

$$dF_x = (-\sigma d\varepsilon + d\sigma)A . \tag{3.13}$$

Since dF_x becomes zero when the necking starts, we get the following relationship at the onset of necking:

$$\frac{d\sigma}{d\varepsilon} = \sigma . \tag{3.14}$$

Thus, the point on the graph of σ vs ε at which the necking starts is characterized by the condition that the slope of the tangent at that point is equal to the ordinate of the point. This point (labeled as A) is shown in Figure 3.5.

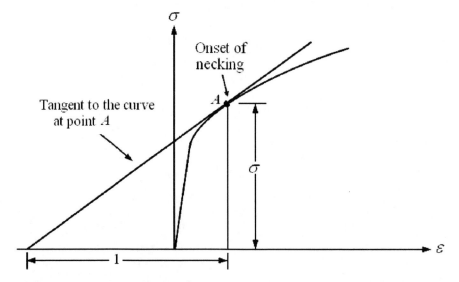

Figure 3.5. True stress vs logarithmic strain diagram in tension test showing the onset of necking

In this book, we shall not consider any more cases of plastic instability. The purpose of the above discussion is to give a brief idea about the phenomenon of plastic instability.

The above experimental observations are based on the tension (and compression) test. Some additional observations on plasticity which are based on experiments other than the tension (and compression) test are as follows.

- Effect of hydrostatic stress

It is observed that the yield stress is unaffected by the hydrostatic part of the stress tensor. See [4] for the original references on this observation. Thus, yielding is essentially caused by the deviatoric part of the stress tensor. This observation has been used in developing the yield criteria in next section.

- Anisotropy

Microstructure of metals is crystalline in nature. In an annealed metal, the crystallographic directions are randomly oriented. At macroscopic level, this means, there is no preferred direction. Thus, an annealed metal is isotropic at the macroscopic level. However, when it is subjected to cold forming processes like drawing, extrusion, rolling *etc.*, the crystallographic directions gradually rotate towards a common axis thus creating a preferred direction. Therefore, after cold forming, the metal usually becomes anisotropic in nature. When this metal is subjected to further forming processes without annealing, the yield criteria and the plastic stress-strain relations used for the analysis of these processes should incorporate the anisotropy. Anisotropic yield criteria and the corresponding plastic stress-strain relations are discussed in next chapter.

3.3 Criteria for Initial Yielding of Isotropic Materials

A law defining the limit of elastic behavior is known as the yield condition or *yield criterion*. Equation 3.7 is the criterion for initial yielding for the one-dimensional state of stress, where only one stress component is non-zero. In a three-dimensional state of stress, where normally all the stress components are non-zero, this condition can be generalized as

$$f(\sigma_{ij}) = 0 ,$$
(3.15)

where σ_{ij} are the components of stress tensor σ with respect to a coordinate system (x, y, z). The function f is called the *yield function*.

As stated earlier, yielding depends only on the deviatoric part of stress tensor. Let σ'_{ij} be the components of the deviatoric part of σ with respect to the coordinate system (x, y, z). Then, in Equation 3.15, f should be a function of σ'_{ij}. Thus, the yield criterion becomes

$$f(\sigma'_{ij}) = 0 .$$
(3.16)

For isotropic materials, the yield function should remain unaffected by a change of the coordinate system. It means the function f should be an invariant of σ'. Since every invariant of a tensor can be expressed in terms of the three principal invariants, f should be a function of the three principal invariants J_1, J_2 and J_3 of σ'. But, since $tr\sigma' = 0$, the first invariant J_1 is zero (Equation 2.104). Therefore, the yield criterion becomes

$$f(J_2, J_3) = 0.$$

(3.17)

At the initial yielding, the yield stress in compression is found to be numerically equal to σ_Y. When we generalise this observation to the 3D case, we expect f to be unaffected if σ'_{ij} is replaced by $-\sigma'_{ij}$. It implies that f should be an even function of σ'_{ij}. Note that J_2 is an even function of σ'_{ij} (Equation 2.105), but J_3 is an odd function of σ'_{ij} (Equation 2.103). Therefore, in Equation 3.17, f should be an *even* function of J_3, but can be any function of J_2.

To determine the specific dependence of f on J_2 and J_3, one has to make a hypothesis and test it against the experimental results. We discuss here two such hypotheses and their experimental validation: (i) von Mises yield criterion and (ii) Tresca yield criterion.

3.3.1 von Mises Yield Criterion

In von Mises yield criterion, the yield function f is assumed to be a linear function of J_2 and independent of J_3. Therefore, the von Mises yield criterion (henceforth simply called the Mises criterion) can be stated as

$$f(J_2, J_3) \equiv J_2 - k = 0.$$

(3.18)

This criterion was proposed by von Mises in 1913, but was anticipated by Huber in 1904. Its physical interpretation was provided by Hencky in 1924 and Nadai in 1933. Hencky showed that J_2 is related to the distortion strain energy density. As stated in Section 2.5.1, the strain energy density is the work done (per unit volume) by σ_{ij} during the deformation and is given by $\int \sigma_{ij} d\varepsilon_{ij}$. The distortion strain energy density is the work done (per unit volume) by the deviatoric part σ'_{ij}. As *per* Hencky's interpretation, Equation 3.18 states that yielding occurs whenever the distortion strain energy density reaches a critical value. Nadai related J_2 to the octahedral shear stress τ_{oct} (Equation 2.98). Therefore, as *per* Nadai's interpretation, Equation 3.18 states that yielding occurs if the octahedral shear stress reaches a critical value.

The material constant k can be determined easily from the one-dimensional experimental results of Section 3.2. In tension test, matrix of the stress components with respect to (x, y, z) coordinate system is given by

$$[\sigma] = \begin{bmatrix} \sigma & 0 & 0 \\ 0 & 0 & 0 \\ 0 & 0 & 0 \end{bmatrix}, \tag{3.19}$$

where σ is given by Equation 3.4. From the above equation, we get $tr\sigma = \sigma$. Then, using Equation 2.100, the matrix of the deviatoric part can be written as

$$[\sigma'] = \begin{bmatrix} \dfrac{2}{3}\sigma & 0 & 0 \\ 0 & -\dfrac{1}{3}\sigma & 0 \\ 0 & 0 & -\dfrac{1}{3}\sigma \end{bmatrix}. \tag{3.20}$$

From this, J_2 is calculated as (Equation 2.105)

$$J_2 = \frac{1}{2}\sigma'_{ij}\sigma'_{ij} = \frac{1}{3}\sigma^2. \tag{3.21}$$

At the initial yielding, σ is equal to σ_Y, and therefore, J_2 becomes $(1/3)\sigma_Y^2$. Substituting the value of J_2 at the initial yielding in Equation 3.18, we get $(1/3)\sigma_Y^2$ as the value of k. Then, the Mises criterion can be stated as

$$J_2 - \frac{1}{3}\sigma_Y^2 = 0. \tag{3.22}$$

There are alternative expressions of the Mises criterion. To develop the first alternative expression, we define the following invariant of σ':

$$\sigma_{eq} = (3J_2)^{1/2} = \left(\frac{3}{2}\sigma'_{ij}\sigma'_{ij}\right)^{1/2}. \tag{3.23}$$

It can easily be shown from Equation 3.20 that, in tension test, σ_{eq} is equal to σ. Therefore, σ_{eq} is called the *equivalent stress*. It is also called the *effective or generalized stress*. Since, at the initial yielding, σ is equal to σ_Y, the Mises criterion in terms of σ_{eq} becomes

$$\sigma_{eq} - \sigma_Y = 0. \tag{3.24}$$

Next, we develop an expression for the Mises criterion in terms of the principal stresses σ_i. In the coordinate system of principal directions, we decompose the matrix of σ (Equation 2.78) into the hydrostatic and deviatoric parts. The matrix of the deviatoric part σ' becomes

$$[\sigma'] = \begin{bmatrix} \sigma_1 - \dfrac{1}{3}(\sigma_1 + \sigma_2 + \sigma_3) & 0 & 0 \\ 0 & \sigma_2 - \dfrac{1}{3}(\sigma_1 + \sigma_2 + \sigma_3) & 0 \\ 0 & 0 & \sigma_3 - \dfrac{1}{3}(\sigma_1 + \sigma_2 + \sigma_3) \end{bmatrix}. \tag{3.25}$$

Then, J_2 can be calculated as

$$J_2 = \frac{1}{2}\sigma'_{ij}\sigma'_{ij} = \frac{1}{2}\Big\{ [\sigma_1 - \frac{1}{3}(\sigma_1 + \sigma_2 + \sigma_3)]^2 + [\sigma_2 - \frac{1}{3}(\sigma_1 + \sigma_2 + \sigma_3)]^2$$
$$+ [\sigma_3 - \frac{1}{3}(\sigma_1 + \sigma_2 + \sigma_3)]^2 \Big\} = \frac{1}{6}[(\sigma_1 - \sigma_2)^2 + (\sigma_2 - \sigma_3)^2 + (\sigma_3 - \sigma_1)^2]. \tag{3.26}$$

Substituting this value of J_2 in Equation 3.22, we get the following expression for the Mises criterion:

$$[(\sigma_1 - \sigma_2)^2 + (\sigma_2 - \sigma_3)^2 + (\sigma_3 - \sigma_1)^2] - 2\sigma_Y^2 = 0. \tag{3.27}$$

The principal stresses are invariants of the stress tensor as they are the roots of an equation involving the invariants (Equation 2.81). Therefore, one can express the yield criterion in terms of the principal stresses only for isotropic materials.

3.3.2 Tresca Yield Criterion

The Tresca yield criterion (henceforth simply called Tresca criterion) was proposed in 1864. However, it was not developed on the basis of Equation 3.17. It was proposed on the basis of experimental observations on extrusion of metals through dies of various shapes. The Tresca criterion states that, whenever the maximum shear stress at a point reaches the critical value, yielding occurs at that point. The maximum shear stress at a point is given by Equation 2.92 when the principal stresses are ordered. However, when the principal stresses are not ordered, the maximum shear stress at a point can be expressed as

$$|\sigma_s|_{\max} = \frac{1}{2}\max\left\{|\sigma_1 - \sigma_2|, |\sigma_2 - \sigma_3|, |\sigma_3 - \sigma_1|\right\}. \tag{3.28}$$

If the critical value of the maximum shear stress (*i.e.*, its value at yielding) is k_1, then the Tresca criterion can be written as

$$[(\sigma_1 - \sigma_2)^2 - 4k_1^2][(\sigma_2 - \sigma_3)^2 - 4k_1^2][(\sigma_3 - \sigma_1)^2 - 4k_1^2] = 0. \tag{3.29}$$

The value of k_1 can be evaluated from the one-dimensional experimental results of Section 3.2. From Equation 3.19, we get the following values of the principal stresses in tension test:

$$\sigma_1 = \sigma, \sigma_2 = 0, \sigma_3 = 0, \tag{3.30}$$

where σ is given by Equation 3.4. Further, at the initial yielding,

$$\sigma = \sigma_Y. \tag{3.31}$$

Substituting Equations 3.30 and 3.31 into Equation 3.29, we get $\sigma_Y/2$ as the value of k_1. Then the Tresca criterion (Equation 3.29) becomes

$$[(\sigma_1 - \sigma_2)^2 - \sigma_Y^2][(\sigma_2 - \sigma_3)^2 - \sigma_Y^2][(\sigma_3 - \sigma_1)^2 - \sigma_Y^2] = 0. \tag{3.32}$$

In terms of the invariants J_2 and J_3, the above expression becomes [3]

$$f(J_2, J_3) \equiv 4\left(J_2 - \frac{\sigma_Y^2}{4}\right)(J_2 - \sigma_Y^2)^2 - 27J_3^2 = 0. \tag{3.33}$$

But the yield function of Tresca criterion in terms of J_2 and J_3 is complicated for application purpose.

3.3.3 Geometric Representation of Yield Criteria

Using Equations 3.27 and 3.32, one can represent the Mises and Tresca criteria geometrically as surfaces in a three-dimensional stress space of $(\sigma_1, \sigma_2, \sigma_3)$. Figure 3.6 shows these surfaces. The *Mises surface is a right circular cylinder* of radius $(\sqrt{2/3})\sigma_Y$, while the *Tresca surface is a right (regular) hexagonal prism* completely inscribed in the Mises cylinder [1,3]. The axis of the cylinder and prism is along the line $\sigma_1 = \sigma_2 = \sigma_3$. When all the principal stresses are equal, the state of stress is purely hydrostatic. Therefore, this line is known as the *hydrostatic line*. Further, this axis is perpendicular to the plane $\sigma_1 + \sigma_2 + \sigma_3 = 0$. Since *trσ* is zero in a purely deviatoric state of stress, this plane is known as the *deviatoric* or *π plane*.

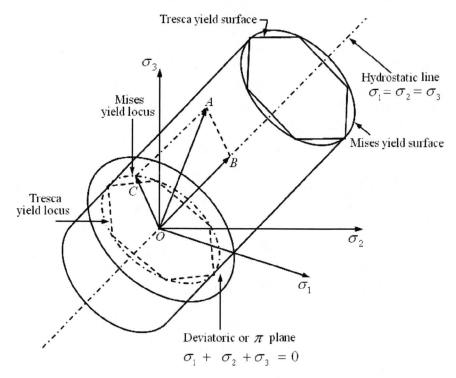

Figure 3.6. Geometric representation of the yield criteria in the stress space of $(\sigma_1, \sigma_2, \sigma_3)$

One can use the yield surfaces of Figure 3.6 to find out graphically when the state of stress at a material particle will reach the yield level. For this purpose, express the state of stress at the material particle in terms of the principal stresses $(\sigma_1, \sigma_2, \sigma_3)$. Then locate the point in the stress space with the coordinates $(\sigma_1, \sigma_2, \sigma_3)$. Let the stress level at the material particle be such that its behavior is still elastic. Then, the point will be inside the yield surfaces. Let us denote this point by A. Then, the vector $\overrightarrow{\mathbf{OA}}$ represents the state of stress at the material particle. This vector can be decomposed into two components: (i) the component OB along the hydrostatic line and (ii) the component OC along the deviatoric plane. The component OB represents the hydrostatic part of the stress while the component OC represents the deviatoric part. Now, let there be an increase in the stress level at the material particle. Further, let the increase be such that only the hydrostatic component OB increases. Then the point A can never reach the yield surfaces and hence there can never be any yielding. This is because the hydrostatic part of stress has no effect on yielding. On the other hand, if the increase in the stress level is such that the deviatoric component OC or both the components increase sufficiently, then the point A can reach the yield surfaces. When that happens, there will be yielding at the material particle. Note that the Tresca prism is completely inside the Mises cylinder except at the six edges. Therefore, if the point A reaches any one of these six edges, then the yielding will occur according to both the Tresca and Mises criteria. Otherwise, the point A will reach the Tresca prism first indicating yielding according to the Tresca criterion.

When the state of stress at a particle is such that its hydrostatic part is zero, then the geometrical representation of the yield criteria reduces to curves: (i) circle of radius $(\sqrt{2/3})\sigma_Y$ for the Mises criterion and (ii) a regular hexagon for the Tresca criterion which is completely inscribed in the Mises circle. These curves are the intersections of the yield surfaces with the deviatoric plane. They are called *yield loci on the deviatoric plane* and are shown separately in Figure 3.7.

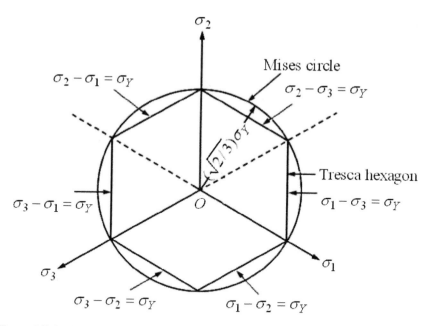

Figure 3.7. Loci of the Mises and Tresca yield surfaces on the deviatoric plane. The axes $(\sigma_1, \sigma_2, \sigma_3)$ are not in the deviatoric plane

On the other hand, if the state of stress at a particle is of *plane stress* type (*i.e.*, if one of the three principal stresses is zero at the particle), then the geometrical representation of the yield criteria reduces to different curves. We can obtain the equations of these curves as follows. If we assume that the principal stresses are not ordered and σ_3 is zero, then Equations 3.27 and 3.32 reduce to

$$\sigma_1^2 - \sigma_1\sigma_2 + \sigma_2^2 - \sigma_Y^2 = 0 \qquad \text{(Mises criterion)}, \tag{3.34}$$

$$[(\sigma_1 - \sigma_2)^2 - \sigma_Y^2][\sigma_2^2 - \sigma_Y^2][\sigma_1^2 - \sigma_Y^2] = 0 \quad \text{(Tresca criterion)}. \tag{3.35}$$

Equation 3.34 of the Mises criterion represents an ellipse in two-dimensional stress plane of (σ_1, σ_2). On the other hand, Equation 3.35 for the Tresca criterion represents a hexagon (but not regular). Both these curves are shown in Figure. 3.8.

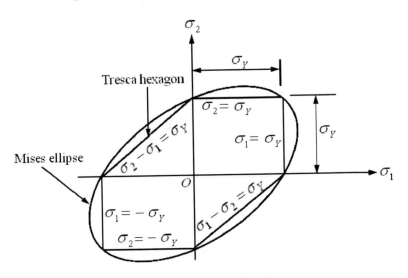

Figure 3.8. Loci of the Mises and Tresca yield surfaces on the plane $\sigma_3 = 0$

Here also, the Tresca hexagon is completely inscribed in the Mises ellipse. These curves are the intersections of the yield surfaces with the plane $\sigma_3 = 0$. Therefore, they are known as the *yield loci on the plane* $\sigma_3 = 0$.

3.3.4 Convexity of Yield Surfaces

It is observed that the regions bounded by the yield surfaces of both the Mises and Tresca criteria (Figure 3.6) are *convex*. A region is defined as convex if a straight line segment joining any two points of the region lies completely inside the region. Note that the regions bounded by the yield loci of Figures 3.7 and 3.8 are also convex. When anisotropic yield criteria are formulated, one of the requirements of these criteria is that their geometric representations must lead to convex yield surfaces.

3.3.5 Experimental Validation

There have been quite a few attempts to compare the predictions of the Mises and Tresca criteria with experimental results on yielding. Notable amongst these are the experiments of Lode [5] and Taylor and Quinney [6].

3.3.5.1 Lode's Experiments
Lode [5], in 1925, conducted experiments on thin tubes subjected to internal pressure as well as axial force (Figure 3.9). The tube material was iron, copper and nickel. Besides comparing his experimental results on yielding with the predictions of the yield criteria, he also studied the influence of the intermediate principal stress on yielding.

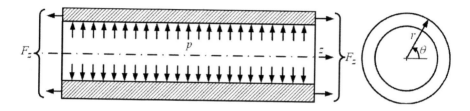

Figure 3.9. Thin tube subjected to internal pressure p and axial force F_z (tensile or compressive) in Lode's experiment

It is convenient to use the cylindrical polar coordinates (r, θ, z) shown in Figure 3.9. With respect to this coordinate system, the matrix of stress components can be expressed as

$$[\sigma] = \begin{bmatrix} \sigma_{rr} & \sigma_{r\theta} & \sigma_{zr} \\ \sigma_{r\theta} & \sigma_{\theta\theta} & \sigma_{\theta z} \\ \sigma_{zr} & \sigma_{\theta z} & \sigma_{zz} \end{bmatrix}. \tag{3.36}$$

The geometry and loading are such that, the stress components in the tube are the same at every point. The normal stress components are given by

$$\sigma_{rr} \approx 0, \; \sigma_{\theta\theta} = \frac{p\,r_i}{t}, \; \sigma_{zz} = \frac{F_z}{A}, \tag{3.37}$$

where p is the internal pressure in the tube, r_i is the inner radius of the tube, t is the wall thickness of the tube, F_z is the axial force (tensile or compressive) acting on the tube and A is the area of the cross-section of the tube. Further, the shear stress components in the tube are zero. Therefore, σ_{rr}, $\sigma_{\theta\theta}$ and σ_{zz} are the principal stresses. Note that whereas $\sigma_{\theta\theta}$ is always tensile, σ_{zz} may be tensile or compressive. Let us order these principal stresses and use the usual notation for them: $\sigma_1 \geq \sigma_2 \geq \sigma_3$. Lode [5] introduced a parameter denoted by μ (not to be confused with the shear modulus) and defined by

$$\mu = \frac{2\sigma_2 - \sigma_3 - \sigma_1}{\sigma_1 - \sigma_3}. \tag{3.38}$$

It is called the *Lode parameter*. Using the definition of Lode parameter (Equation 3.38), Equation 3.27 for the Mises criterion becomes

$$\frac{\sigma_1 - \sigma_3}{\sigma_Y} = \frac{2}{\sqrt{3 + \mu^2}}. \tag{3.39}$$

Further, since the principal stresses are ordered, the Tresca criterion can be expressed as

$$\frac{\sigma_1 - \sigma_3}{\sigma_Y} = 1. \tag{3.40}$$

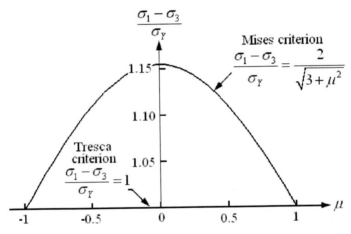

Figure 3.10. Mises and Tresca curves in Lode's experiments (experimental points are excluded)

The plots of both the Mises and Tresca criteria are shown in Figure 3.10. The axes used are $(\sigma_1 - \sigma_3)/\sigma_Y$ and μ. Lode [5] found the experimental values of $(\sigma_1 - \sigma_3)/\sigma_Y$ and μ at yielding by varying the internal pressure p and the axial force F_z. The experimental points are *not* shown in the figure. However they fall between the Mises and Tresca curves and are closer to the Mises curve.

3.3.5.2 Experiments of Taylor and Quinney

Taylor and Quinney [6], in 1931, also conducted experiments on thin tubes subjected to axial force. But the other loading was twisting moment instead of the internal pressure. The tube material was mild steel, copper and aluminum.

Again, it is convenient to use the cylindrical polar coordinates. Here, the normal stress due to axial force is constant, but the shear stress due to twisting moment increases in the radial direction and attains the maximum value at the outer tube surface. Therefore, yielding will take place at the outer surface. The non-zero stress components at the outer surface are

$$\sigma_{\theta z} = \frac{M_z d}{2 I_{zz}}, \quad \sigma_{zz} = \frac{F_z}{A}, \tag{3.41}$$

where M_z is the twisting moment acting on the tube, d is the outer diameter of the tube and I_{zz} is the moment of inertia of the tube cross-section about the z-axis (also called the polar moment of inertia). The symbols F_z and A have been defined earlier. The principal stresses in the tube are given by

$$\sigma_1, \sigma_3 = \frac{\sigma_{zz}}{2} \pm \left\{ \left(\frac{\sigma_{zz}}{2} \right)^2 + \sigma_{\theta z}^2 \right\}^{1/2}, \sigma_2 = 0. \tag{3.42}$$

Substituting Equation 3.42, Equation 3.27 for the Mises criteria becomes

$$\left(\frac{\sigma_{zz}}{\sigma_Y} \right)^2 + \left(\frac{\sqrt{3}\sigma_{\theta z}}{\sigma_Y} \right)^2 = 1. \tag{3.43}$$

Since the principal stresses are ordered, the Tresca criterion is given by Equation 3.40. Substituting Equation 3.42 in Equation 3.40, the Tresca criterion can be written as

$$\left(\frac{\sigma_{zz}}{\sigma_Y} \right)^2 + \left(\frac{2\sigma_{\theta z}}{\sigma_Y} \right)^2 = 1. \tag{3.44}$$

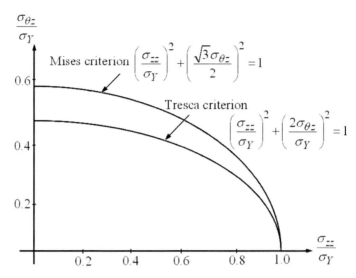

Figure 3.11. Mises and Tresca curves in the experiments of Taylor and Quinney (experimental points are excluded)

The plots of both the Mises and Tresca criteria are shown in Figure 3.11. The axes used are $\sigma_{\theta z}/\sigma_Y$ and σ_{zz}/σ_Y. Taylor and Quinney [6] found the experimental values of $\sigma_{\theta z}/\sigma_Y$ and σ_{zz}/σ_Y at yielding by varying the twisting moment M_z and the axial force F_z. The experimental points are *not* shown in the figure. But they lie closer to the Mises curve.

So both the experiments indicate that the Mises criterion is in better agreement with experimental results. Henceforth, we shall use only the Mises criterion. The experiments further indicate that the Tresca criterion is conservative as far as prediction of yielding is concerned. Therefore, it is preferred in the design of structures and machine elements where the objective is to avoid yielding. The Tresca yield surface is not smooth like the Mises yield surface (Figure 3.6). Normal to the Tresca yield surface does not exist along the six edges of the hexagonal prism. This creates difficulties in applying the plastic stress-strain relations along the edges as these relations depend on the normal. This is another reason why we use only the Mises criterion hereafter.

Example 3.1: Figure 3.12 shows axisymmetric drawing of steel wire. The yield stress of the material is $\sigma_Y = 360$ MPa. Matrices of the stress tensor σ at points A and B, with respect to (r, θ, z) coordinate system, are given by

$$[\sigma]_A = \begin{bmatrix} -80 & 0 & 160 \\ 0 & -80 & 0 \\ 160 & 0 & 70 \end{bmatrix} \text{MPa}, \quad [\sigma]_B = \begin{bmatrix} -100 & 0 & 180 \\ 0 & -100 & 0 \\ 180 & 0 & 80 \end{bmatrix} \text{MPa}. \quad (3.45)$$

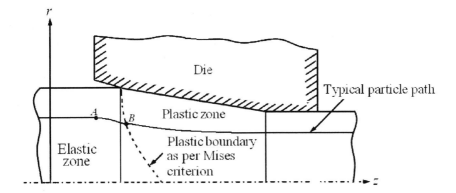

Figure 3.12. Typical plastic boundary in axisymmetric drawing

(a) Find the matrices of the deviatoric part σ' at points A and B with respect to (r, θ, z) coordinate system.

(b) Find the equivalent stress σ_{eq} at points A and B. Further, using the Mises criterion, show that point A lies in the elastic zone whereas point B lies on the plastic boundary.

(c) Find the invariants J_2 and J_3 of the deviatoric part σ' at point B. Using the Tresca criterion in terms of the invariants (Equation 3.33), check whether yielding occurs at point B.

Solution: (a) From Equation 2.100, we get the matrix of σ' with respect to any coordinate system as

$$[\sigma'] = [\sigma] - \left(\frac{1}{3}\sigma_{ii}\right)[1]. \tag{3.46}$$

Substituting the values of σ_{ij} at points A and B from Equation 3.45, we get

$$[\sigma']_A = \begin{bmatrix} -50 & 0 & 160 \\ 0 & -50 & 0 \\ 160 & 0 & 100 \end{bmatrix} \text{MPa}, \quad [\sigma']_B = \begin{bmatrix} -60 & 0 & 180 \\ 0 & -60 & 0 \\ 180 & 0 & 120 \end{bmatrix} \text{MPa}. \tag{3.47}$$

(b) Substituting the values of σ'_{ij} from part (a) in Equation 3.23, we get the following values of the equivalent stress at points A and B:

At point A,

$$\sigma_{eq} = \left(\frac{3}{2}\sigma_{ij}'\sigma_{ij}'\right)^{1/2} = \left\{\frac{3}{2}[2(-50)^2 + 2(160)^2 + (100)^2 + 4(0)^2]\right\}^{1/2} = 315.2\,\text{MPa}.$$

(3.48a)

At point B,

$$\sigma_{eq} = \left(\frac{3}{2}\sigma_{ij}'\sigma_{ij}'\right)^{1/2} = \left\{\frac{3}{2}[2(-60)^2 + 2(180)^2 + (120)^2 + 4(0)^2]\right\}^{1/2} = 360\,\text{MPa}.$$

(3.48b)

For the given material, $\sigma_Y = 360\,\text{MPa}$. Since, $\sigma_{eq} < \sigma_Y$ at point A, yielding has not taken place at this point according to the Mises criterion. Therefore, point A lies in the elastic zone. But $\sigma_{eq} = \sigma_Y$ at point B. Therefore, yielding has taken place at this point according to the Mises criterion. Therefore, point B lies on the plastic boundary.

(c) Substituting the values of σ_{ij}' at point B from part (a) in Equations 2.105 and 2.103, we get the following values of the invariants at point B:

$$J_2 = \frac{1}{2}\sigma_{ij}'\sigma_{ij}' = \frac{1}{2}[2(-60)^2 + 2(180)^2 + (120)^2 + 4(0)^2],$$
$$= 4.32 \times 10^4 \,(\text{MPa})^2.$$

(3.49a)

$$J_3 = \in_{ijk} \sigma_{1i}' \sigma_{2j}' \sigma_{3k}',$$
$$= \sigma_{11}'(\sigma_{22}'\sigma_{33}' - \sigma_{23}'\sigma_{32}') + \sigma_{12}'(\sigma_{23}'\sigma_{31}' - \sigma_{21}'\sigma_{33}') + \sigma_{13}'(\sigma_{21}'\sigma_{32}' - \sigma_{22}'\sigma_{31}'),$$
$$= (-60)[(-60)\times(120) - 0\times 0] + 0[0\times 180 - 0\times 120] + 180[0\times 0 - (-60)\times 180],$$
$$= 2.376 \times 10^6 \,(\text{MPa})^3.$$

(3.49b)

Substituting the above values of J_2 and J_3 at point B and $\sigma_Y = 360\,\text{MPa}$ in the left side of Equation 3.33, we get

$$4\left(J_2 - \frac{\sigma_Y^2}{4}\right)(J_2 - \sigma_Y^2)^2 - 27J_3^2,$$

$$= 4\left[4.32\times 10^4 - \frac{1}{4}\times(3.6)^2 \times 10^4\right][4.32\times 10^4 - (3.6)^2 \times 10^4]^2 - 27\times(2.376)^2 \times 10^{12},$$

$$= 1.7006 \times 10^{14}.$$

(3.50)

Since, at point B, the left side of Equation 3.33 is more than zero, yielding has already taken place at this point according to the Tresca criterion.

Therefore, if the plastic boundary is constructed in Figure 3.12 according to the Tresca criterion, it will lie to the left of the Mises plastic boundary.

If we find the principal stresses of σ using the matrices given by Equation 3.45 and then use the Mises and Tresca criteria in terms of the principal stresses (Equations 3.27 and 3.32), we will get exactly the same results.

3.4 Incremental Strain and Strain Rate Measures

As indicated in Section 3.2, the plastic deformation is usually large. Therefore, we cannot use the linear or infinitesimal strain tensor ε as the measure of plastic deformation. While developing a measure of plastic deformation, we need to take into account the fact that, in plastic behavior, the stress depends on the history of deformation. To express this behavior mathematically, it would be convenient to have a measure of plastic deformation either in the incremental form or in the rate form. Normally, the following two measures are used to represent the incremental plastic deformation.

3.4.1 Incremental Linear Strain Tensor

Figure 3.13 shows three configurations in a forging process: (i) *initial configuration*, (ii) deformed configuration at current time t (known as the *current configuration*) and (iii) deformed configuration at time $t+dt$ where dt is the time increment (known as the *incremental configuration*). A material particle occupies position P_0 in the initial configuration, position P in the current configuration and position P' in the incremental configuration. Let

$$x = x\hat{i} + y\hat{j} + z\hat{k} \tag{3.51}$$

be the current position vector of the particle (*i.e.*, the position vector of point P) and

$$du = du_x\hat{i} + du_y\hat{j} + du_z\hat{k} \tag{3.52}$$

be the incremental displacement vector (of point P) in time dt. Here, $(\hat{i}, \hat{j}, \hat{k})$ are the unit vectors along the $(x,\ y,\ z)$ axes.

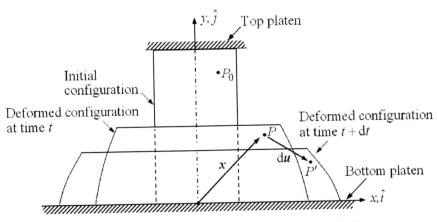

Figure 3.13. Incremental deformation in the time interval dt in forging process

We have seen in Subsection 2.4.1.1 that gradient of a vector is a tensor. Therefore, the quantity $\nabla(d\boldsymbol{u})$, which is a gradient of the vector $d\boldsymbol{u}$ with respect to the current position vector $\boldsymbol{x}$, is a tensor. Since the gradient is not with respect to the initial configuration, the operator ∇ here does not have the subscript zero like the gradient symbol of Subsection 2.4.1.1. The components of $\nabla(d\boldsymbol{u})$ with respect to (x, y, z) coordinate system are given by

$$[\nabla(du)] = \begin{bmatrix} \dfrac{\partial(du_x)}{\partial x} & \dfrac{\partial(du_x)}{\partial y} & \dfrac{\partial(du_x)}{\partial z} \\[2ex] \dfrac{\partial(du_y)}{\partial x} & \dfrac{\partial(du_y)}{\partial y} & \dfrac{\partial(du_y)}{\partial z} \\[2ex] \dfrac{\partial(du_z)}{\partial x} & \dfrac{\partial(du_z)}{\partial y} & \dfrac{\partial(du_z)}{\partial z} \end{bmatrix}. \tag{3.53}$$

We have seen in Subsection 2.4.1.1 that if $\boldsymbol{u}$ is the displacement from the initial configuration to a deformed configuration, then the symmetric part of the gradient of $\boldsymbol{u}$ (with respect to the position vector in the initial configuration) can be chosen as a measure of that deformation, provided the deformation is small. Now, *assume that the incremental deformation from the current configuration is small.* (Note that, mathematically, this assumption means the components of the tensor $\nabla(d\boldsymbol{u})$ are small compared to 1.) Since $d\boldsymbol{u}$ is the incremental displacement from the current configuration, the symmetric part of the tensor $\nabla(d\boldsymbol{u})$ can be selected as a measure of the incremental deformation. Thus, our measure of incremental deformation is the tensor $d\boldsymbol{\varepsilon}$, which is the symmetric part of $\nabla(d\boldsymbol{u})$. We call it the *incremental linear strain tensor.*

We can write the relation between $d\boldsymbol{\varepsilon}$ and $d\boldsymbol{u}$, in tensor notation, as

$$d\boldsymbol{\varepsilon} = \frac{1}{2}\left(\nabla(d\boldsymbol{u}) + (\nabla(d\boldsymbol{u}))^{\mathrm{T}}\right),$$
(3.54)

and in index notation as

$$d\varepsilon_{ij} = \frac{1}{2}\left(\frac{\partial(du_i)}{\partial x_j} + \frac{\partial(du_j)}{\partial x_i}\right) = \frac{1}{2}(du_{i,j} + du_{j,i}),$$
(3.55)

where it is understood that the comma indicates differentiation with respect to the current coordinates. Let the components of $d\boldsymbol{\varepsilon}$ with respect to (x,y,z) coordinate system be

$$[d\varepsilon] = \begin{bmatrix} d\varepsilon_{xx} & d\varepsilon_{xy} & d\varepsilon_{zx} \\ d\varepsilon_{xy} & d\varepsilon_{yy} & d\varepsilon_{yz} \\ d\varepsilon_{zx} & d\varepsilon_{yz} & d\varepsilon_{zz} \end{bmatrix}.$$
(3.56)

Then using Equation 3.53, the component form of Equation 3.54 can be written as

$$d\varepsilon_{xx} = \frac{\partial(du_x)}{\partial x}, \quad d\varepsilon_{yy} = \frac{\partial(du_y)}{\partial y}, \quad d\varepsilon_{zz} = \frac{\partial(du_z)}{\partial z},$$

$$d\varepsilon_{xy} = \frac{1}{2}\left(\frac{\partial(du_x)}{\partial y} + \frac{\partial(du_y)}{\partial x}\right),$$

$$d\varepsilon_{yz} = \frac{1}{2}\left(\frac{\partial(du_y)}{\partial z} + \frac{\partial(du_z)}{\partial y}\right),$$
(3.57)

$$d\varepsilon_{zx} = \frac{1}{2}\left(\frac{\partial(du_z)}{\partial x} + \frac{\partial(du_x)}{\partial z}\right).$$

Equations 3.54 or 3.55 or 3.57 are called the *incremental strain-displacement relations*.

To find the incremental displacements, incremental strains and incremental stresses in a deformable body, one needs to solve three sets of incremental equations in the current configuration. Such a formulation is called the *updated Lagrangian formulation*, which is described in more detail later. The above equation is a *first set of governing equations* for this formulation when the incremental deformation is small. We shall discuss in the next chapter the modifications required for the large incremental deformation.

The physical interpretation of the components of $d\boldsymbol{\varepsilon}$ is similar to that of the components of the linear strain tensor. The component $d\varepsilon_{xx}$ represents the *change in current length per unit current length* along the direction which is currently along x-direction. The components $d\varepsilon_{yy}$ and $d\varepsilon_{zz}$ have similar interpretation. The component $d\varepsilon_{xy}$ represents *half the change in angle* between the directions which are currently along x and y directions. The components $d\varepsilon_{yz}$ and $d\varepsilon_{zx}$ have similar interpretation. The sign convention for the components of $d\varepsilon$ is the same as that of the components of the linear strain tensor.

Just like the linear strain tensor, the tensor $d\varepsilon$ has the principal values, principal directions, principal invariants and the hydrostatic and deviatoric parts. They are defined similarly. The incremental volumetric strain $d\varepsilon_v$, when the incremental deformation is small, is defined by an equation similar to Equation 2.178:

$$d\varepsilon_v = d\varepsilon_{ii}.$$
(3.58)

Define the tensor $d\omega$ as the antisymmetric part of $\nabla(du)$:

$$d\omega = \frac{1}{2}\left(\nabla(du) - (\nabla(du))^{\mathrm{T}}\right).$$
(3.59)

It can be shown that, the tensor $d\omega$ represents the incremental rotation of a neighborhood of the particle in time dt, if the rotation is small (*i.e.*, if the components of the tensor $\nabla(du)$ are small compared to 1). It is called the *incremental infinitesimal rotation tensor*. In index notation, Equation 3.59 can be written as

$$d\omega_{ij} = \frac{1}{2}\left(\frac{\partial(du_i)}{\partial x_j} - \frac{\partial(du_j)}{\partial x_i}\right) = \frac{1}{2}\left(du_{i,j} - du_{j,i}\right),$$
(3.60)

where it is understood that the comma indicates the differentiation with respect to the current coordinates. Let the components of $d\omega$ with respect to (x, y, z) coordinate system be

$$[d\omega] = \begin{bmatrix} d\omega_{xx} & d\omega_{xy} & d\omega_{xz} \\ d\omega_{yx} & d\omega_{yy} & d\omega_{yz} \\ d\omega_{zx} & d\omega_{zy} & d\omega_{zz} \end{bmatrix}.$$
(3.61)

Then using Equation 3.53, the component form of Equation 3.59 can be written as

$$d\omega_{xx} = d\omega_{yy} = d\omega_{zz} = 0,$$

$$d\omega_{zy} = -d\omega_{yz} = \frac{1}{2}\left(\frac{\partial(du_z)}{\partial y} - \frac{\partial(du_y)}{\partial z} \right),$$

$$d\omega_{xz} = -d\omega_{zx} = \frac{1}{2}\left(\frac{\partial(du_x)}{\partial z} - \frac{\partial(du_z)}{\partial x} \right),$$

$$d\omega_{yx} = -d\varepsilon_{xy} = \frac{1}{2}\left(\frac{\partial(du_y)}{\partial x} - \frac{\partial(du_x)}{\partial y} \right).$$

(3.62)

The components $d\omega_{zy}$, $d\omega_{xz}$ and $d\omega_{yx}$ represent the angle of incremental rotation respectively about x, y and z axes. Their sign convention is similar to that for the components of the infinitesimal rotation tensor.

3.4.2 Strain Rate Tensor

While analyzing the forming processes like rolling, drawing, extrusion *etc.*, the domain used is usually a region fixed in space (called the *control volume*) rather than a fixed set of material particles. Such a formulation is called the *Eulerian formulation*, which is described in more detail later. In this formulation, it is not convenient to analyze the deformation increment by increment. Instead, it is easy to study the deformation of the whole control volume simultaneously. This becomes possible by choosing the velocity as a primary variable (instead of the incremental displacement). Further, in this case, it is the rate of deformation which is a more relevant secondary variable than the deformation itself.

To develop a measure of the rate of deformation, we consider the control volume for a drawing process shown in Figure 3.14. Here, point P is a location of some material particle at time t. Typical path line of the material particle is also shown in the figure. Let x be the position vector of point P and

$$v = v_x \hat{i} + v_y \hat{j} + v_z \hat{k}$$

(3.63)

be its velocity. The expression for x in terms of $(\hat{i}, \hat{j}, \hat{k})$ is given by Equation 3.51. We define the tensor ∇v (at point P) as the gradient of v with respect to the position vector x. It is called the *velocity gradient tensor*. Its components with respect to (x, y, z) coordinate system are given by

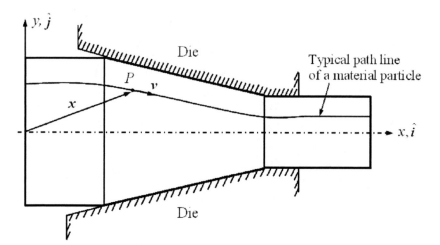

Figure 3.14. Control volume for drawing process

$$[\nabla v] = \begin{bmatrix} \dfrac{\partial v_x}{\partial x} & \dfrac{\partial v_x}{\partial y} & \dfrac{\partial v_x}{\partial z} \\[2mm] \dfrac{\partial v_y}{\partial x} & \dfrac{\partial v_y}{\partial y} & \dfrac{\partial v_y}{\partial z} \\[2mm] \dfrac{\partial v_z}{\partial x} & \dfrac{\partial v_z}{\partial y} & \dfrac{\partial v_z}{\partial z} \end{bmatrix}. \tag{3.64}$$

In books on continuum mechanics, the velocity gradient tensor is denoted by $\boldsymbol{L}$.

We decompose the tensor ∇v into a sum of symmetric and antisymmetric parts. Define the tensor $\dot{\varepsilon}$ as the symmetric part of ∇v :

$$\dot{\varepsilon} = \frac{1}{2}\left(\nabla v + (\nabla v)^{\mathrm{T}}\right). \tag{3.65}$$

It can be shown that the tensor $\dot{\varepsilon}$ completely describes the rate of deformation at a point. That is, given the tensor $\dot{\varepsilon}$ at a point, we can find the *rate* of change of length per unit length in any direction at that point. Further, we can also find the *rate* of change of angle between any pair of perpendicular directions at that point [2,3,7]. The tensor $\dot{\varepsilon}$ is called the *strain rate tensor*. Note that the tensor $\dot{\varepsilon}$ represents the rate of deformation at a point irrespective of whether the rate of deformation at that point is small or large. Further, even though we use the symbol $\dot{\varepsilon}$, commonly employed in metal forming literature for this tensor, it is to be noted that this tensor is not the time derivative of the linear strain tensor ε. In books on continuum mechanics, this tensor is usually denoted by $\boldsymbol{D}$ and is called the *rate of deformation tensor*.

In index notation, Equation 3.65 can be written as

$$\dot{\varepsilon}_{ij} = \frac{1}{2}\left(\frac{\partial v_i}{\partial x_j} + \frac{\partial v_j}{\partial x_i}\right) = \frac{1}{2}(v_{i,j} + v_{j,i}), \qquad (3.66)$$

where it is understood that the comma indicates differentiation with respect to the components of *x*. Let the components of $\dot{\varepsilon}$ with respect to (x, y, z) coordinate system be

$$[\dot{\varepsilon}] = \begin{bmatrix} \dot{\varepsilon}_{xx} & \dot{\varepsilon}_{xy} & \dot{\varepsilon}_{zx} \\ \dot{\varepsilon}_{xy} & \dot{\varepsilon}_{yy} & \dot{\varepsilon}_{yz} \\ \dot{\varepsilon}_{zx} & \dot{\varepsilon}_{yz} & \dot{\varepsilon}_{zz} \end{bmatrix}. \qquad (3.67)$$

Then using Equation 3.64, the component form of Equation 3.65 can be written as

$$\dot{\varepsilon}_{xx} = \frac{\partial v_x}{\partial x}, \; \dot{\varepsilon}_{yy} = \frac{\partial v_y}{\partial y}, \; \dot{\varepsilon}_{zz} = \frac{\partial v_z}{\partial z},$$

$$\dot{\varepsilon}_{xy} = \frac{1}{2}\left(\frac{\partial v_x}{\partial y} + \frac{\partial v_y}{\partial x}\right),$$

$$\dot{\varepsilon}_{yz} = \frac{1}{2}\left(\frac{\partial v_y}{\partial z} + \frac{\partial v_z}{\partial y}\right), \qquad (3.68)$$

$$\dot{\varepsilon}_{zx} = \frac{1}{2}\left(\frac{\partial v_z}{\partial x} + \frac{\partial v_x}{\partial z}\right).$$

Equations 3.65 or 3.66 or 3.68 are called the *strain rate– velocity relations.*

To find the velocities, strain rates and stresses in a control volume, one needs to solve three sets of governing equations for the Eulerian formulation. The above equation is a *first set of governing equations* for this formulation.

The physical interpretation of the components of $\dot{\varepsilon}$ is similar to that of the components of the linear strain tensor. The component $\dot{\varepsilon}_{xx}$ represents the *rate of change of current length per unit current length* along the direction which is currently along *x*-direction. The components $\dot{\varepsilon}_{yy}$ and $\dot{\varepsilon}_{zz}$ have similar interpretation. These components are called the *normal* strain rate components. The component $\dot{\varepsilon}_{xy}$ represents *half the rate of change of angle* between the directions which are currently along *x* and *y* directions. The components $\dot{\varepsilon}_{yz}$ and $\dot{\varepsilon}_{zx}$ have similar interpretation. These components are called the *shear* strain rate components. The sign convention for the components of $\dot{\varepsilon}$ is similar to that of the components of the linear strain tensor.

Analysis of the tensor $\dot\varepsilon$ at a point can be carried out in a manner similar to the analysis of linear strain tensor. Let $\dot\varepsilon_i$ be the eigenvalues and $\hat e_i$ be the orthonormal eigenvectors of the tensor $\dot\varepsilon$. Further, assume that $\dot\varepsilon_i$ have been arranged in decreasing order and the senses of $\hat e_i$ have been chosen so as form a right-handed system. It can be shown that the rates of change of angles associated with the directions $\hat e_i$ are zero. Therefore, these directions are called the *principal directions*. Further, $\dot\varepsilon_i$ are called the *principal values* of $\dot\varepsilon$. The matrix of $\dot\varepsilon$ with respect to the principal directions becomes a diagonal matrix with $\dot\varepsilon_i$ as the diagonal components. It is observed that $\dot\varepsilon_1$ represents the maximum rate of change of length per unit length which occurs in the direction $\hat e_1$. Similarly, $\dot\varepsilon_3$ represents the minimum rate of change of length per unit length which occurs in the direction $\hat e_3$. Further, the maximum rate of change of angle is given by $\dot\varepsilon_1 - \dot\varepsilon_3$. The directions which undergo this *maximum rate of change of angle* are given by $(\pm 1/\sqrt2)(\hat e_1 \pm \hat e_3)$. The coefficients in the eigenvalue equation are the three *principal invariants* of the tensor $\dot\varepsilon$ which are defined by equations similar to Equations 2.172–2.174. Just like the linear strain tensor, the tensor $\dot\varepsilon$ can also be decomposed into a sum of hydrostatic and deviatoric parts. The hydrostatic part represents the rate of change of current volume (without change of current shape) and the deviatoric part represents the rate of change of current shape (without change of current volume). The rate of change of current volume per unit current volume ($\dot\varepsilon_v$) is defined by

$$\dot\varepsilon_v = \dot\varepsilon_{ii} = v_{i,i}. \tag{3.69}$$

It is called volumetric strain rate.

Define the tensor $\dot\omega$ as the antisymmetric part of ∇v :

$$\dot\omega = \frac{1}{2}\left(\nabla v - (\nabla v)^{\mathrm T}\right). \tag{3.70}$$

It can be shown that the tensor $\dot\omega$ represents the angular velocity of a neighborhood of the point. This result is valid irrespective of whether the angular velocity at that point is small or large. The tensor $\dot\omega$ is called the *spin or vorticity tensor*. In index notation, Equation 3.70 can be written as

$$\dot\omega_{ij} = \frac{1}{2}\left(\frac{\partial v_i}{\partial x_j} - \frac{\partial v_j}{\partial x_i}\right) = \frac{1}{2}(v_{i,j} - v_{j,i}), \tag{3.71}$$

where it is understood that the comma indicates the differentiation with respect to the components of $\boldsymbol{x}$. Let the components of $\dot{\omega}$ with respect to (x,y,z) coordinate system be

$$
[\dot{\omega}] = \begin{bmatrix} \dot{\omega}_{xx} & \dot{\omega}_{xy} & \dot{\omega}_{xz} \\ \dot{\omega}_{yx} & \dot{\omega}_{yy} & \dot{\omega}_{yz} \\ \dot{\omega}_{zx} & \dot{\omega}_{zy} & \dot{\omega}_{zz} \end{bmatrix}.
\tag{3.72}
$$

Then using Equation 3.64, the component form of Equation 3.70 can be written as

$$
\dot{\omega}_{xx} = \dot{\omega}_{yy} = \dot{\omega}_{zz} = 0,
$$

$$
\dot{\omega}_{zy} = -\dot{\omega}_{yz} = \frac{1}{2}\left(\frac{\partial v_z}{\partial y} - \frac{\partial v_y}{\partial z} \right),
$$

$$
\dot{\omega}_{xz} = -\dot{\omega}_{zx} = \frac{1}{2}\left(\frac{\partial v_x}{\partial z} - \frac{\partial v_z}{\partial x} \right),
\tag{3.73}
$$

$$
\dot{\omega}_{yx} = -\dot{\varepsilon}_{xy} = \frac{1}{2}\left(\frac{\partial v_y}{\partial x} - \frac{\partial v_x}{\partial y} \right).
$$

The quantities $\dot{\omega}_{zy}$, $\dot{\omega}_{xz}$ and $\dot{\omega}_{yx}$ represent the components of angular velocity respectively about x, y and z axes. Their sign convention is similar to that for the components of the infinitesimal rotation tensor. Since, $\dot{\omega}$ is an antisymmetric tensor, we can associate the following vector with it:

$$
\begin{aligned}
\dot{\omega}_{zy}\,\hat{\boldsymbol{i}} + \dot{\omega}_{xz}\,\hat{\boldsymbol{j}} + \dot{\omega}_{yx}\,\hat{\boldsymbol{k}} &= \frac{1}{2}\left[\left(\frac{\partial v_z}{\partial y} - \frac{\partial v_y}{\partial z} \right)\hat{\boldsymbol{i}} + \left(\frac{\partial v_x}{\partial z} - \frac{\partial v_z}{\partial x} \right)\hat{\boldsymbol{j}} + \left(\frac{\partial v_y}{\partial x} - \frac{\partial v_x}{\partial y} \right)\hat{\boldsymbol{k}} \right], \\
&= \frac{1}{2}\,\epsilon_{ijk}\,\frac{\partial v_k}{\partial x_j}\,\hat{\boldsymbol{i}}_i, \\
&= \frac{1}{2}\nabla \times \boldsymbol{v}.
\end{aligned}
\tag{3.74}
$$

The vector $(1/2)\nabla \times \boldsymbol{v}$ is indeed the angular velocity of a neighborhood of the point.

3.4.3 Relation Between Incremental Linear Strain Tensor and Strain Rate Tensor

Note that the symbol $\mathbf{d}u$ in Section 3.4.1 does not represent *the relative displacement* of a particle with respect to its neighboring particle. Instead, it is an *infinitesimal change in the displacement of a single particle in time* $\mathrm{d}t$ (Figure 3.13). Then, the velocity v of the particle would be $\mathbf{d}u / \mathrm{d}t$. In view of this relation between $\mathbf{d}u$ and v, comparison of Equations 3.54 and 3.65 gives the following relation between the incremental linear strain tensor $\mathbf{d}\varepsilon$ and the strain rate tensor $\dot{\varepsilon}$:

$$\mathbf{d}\varepsilon = \dot{\varepsilon}\mathrm{d}t , \tag{3.75}$$

or in index notation

$$\mathrm{d}\varepsilon_{ij} = \dot{\varepsilon}_{ij}\mathrm{d}t . \tag{3.76}$$

Further, we get a similar relation between the incremental infinitesimal rotation tensor $\mathbf{d}\omega$ and the spin tensor $\dot{\omega}$:

$$\mathrm{d}\omega = \dot{\omega}\mathrm{d}t , \tag{3.77}$$

or

$$\mathrm{d}\omega_{ij} = \dot{\omega}_{ij}\mathrm{d}t . \tag{3.78}$$

We use Equation 3.76 to find out what happens when we integrate the strain rate tensor $\dot{\varepsilon}$. For this purpose, we consider a very simple deformation. The deformation in the control volume of Figure 3.15 is such that the particle paths are straight lines parallel to x-axis. It means that the particles do not rotate, and therefore both $\dot{\omega}$ and $\mathbf{d}\omega$ are zero at every point of the control volume. Consider a typical particle path where the length of the particle is ℓ_0 when it enters the control volume (*i.e.*, at time $t = t_0$) and ℓ_f when it leaves the control volume (*i.e.*, at time $t = t_f$).

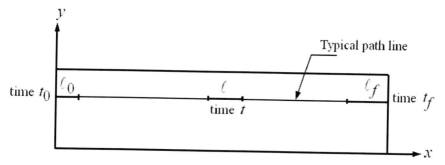

Figure 3.15. Simple deformation in which particle path-lines are parallel to x-axis

Further, let the length at time t be ℓ and the length increment in time dt be dℓ. Then, from Equation 3.76 and the definition of $d\varepsilon_{xx}$, we get

$$\dot{\varepsilon}_{xx}dt = d\varepsilon_{xx} = \frac{d\ell}{\ell}. \tag{3.79}$$

Integration of $\dot{\varepsilon}_{xx}$ from $t = t_0$ to $t = t_f$ gives

$$\int_{t_0}^{t_f} \dot{\varepsilon}_{xx}dt = \int_{t_0}^{t_f} d\varepsilon_{xx} = \int_{\ell_0}^{\ell_f} \frac{d\ell}{\ell} = \ln\frac{\ell_f}{\ell_0} = \varepsilon_f \neq \varepsilon_{xx}. \tag{3.80}$$

Here, ε_f is the value of logarithmic strain ε (defined by Equation 3.1) at $t = t_f$. Thus, when we integrate a normal strain rate component, we get the *logarithmic strain* in that direction, and *not* the corresponding component of the linear strain tensor. This result is based on the assumption that there is *no rotation* of particles. When the rotation of particles is present, no physically meaningful quantity emerges from the integration of a strain rate component [7]. Generalizing this result to the three-dimensional case, we can state that integration of the strain rate tensor $\dot{\varepsilon}$ gives the logarithmic strain tensor when there is no rotation of the particles. When rotation of the particles is present, the integration of $\dot{\varepsilon}$ does not give any physically meaningful tensor.

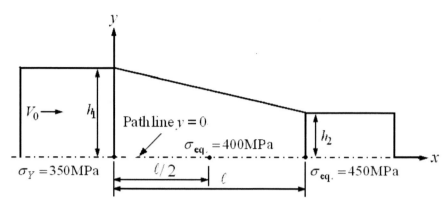

Figure 3.16. Extrusion of a steel sheet

Example 3.2: Figure 3.16 shows extrusion of a sheet. Components of the incremental displacement vector **du** with respect to (x, y, z) coordinate system are given by

$$du_x = v_x dt, \; du_y = v_y dt, \; du_z = v_z dt , \tag{3.81}$$

where dt is the time increment and the components (v_x, v_y, v_z) of the velocity vector **v** are

$$v_x = \frac{V_0 \ell}{[(\ell - x) + (h_2 / h_1)x]}, \; v_y = -\frac{V_0[1 - (h_2 / h_1)]\ell}{[(\ell - x) + (h_2 / h_1)x]^2} y, \; v_z = 0 . \tag{3.82}$$

Here, V_0 is the ram speed and ℓ, h_1 and h_2 are the geometric dimensions shown in the figure.
(a) Find the components of the strain rate tensor $\dot{\varepsilon}$ and the incremental linear strain tensor $d\varepsilon$.

(b) Find the volumetric strain rate $\dot{\varepsilon}_v$.

Solution: (a) Substituting the values of v_i from Equation 3.82 in the strain rate–velocity relations (Equation 3.68), we get

$$\dot{\varepsilon}_{xx} = \frac{\partial v_x}{\partial x} = +\frac{V_0[1-(h_2/h_1)]\ell}{[(\ell-x)+(h_2/h_1)x]^2},$$

$$\dot{\varepsilon}_{yy} = \frac{\partial v_y}{\partial y} = -\frac{V_0[1-(h_2/h_1)]\ell}{[(\ell-x)+(h_2/h_1)x]^2},$$

$$\dot{\varepsilon}_{zz} = \frac{\partial v_z}{\partial z} = 0,$$

$$\dot{\varepsilon}_{xy} = \frac{1}{2}\left(\frac{\partial v_x}{\partial y} + \frac{\partial v_y}{\partial x}\right) = -\frac{V_0[1-(h_2/h_1)]^2\ell}{[(\ell-x)+(h_2/h_1)x]^3}y, \qquad (3.83)$$

$$\dot{\varepsilon}_{yz} = \frac{1}{2}\left(\frac{\partial v_y}{\partial z} + \frac{\partial v_z}{\partial y}\right) = 0,$$

$$\dot{\varepsilon}_{zx} = \frac{1}{2}\left(\frac{\partial v_z}{\partial x} + \frac{\partial v_x}{\partial z}\right) = 0.$$

Substituting the values of du_i from Equations 3.81 and 3.82 in the incremental strain-displacement relations (Equation 3.57), we get

$$d\varepsilon_{xx} = \frac{\partial(du_x)}{\partial x} = +\frac{V_0[1-(h_2/h_1)]\ell}{[(\ell-x)+(h_2/h_1)x]^2}dt,$$

$$d\varepsilon_{yy} = \frac{\partial(du_y)}{\partial y} = -\frac{V_0[1-(h_2/h_1)]\ell}{[(\ell-x)+(h_2/h_1)x]^2}dt,$$

$$d\varepsilon_{zz} = \frac{\partial(du_z)}{\partial z} = 0,$$

$$d\varepsilon_{xy} = \frac{1}{2}\left(\frac{\partial(du_x)}{\partial y} + \frac{\partial(du_y)}{\partial x}\right) = -\frac{V_0[1-(h_2/h_1)]^2\ell}{[(\ell-x)+(h_2/h_1)x]^3}ydt, \qquad (3.84)$$

$$d\varepsilon_{yz} = \frac{1}{2}\left(\frac{\partial(du_y)}{\partial z} + \frac{\partial(du_z)}{\partial y}\right) = 0,$$

$$d\varepsilon_{zx} = \frac{1}{2}\left(\frac{\partial(du_z)}{\partial x} + \frac{\partial(du_x)}{\partial z}\right) = 0.$$

(b) Substituting the values of $\dot{\varepsilon}_{ij}$ from part (a) in Equation 3.69, we get the following expression for the volumetric strain rate:

$$\dot{\varepsilon}_v = \dot{\varepsilon}_{ii} = \frac{V_0[1-(h_2/h_1)]\ell}{[(\ell-x)+(h_2/h_1)x]^2}(1-1+0) = 0. \qquad (3.85)$$

Note that when the strain rate components $\dot{\varepsilon}_{xz}$, $\dot{\varepsilon}_{yz}$ and $\dot{\varepsilon}_{zz}$ are zero at a point, the state of deformation is called the *state of plane strain (at a point) in* $x - y$ *plane*. When these strain rate components are zero at every point of the control volume and if, additionally, the remaining strain rate components $\dot{\varepsilon}_{xx}$, $\dot{\varepsilon}_{yy}$ and $\dot{\varepsilon}_{xy}$ are independent of z, it is called the *state of plane strain (in a body) in* $x - y$ *plane*. It can be shown that the state of deformation described by Equation 3.83 is of this type.

3.5 Modeling of Isotropic Hardening or Criterion for Subsequent Isotropic Yielding

Figure 3.2 represents one-dimensional experimental result for subsequent yielding. It shows the variation of yield stress with the level of plastic deformation. Mathematical expression of this variation is given by Equation 3.8, where a specific form of hardening function h could be Equation 3.9 or 3.10 or 3.11. To generalize Equation 3.8 to three-dimensional state of stress, we follow a somewhat different approach to that adopted for the generalization of initial yielding. To gain some insight into the phenomenon of subsequent yielding, we consider graphical representation of the initial yield criteria, rather than its mathematical expression. Figure 3.6 shows yield surfaces corresponding to the initial yield criteria. However, since yielding does not depend on the hydrostatic part of stress, it is simpler to consider a two-dimensional curve called yield locus (Figure 3.7), which is the intersection of the yield surface and the deviatoric plane. We expect that, in general, both the size and shape of the initial yield locus would change with the complete history of plastic deformation since the last annealing. Further, due to the Bauschinger effect, the center of the yield locus would move away from the origin. However, if the Bauschinger effect is to be neglected, the center of the yield locus must coincide with the origin.

To simplify the development of criterion for subsequent yielding, we assume that the material remains *isotropic during hardening*. We shall discuss the validity of this assumption at the end of this section. Note that the assumption of isotropy automatically excludes the Bauschinger effect. Further, the assumption of isotropy means the following:

- During subsequent yielding, only the size of the initial yield locus changes, but its shape remains unchanged.
- The change in size depends on the invariants of a tensor describing the history of plastic deformation.

The first step in the development of criterion for subsequent yielding is the evaluation of the invariants. The measure of deformation developed in Section 3.4, namely the incremental linear strain tensor, is only for an incremental deformation. Further, because of the rotation of the particles, its integration up to current time does not give a physically meaningful measure of the history of deformation. Therefore, we first find the invariants of the incremental linear strain tensor and then integrate them, along the path of deformation, to get scalar measures of the history of deformation. Since invariants are coordinate-independent functions, rotation of the particles does not affect their values during the integration process.

Another thing to be noted is that the measure of incremental deformation developed in Section 3.4 contains both the elastic and plastic parts. Change in the size of yield locus depends only on the plastic part of history of deformation. Therefore, we need to separate the plastic part from the incremental linear strain tensor. For this purpose, we assume that the elastic and plastic parts (of the incremental linear strain tensor) are additive. The elastic and plastic parts were found to be additive for the one-dimensional case (Figure 3.1), not just for incremental deformation but also for total deformation. Therefore, we expect them to be additive for the three-dimensional case, at least for the case of small incremental deformation. Thus, we assume that

$$d\varepsilon_{ij} = d\varepsilon_{ij}^{e} + d\varepsilon_{ij}^{p}, \tag{3.86}$$

where $d\varepsilon_{ij}^{e}$ and $d\varepsilon_{ij}^{p}$ are respectively the elastic and plastic parts of the incremental linear strain tensor. We shall discuss the validity of this assumption later when we discuss the measures of finite incremental deformation in the next chapter. We can now define the principal invariants $I_{d\varepsilon^{p}}$, $II_{d\varepsilon^{p}}$ and $III_{d\varepsilon^{p}}$ of the tensor $d\varepsilon_{ij}^{p}$ by equations similar to Equations 2.172–2.174:

$$I_{d\varepsilon^{p}} = d\varepsilon_{ii}^{p}, \tag{3.87}$$

$$II_{d\varepsilon^{p}} = \frac{1}{2}(d\varepsilon_{ij}^{p}d\varepsilon_{ij}^{p} - d\varepsilon_{ii}^{p}d\varepsilon_{jj}^{p}), \tag{3.88}$$

$$III_{d\varepsilon^{p}} = \in_{ijk} d\varepsilon_{1i}^{p}d\varepsilon_{2j}^{p}d\varepsilon_{3k}^{p}. \tag{3.89}$$

Incremental volumetric strain corresponding to $d\varepsilon_{ij}^{p}$ is defined by an equation similar to Equation 3.58:

$$d\varepsilon_{v}^{p} = d\varepsilon_{ii}^{p}. \tag{3.90}$$

Since, there is no change in volume during plastic deformation, $d\varepsilon_{v}^{p}$ is zero. Therefore, $d\varepsilon_{ii}^{p}$ is zero. Because of this, Equations 3.87 and 3.88 for the first two invariants get modified as follows:

$$I_{d\varepsilon^{p}} = 0, \tag{3.91}$$

$$II_{d\varepsilon^{p}} = \frac{1}{2}d\varepsilon_{ij}^{p}d\varepsilon_{ij}^{p}. \tag{3.92}$$

Note that Equations 3.87 and 3.91 show that the hydrostatic part of $d\varepsilon_{ij}^P$ is zero and therefore its deviatoric part is equal to itself. Since $I_{d\varepsilon^P}$ is zero, the size of the yield locus (for isotropic hardening) should depend only on $II_{d\varepsilon^P}$ and $III_{d\varepsilon^P}$ or rather on their integrals along the path of deformation.

Regarding the dependence of the size of the yield locus on the integrals of $II_{d\varepsilon^P}$ and $III_{d\varepsilon^P}$, there are two hypotheses: (i) *strain hardening hypothesis* and (ii) *work hardening hypothesis*. We consider both of these one by one. Regarding the choice of the initial yield locus, we do not choose the most general yield locus. Instead, we choose the Mises yield locus, because we have decided to use the Mises criterion for initial yielding in the rest of the book.

3.5.1 Strain Hardening Hypothesis

In strain hardening hypothesis, the size of the yield locus is assumed to be independent of the third invariant $III_{d\varepsilon^P}$. Further, it is assumed to be a function not of the integral of the second invariant $II_{d\varepsilon^P}$ but of the following invariant:

$$d\varepsilon_{eq}^P = \left(\frac{4}{3}II_{d\varepsilon^P}\right)^{1/2} = \left(\frac{2}{3}d\varepsilon_{ij}^P d\varepsilon_{ij}^P\right)^{1/2}. \tag{3.93}$$

The choice of this invariant makes it convenient to evaluate the hardening function of three-dimensional criterion from the results of tension test. It is so because, in tension test, this invariant is equal to the plastic part of the incremental axial strain. To see this, note that, the shear strain components are zero in tension test as stated in Subsection 2.5.1.1. Therefore,

$$d\varepsilon_{xy}^P = d\varepsilon_{yz}^P = d\varepsilon_{zx}^P = 0. \tag{3.94}$$

Further, the condition of no volume change in plastic deformation and Equation 3.90 imply that

$$d\varepsilon_{xx}^P + d\varepsilon_{yy}^P + d\varepsilon_{zz}^P = 0. \tag{3.95}$$

Additionally, because of transverse symmetry of the rod, we get

$$d\varepsilon_{yy}^P = d\varepsilon_{zz}^P. \tag{3.96}$$

In view of Equations 3.94–3.96, the quantity $d\varepsilon_{eq}^P$, in tension test, reduces to $d\varepsilon_{xx}^P$. Therefore, $d\varepsilon_{eq}^P$ is called the *equivalent plastic strain increment*. It is also called the *effective or generalized plastic strain increment*. Integration of $d\varepsilon_{eq}^{p}$ along the path of deformation gives

$$\varepsilon_{eq}^{p} = \int d\varepsilon_{eq}^{p} . \tag{3.97}$$

We call the quantity ε_{eq}^{p} the *equivalent* (or *effective* or *generalized*) *plastic strain*.

Now we apply the strain-hardening hypothesis to the Mises yield locus on the deviatoric plane. Radius of the Mises yield locus, in initial yielding, is $(\sqrt{2/3})\sigma_{Y}$. Using Equation 3.24, the expression for the radius can be written as $(\sqrt{2/3})\sigma_{eq}$. If we take σ_{eq} as a measure of the size of the Mises yield locus, the strain hardening hypothesis can be expressed mathematically as

$$\sigma_{eq} = H(\varepsilon_{eq}^{p}), \tag{3.98}$$

where H, known as the hardening function, depends on the material. This equation states that the amount of hardening depends on the sum of all infinitesimal plastic increments leading to the final deformation of the particle since last annealing and not just on the difference between the final and initial deformations of the particle.

The function H is evaluated from the one-dimensional stress-strain curve in tension test. For this purpose, we show that the equivalent plastic strain ε_{eq}^{p}, in tension test, is equal to ε^{p} (the plastic part of the logarithmic strain). To show this, we proceed as follows. Earlier in this section, we have shown that the equivalent plastic strain increment $d\varepsilon_{eq}^{p}$ is equal to $d\varepsilon_{xx}^{p}$ (plastic part of the incremental axial strain) in tension test. Further, Equation 3.80 shows that the integral of $d\varepsilon_{xx}$ is the logarithmic strain ε in x-direction provided there is no rotation of the particle. The condition of no rotation holds good for tension test. Therefore, the integral of $d\varepsilon_{xx}$ will be ε. Further, in tension test, the elastic and plastic parts of the strain are additive. Therefore, the integral of the plastic part of $d\varepsilon_{xx}$ will be equal to the plastic part of ε. Thus, in tension test, we get

$$\varepsilon_{eq}^{p} = \int d\varepsilon_{eq}^{p} = \int d\varepsilon_{xx}^{p} = \varepsilon^{p} . \tag{3.99}$$

Further, as already shown in Subsection 3.3.1, the equivalent stress σ_{eq} is equal to the axial stress σ in tension test. Therefore, in tension test, the three-dimensional hardening function H reduces to a function between σ and ε^{p}. In other words, for isotropic hardening, the function H is the same as the one-dimensional hardening function h describing the stress-strain curve of Figure 3.2. As stated earlier, this curve is commonly represented by Equation 3.9 or 3.10 or 3.11.

3.5.2 Work Hardening Hypothesis

In work hardening hypothesis, the size of the yield locus is assumed to depend on the total plastic work done (per unit volume) to achieve the present state of plastic deformation since last annealing. This quantity is the integral of the incremental plastic work (per unit volume) along the path of deformation:

$$W^P = \int dW^P = \int \sigma_{ij} d\varepsilon_{ij}^p . \qquad (3.100)$$

Note that $dW^P = \sigma_{ij} d\varepsilon_{ij}^p$ is an invariant of the tensor $d\varepsilon_{ij}^p$ as the work is a coordinate-independent quantity. Now we apply the work hardening hypothesis to the Mises yield locus on the deviatoric plane. Since σ_{eq} is a measure of the size of the Mises yield locus, the work hardening hypothesis can be expressed mathematically as

$$\sigma_{eq} = F(W^P), \qquad (3.101)$$

where the function F depends on the material.

The function F is evaluated from the one-dimensional stress-strain curve in tension test. As stated earlier, the equivalent stress σ_{eq} is equal to the axial stress σ in tension test. Further, the only non-zero stress component in tension test is the axial stress σ. Therefore, the expression for W^P in tension test reduces to

$$W_{tension-test}^P = \int \sigma d\varepsilon^P . \qquad (3.102)$$

Thus, for isotropic hardening, the function F reduces to a function between σ and $W_{tension-test}^P$ in tension test. We construct this function as follows. First, using Equation 3.102, we find $W_{tension-test}^P$ corresponding to all values of σ greater than σ_Y from Figure 3.2. Then we plot a graph of σ vs $W_{tension-test}^P$. Finally, we fit a mathematical equation similar to Equations 3.9–3.11 to this graph.

It can be shown that the work hardening hypothesis is equivalent to the strain hardening hypothesis for the materials whose plastic potential is the same as the Mises yield function. The plastic potential is defined in the next section.

3.5.3 Experimental Validation

There have been very few attempts to study the change in yield locus or yield surface due to hardening. Naghdi et al. [8] studied the subsequent yield loci through tension-torsion tests on thin aluminum tubes. (Their experiments may be considered as extensions of experiments carried out by Taylor and Quinney [6].) Naghdi et al. [8] plotted the initial and subsequent yield loci using the normal and shear stress components as the axes. Their results show that the initial yield locus, which is symmetric about the normal stress axis, becomes unsymmetric by getting shifted along the shear stress axis. This is due to the Bauschinger effect. Experiments of Philips and Das [9] show that, during subsequent yielding, the yield surfaces expand, distort and translate. (They also studied the effect of temperature on yield surfaces.) Thus, the assumption of isotropic hardening does not seem to have experimental support. However, the predictions of strain hardening hypothesis do agree with experimental results when the *loading is proportional* [10].

To express the change of shape of yield locus in a mathematical form, more experimental data than presently available is needed. In the absence of such data, we shall use the strain hardening hypothesis for the purpose of our analysis. Further, we shall use the power law similar to Equation 3.9 to represent the stress-strain curve of Figure 3.2. With these choices, the *three-dimensional hardening function H* can be written as

$$\sigma_{eq} = H(\varepsilon_{eq}^{p}) \equiv \sigma_Y + K(\varepsilon_{eq}^{p})^n. \tag{3.103}$$

Further, using Equations 3.22, 3.24 and 3.98, the *criterion for subsequent yielding* for the Mises material can now be expressed as

$$f(J_2, J_3; \varepsilon_{eq}^{p}) \equiv J_2 - \frac{1}{3}H^2(\varepsilon_{eq}^{p}) = 0. \tag{3.104}$$

The function f is called the *yield function*. For subsequent yielding, the yield function, besides being a function of the invariants J_2 and J_3, also depends on the equivalent plastic strain (called the hardening parameter). For initial yielding, the value of H reduces to σ_Y and thus, Equation 3.104 reduces to Equation 3.22.

Example 3.3: Consider the sheet extrusion of Figure 3.16.
(a) Assume that the elastic part of the incremental linear strain tensor is zero (*i.e.*, $d\varepsilon_{ij}^{p} = d\varepsilon_{ij}$). Then, using the expressions of $d\varepsilon_{ij}$ from Example 3.2 (Equation 3.84), find the equivalent plastic strain increment $d\varepsilon_{eq}^{p}$ along the path $y = 0$.

(b) Using the expression of $d\varepsilon_{eq}^{p}$ from part (a), calculate the values of equivalent plastic strain ε_{eq}^{p} at points $x = \ell/2$ and $x = \ell$ along the path $y = 0$. Use the following values of the geometric parameters:

$$h_1 = 10 \, \text{mm}, \ h_2 = 5 \, \text{mm}, \ \ell = 20 \, \text{mm}. \tag{3.105}$$

(c) Use the values of ε_{eq}^{p} at $x = \ell/2$ and $x = \ell$ from part (b) to determine the material constants K and n of the hardening relation given by Equation 3.103.

Use the following values:

$$\sigma_Y = 360 \text{ MPa}, \ \sigma_{eq} = 400 \text{ MPa at } x = \ell/2, \sigma_{eq} = 450 \text{ at } x = \ell. (3.106)$$

Solution: (a) It is given that $d\varepsilon_{ij}^p = d\varepsilon_{ij}$. Then, the expression for the equivalent plastic strain increment $d\varepsilon_{eq}^p$ (Equation 3.93) becomes

$$d\varepsilon_{eq}^p = \left(\frac{2}{3} d\varepsilon_{ij} d\varepsilon_{ij}\right)^{1/2}. \tag{3.107}$$

The expressions for $d\varepsilon_{ij}$ given in Example 3.2 (Equation 3.84) are such that the components $d\varepsilon_{zz}$, $d\varepsilon_{yz}$ and $d\varepsilon_{zx}$ are zero everywhere. Further, along the path $y = 0$, the component $d\varepsilon_{xy}$ becomes zero. Substituting the expressions for the components $d\varepsilon_{xx}$ and $d\varepsilon_{yy}$ in Equation 3.107, we get the following expression for $d\varepsilon_{eq}^p$ along the path $y = 0$:

$$
\begin{aligned}
d\varepsilon_{eq}^p &= \left\{\frac{2}{3}[(d\varepsilon_{xx})^2 + (d\varepsilon_{yy})^2]\right\}^{1/2}, \\
&= \frac{2}{\sqrt{3}} \frac{V_0[1-(h_2/h_1)]\ell}{[(\ell-x)+(h_2/h_1)x]^2} dt.
\end{aligned} \tag{3.108}
$$

(b) The equivalent plastic strain ε_{eq}^p is given by Equation 3.97. To carry out the integration of Equation 3.97, we change the variable of integration from t to x using the following definition of the velocity component v_x:

$$v_x = dx/dt. \tag{3.109}$$

Substituting the above equation along with Equation 3.82 for v_x, the expression for $d\varepsilon_{eq}^p$ along the path $y = 0$ (Equation 3.108) becomes

$$d\varepsilon_{eq}^p = \frac{2}{\sqrt{3}} \frac{[1-(h_2/h_1)]}{[(\ell-x)+(h_2/h_1)x]} dx. \tag{3.110}$$

Before integration, we simplify this expression by substituting the values of geometric parameters from Equation 3.105. Then, we get

$$d\varepsilon_{eq}^p = \frac{1}{\sqrt{3}(20-0.5x)} dx. \tag{3.111}$$

Now, substituting the above expression in Equation 3.97, we get the following expression for ε_{eq}^p:

$$\varepsilon_{eq}^p = \int d\varepsilon_{eq}^p = \int_0^x \frac{dx}{\sqrt{3}\,(20-0.5x)} = \frac{2}{\sqrt{3}} \ln\left[\frac{20}{(20-0.5x)}\right]. \tag{3.112}$$

Evaluating the above expression at $x = \ell/2$ and $x = \ell$, we get the following values of ε_{eq}^p:

$$\varepsilon_{eq}^p = \frac{2}{\sqrt{3}} \ln\frac{4}{3} = 0.332, \quad \text{at } x = \frac{\ell}{2} = 10\text{mm},$$

$$\varepsilon_{eq}^p = \frac{2}{\sqrt{3}} \ln 2 = 0.800, \quad \text{at } x = \ell = 20\text{mm}. \tag{3.113}$$

(c) Taking the natural logarithm of Equation 3.103 and substituting the value of σ_Y from Equation 3.106, we get

$$\ln(\sigma_{eq} - 360) = \ln K + n\ln \varepsilon_{eq}^p. \tag{3.114}$$

Substituting the values of $(\sigma_{eq}, \varepsilon_{eq}^p)$ at $x = \ell/2$ and $x = \ell$ from Equations 3.106 and 3.113, we get the following two equations for the unknowns K and n:

$$\ln(40) = \ln K + n\ln(0.332),$$
$$\ln(90) = \ln K + n\ln(0.800). \tag{3.115}$$

Solving these two equations, we get the following values of the hardening parameters:

$$K = 110.565 \text{ MPa}, \quad n = 0.922. \tag{3.116}$$

3.6 Plastic Stress-Strain and Stress-Strain Rate Relations for Isotropic Materials

In Chapter 2, we obtained three-dimensional stress-strain relations for linearly elastic material by generalizing one-dimensional experimental observations. Such an approach is not feasible for developing three-dimensional plastic stress-strain relations. Earlier attempts at developing three-dimensional plastic stress-strain relations were based on the Saint-Venant's proposal that, for isotropic materials, the principal directions of the plastic part of the incremental linear strain tensor coincide with the principal directions of the stress tensor [4]. Based on this proposal, Levy and Mises independently proposed three-dimensional stress-strain relations for rigid perfectly plastic material whereas Prandtl and Reuss independently developed them for elastic perfectly plastic material. But, these relations are not applicable to hardening materials.

Two approaches have emerged for developing the three-dimensional stress strain relations for hardening materials: (i) approach based on *Drucker's postulate for stable plastic material* [7] and (ii) approach based on the *postulate of plastic potential*. In this book, we follow the second approach because it is less mathematical. But the first approach is discussed very briefly in the next paragraph.

In tension testing, beyond initial yielding, the stress increases monotonically with the strain (Figure 3.1). Drucker calls this type of hardening material stable plastic material. To generalize this concept, Drucker introduces an external agency (distinct from the agency which causes the existing state of stress) which slowly applies a set of self-equilibrating forces on the body in a state of equilibrium and then slowly removes them. The original configuration may or may not be restored at the end of the cycle, but the state of stress is returned to the original equilibrium state. As a generalization of the hardening behavior in tension test, Drucker makes the following postulate for stable plastic material in three-dimensional state of stress. In a *stable plastic material*, (i) the plastic work done by the external agency during the application of the additional stresses is positive and (ii) the net plastic work performed by the external energy during the cycle of application and removal of the forces is non-negative. One of the consequences of the Drucker's stability postulate is that the direction of the geometrical vector representation of the plastic part of incremental linear strain tensor is along the normal to the yield surface [2, 7]. The plastic stress strain relations are then just a mathematical expression of this statement.

In the second approach, we postulate the existence of a scalar function of the stress tensor called *plastic potential*. We denote it by $g(\sigma_{ij})$. We define it as a function whose derivatives with respect to σ_{ij} specify the ratios of the components of the plastic part of incremental linear strain tensor. However, they do not define the magnitude of this tensor. Thus, we have

$$d\varepsilon_{ij}^{p} = d\lambda \frac{\partial g}{\partial \sigma_{ij}}, \tag{3.117}$$

where $d\lambda$ is a positive scalar whose value depends on the stress increment, the hardening relation of the material and possibly the state of stress. The concept of plastic potential was proposed by Mises in 1928. Its existence has been justified by Hill [4]. Note that, here, the word potential is used in a mathematical sense. Mathematically, if a vector or a tensor valued function is expressed as a vector or a tensor derivative of a scalar function, then that scalar function is known as the potential. However, the plastic potential does not have any physical meaning in the sense of gravitational or electromagnetic potential. Equation 3.117 is called the *flow rule*.

The plastic potential has the following properties:

- Since yielding depends only on the deviatoric part of stress tensor, we expect that $d\varepsilon_{ij}^p$ should depend only on the deviatoric part of σ_{ij}. Therefore, the function g should depend only on σ_{ij}', the deviatoric part of σ_{ij}.

- For isotropic materials, g should be a coordinate-independent function, i.e., it should be an invariant of σ_{ij}'. Since every invariant of σ_{ij}' can be expressed in terms of the principal invariants J_2 and J_3 (as J_1 is zero), g should be a function of J_2 and J_3.

- We have decided to neglect the Bauschinger effect. Therefore, if $d\varepsilon_{ij}^p$ is the plastic part of the incremental linear strain tensor corresponding to σ_{ij}', then corresponding to $-\sigma_{ij}'$, it should be $-d\varepsilon_{ij}^p$. This means g should be an even function of J_3.

Thus, we see that the plastic potential g and the yield function f (defined in Sections 3.3 and 3.5) have similar properties. Even though it may be possible to relate g and f on the basis of microstructural observations, the form of potential function is not known for any material [4]. Therefore, based on the above observations, we *assume* that the plastic potential g is identical to the yield function f. Note that, if we start from the Drucker's stability postulate, then one of its consequences is that the plastic part of incremental linear strain tensor is proportional to the derivative of the yield function with respect to the stress tensor. Thus, the first approach of developing the plastic stress strain relations seems to be better, as we need to have only one postulate, namely the Drucker's postulate, whereas in the second approach, besides postulating the existence of the plastic potential, we also need to assume that it is identical to the yield function.

3.6.1 Associated Flow Rule

When we assume that $g = f$, Equation 3.117 becomes

$$d\varepsilon_{ij}^p = d\lambda \frac{\partial f}{\partial \sigma_{ij}}.$$

(3.118)

This flow rule is called the *associated flow rule*. The yield function f depends on the stress tensor σ_{ij} through the invariants J_2 and J_3 of its deviatoric part. Using the chain rule, we can write the associated flow rule (Equation 3.118) as

$$d\varepsilon_{ij}^p = d\lambda \left(\frac{\partial f}{\partial J_2} \frac{\partial J_2}{\partial \sigma_{ij}} + \frac{\partial f}{\partial J_3} \frac{\partial J_3}{\partial \sigma_{ij}} \right).$$

(3.119)

To evaluate the derivatives of the invariants J_2 and J_3 with respect to σ_{ij}, we again use the chain rule:

$$\frac{\partial J_2}{\partial \sigma_{ij}} = \frac{\partial J_2}{\partial \sigma'_{pq}} \frac{\partial \sigma'_{pq}}{\partial \sigma_{ij}}, \tag{3.120}$$

$$\frac{\partial J_3}{\partial \sigma_{ij}} = \frac{\partial J_3}{\partial \sigma'_{pq}} \frac{\partial \sigma'_{pq}}{\partial \sigma_{ij}}. \tag{3.121}$$

First, we determine the derivate of σ'_{pq} with respect to σ_{ij}. For this, we use Equation 2.100 after changing the indices i and j to p and q. Then, we use Equation 2.35 to express the derivatives of σ_{pq} and σ_{kk} with respect to σ_{ij} in terms of δ. Then, we simplify the resulting expression using the identity at Equation 2.15. Thus, we get

$$\frac{\partial \sigma'_{pq}}{\partial \sigma_{ij}} = \frac{\partial \left(\sigma_{pq} - \frac{1}{3} \sigma_{kk} \delta_{pq} \right)}{\partial \sigma_{ij}},$$

$$= \frac{\partial \sigma_{pq}}{\partial \sigma_{ij}} - \frac{1}{3} \frac{\partial \sigma_{kk}}{\partial \sigma_{ij}} \delta_{pq}, \tag{3.122}$$

$$= \delta_{pi} \delta_{qj} - \frac{1}{3} \delta_{ki} \delta_{kj} \delta_{pq},$$

$$= \delta_{pi} \delta_{qj} - \frac{1}{3} \delta_{ij} \delta_{pq}.$$

Next, we estimate the derivative of J_2 with respect to σ'_{pq}. For this, we use Equation 2.105 after changing the dummy indices from i and j to m and n. Then, we use Equation 2.35 to express the derivative of σ'_{mn} with respect to σ'_{pq} in terms of δ. Then, as before, we simplify the resulting expression using the identity at Equation 2.15. Thus, we obtain

$$\frac{\partial J_2}{\partial \sigma'_{pq}} = \frac{1}{2} \frac{\partial (\sigma'_{mn} \sigma'_{mn})}{\partial \sigma'_{pq}} = \frac{1}{2} \left(2\sigma'_{mn} \frac{\partial \sigma'_{mn}}{\partial \sigma'_{pq}} \right),$$

$$= \frac{1}{2} (2\sigma'_{mn} \delta_{mp} \delta_{nq}) = \sigma'_{pq}. \tag{3.123}$$

Finally, we evaluate the derivative of J_3 with respect to σ'_{pq}. For this, instead of using Equation 2.103, we use an alternate expression for the determinant of a tensor given by Equation 2.25. Further, since the tensor σ' is symmetric and the sum of its diagonal terms is zero, the result of Example 2.4 implies that this expression reduces to

$$J_3 \equiv \det \sigma' = \frac{1}{3}\sigma'_{lm}\sigma'_{mn}\sigma'_{nl}. \tag{3.124}$$

We use this expression to determine the derivative of J_3 with respect to σ'_{pq}. We use Equation 2.35 to express the derivatives of $\sigma'_{lm}, \sigma'_{mn}$ and σ'_{nl} with respect to σ'_{pq} in terms of δ. Then, as before, we simplify the resulting expression using the identity at Equation 2.15. Then, we reshuffle the terms, interchange the indices using the symmetry of σ' and change the dummy indices to m to make all the three terms identical. Finally, we combine the three terms. Thus, we get

$$\begin{aligned}
\frac{\partial J_3}{\partial \sigma'_{pq}} &= \frac{1}{3}\frac{\partial}{\partial \sigma'_{pq}}(\sigma'_{lm}\sigma'_{mn}\sigma'_{nl}), \\
&= \frac{1}{3}(\delta_{lp}\delta_{mq}\sigma'_{mn}\sigma'_{nl} + \sigma'_{lm}\delta_{mp}\delta_{nq}\sigma'_{nl} + \sigma'_{lm}\sigma'_{mn}\delta_{np}\delta_{lq}), \\
&= \frac{1}{3}(\sigma'_{qn}\sigma'_{np} + \sigma'_{lp}\sigma'_{ql} + \sigma'_{qm}\sigma'_{mp}), \\
&= \frac{1}{3}(\sigma'_{pn}\sigma'_{nq} + \sigma'_{pl}\sigma'_{lq} + \sigma'_{pm}\sigma'_{mq}), \\
&= \sigma'_{pm}\sigma'_{mq}.
\end{aligned} \tag{3.125}$$

Now, to get the derivative of J_2 with respect to σ_{ij}, we substitute Equations 3.123 and 3.122 in Equation 3.120. We simplify the resulting expression using the identity at Equation 2.15 and the condition that σ'_{pp} is zero (Equation 2.100). Thus, we obtain

$$\frac{\partial J_2}{\partial \sigma_{ij}} = (\sigma'_{pq})\left(\delta_{pi}\delta_{qj} - \frac{1}{3}\delta_{ij}\delta_{pq}\right) = \sigma'_{ij} - \frac{1}{3}\sigma'_{pp}\delta_{ij} = \sigma'_{ij}. \tag{3.126}$$

Further, to get the derivative of J_3 with respect to σ_{ij}, we substitute Equations 3.125 and 3.122 in Equation 3.121. As before, we simplify the resulting expression using the identity at equation 2.15. Thus, we get

$$\frac{\partial J_3}{\partial \sigma_{ij}} = (\sigma'_{pm}\sigma'_{mq})\left(\delta_{pi}\delta_{qj} - \frac{1}{3}\delta_{ij}\delta_{pq}\right) = \sigma'_{im}\sigma'_{mj} - \frac{1}{3}\sigma'_{pm}\sigma'_{mp}\delta_{ij}. \tag{3.127}$$

Let p be the deviatoric part of σ'^2 (*i.e.*, square of σ'). Then, similar to Equation 2.100, we can write

$$p = \sigma'^2 - \frac{1}{3}tr(\sigma'^2)\mathbf{1},$$ (3.128a)

or in index notation

$$p_{ij} = \sigma'_{im}\sigma'_{mj} - \frac{1}{3}(\sigma'_{pm}\sigma'_{mp})\delta_{ij}.$$ (3.128b)

Comparing Equations 3.127 and 3.128b, we get

$$\frac{\partial J_3}{\partial \sigma_{ij}} = p_{ij}.$$ (3.129)

Substituting Equations 3.126 and 3.129 in Equation 3.119, the associated flow rule becomes

$$d\varepsilon^p_{ij} = d\lambda\left(\frac{\partial f}{\partial J_2}\sigma'_{ij} + \frac{\partial f}{\partial J_3}p_{ij}\right),$$ (3.130)

where the tensor p is defined by Equation 3.128. For a specific yield function (for example the Mises or the Tresca yield function), we can evaluate the derivatives of f with respect to J_2 and J_3 and then simplify the expression (Equation 3.130) of the associated flow rule. But, before that, we shall discuss some consequences of Equation 3.130. This is done in the next few paragraphs.

The first consequence of Equation 3.130 is that the principal directions of the plastic part $d\varepsilon^p$ of the incremental linear strain tensor are the same as those of the stress tensor σ. To show this, we use the following two results.

Result 1: Principal directions of the deviatoric part of a tensor are the same as those of the tensor itself. This can be shown by combining the equation governing the eigenvalues and eigenvectors of a tensor (Equation 2.80) with the one which defines the deviatoric part of a tensor (Equation 2.100).

Result 2: Principal directions of the square of a tensor are the same as those of the tensor itself. This again can be shown using Equation 2.80.

Let $\hat{e}_i$ be the principal directions of the stress tensor σ. Then, result 1 implies that $\hat{e}_i$ would be the principal directions of σ' also. Further, using result 2, we can show that the principal directions of σ'^2 also would be $\hat{e}_i$. Again, using result 1, we can show that the principal directions of p (*i.e.*, the deviatoric part of σ'^2) would also be $\hat{e}_i$. As *per* Equation 3.130, the tensor $d\varepsilon^p$ is a linear combination of the tensors

σ' and p. Since the principal directions of σ' and p are identical and equal to $\hat{e}_i$, the principal directions of $d\varepsilon^p$ also would be equal to $\hat{e}_i$. Thus, we get the result that the principal directions of $d\varepsilon^p$ coincide with that of σ. This was proposed by Saint-Venant in 1870–1871.

To discuss the second consequence of the associated flow rule, we look at its geometric interpretation. We assume that $\hat{e}_i$ (*i.e.*, the principal directions of σ and $d\varepsilon^p$) form a right-handed triad. When we use $\hat{e}_i$ as the coordinate axes, the matrices of σ and $d\varepsilon^p$ become

$$[\sigma] = \begin{bmatrix} \sigma_1 & 0 & 0 \\ 0 & \sigma_2 & 0 \\ 0 & 0 & \sigma_3 \end{bmatrix}, [d\varepsilon^p] = \begin{bmatrix} d\varepsilon_1^p & 0 & 0 \\ 0 & d\varepsilon_2^p & 0 \\ 0 & 0 & d\varepsilon_3^p \end{bmatrix}, \qquad (3.131)$$

where σ_i and $d\varepsilon_i^p$ are the principal values of the tensors σ and $d\varepsilon^p$ respectively. (The σ_i are also called as the principal stresses). Since, these matrices have only three non-zero components, the tensors σ and $d\varepsilon^p$ can be represented as vectors in the combined three-dimensional space of $(\sigma_1, \sigma_2, \sigma_3)$ and $(d\varepsilon_1^p, d\varepsilon_2^p, d\varepsilon_3^p)$. Further, by evaluating J_2 and J_3 in terms of σ_i, we can express the yield function f as a function of σ_i and the hardening parameter ε_{eq}^p : $f(\sigma_1, \sigma_2, \sigma_3; \varepsilon_{eq}^p)$. (For initial yielding, it would be a function of only σ_i.) Then, the yield criterion $f = 0$ can be represented as a surface, called the yield surface, in the three-dimensional stress space of $(\sigma_1, \sigma_2, \sigma_3)$. (We have seen this in Section 3.3 for the special cases of Mises and Tresca criteria for initial yielding.) Now, consider the gradient of the yield function with respect to σ_i. The array of its components with respect to $\hat{e}_i$ is given by

$$\{\nabla f\}^T = \left\{ \frac{\partial f}{\partial \sigma_1}, \frac{\partial f}{\partial \sigma_2}, \frac{\partial f}{\partial \sigma_3} \right\}. \qquad (3.132)$$

Note that this vector is normal to the yield surface. Now, we define another vector whose array of the components is

$$\{d\varepsilon^p\}^T \equiv \{d\varepsilon_1^p, d\varepsilon_2^p, d\varepsilon_3^p\}. \qquad (3.133)$$

To see the relation between these two vectors, we write the associated flow rule (Equation 3.118) in the component form:

$$d\varepsilon_1^P = d\lambda \frac{\partial f}{\partial \sigma_1}, \ d\varepsilon_2^P = d\lambda \frac{\partial f}{\partial \sigma_2}, \ d\varepsilon_3^P = d\lambda \frac{\partial f}{\partial \sigma_3}, \tag{3.134a}$$

or

$$\{d\varepsilon^P\} = d\lambda\{\nabla f\}. \tag{3.134b}$$

This equation states that, in the combined three-dimensional space of $(\sigma_1, \sigma_2, \sigma_3)$ and $(d\varepsilon_1^P, d\varepsilon_2^P, d\varepsilon_3^P)$, the vectors $\{d\varepsilon^P\}$ and $\{\nabla f\}$ have the same direction. Since, $\{\nabla f\}$ is normal to the yield surface, the vector $\{d\varepsilon^P\}$ is also along the normal to the yield surface. This is shown in Figure 3.17 for the Mises and Tresca yield loci on the deviatoric plane. Since the vector $\{d\varepsilon^P\}$ is parallel to the deviatoric plane, three-dimensional graphical representation is not necessary. Therefore, the graphical representation of Figure 3.17 is confined to the deviatoric plane only.

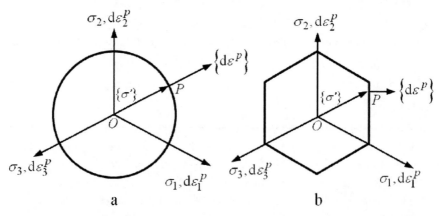

Figure 3.17. Normality rule or the graphical representation of the associated flow rule for the Mises and Tresca yield functions. **a** Mises yield locus on the deviatoric plane. **b** Tresca yield locus on the deviatoric plane

As stated in Section 3.3, the vector $\overrightarrow{OP}$ in Figure 3.17 represents the deviatoric part of the stress tensor. Thus, the components of the vector $\overrightarrow{OP}$ are $(\sigma_1', \sigma_2', \sigma_3')$ where σ_i' are the principal values of σ'. This vector is denoted as $\{\sigma'\}$ in Figure 3.17. Note that the vector $\{d\varepsilon^P\}$ is parallel to the vector $\{\sigma'\}$ for the Mises material but not for the Tresca material. As a result, the tensor $d\varepsilon_{ij}^P$ is a scalar multiple of σ_{ij}' for the Mises material but not for the Tresca material. This is also expected from Equation 3.130, where the derivative $\partial f / \partial J_3$ is zero for the Mises material but not for the Tresca material. There is another difference between the Mises and Tresca yield criterion as far as the evaluation of $d\varepsilon_{ij}^p$ is concerned. For the Mises surface, the normal is defined at every point of the yield surface. But for the Tresca surface, it is not defined along the six edges of the prism (*i.e.*, at the corners of the yield loci of Figure 3.17). Therefore, the vector $\{d\varepsilon^p\}$ is not uniquely determined at the edges of the Tresca surface. This is one of the reasons for not using the Tresca yield criterion in our analysis.

The statement that the vector $\{d\varepsilon^p\}$ is normal to the yield surface is known as the *normality rule*. In our approach, it follows from the assumption that the plastic potential and the yield function are identical. But, in the other approach, it follows from the Drucker's stability postulate without any additional assumptions.

Now, we illustrate the procedure for determining the scalar $d\lambda$ appearing in the associated flow rule for the case of materials which obey the strain hardening hypothesis. For this case, the yield function can be expressed as

$$f(\sigma_{ij}; \varepsilon_{eq}^p) = 0. \tag{3.135}$$

Setting the differential of f to zero, we get

$$df \equiv \frac{\partial f}{\partial \sigma_{ij}} d\sigma_{ij} + \frac{\partial f}{\partial \varepsilon_{eq}^p} d\varepsilon_{eq}^p = 0. \tag{3.136}$$

This is called the consistency condition. It means the state $\left(\sigma_{ij} + d\sigma_{ij}, \; \varepsilon_{eq}^p + d\varepsilon_{eq}^p\right)$ also lies on the yield surface. Combining the definition of $d\varepsilon_{eq}^p$ (Equation 3.93) and the associated flow rule (Equation 3.118), we get

$$d\varepsilon_{eq}^p = \left(\frac{2}{3} d\varepsilon_{ij}^p d\varepsilon_{ij}^p\right)^{1/2} = \sqrt{\frac{2}{3}} \left(\frac{\partial f}{\partial \sigma_{ij}} \frac{\partial f}{\partial \sigma_{ij}}\right)^{1/2} d\lambda. \tag{3.137}$$

Substituting the above expression in Equation 3.136, we get the following expression for $d\lambda$:

$$d\lambda = -\sqrt{\frac{3}{2}} \left(\frac{\dfrac{\partial f}{\partial \sigma_{ij}} d\sigma_{ij}}{\dfrac{\partial f}{\partial \varepsilon_{eq}^p} \left(\dfrac{\partial f}{\partial \sigma_{ij}} \dfrac{\partial f}{\partial \sigma_{ij}}\right)^{1/2}} \right). \tag{3.138}$$

Since we have decided to use the Mises yield criterion in the remainder of the book, it is imperative to use the corresponding yield function in the associated flow rule. The Mises yield function is given by Equation 3.104. Differentiating Equation 3.104 with respect to σ_{ij} and ε_{eq}^p and substituting the expression for the derivative of J_2 (Equation 3.126), we get

$$\frac{\partial f}{\partial \sigma_{ij}} = \frac{\partial J_2}{\partial \sigma_{ij}} = \sigma_{ij}', \qquad (3.139)$$

$$\frac{\partial f}{\partial \varepsilon_{eq}^p} = -\frac{1}{3}(2HH'). \qquad (3.140)$$

Here, H' is the derivative of the hardening function (Equation 3.103) with respect to the equivalent plastic strain. We substitute the above expressions in Equation 3.138 along with the expression for σ_{eq} (Equation 3.23) and the equality $H = \sigma_{eq}$. This leads to

$$d\lambda = \frac{9}{4} \frac{\sigma_{ij}' d\sigma_{ij}}{H' \sigma_{eq}^2}. \qquad (3.141)$$

This is the expression for $d\lambda$ for the Mises material.

Now, we obtain the associated follow rule for the Mises material. The Mises yield function (Equation 3.104) is linear in J_2 and independent of J_3. Therefore, the associated flow rule (Equation 3.130) becomes

$$d\varepsilon_{ij}^p = d\lambda \sigma_{ij}'. \qquad (3.142)$$

Thus, as stated earlier, the tensor $d\varepsilon_{ij}^p$ is a scalar multiple of σ_{ij}' for the Mises material. This is also expressed graphically in Figure 3.17a. Substituting the expression for $d\lambda$ (Equation 3.141) and changing the dummy indices from i and j to k and l, we get the following expression for the associated flow rule of the Mises material:

$$d\varepsilon_{ij}^p = \frac{9}{4} \frac{\sigma_{ij}' \sigma_{kl}'}{H' \sigma_{eq}^2} d\sigma_{kl}. \qquad (3.143)$$

Note that, in the above expression, the differential is *not of the deviatoric part but of the whole stress tensor*.

We modify the above expression so as to bring in the term $d\varepsilon_{eq}^p$ on the right hand side. This needs to be done to make it useful for the Eulerian formulation.

First, we decompose $\mathrm{d}\sigma_{kl}$ into the hydrostatic and deviatoric parts. Then we simplify the product $\sigma'_{kl}\mathrm{d}\sigma_{kl}$ by using the identity at Equation 2.15 and the condition $\sigma'_{kk} = 0$. Thus, we get

$$\sigma'_{kl}\mathrm{d}\sigma_{kl} = \sigma'_{kl}\left(\frac{1}{3}\mathrm{d}\sigma_{mm}\delta_{kl} + \mathrm{d}\sigma'_{kl}\right) = \frac{1}{3}\sigma'_{kk}\mathrm{d}\sigma_{mm} + \sigma'_{kl}\mathrm{d}\sigma'_{kl},$$
$$= \sigma'_{kl}\mathrm{d}\sigma'_{kl}. \tag{3.144}$$

Next, we square both sides of Equation 3.23 and differentiate them. We also change the dummy indices from i and j to k and l. Then we obtain

$$2\sigma_{eq}\mathrm{d}\sigma_{eq} = \frac{3}{2}(2\sigma'_{kl}\mathrm{d}\sigma'_{kl}). \tag{3.145}$$

Further, by differentiating the hardening function (Equation 3.103), we get

$$\mathrm{d}\sigma_{eq} = H'\mathrm{d}\varepsilon^p_{eq}. \tag{3.146}$$

Eliminating $\sigma'_{kl}\mathrm{d}\sigma'_{kl}$ and $\mathrm{d}\sigma_{eq}$ from the above three equations, we get

$$\sigma'_{kl}\mathrm{d}\sigma_{kl} = \frac{2}{3}\sigma_{eq}H'\mathrm{d}\varepsilon^p_{eq}. \tag{3.147}$$

Finally, we substitute the above expression for $\sigma'_{kl}\mathrm{d}\sigma_{kl}$ in Equation 3.143 and cancel the factor $H'\sigma_{eq}$ from the numerator and the denominator. Then, we obtain

$$\mathrm{d}\varepsilon^p_{ij} = \frac{3}{2}\frac{\mathrm{d}\varepsilon^p_{eq}}{\sigma_{eq}}\sigma'_{ij}. \tag{3.148}$$

Starting from Equations 3.143 and 3.148, we now derive the following two constitutive relations for the Mises material in the next two subsections: (i) elastic-plastic incremental stress-strain relation to be used for the updated Lagrangian formulation and (ii) elastic-plastic stress–strain rate relation for the Eulerian formulation.

3.6.2 Elastic-Plastic Incremental Stress-Strain Relation for Mises Material

The constitutive equation (Equation 3.143) is a relationship between the incremental stress tensor $\mathrm{d}\sigma_{kl}$ and only the plastic part of incremental linear strain tensor $\mathrm{d}\varepsilon_{ij}$. However, we need a relationship between $\mathrm{d}\sigma_{kl}$ and the whole of $\mathrm{d}\varepsilon_{ij}$.

To develop this relationship, we first need to relate $d\sigma_{kl}$ with the elastic part of $d\varepsilon_{ij}$ and then combine this relationship with Equation 3.143.

To relate $d\sigma_{kl}$ with $d\varepsilon_{ij}^{e}$, we proceed as follows. Note that we have decided to neglect the hysteresis in tension test. Then, the slope of the unloading path becomes equal to the slope of the elastic path (line OY of Figure 3.1). This means, in a three-dimensional elastic-plastic state of deformation, the relation between $d\sigma_{kl}$ and $d\varepsilon_{ij}^{e}$ is the same as the constitutive equation of the linearly elastic material (Equations 2.213 or 2.218 or 2.224 and 2.225). To relate $d\varepsilon_{ij}^{e}$ with $d\sigma_{kl}$, we use the inverse stress-strain relationship (Equation 2.218):

$$d\varepsilon_{ij}^{e} = \frac{1}{E}[-v d\sigma_{kk}\delta_{ij} + (1+v)d\sigma_{ij}].$$
(3.149)

Using the identity at Equation 2.15, we can rewrite the above equation as

$$d\varepsilon_{ij}^{e} = \frac{1}{E}[-v\delta_{kl}\delta_{ij} + (1+v)\delta_{ik}\delta_{jl}]d\sigma_{kl}.$$
(3.150)

Next, we combine the elastic and plastic parts of the constitutive equation. For this, as before, we assume that the elastic and plastic parts of the incremental linear strain tensor are additive (Equation 3.86). It seems justified for the case of small incremental deformation. As stated earlier, we shall discuss the validity of this assumption in the next chapter. By combining Equations 3.143 and 3.150, we obtain

$$d\varepsilon_{ij} = d\varepsilon_{ij}^{e} + d\varepsilon_{ij}^{p},$$

$$= \left\{ \frac{1}{E}[-v\delta_{kl}\delta_{ij} + (1+v)\delta_{ik}\delta_{jl}] + \frac{9}{4}\frac{\sigma_{ij}'\sigma_{kl}'}{H'\sigma_{eq}^{2}} \right\} d\sigma_{kl}.$$
(3.151)

Inverting this relationship [3], we get

$$d\sigma_{ij} = C_{ijkl}^{EP} d\varepsilon_{kl},$$
(3.152)

where the *fourth order elastic- plastic tensor* C_{ijkl}^{EP} is given by

$$C_{ijkl}^{EP} = 2\mu \left[\frac{v}{1-2v}\delta_{ij}\delta_{kl} + \delta_{ik}\delta_{jl} - \frac{9\mu}{2}\frac{\sigma_{ij}'\sigma_{kl}'}{(H'+3\mu)\sigma_{eq}^{2}} \right].$$
(3.153)

Note that, the tensor C_{ijkl}^{EP} depends on (i) the elastic material constants μ and ν (ii) the material hardening curve through $\sigma_{eq} = H(\varepsilon_{eq}^p)$ and its slope H' and (iii) the current stress through σ'. Equations 3.152 and 3.153 are a *second set of governing equations* for the *updated Lagrangian formulation* when the incremental deformation is small. We shall discuss the modifications required for the large incremental deformation in the next chapter.

The incremental stress in Equation 3.152 has to be *objective*. The definition of objectivity and an objective incremental stress tensor is discussed in the next section.

3.6.3 Elastic-Plastic Stress-Strain Rate Relation for Mises Material

As stated in Subsection 3.4.2, in Eulerian formulation, we use the strain rate tensor $\dot{\varepsilon}_{ij}$ as the measure of deformation. Therefore, for Eulerian formulation, we need to develop the constitutive relation in terms of $\dot{\varepsilon}_{ij}$ rather than the incremental linear strain tensor. We begin by separating $\dot{\varepsilon}_{ij}$ into the elastic and plastic parts. For this purpose, like in the case of incremental linear strain tensor, we assume that it is possible to decompose $\dot{\varepsilon}_{ij}$ additively into the elastic and plastic parts. (We shall show in the next chapter that this is only approximately true when the rotation is small.) Therefore, we write

$$\dot{\varepsilon}_{ij} = \dot{\varepsilon}_{ij}^e + \dot{\varepsilon}_{ij}^p .$$

(3.154)

Next, we obtain the constitutive equation for $\dot{\varepsilon}_{ij}^p$ from the associated flow rule. For this purpose, we develop a relationship between $\dot{\varepsilon}_{ij}^p$ and $d\varepsilon_{ij}^p$. We substitute the decompositions of $d\varepsilon_{ij}$ and $\dot{\varepsilon}_{ij}$ (Equations 3.86 and 3.154) in the equation relating these two quantities (Equation 3.76) and equate the plastic parts of both sides. Then, we get

$$d\varepsilon_{ij}^p = \dot{\varepsilon}_{ij}^p dt .$$

(3.155)

Note that the associated flow rule (Equation 3.148) also involves the equivalent plastic strain increment $d\varepsilon_{eq}^p$. To express it in terms of $\dot{\varepsilon}_{ij}^p$, we define the *equivalent, or effective or generalized plastic strain rate* as

$$\dot{\varepsilon}_{eq}^p = \left(\frac{2}{3} \dot{\varepsilon}_{ij}^p \dot{\varepsilon}_{ij}^p \right)^{1/2} .$$

(3.156)

It is an invariant of the tensor $\dot{\varepsilon}_{ij}^p$. Further, it can be shown to be equal to the plastic part of the axial strain rate in tension test. Substituting Equation 3.155 in the expression for $d\varepsilon_{eq}^p$ (Equation 3.93) and using the definition of $\dot{\varepsilon}_{eq}^p$ (Equation 3.156), we obtain $d\varepsilon_{eq}^p$ in terms of $\dot{\varepsilon}_{eq}^p$:

$$d\varepsilon_{eq}^p = \dot{\varepsilon}_{eq}^p dt \ . \tag{3.157}$$

Before we proceed further, we express the equivalent plastic strain ε_{eq}^p (the argument of the hardening function H) in terms of $\dot{\varepsilon}_{eq}^p$ by substituting the above expression in Equation 3.97:

$$\varepsilon_{eq}^p = \int \dot{\varepsilon}_{eq}^p dt \ . \tag{3.158}$$

Here, the integration is to be carried out along the path line of the material particle.

To express the associated flow rule (Equation 3.148) in terms of $\dot{\varepsilon}_{ij}^p$, we substitute the expressions for $d\varepsilon_{ij}^p$ and $d\varepsilon_{eq}^p$ (Equations 3.155 and 3.157) in the above equation and cancel the factor dt from both the sides. This leads to

$$\dot{\varepsilon}_{ij}^p = \frac{3}{2} \frac{\dot{\varepsilon}_{eq}^p}{\sigma_{eq}} \sigma_{ij}' \ . \tag{3.159}$$

Taking the trace of both sides, and using the condition that the trace of σ' is zero (Equation 2.99), we get

$$\dot{\varepsilon}_{kk}^p = 0 \ . \tag{3.160a}$$

Note that, similar to Equation 3.69, $\dot{\varepsilon}_{kk}^p$ represents the volumetric strain rate corresponding to the plastic part of the strain rate tensor. Thus, Equation 3.160a is consistent with the observation that there is no change in volume corresponding to the plastic part of the deformation. Equation 3.160a also states that the hydrostatic part of $\dot{\varepsilon}_{ij}^p$ is zero. Therefore, the whole of $\dot{\varepsilon}_{ij}^p$ is equal to its deviatoric part. Thus,

$$\dot{\varepsilon}_{ij}'^p = \frac{3}{2} \frac{\dot{\varepsilon}_{eq}^p}{\sigma_{eq}} \sigma_{ij}' \ . \tag{3.160b}$$

Next, we get the constitutive equation for $\dot{\varepsilon}_{ij}^e$. For this purpose, as before, we assume that the elastic part of strain is related to the stress by the constitutive equation of the linearly elastic material (Equations 2.213 or 2.218 or 2.224 and 2.225). To relate $\dot{\varepsilon}_{ij}^e$ with σ_{ij}, we first relate $d\varepsilon_{ij}^e$ with $d\sigma_{ij}$ using the third form of the stress strain relationship (Equations 2.224 and 2.225). We interchange the sides so as to write the incremental strain in terms of the incremental stress. Further, using Equation 2.229, we replace $(3\lambda + 2\mu)$ with $3K$. Then, we get

$$d\varepsilon_{kk}^e = \frac{d\sigma_{kk}}{3K}, \tag{3.161a}$$

$$d\varepsilon_{ij}^{\prime e} = \frac{d\sigma_{ij}'}{2\mu}. \tag{3.161b}$$

Now, we express the strain increments in terms of the strain rates (using Equation 3.76) and the stress increments in terms of the stress rates:

$$\dot{\varepsilon}_{kk}^e \, dt = \frac{\dot{\sigma}_{kk}}{3K} \, dt, \tag{3.162a}$$

$$\dot{\varepsilon}_{ij}^{\prime e} \, dt = \frac{\dot{\sigma}_{ij}'}{2\mu} \, dt. \tag{3.162b}$$

Here, $\dot{\sigma}_{kk}$ is the time rate of σ_{kk} and $\dot{\sigma}_{ij}'$ is the time rate of σ_{ij}'. Finally, we add the elastic as well as the plastic contributions of the hydrostatic and the deviatoric parts of $\dot{\varepsilon}_{ij}$ separately. Thus, adding (a) as well as (b) parts of Equations 3.160 and 3.162, we obtain

$$\dot{\varepsilon}_{kk} = \dot{\varepsilon}_{kk}^e + \dot{\varepsilon}_{kk}^p = \frac{\dot{\sigma}_{kk}}{3K} + 0 = \frac{\dot{\sigma}_{kk}}{3K}, \tag{3.163a}$$

$$\dot{\varepsilon}_{ij}' = \dot{\varepsilon}_{ij}^{\prime e} + \dot{\varepsilon}_{ij}^{\prime p} = \frac{1}{2\mu}\dot{\sigma}_{ij}' + \frac{3\dot{\varepsilon}_{eq}^p}{2\sigma_{eq}}\sigma_{ij}'. \tag{3.163b}$$

This is the constitutive relation for the Eulerian formulation. Note that, in this relation, $\dot{\varepsilon}_{ij}$ depends on (i) the elastic material constants K and μ, (ii) the material hardening curve through $\sigma_{eq} = H(\int \dot{\varepsilon}_{eq}^p dt)$, (iii) the current stress through σ' and (iv) the current stress rate through $\dot{\sigma}$. Equations 3.163a, b are a *second set of governing equations* for the *Eulerian formulation* when the rotation is small.

Modifications required when this assumption is not valid are discussed in the next chapter.

The stress rates in Equations 3.163a,b have to be *objective*. The definition of objectivity and an objective stress rate tensor, namely the Jaumann stress rate tensor, are discussed in the next subsection.

Now, we consider two special cases of the constitutive relation (Equations 3.163a,b). First case arises if we neglect hardening. For *non-hardening materials*, the size of the initial yield surface does not change with the plastic deformation. Therefore, as *per* Equation 3.103, the equivalent stress σ_{eq} remains constant at the value of yield stress σ_Y. When we substitute $\sigma_{eq} = \sigma_Y$ in Equations 3.163a, b, the resulting equations for non-hardening materials are called the *Prandtl-Reuss equations*. They were proposed by Prandtl in 1924 for plane problems and by Reuss in 1930 for general case based on Saint Venant's proposal. The second case arises if we also neglect the elastic deformation. Note that, Equations 3.163a, b involve both the stress and stress rate, and therefore the resulting problem becomes quite difficult to solve. This difficulty can be circumvented if we neglect $\dot{\varepsilon}_{ij}^e$ compared to $\dot{\varepsilon}_{ij}^p$. This is justified as the elastic deformation in metals is normally quite small. Note that, when we neglect the elastic deformation, we are assuming the material to be *rigid-plastic material*. When we neglect $\dot{\varepsilon}_{ij}^e$, Equation 3.163a becomes meaningless and we are left with only the following modified version of Equation 3.163b:

$$\dot{\varepsilon}_{ij} \cong \frac{3\dot{\varepsilon}_{eq}}{2\sigma_{eq}}\sigma'_{ij} .$$
(3.164)

As stated before, for non-hardening materials, the equivalent stress σ_{eq} is equal to the yield stress σ_Y. When we substitute $\sigma_{eq} = \sigma_Y$ in Equation 3.164, the resulting equation for *rigid-plastic non-hardening materials* is called the *Levy-Mises equation*. It was proposed independently by Levy in 1871 and by Mises in 1913 based on Saint Venant's proposal.

Equation 3.164 is usually written in the transposed form:

$$\sigma'_{ij} = \frac{2\sigma_{eq}}{3\dot{\varepsilon}_{eq}}\dot{\varepsilon}_{ij} .$$
(3.165)

This equation states that, for the rigid-plastic materials, we can determine only the deviatoric part of stress from the (plastic) deformation, and the hydrostatic part of stress remains constitutively indeterminate. In elastic plastic materials, there is a small change in volume due to the elastic part of the deformation. But the rigid plastic materials are incompressible as the plastic deformation cannot produce any change in volume. The hydrostatic part of stress, in rigid plastic materials, arises as a reaction to this incompressibility constraint. Therefore, it can be determined from the condition that the volumetric strain rate is zero:

$$\dot{\varepsilon}_{kk} = 0.$$

(3.166)

3.6.4 Viscoplasticity and Temperature Softening

Suppose the material exhibits viscoplasticity as well as strain hardening. Then, as in the case of the strain hardening postulate, we assume that:

- During subsequent yielding, only the size of the yield locus changes with the strain rate tensor. The shape or the center of the yield locus remains unchanged.
- The change in size depends on the equivalent plastic strain rate $\dot{\varepsilon}_{eq}^{p}$, an invariant of the plastic part of the strain rate tensor.

For the Mises material, σ_{eq} represents the size of the yield locus. Therefore, for the material which exhibits both strain hardening and viscoplasticity, we can express σ_{eq} as a function of ε_{eq}^{p} and $\dot{\varepsilon}_{eq}^{p}$:

$$\sigma_{eq} = H(\varepsilon_{eq}^{p}, \dot{\varepsilon}_{eq}^{p}).$$

(3.167)

Like the dependence of function H on ε_{eq}^{p}, the dependence on $\dot{\varepsilon}_{eq}^{p}$ is also determined from the tension test. In tension test, σ_{eq} is equal to the axial stress and $\dot{\varepsilon}_{eq}^{p}$ is equal to the plastic part of the axial strain rate. Therefore, the dependence of H on $\dot{\varepsilon}_{eq}^{p}$ is found from the graph of axial stress vs the plastic part of the axial strain rate.

Suppose the material exhibits temperature softening and strain hardening. Then we assume that, during subsequent yielding, the shape or the center of the yield locus remains unchanged. Further, the size of the yield locus depends on the temperature T. Therefore, for the material which exhibits both strain hardening and temperature softening, we can express σ_{eq} as a function of ε_{eq}^{p} and T:

$$\sigma_{eq} = H(\varepsilon_{eq}^{p}, T).$$

(3.168)

We determine the dependence of H on T from the graph of axial stress vs temperature in tension testing.

If the material exhibits strain hardening and viscoplasticity as well as temperature softening, then we can express σ_{eq} as a function of ε_{eq}^p, $\dot{\varepsilon}_{eq}^p$ and T:

$$\sigma_{eq} = H(\varepsilon_{eq}^p, \dot{\varepsilon}_{eq}^p, T). \tag{3.169}$$

For the materials which exhibit viscoplasticity or/and temperature softening besides strain hardening, the equivalent stress (σ_{eq}) in the constitutive equations (Equations 3.152, 3.153 and 3.163) should be evaluated from Equations 3.167 or 3.168 or 3.169 as the case may be. Similar to Equation 3.103, the function H in these expressions (Equations 3.167–3.169) can be approximated as a power law in (ε_{eq}^p, $\dot{\varepsilon}_{eq}^p$) or (ε_{eq}^p, T) or (ε_{eq}^p, $\dot{\varepsilon}_{eq}^p$, T). Further, H' in Equation 3.153 should be interpreted as $\partial H / \partial \varepsilon_{eq}^p$.

Example 3.4: In plane strain rolling of a sheet, matrix of the stress tensor σ with respect to (x, y, z) coordinate system, at point A, is given by

$$[\sigma] = \begin{bmatrix} 0.21 & 0 & 0 \\ 0 & -0.09 & 0 \\ 0 & 0 & 0.06 \end{bmatrix} \text{GPa.} \tag{3.170}$$

(a) The state of stress at point A lies on the initial Tresca yield surface (*i.e.*, the Tresca yield surface corresponding to zero equivalent plastic strain). On which side of the yield surface does it lie ?

(b) Find the matrices of the tensors σ', σ'^2 and p at point A.

(c) Find $d\varepsilon^p$ in terms of $d\lambda$, at point A, from the associated flow rule. Use the Tresca yield function as the plastic potential.

Solution: (a) The stress matrix of Equation 3.170 is already in the diagonal form. Therefore, $(\hat{i}, \hat{j}, \hat{k})$, the unit vectors along (x, y, z) axes, are the principal directions at point A. Let the labeling of the principal directions be as follows:

$$\hat{e}_1 = \hat{i}, \hat{e}_2 = \hat{j}, \hat{e}_3 = \hat{k}. \tag{3.171}$$

Then the principal stresses at point A are

$$\sigma_1 = 0.21 \text{GPa}, \sigma_2 = -0.09 \text{GPa}, \sigma_3 = 0.06 \text{GPa}. \tag{3.172}$$

Note that these principal stresses are not ordered. The maximum principal stress difference at point A is

$$\max[|\sigma_1 - \sigma_2|, |\sigma_2 - \sigma_3|, |\sigma_3 - \sigma_1|] = (\sigma_1 - \sigma_2). \tag{3.173}$$

Therefore, the state of stress at point A lies on the side $\sigma_1 - \sigma_2 = \sigma_Y$. Further, the value of the yield stress of the material becomes

$$\sigma_Y = \sigma_1 - \sigma_2 = 0.21 - (-0.09) = 0.3\,\text{GPa}\,. \tag{3.174}$$

Note that, if the labeling of the principal directions at point A is changed, the side on which the state of stress lies will also be changed.

(b) From Equation 2.100, we get the matrix of σ' with respect to any coordinate system as

$$[\sigma'] = [\sigma] - \left(\frac{1}{3}\sigma_{ii}\right)[1]\,. \tag{3.175}$$

Substituting the values of σ_{ij} from Equation 3.170, we get the matrix of σ' at point A as

$$[\sigma'] = \begin{bmatrix} 0.15 & 0 & 0 \\ 0 & -0.15 & 0 \\ 0 & 0 & 0 \end{bmatrix}\text{GPa}\,. \tag{3.176}$$

Multiplication of the matrix of Equation 3.176 with itself gives the matrix of σ'^2 at point A. Thus, we get

$$[\sigma']^2 = \begin{bmatrix} 0.15 & 0 & 0 \\ 0 & -0.15 & 0 \\ 0 & 0 & 0 \end{bmatrix} \begin{bmatrix} 0.15 & 0 & 0 \\ 0 & -0.15 & 0 \\ 0 & 0 & 0 \end{bmatrix},$$

$$= \begin{bmatrix} 0.0225 & 0 & 0 \\ 0 & 0.0225 & 0 \\ 0 & 0 & 0 \end{bmatrix}(\text{GPa})^2\,. \tag{3.177}$$

The tensor p is the deviatoric part of σ'^2. Therefore, the matrix of p is given by

$$[p] = [\sigma']^2 - \left(\frac{1}{3}tr\sigma'^2\right)[1]\,. \tag{3.178}$$

Substituting the values of $\sigma_{ij}'^2$ from Equation 3.177, we get the matrix of p at point A as

$$[p] = \frac{1}{3} \begin{bmatrix} 0.0225 & 0 & 0 \\ 0 & 0.0225 & 0 \\ 0 & 0 & -0.0450 \end{bmatrix} (\text{GPa})^2. \tag{3.179}$$

(c) The Tresca yield function, for initial yielding, is given by Equation 3.33. Expanding this expression, we get

$$f(J_2, J_3) = (4)J_2^3 - (9\sigma_Y^2)J_2^2 + (6\sigma_Y^4)J_2 - \sigma_Y^6 - (27)J_3^2. \tag{3.180}$$

Differentiating the above expression partially with respect to J_2 and J_3, we obtain

$$\frac{\partial f}{\partial J_2} = (12)J_2^2 - (18\sigma_Y^2)J_2 + (6\sigma_Y^4),$$
$$\frac{\partial f}{\partial J_3} = -(54)J_3. \tag{3.181}$$

To evaluate these derivatives at point A, we first calculate J_2 and J_3 at point A. This is done by substituting the values of σ_{ij}' from Equation 3.176 into Equations 2.105 and 2.103. Thus, we get

$$J_2 = \frac{1}{2}\sigma_{ij}'\sigma_{ij}' = \frac{1}{2}[(0.15)^2 + (-0.15)^2 + 7(0)^2] = 0.0225(\text{GPa})^2; \tag{3.182a}$$

$$J_3 = \epsilon_{ijk}\,\sigma_{1i}'\sigma_{2j}'\sigma_{3k}',$$
$$= \sigma_{11}'(\sigma_{22}'\sigma_{33}' - \sigma_{23}'\sigma_{32}') + \sigma_{12}'(\sigma_{23}'\sigma_{31}' - \sigma_{21}'\sigma_{33}') + \sigma_{13}'(\sigma_{21}'\sigma_{32}' - \sigma_{22}'\sigma_{31}'),$$
$$= (0.15)[(-0.15)\times 0 - 0\times 0] + 0[0\times 0 - 0\times 0] + 0[0\times 0 - (-0.15)\times 0],$$
$$= 0(\text{GPa})^3. \tag{3.182b}$$

Substituting the values of σ_Y, J_2 and J_3 from Equations 3.174, 3.182a and 3.182b into Equation 3.181, we obtain the following values of the derivatives at point A:

$$\frac{\partial f}{\partial J_2} = (12)J_2^2 - (18\sigma_Y^2)J_2 + (6\sigma_Y^4),$$
$$= 12(0.0225)^2 - 18\times(0.3)^2(0.0225) + 6(0.3)^4, \tag{3.183a}$$
$$= 0.018225;$$

$$\frac{\partial f}{\partial J_3} = -(54)J_3 = 0. \tag{3.183b}$$

Next, we get the expression for $d\varepsilon_{ij}^{p}$ (at point A) in terms of σ_{ij}' by substituting the above values in the associated flow rule (Equation 3.130):

$$d\varepsilon_{ij}^{p} = d\lambda \left(\frac{\partial f}{\partial J_2} \sigma_{ij}' + \frac{\partial f}{\partial J_3} p_{ij} \right) = d\lambda[(0.018225)\sigma_{ij}' + (0)p_{ij}] = (0.018225 d\lambda)\sigma_{ij}'.$$

(3.184)

Finally, we evaluate $d\varepsilon_{ij}^{p}$ at point A (in terms of $d\lambda$); by substituting the values of σ_{ij}' from Equation 3.176, we obtain

$$d\varepsilon_{xx}^{p} = (0.018225 d\lambda)(0.15) = (0.00273375)d\lambda,$$
$$d\varepsilon_{yy}^{p} = (0.018225 d\lambda)(-0.15) = -(0.00273375)d\lambda,$$

(3.185)

$$d\varepsilon_{ij}^{p} = 0 \text{ for other values of } (i, j).$$

3.7 Objective Stress Rate and Objective Incremental Stress Tensors

The stress rate tensor in Equations 3.163a, b and the incremental stress tensor in Equation 3.152 have to be *objective* tensors. It means these tensors have to be frame-invariant or invariant under a change of reference frame.

We explain the concept of frame-invariance as follows. A position vector of point P, when described with respect to a moving frame, has a different mathematical representation than when it is described with respect to a fixed frame. Let x and x^{*} be the mathematical representations of the position vector with respect to fixed and moving frames respectively. The quantities x and x^{*} are related as follows. The distance between two arbitrary points in space is represented as a magnitude of the difference of the position vectors of these two points. The requirement that this distance should be invariant under a change of frame leads to the following relation between x and x^{*} [11]:

$$x^{*} = c(t) + Q(t)x,$$

(3.186)

where the vector $c(t)$ represents the translation and the tensor $Q(t)$ represents the rotation of the moving frame at time t with respect to the fixed frame. Here, the tensor $Q(t)$ is an *orthogonal* tensor. That is,

$$Q(t)Q^{T}(t) = Q^{T}(t)Q(t) = 1.$$

(3.187)

Further, the determinant of $Q(t)$ is +1. Therefore $Q(t)$ is called the *proper* orthogonal tensor. In Subsection 2.3.1.3 we talked about an orthogonal matrix. It is related to the orthogonal tensor as follows. The matrix of an orthogonal tensor is orthogonal matrix in every coordinate system. In the next paragraph, we shall use Equation 3.186 to derive the frame-invariance of vector and tensor quantities.

Now, we *postulate* that the direction of the stress vector (at every point on every plane) should be invariant under a change of frame. This requirement leads to a relation between their mathematical representations with respect to fixed and moving frames. To work out this relation, let us denote the mathematical representations of the stress vector as t_n and t_n^*. Since the direction of a vector is related to the difference of the position vectors of its end points, it can easily be shown, using Equation 3.186, that

$$t_n^* = Q(t)t_n. \tag{3.188}$$

Vectors which satisfy this relation are called *objective* vectors. Note that all vectors are not objective. It is well known that vectors like velocity and acceleration change with the frame. Next, we discus the frame-invariance of the Cauchy stress tensor. Let σ and $\hat{n}$ be the mathematical representations of the stress tensor and the normal vector (to the plane on which the stress vector acts) respectively with respect to the fixed frame. Further, let σ^* and $\hat{n}^*$ be the representations with respect to the moving frame. Note that σ and $\hat{n}$ are related by Equation 2.65. Therefore, in the moving frame, σ^* and $\hat{n}^*$ will be related by

$$t_n^* = \sigma^* \hat{n}^*. \tag{3.189}$$

Note that the normal vector also has to be objective. Therefore,

$$\hat{n}^* = Q(t)\hat{n}. \tag{3.190}$$

Combining Equations 2.65 and 3.188–3.190 and using the orthogonality of $Q(t)$, we get the following relation between σ and σ^*:

$$\sigma^* = Q(t)\sigma Q^{\mathrm{T}}(t). \tag{3.191}$$

The tensors which satisfy the above relation are known as the *objective* tensors. Thus, the Cauchy stress tensor is objective. But its rate is not objective as can be seen by taking the time derivative of Equation 3.191:

$$\dot{\sigma}^* = \dot{Q}(t)\sigma Q^{\mathrm{T}}(t) + Q(t)\dot{\sigma} Q^{\mathrm{T}}(t) + Q(t)\sigma \dot{Q}^{\mathrm{T}}(t). \tag{3.192}$$

3.7.1 Jaumann Stress Rate and Associated Objective Incremental Stress Tensor

To develop an objective stress rate tensor, we proceed as follows. In general, in a time increment, a material particle gets both deformed and rotated. *Assume that, in a current time increment, the particle does not deform, but only rotates.* Then, the strain rate tensor $\dot{\varepsilon}$ will be zero at the present time t but not the spin tensor $\dot{\omega}$. In this case, we expect the objective stress rate tensor also to be zero as we do not expect it to depend on the rotation. However, the Cauchy stress rate does depend on the rotation. If we define a rate of the Cauchy stress tensor which consists of the total rate minus the rate only due to rotation, then we expect this stress rate tensor to be objective. As a first step in the development of such an objective stress rate tensor, we begin with the calculation of the Cauchy stress rate only due to rotation.

To find the change in the Cauchy stress tensor in the time interval dt due to rotation, consider two deformed configurations as shown in Figure 3.18: (i) deformed configuration at time t and (ii) deformed configuration at time t+dt. Further, consider a parallelepiped shaped element around point P of the first configuration. Now, choose two frames: (i) a fixed frame and (ii) a moving frame which rotates with the particle but coincides with the fixed frame at time t. Such a frame is called the *co-rotational* or *material* frame. Coordinate axes of the fixed frame as well as the moving frame (both at time t and t+dt) are shown in the figure.

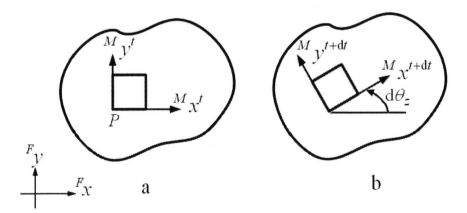

Figure 3.18. Small rotation of the particle in time interval dt. The meaning of the superscripts on the coordinate axes is as follows. The *left superscript* refers to the frame: F for fixed frame and M for material frame. For the material frame, the *right superscript* denotes the time. **a** Deformed configuration at time t. **b** Deformed configuration at time t+dt

Let $[^{F}\sigma^{t}]$ be the matrix of (the Cauchy) stress components at time t with respect to the fixed frame. Since at time t the material frame coincides with the fixed frame, matrix $[^{M}\sigma^{t}]$ of the stress components with respect to the material frame at time t will also be equal to $[^F\sigma^t]$. We are assuming that the element only rotates and does not deform during the time interval dt. Therefore, there is no change in the stress components during the interval dt with respect to the material frame. Then, the matrix $[^M\sigma^{t+dt}]$ of the stress components with respect to the material frame at time $t+dt$ will also be equal to $[^M\sigma^t]$ and hence equal to $[^F\sigma^t]$. To get the matrix $[^F\sigma^{t+dt}]$ of the stress components at time $t+dt$ with respect to the fixed frame, we carry out the component transformation from $^Mx_i^{t+dt}$ to ^Fx_i, the corresponding transformation matrix being $[Q]$. Note that, even though this $[Q]$ is related to the rotation of the moving frame with respect to the fixed frame, it does not directly represent the matrix form of the tensor $Q(t)$ of Equation 3.186 and the related equations.

To find $[Q]$, we proceed as follows. If the particle rotation in the interval dt is small, then it is described by the incremental infinitesimal rotation tensor $d\omega$ (Equation 3.59) which, by Equation 3.77, is equal to $\dot{\omega}dt$. Further, the angles $(d\theta_x, d\theta_y, d\theta_z)$ through which the material axes $(^Mx^{t+dt}, {}^My^{t+dt}, {}^Mz^{t+dt})$ rotate about the fixed axes $(^Fx, {}^Fy, {}^Fz)$ during dt are given by

$$d\theta_x = {}^F\dot{\omega}_{zy}dt = -{}^F\dot{\omega}_{yz}dt , \tag{3.193a}$$

$$d\theta_y = {}^F\dot{\omega}_{xz}dt = -{}^F\dot{\omega}_{zx}dt , \tag{3.193b}$$

$$d\theta_z = {}^F\dot{\omega}_{yx}dt = -{}^F\dot{\omega}_{xy}dt , \tag{3.193c}$$

where the left superscript F means the components of $\dot{\omega}$ are taken with respect to the fixed frame. The angle $d\theta_z$ is shown in Figure 3.18. Since small rotation is a vector, we can consider the total rotation of the $(^Mx^{t+dt}, {}^My^{t+dt}, {}^Mz^{t+dt})$ axes in time dt as the sum of three rotations: (i) rotation in which only $d\theta_x$ is non-zero, (ii) rotation in which only $d\theta_y$ is non-zero, and (iii) rotation in which only $d\theta_z$ is non-zero. Since the angles $(d\theta_x, d\theta_y, d\theta_z)$ are small, while obtaining the transformation matrix, we make the simplification that cosines of these angles are 1 while the sines are equal to the angles themselves. Corresponding to the third rotation, the transformation matrix $[Q]^3$ obtained from Figure 3.18 and Equation 2.54 is given by

$$[Q]^3 = \begin{bmatrix} 1 & -d\theta_z & 0 \\ d\theta_z & 1 & 0 \\ 0 & 0 & 1 \end{bmatrix}.$$

(3.194)

Here, the i-th row of $[Q]^3$ contains the direction cosines of the fixed axes ${}^F x_i$ with respect to the material axes $({}^M x^{t+dt}, {}^M y^{t+dt}, {}^M z^{t+dt})$ axes. The negative sign in second column of the first row is because the axis ${}^F x$ is clockwise from the ${}^M x^{t+dt}$ axis. Similarly, we obtain the transformation matrices $[Q]^1$ and $[Q]^2$ corresponding to the first and second rotations. Adding the three transformation matrices and using Equation 3.193, the net transformation matrix becomes

$$[Q] = [Q]^1 + [Q]^2 + [Q]^3,$$
$$= \begin{bmatrix} 1 & -d\theta_z & d\theta_y \\ d\theta_z & 1 & -d\theta_x \\ -d\theta_y & d\theta_x & 1 \end{bmatrix},$$

(3.195a)

or

$$[Q] = [1] + [{}^F \dot{\omega}]dt ,$$

(3.195b)

where $[{}^F \dot{\omega}]$ is the matrix of the spin tensor $\dot{\omega}$ with respect to the fixed frame.

Now, using the tensor transformation relation (Equation 2.55), we get the matrix $[{}^F \sigma^{t+dt}]$ of the stress components at time $t+dt$ with respect to the fixed frame as

$$[{}^F \sigma^{t+dt}] = [Q][{}^M \sigma^{t+dt}][Q]^T = \left([1] + [{}^F \dot{\omega}]dt\right)[{}^M \sigma^{t+dt}]\left([1]^T + [{}^F \dot{\omega}]^T dt\right).$$

(3.196)

Since the matrix $[{}^M \sigma^{t+dt}]$ is equal to $[{}^F \sigma']$, the above equation becomes

$$[{}^F \sigma^{t+dt}] = \left([1] + [{}^F \dot{\omega}]dt\right)[{}^F \sigma']\left([1]^T + [{}^F \dot{\omega}]^T dt\right),$$
$$= \left([1] + [{}^F \dot{\omega}]dt\right)\left([{}^F \sigma'] + [{}^F \sigma'][{}^F \dot{\omega}]^T dt\right),$$
$$= [{}^F \sigma'] + [{}^F \dot{\omega}][{}^F \sigma']dt + [{}^F \sigma'][{}^F \dot{\omega}]^T dt + [{}^F \dot{\omega}][{}^F \sigma'][{}^F \dot{\omega}]^T (dt)^2.$$

(3.197)

From this, we get the following expression for the rate of change of $[^F\sigma^t]$:

$$[^F\dot{\sigma}^t] = \lim_{dt \to 0} \frac{[^F\sigma^{t+dt}]-[^F\sigma^t]}{dt} = [^F\dot{\omega}][^F\sigma^t] + [^F\sigma^t][^F\dot{\omega}]^T . \qquad (3.198)$$

Since the unit vectors of the fixed frame do not change the directions with time, the above equation can be written in a tensor form. While writing the tensor form, we omit the right superscript t for the stress matrix. Then, we get

$$\dot{\sigma} = \dot{\omega}\sigma + \sigma\dot{\omega}^T . \qquad (3.199)$$

Note that the above expression of the Cauchy stress rate tensor is based on the assumption that the time increment consists of pure rotation and no deformation. When, the time increment consists of both the deformation and rotation, we define a rate of Cauchy stress tensor which consists of the total rate $\dot{\sigma}$ minus the above rate due to rotation. We denote it by $\overset{o}{\sigma}$. Thus, $\overset{o}{\sigma}$ is given by

$$\overset{o}{\sigma} = \dot{\sigma} - (\dot{\omega}\sigma + \sigma\dot{\omega}^T) . \qquad (3.200)$$

The stress rate $\overset{o}{\sigma}$ is called the *Jaumann stress rate*. It can be shown that $\overset{o}{\sigma}$ is an objective tensor. To show it, we need the expression for the spin tensor in a moving frame. To obtain this expression, we proceed as follows.

Let $\mathbf{v}$, ∇v and ω be the mathematical representations of the velocity vector, the velocity gradient tensor and the spin tensor respectively with respect to a fixed frame. Further, let v^*, $\nabla^* v^*$ and ω^* be the representations with respect to a moving frame. Note that the velocity vector is the rate of change of the position vector with time. Differentiating Equation 3.186 with time, we get the expression for v^*:

$$v^* = \dot{c}(t) + \dot{Q}(t)x + Q(t)v . \qquad (3.201)$$

Next, we obtain the expression for $\nabla^* v^*$ using the chain rule and differentiating Equation 3.201 with x and Equation 3.186 with x^*. We modify this expression by decomposing ∇v into the symmetric and antisymmetric parts. Thus, we get

$$\nabla^* v^* = \frac{\partial v^*}{\partial x} \frac{\partial x}{\partial x^*},$$

$$= \left(\dot{\boldsymbol{Q}}(t) + \boldsymbol{Q}(t)\nabla v \right) \boldsymbol{Q}^T (t), \tag{3.202}$$

$$= \dot{\boldsymbol{Q}}(t)\boldsymbol{Q}^T (t) + \boldsymbol{Q}(t)\dot{\varepsilon}\boldsymbol{Q}^T (t) + \boldsymbol{Q}(t)\dot{\omega}\boldsymbol{Q}^T (t).$$

Finally, we obtain $\dot{\omega}^*$ as the antisymmetric part of $\nabla^* v^*$:

$$\dot{\omega}^* = \frac{1}{2}\left(\nabla^* v^* - (\nabla^* v^*)^T \right),$$

$$= \frac{1}{2}\left(\dot{\boldsymbol{Q}}(t)\boldsymbol{Q}^T (t) + \boldsymbol{Q}(t)\dot{\varepsilon}\boldsymbol{Q}^T (t) + \boldsymbol{Q}(t)\dot{\omega}\boldsymbol{Q}^T (t) \right. \tag{3.203}$$

$$\left. - \boldsymbol{Q}(t)\dot{\boldsymbol{Q}}^T (t) - \boldsymbol{Q}(t)\dot{\varepsilon}^T \boldsymbol{Q}^T (t) - \boldsymbol{Q}(t)\dot{\omega}^T \boldsymbol{Q}^T (t) \right).$$

This expression can be simplified by using the orthogonality of $\boldsymbol{Q}(t)$. Differentiating Equation 3.187 with time, we get

$$\dot{\boldsymbol{Q}}(t)\boldsymbol{Q}^T (t) = -\boldsymbol{Q}(t)\dot{\boldsymbol{Q}}^T (t). \tag{3.204}$$

Substituting Equation 3.204 into Equation 3.203 and using the symmetry of $\dot{\varepsilon}$ and antisymmetry of $\dot{\omega}$, we get the following expression for $\dot{\omega}^*$:

$$\dot{\omega}^* = \dot{\boldsymbol{Q}}(t)\boldsymbol{Q}^T (t) + \boldsymbol{Q}(t)\dot{\omega}\boldsymbol{Q}^T (t). \tag{3.205}$$

Using Equation 3.200, the mathematical representation of the Jaumann stress rate with respect to the moving frame can be written as

$$\overset{\circ}{\sigma}^* = \dot{\sigma}^* - \left(\dot{\omega}^* \sigma^* + \sigma^* \dot{\omega}^{*T} \right), \tag{3.206}$$

where $\dot{\sigma}^*$, $\dot{\omega}^*$ and σ^* are given by Equations 3.192, 3.205 and 3.191 respectively.

Using the orthogonality of $\boldsymbol{Q}(t)$, it can be shown that $\overset{\circ}{\sigma}^*$ satisfies the following relation [7, 12]:

$$\overset{\circ}{\sigma}^* = \boldsymbol{Q}(t)\overset{\circ}{\sigma}\boldsymbol{Q}^T (t). \tag{3.207}$$

This shows that the Jaumann stress rate is objective.

Quite a few other objective stress rate measures like the Truesdell rate, Green-Naghdi rate (also called as Green-McInnis rate), Metzger-Dubey rate *etc.* have been proposed. The Jaumman stress rate is based on the spin tensor as the rate of rotation. The rates of rotation used in developing other objective stress measures are related to the quantities associated with the kinematics of finite deformation. Therefore, these measures will be described in the next chapter. Compared to other objective stress rates, the Jaumann stress rate is simpler to implement in a numerical scheme. Therefore, we use the Jaumann stress rate in our analysis.

Thus, in the constitutive equation for the Eulerian formulation (Equations 3.163a, b), we replace $\dot{\sigma}$ by the Jaumann stress rate $\overset{\text{o}}{\sigma}$. Further, in the constitutive equations for the updated Lagrangian formulation (Equation 3.152), we replace $\mathbf{d}\sigma$ by $\mathbf{d}\overset{\text{o}}{\boldsymbol{\sigma}}$:

$$\mathbf{d}\overset{\text{o}}{\boldsymbol{\sigma}} = \overset{\text{o}}{\boldsymbol{\sigma}}\,\mathbf{d}t\,, \qquad\qquad (3.208)$$

the product of the Jaumann stress rate and the time increment $\mathbf{d}t$. This is equivalent to assuming that the incremental rotation is small and is given by $\dot{\omega}\mathbf{d}t$. Other objective incremental stress measures will be discussed in the next chapter.

3.8 Unloading Criterion

Stress-strain curve (Figure 3.1) of tension test shows that, in a one-dimensional state of stress, unloading (at a point on the curve YF) occurs if the stress decreases, *i.e.*, if $\mathrm{d}\sigma < 0$. However, this is not true for the compression test, where σ itself is negative and therefore, unloading occurs if $\mathrm{d}\sigma > 0$. Thus, the unloading criterion which is valid for both tension and compression tests can be stated as

$$\sigma \mathrm{d}\sigma < 0. \qquad\qquad (3.209)$$

To extend this criterion to a three-dimensional state of stress, we turn to the graphical representation of the yield criteria. In a one-dimensional state of stress, graphical representation of the yield criterion is just a pair of points $\sigma = \pm h(\varepsilon^P)$ on the σ-axis. Equation 3.209 means, during unloading, direction of the stress increment $\mathrm{d}\sigma$ is inward from the yield points (*i.e.*, towards the origin of the σ-axis). In a three-dimensional state of stress, graphical representation of the yield criterion is a surface in the stress space of the principal stresses $(\sigma_1, \sigma_2, \sigma_3)$. However, for convenience, we consider its locus on the deviatoric plane.

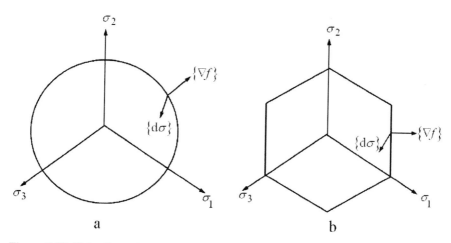

Figure 3.19. Unloading criteria. The vector {dσ} points inward from the yield locus and makes an obtuse angle with the outward normal. **a** Mises yield locus on the deviatoric plane. **b**. Tresca yield locus on the deviatoric plane

Let {dσ} be the array of the principal values of the stress increment $d\sigma_{ij}$. Then, it can be represented as a vector in the stress space. We expect that, during unloading, the vector {dσ} will be pointing inward from the yield locus as shown in Figure 3.19. Then, it will make an obtuse angle with the vector {∇f} which is normal to the yield locus. This condition can be stated as

$$\{\nabla f\}^T \{d\sigma\} < 0.$$
(3.210)

In index notation, this condition can be expressed as [3]

$$\frac{\partial f}{\partial \sigma_{ij}} d\sigma_{ij} < 0.$$
(3.211)

For the Mises material, the derivative of the yield function f with respect σ_{ij} is given by Equation 3.139. Using this result, the unloading criterion (Equation 3.211) becomes

$$\sigma_{ij}' d\sigma_{ij} < 0.$$
(3.212)

3.9 Eulerian and Updated Lagrangian Formulations for Metal Forming Processes

Now, we put together all the three governing equations to make a mathematical model of the metal forming processes. As stated in the introduction, Eulerian formulation is convenient for processes like rolling, drawing, extrusion *etc.* whereas the updated Lagrangian formulation is convenient for the processes like forging, deep drawing, sheet bending *etc.* In this section, we develop these two formulations. Since, some of the governing equations are differential equations in space and time variables, we need the boundary and initial conditions. They are also discussed in this section. One of the governing equations, namely the equation of motion (Equation 2.129), is in terms of the acceleration vector *a*. Since the primary variable is the velocity vector for the Eulerian formulation, we need to express *a* in terms of the velocity vector. Further, for the updated Lagrangian formulation, we need to put the equation of motion in the incremental form. These things are discussed first before developing the two formulations.

3.9.1 Equation of Motion in Terms of Velocity Derivatives

As stated earlier, in Eulerian formulation, the primary variable of the problem is the velocity vector v. The acceleration vector a is the time rate of the velocity vector. Note that the velocity vector v of a particle, besides depending explicitly on time t, also depends implicitly on t through its position vector x. Therefore, using the chain rule, the acceleration vector a can be expressed as

$$a_i \equiv \frac{dv_i}{dt} = \frac{\partial v_i}{\partial t} + \frac{\partial v_i}{\partial x_j}\frac{dx_j}{dt} = \frac{\partial v_i}{\partial t} + (v_{i,j})v_j. \qquad (3.213)$$

Here, the last equality follows from the definition of the velocity vector v, that it is the rate of change of the position vector x of the particle. The comma in the second term of the last equality indicates the derivative with respect to the components of the position vector x.

In tensor notation, the above equation can be written as

$$a \equiv \frac{dv}{dt} = \frac{\partial v}{\partial t} + (\nabla v)v, \qquad (3.214)$$

where the velocity gradient tensor ∇v has been defined in Subsection 3.4.2. The derivative dv/dt is called the *material time derivative* of the velocity vector. The physical interpretation of the two parts of dv/dt is as follows. The first part consists of the partial derivative of v with respect to time. It represents the change in velocity vector of the point of the control volume which the particle occupies at time t. It is called the *unsteady* term of the material time derivative. Since the particle continues to change its position with time, the second part consists of the partial derivative of v with respect to x. It represents the change in velocity vector due to the change in its position. This term is called the *convective* term of the material time derivative. Equation 3.214 shows that the acceleration vector a is a non-linear function of the velocity vector v. Because of this, in Eulerian formulation, the equation of motion becomes a *non-linear* equation.

Substituting the expression for the acceleration vector (Equation 3.213), the equation of motion (Equation 2.131) now becomes

$$\rho \left(\frac{\partial v_i}{\partial t} + v_{i,j} v_j \right) = \rho b_i + \sigma_{ij,j} .$$

(3.215)

For a *steady process*, the first part of the acceleration vector, namely $\partial v_i / \partial t$, is zero.

For a *rigid-plastic material*, it is convenient to decompose the last term of the equation of motion into the hydrostatic and deviatoric parts. In metal forming and machining literature, the negative of the hydrostatic part is often called *pressure* and is denoted by the letter p:

$$p = -\frac{1}{3} \sigma_{kk} .$$

(3.216)

Note that, unlike in fluids, the hydrostatic part in solids is sometimes tensile (*i.e.*, p is sometimes negative). However, whenever the hydrostatic part becomes tensile at a point, there is likelihood of material separation at that point. This aspect will be dealt with in the next chapter, when we discuss the theories of fracture. Substituting Equation 3.216 in the expression for decomposition of the stress tensor (Equation 2.100), we get

$$\sigma_{ij} = -p\delta_{ij} + \sigma'_{ij} .$$

(3.217)

Now, we evaluate the divergence of the first term on the right side. Using the product rule and the identity at Equation 2.15 and noting that δ is a constant, we obtain

$$(p\delta_{ij})_{,j} = p_{,j}\delta_{ij} + p\delta_{ij,j} = p_{,i} + 0 = p_{,i} .$$

(3.218)

Taking the divergence of each side of Equation 3.217 and using Equation 3.218, we obtain

$$\sigma_{ij,j} = -p_{,i} + \sigma'_{ij,j} .$$

(3.219)

Substituting Equation 3.219, the equation of motion (Equation 3.215) now becomes

$$\rho\left(\frac{\partial v_i}{\partial t}+v_{i,j}v_j\right)=\rho b_i - p_{,i}+\sigma'_{ij,j}.$$
(3.220)

3.9.2 Incremental Equation of Motion

For updated Lagrangian formulation, we need to express the equation of motion in an incremental form. This can be done as follows. Let a_i and σ_{ij} be the components of the acceleration vector and the stress tensor respectively at a particle at time t. Then, a_i and σ_{ij} will satisfy the equation of motion given by Equation 2.131 in the deformed configuration at time t (called the *current configuration*). Let $\mathrm{d}a_i$ and $\mathrm{d}\sigma_{ij}$ be the increments in the acceleration vector and the stress tensor at the particle during the time increment $\mathrm{d}t$. Then, $a_i+\mathrm{d}a_i$ and $\sigma_{ij}+\mathrm{d}\sigma_{ij}$ will satisfy the equation of motion in the deformed configuration at time $t+\mathrm{d}t$. This equation will be similar to Equation 2.131 except that the derivatives will now be with respect to the position vector of the particle at time $t+\mathrm{d}t$. However, the deformed configuration at time $t+\mathrm{d}t$ is not known and, therefore, the position vector of the particle at time $t+\mathrm{d}t$ is also unknown. Note that while developing a measure of incremental deformation, *we have assumed that the incremental deformation during the time interval dt is small.* It means the deformed configuration at time t does not change much geometrically during the time interval $\mathrm{d}t$. Therefore, the derivative with respect to the position vector (of a particle) at time $t+\mathrm{d}t$ will be approximately equal to the derivative with respect to the position vector at time t. Then, the approximate equation of motion at time $t+\mathrm{d}t$ will be

$$\rho(a_i+\mathrm{d}a_i)=\rho(b_i+\mathrm{d}b_i)+(\sigma_{ij}+\mathrm{d}\sigma_{ij})_{,j},$$
(3.221)

where $\mathrm{d}b_i$ is the body force increment (per unit mass) in the time interval $\mathrm{d}t$ and the comma denotes the derivative with respect to components of the position vector at time t. Subtracting Equation 2.131 from the above equation, we get the following form of the incremental equation of motion:

$$\rho\,\mathrm{d}a_i = \rho\,\mathrm{d}b_i + \mathrm{d}\sigma_{ij,j}.$$
(3.222)

   Hill [4] has derived the incremental *equilibrium* equation taking into account the change in the position vector during the time interval $\mathrm{d}t$. One can extend this derivation to obtain the incremental *equation of motion*. The incremental equation of motion is not really convenient for the finite element formulation of the problem. Therefore, Hill's incremental equilibrium equation or the corresponding incremental equations of motion are not presented here.

3.9.3 Eulerian Formulation for Metal Forming Problems

As stated earlier, the Eulerian formulation is convenient for the analysis of metal forming processes like rolling, drawing, extrusion *etc*. In this formulation, a region fixed in space (called the *control volume*) is chosen as the domain for the analysis. A possible control volume for *wire drawing* is shown in Figure 3.20. Because of *symmetry*, only *half* the wire is considered. While choosing the control volume, the boundaries AB and EF are placed sufficiently away from the die interface CD so as to simplify the boundary conditions on these boundaries by taking advantage of the uniform velocity fields existing there. The figure also shows the possible *plastic boundaries*. These boundaries are not known *a priori* but have to be determined as a part of the solution.

In any metal forming process, there is always some temperature change due to the dissipation of mechanical energy into heat. However, for slow processes, the temperature change is small and therefore, these processes can be approximated as *isothermal*. For an isothermal process, the velocity field v_i, the strain rate field $\dot{\varepsilon}_{ij}$ and the stress field σ_{ij} in the control volume are governed by the following equations. For the sake of completeness, these equations have been reproduced below.

Governing Equations

(i) Strain rate – velocity relations (Equation 3.66), six scalar equations:

$$\dot{\varepsilon}_{ij} = \frac{1}{2}(v_{i,j} + v_{j,i}).$$
(3.223)

(ii) Elastic-plastic stress-strain rate relations (Equations 3.163a, b, 3.103 and 3.162), six scalar equations:
In Plastic Zone:

$$\dot{\varepsilon}_{kk} = \frac{\overset{o}{\sigma}_{kk}}{3K}, \quad \dot{\varepsilon}'_{ij} = \frac{1}{2\mu}\overset{o}{\sigma}'_{ij} + \frac{3\dot{\varepsilon}^p_{eq}}{2\sigma_{eq}}\sigma'_{ij},$$
(3.224a)

where

$$\sigma_{eq} = \sigma_Y + K(\varepsilon^p_{eq})^n.$$
(3.224b)

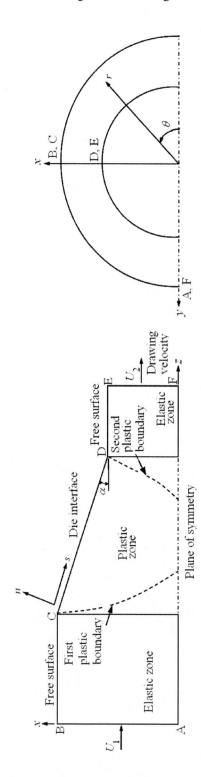

Figure 3.20. Domain for the Eulerian formulation of wire drawing. It is the control volume consisting of half the wire

In elastic zone:

$$\dot{\varepsilon}_{kk} = \frac{\overset{o}{\sigma}_{kk}}{3K}, \quad \dot{\varepsilon}'_{ij} = \frac{1}{2\mu}\overset{o}{\sigma}'_{ij}.$$
(3.224c)

Here, the superscript $\circ$ denotes that it is the *Jaumann stress rate*. The Jaumann stress rate is related to the Cauchy stress rate through spin tensor by Equation 3.200. The spin tensor is given by Equation 3.71. Thus

$$\overset{o}{\sigma}_{kk} = \dot{\sigma}_{kk} - (\dot{\omega}_{kl}\sigma_{lk} + \sigma_{kl}\dot{\omega}^T_{lk}),$$
(3.224d)

and

$$\overset{o}{\sigma}'_{ij} = \dot{\sigma}'_{ij} - (\dot{\omega}_{il}\sigma'_{lj} + \sigma'_{il}\dot{\omega}^T_{lj}),$$
(3.224e)

where

$$\dot{\omega}_{ij} = \frac{1}{2}(v_{i,j} - v_{j,i}).$$
(3.224f)

Note that the time derivative of the Cauchy stress in Equations 3.224d, e has to be the material time derivative.

(iii) Equations of motion (Equation 3.215), three scalar equations:

$$\rho\left(\frac{\partial v_i}{\partial t} + v_{i,j}v_j\right) = \rho b_i + \sigma_{ij,j}.$$
(3.225)

We need the equation of conservation of mass (also known as the continuity equation) if the density is treated as unknown. However, for isothermal processes, the change in density is very small and ρ can be treated as a constant. Thus, we have 15 scalar equations for 15 unknowns: (i) 3 velocity components v_i, (ii) 6 strain rate components $\dot{\varepsilon}_{ij}$ and (iii) 6 stress components σ_{ij}. To solve these equations for the given material, the material properties have to be supplied: (i) density ρ, (ii) elastic properties K and μ and (iii) the yield stress σ_Y and the hardening parameters K and n. Further, the body force b_i (per unit mass) also has to be specified.

All these 15 equations are differential equations in spatial variables x_j and time t. Therefore, boundary and initial conditions are required for solving these equations.

Boundary Conditions

Typical boundary conditions are as follows. Let the boundary of the domain be denoted by S.

(i) On a part of the boundary (S_v), a velocity vector v is specified. Thus,

$$v_i = v_i^* \qquad \text{on } S_v, \tag{3.226}$$

where v_i^* represents the specified value. This is called the *kinematic* or *velocity* boundary condition.

(ii) On the remaining part of the boundary (S_t), a stress vector $t_n = \sigma \hat{n}$ is specified,. Thus,

$$(t_n)_i \equiv \sigma_{ij} n_j = (t_n^*)_i \qquad \text{on } S_t, \tag{3.227}$$

where $(t_n)_i^*$ represents the specified value. This is called the *stress* or *traction* boundary condition. Note that the parts S_v and S_t have to be disjoint. Further, their union has to be equal to the total boundary S. Thus,

$$S_v \cap S_t = \phi, \; S = S_v \cup S_t. \tag{3.228}$$

The individual parts S_v and S_t may consist of several disjoint segments.

In practice, the boundary conditions differ from those specified by Equations 3.226 and 3.227. Sometimes at a point, all the three components of the velocity vector v or the stress vector t_n may not be known. Instead, only one velocity component and two stress components or two velocity components and one stress component may be known. Such boundary conditions are called *mixed* boundary conditions. Further, on a boundary inclined to coordinate axes or on a boundary where friction is present, individual components of v or t_n may not be known. Instead, a combination of their components is known.

As an illustration of the boundary conditions for a practical problem, now, we write the boundary conditions for the problem of Figure 3.20. Note that the boundaries AB and EF are plane surfaces of semi-circular shape. Further, the boundaries BC and DE represent semi-cylindrical surfaces whereas the boundary CD represents a semi-conical surface. Finally, the boundary AF is again a plane surface consisting of a rectangle joined by a trapezoid followed again by a rectangle. It would be convenient to use the cylindrical polar coordinates (r, θ, z) to write the boundary conditions.

Entry and exit boundaries AB and EF:

As stated earlier, the boundaries AB and EF are chosen sufficiently away from the die interface CD. Therefore, we can assume that, at these boundaries, the velocity vector has only z-component and it is uniform over the whole cross-section of the wire. Let U_2 be the specified *drawing velocity*. Then, the boundary conditions at the exit boundary EF become

$$v_r = 0, \quad v_\theta = 0, \quad v_z = U_2, \tag{3.229}$$

where v_r, v_θ and v_z are the components of the velocity vector v in cylindrical polar coordinates. Let U_1 be the velocity (along z-axis) at the entry boundary AB. It can be expressed in terms of U_2 and the *reduction* r_d using the conservation of mass equation. Since the density is treated as constant, the conservation of mass implies

$$U_1 A_1 = U_2 A_2, \tag{3.230}$$

where A_1 and A_2 are the areas of cross-section of the wire at the entry and exit boundaries respectively. The definition of the reduction r_d is (Equation 1.13):

$$r_d = 1 - \frac{A_2}{A_1}. \tag{3.231}$$

Eliminating A_2 / A_1 from Equations 3.230 and 3.231, we get

$$U_1 = U_2(1 - r_d). \tag{3.232}$$

Now, the boundary conditions at the entry boundary AB can be written as

$$v_r = 0, \; v_\theta = 0, \; v_z = U_2(1 - r_d). \tag{3.233}$$

Stress free boundaries BC and DE:

The boundary BC is a stress-free surface. We assume that the *die does not have the land portion*. Then, the whole of the boundary DE is also a stress-free surface. On the stress-free surfaces, the stress vector is zero at every point. Therefore, the boundary conditions at the boundaries BC and DE can be expressed as

$$t_r = 0, \; t_\theta = 0, \; t_z = 0, \tag{3.234}$$

where t_r, t_θ and t_z are the components of the stress vector t_n in cylindrical polar coordinates. Sometimes an alternate set of boundary conditions is used on these boundaries. This set is as follows. Since the direction of the velocity vector at the boundaries BC and DE is always along z-axis, the boundary condition (Equation 3.234) may be modified to specify v_r and v_θ to be zero instead of t_r and t_θ

being zero. The modified boundary condition is expected to give a more accurate velocity field.

Plane of symmetry AF:

On the plane of symmetry, the normal component of the velocity vector and both the shear components of the stress vector are zero at every point. Therefore, the boundary conditions at the boundary AF can be written as

$$t_r = 0, \; v_\theta = 0, \; t_z = 0. \tag{3.235}$$

Note that these are mixed type of boundary conditions.

Die interface CD:

Let n be the direction normal to the die interface and s be the direction along the interface. At the interface, there cannot be any material flow along the normal direction n. Therefore, component of the velocity vector along the direction n must be zero. Using the die semi-angle α, the normal component of the velocity vector can be expressed as

$$v_n = v_r \cos \alpha + v_z \sin \alpha. \tag{3.236}$$

We assume that the die does not exert any frictional (or shear) stress in the θ-direction. Further, the frictional (or shear) stress exerted by the die in s-direction is assumed to be governed by the *Coulomb's law*:

$$\left| t_s \right| = f \left| t_n \right|, \tag{3.237}$$

where f is the coefficient of friction and t_s and t_n are the components of the stress vector $\boldsymbol{t_n}$ along the directions s and n respectively. Since the material flow at the interface is in the positive s-direction, the frictional stress will be in the opposite direction, *i.e.*, in the negative s-direction. Further, the normal stress exerted by the die is always compressive, *i.e.*, in the negative n-direction. Therefore, both t_s and t_n are negative. Then, Equation 3.237 becomes

$$t_s = f \, t_n. \tag{3.238}$$

Using the die semi-angle α, we can express t_s and t_n in terms of t_r and t_z:

$$t_s = -t_r \sin \alpha + t_z \cos \alpha, \; t_n = t_r \cos \alpha + t_z \sin \alpha. \tag{3.239}$$

Eliminating t_s and t_n from Equations 3.238 and 3.239, the Coulomb's law can be written as

$$-(\sin\alpha + f\cos\alpha)t_r + (\cos\alpha - f\sin\alpha)t_z = 0. \tag{3.240}$$

Now, the boundary conditions at the boundary CD become

$$v_r \cos\alpha + v_z \sin\alpha = 0,$$
$$t_\theta = 0, \tag{3.241}$$
$$-(\sin\alpha + f\cos\alpha)t_r + (\cos\alpha - f\sin\alpha)t_z = 0.$$

These boundary conditions involve combinations of v_r and v_z and t_r and t_z.

The shear stress at the die interface is subject to a constraint that it cannot exceed its maximum value. The maximum value for the Mises material can be found as follows. Maximum value of the shear component of the stress vector at a point *with respect to the orientation of the plane* on which it acts is given by (Equation 2.92)

$$\tau \equiv |\sigma_s|_{max} = \frac{(\sigma_1 - \sigma_3)}{2}, \tag{3.242}$$

where σ_1 and σ_3 are respectively the largest and smallest principal stresses at the point. Note that, the values of σ_1 and σ_3 change from point to point. However, they have to satisfy the Mises yield criterion during the plastic deformation. Using Equations 3.27, 3.24 and 3.98, the Mises criterion for subsequent yielding, in terms of the principal stresses, can be written as

$$[(\sigma_1 - \sigma_2)^2 + (\sigma_2 - \sigma_3)^2 + (\sigma_3 - \sigma_1)^2] - 2\sigma_{eq}^2 = 0,$$
$$\sigma_{eq} = H(\varepsilon_{eq}^p). \tag{3.243}$$

Maximizing τ of Equation 3.242 with respect to σ_1 and σ_3 subject to the constraint of Equation 3.243, we get

$$\tau_{max} = \frac{\sigma_{eq}}{\sqrt{3}}, \quad \sigma_{eq} = H(\varepsilon_{eq}^p). \tag{3.244}$$

Thus, the maximum value of the shear stress on the die interface can not exceed $\sigma_{eq}/\sqrt{3}$ for the Mises material. More details on modeling of friction at the interface are discussed in the next chapter.

Initial Conditions
Note that the governing equations (Equations 3.224d, e and 3.225) involve the *first* (partial) time derivative of the velocity vector as well as of the hydrostatic and deviatoric parts of the stress tensor. Therefore, we need to specify the initial values

of v_i, σ_{kk} and σ'_{ij} at every point of the control volume. Thus, the initial conditions are

$$v_i = v_i^0, \ \sigma_{kk} = \sigma_{kk}^0, \ \sigma'_{ij} = \sigma_{ij}'^0 \ \text{at} \ t = t_0, \tag{3.245}$$

where v_i^0, σ_{kk}^0 and $\sigma_{ij}'^0$ are the specified values at the initial time t_0. For a *steady* process, the partial time derivative (*i.e.*, the unsteady part of the material time derivative) of the velocity vector as well as of the stress tensor is zero. Therefore, we do *not* need the initial conditions.

Governing Equations for Rigid-Plastic Material
Two governing equations (Equations 3.224 and 3.225) are non-linear differential equations which need to be solved by iteration. Many times, the iterative scheme does not converge. In that case, one can still obtain a reasonably accurate solution by simplifying these equations by neglecting the elastic part of the deformation. This amounts to assuming the material to be *rigid-plastic*. The governing equations (Equations 3.223 and 3.225) are applicable for the rigid-plastic material too. However, we replace the equation of motion (Equation 3.225) by an alternate expression (Equation 3.220) which involves the separation of the stress tensor into the hydrostatic and deviatoric parts. Further, the constitutive equation (Equation 3.224) needs to be replaced by the one for the rigid-plastic material (Eq. 3.165). Thus, the governing equations of the rigid-plastic material are
(i) Strain rate — velocity relations (Equation 3.66), six scalar equations:

$$\dot{\varepsilon}_{ij} = \frac{1}{2}(v_{i,j} + v_{j,i}). \tag{3.246}$$

(ii) Rigid-plastic stress-strain rate relations (Equations 3.165 and 3.103), six scalar equations:

$$\sigma'_{ij} = \frac{2\sigma_{eq}}{3\dot{\varepsilon}_{eq}}\dot{\varepsilon}_{ij}, \tag{3.247a}$$

where

$$\sigma_{eq} = \sigma_Y + K(\varepsilon_{eq})^n. \tag{3.247b}$$

Note that, now the constitutive equation *does not* contain the time derivative of the stress tensor.
(iii) Equations of motion (Equation 3.220), three scalar equations:

$$\rho \left(\frac{\partial v_i}{\partial t} + v_{i,j} v_j \right) = \rho b_i - p_{,i} + \sigma'_{ij,j} \, . \tag{3.248}$$

As stated earlier, the hydrostatic part of stress is constitutively indeterminate for the rigid-plastic materials. Therefore, we need an additional equation for its determination. This equation is provided by the constraint that the rate of volume change is zero as the hydrostatic part arises as a reaction to this constraint. Thus, for rigid-plastic materials, we have an *additional governing equation.*
(iv) Incompressibility constraint (Equation 3.166), one scalar equation:

$$\dot{\varepsilon}_{kk} = 0 \, . \tag{3.249}$$

In isothermal and isochoric (*i.e.*, no volume change) process, the density ρ can still depend on the hydrostatic part of stress. However, as before, we assume the change in density to be small and treat ρ as a constant. Thus, now, we have 16 scalar equations for 16 unknowns: (i) 3 velocity components v_i, (ii) 6 strain rate components $\dot{\varepsilon}_{ij}$, (iii) 6 deviatoric stress components σ'_{ij} and (iv) 1 hydrostatic stress component p. Since the elastic deformation has been neglected, we do not need the elastic material properties to solve these equations. The boundary conditions for this problem are as before. However, now, the constitutive equation does not contain the time derivative of the stress tensor. Therefore, we need only one initial condition, namely on the velocity vector.

These governing equations are also non-linear and therefore need to be solved by an iterative scheme. But, they are easier to solve than the governing equations of the elastic-plastic materials.

Location of Plastic Boundaries
The first plastic boundary corresponds to the process of initial yielding at the material particles. The second plastic boundary corresponds to the process of unloading at these particles. In Eulerian formulation, the velocity, strain rate and stress at every point of the control volume are determined in a single step. Thus, it is not an incremental procedure. Therefore, there is no scope for carrying out the elastic analysis first, then applying the initial yield criterion, then carrying out the plastic analysis and finally checking for the unloading. Thus, in Eulerian formulation, it is not possible to locate the plastic boundaries on the basis of initial yielding and unloading criteria.

Instead, in the solution procedure of the Eulerian formulation, we treat the whole control volume as the plastic zone and use only the plastic stress- strain rate relations (for the elastic-plastic as well as for the rigid-plastic materials) without using the initial yield criterion or the unloading criterion. After the analysis, it is observed that the strain rates are very small in the entry and exit regions compared to their values in the middle region. Therefore, the entry and exit regions can be interpreted as either elastic (for the elastic-plastic materials) or rigid (for the rigid-plastic materials). A sufficiently small value of the equivalent plastic strain rate

$\dot{\varepsilon}_{eq}^{p}$ ($\dot{\varepsilon}_{eq}$ for the rigid-plastic materials) can be used as a *cut-off* to demarcate the plastic zone from the rest of the control volume. The cut-off can be a small percent of the maximum value of the equivalent plastic strain rate over the control volume. Accuracy of the plastic boundaries depends on the cut-off value.

Another drawback of this formulation is that the stresses in the entry and exit regions are highly inaccurate since they are determined not by the elastic (or rigid) constitutive equations but by the plastic stress-strain rate relations. Value of the equivalent stress σ_{eq} which we get in the entry region is equal to σ_Y while in the exit region, it is $H(\varepsilon_{eq}^{p})$. Thus, we do not get any reasonable estimates of the residual stresses.

However, accuracy of the solution (*i.e.*, of the deformation and stress fields) in the plastic region is quite good. Further, estimate of the power required to carry out the process is also reasonably accurate.

3.9.4 Updated Lagrangian Formulation for Metal Forming Problems

Processes like forging, deep drawing, sheet bending *etc.* can be analyzed increment by increment. Therefore, updated Lagrangian formulation is convenient for their analysis. We illustrate this formulation through an example of *forging of a cylindrical block*. We assume that *both the platens move with the same velocity but in the opposite direction*. Further, because of *symmetry*, only a *quarter* of the block is considered. The final configuration is assumed to be achieved by moving the platens in an incremental fashion through several increments. Figure 3.21 shows the deformed configuration of the quarter block at the present time t (called the current configuration) obtained after a certain number of increments. The deformation and the stress at time t are completely known through the analysis of earlier increments up to the previous increment. We treat the *current configuration as the reference configuration* for analyzing the incremental displacement, incremental deformation and incremental stress. Let du_i, $d\varepsilon_{ij}$ and $d\sigma_{ij}$ be the components of the incremental displacement vector, the incremental linear strain tensor and the incremental Cauchy stress tensor respectively. As before, we assume that the process is *isothermal*. For an isothermal process, the increments du_i, $d\varepsilon_{ij}$ and $d\sigma_{ij}$ are governed by the following equations. For the sake of completeness, these equations have been reproduced below.

Governing Equations:
(i) Incremental strain — displacement relations (Equation 3.55), six scalar equations:

$$d\varepsilon_{ij} = \frac{1}{2}(du_{i,j} + du_{j,i}). \tag{3.250}$$

(ii) Incremental elastic–plastic stress-strain relations (Equations 3.152, 3.153, 3.103 and 2.217), six scalar equations:

After Yielding:

$$\overset{o}{\mathrm{d}\sigma}_{ij} = C_{ijkl}^{EP}\mathrm{d}\varepsilon_{kl}, \tag{3.251a}$$

where

$$C_{ijkl}^{EP} = 2\mu\left[\frac{\nu}{1-2\nu}\delta_{ij}\delta_{kl} + \delta_{ik}\delta_{jl} - \frac{9\mu}{2}\frac{\sigma_{ij}'\sigma_{kl}'}{(H'+3\mu)\sigma_{eq}^2}\right], \tag{3.251b}$$

$$\sigma_{eq} = \sigma_Y + K(\varepsilon_{eq}^p)^n, \tag{3.251c}$$

$$\nu = \frac{\lambda}{2(\lambda+\mu)}; \tag{3.251d}$$

Before yielding and after unloading:

$$\overset{o}{\mathrm{d}\sigma}_{ij} = C_{ijkl}^{E}\mathrm{d}\varepsilon_{kl}, \tag{3.251e}$$

where

$$C_{ijkl}^{E} = \lambda\delta_{kl}\delta_{ij} + 2\mu\delta_{ik}\delta_{jl}. \tag{3.251f}$$

Here, the superscript ^{o} on the stress increment in Equations 3.251a, e means it is the *product of the Jaumann stress rate and the time increment* (Equation 3.208). The Jaumann stress rate is related to the Cauchy stress rate through spin tensor by Equation 3.200. As *per* Equation 3.78, the product of spin tensor and the time increment is equal to the incremental infinitesimal rotation tensor which is given by Equation (3.60). Thus, we have

$$\overset{o}{\mathrm{d}\sigma}_{ij} = \overset{o}{\sigma}_{ij}\,\mathrm{d}t, \tag{3.251g}$$

where

$$\overset{o}{\sigma}_{ij}\,\mathrm{d}t = \dot{\sigma}_{ij}\mathrm{d}t - \left(\dot{\omega}_{il}\mathrm{d}t\sigma_{lj} + \sigma_{il}\dot{\omega}_{lj}^T\mathrm{d}t\right) = \mathrm{d}\sigma_{ij} - \left(\mathrm{d}\omega_{il}\sigma_{lj} + \sigma_{il}\mathrm{d}\omega_{lj}^T\right), \tag{3.251h}$$

$$\mathrm{d}\omega_{ij} = \frac{1}{2}(\mathrm{d}u_{i,j} - \mathrm{d}u_{j,i}). \tag{3.251i}$$

(iii) Incremental equations of motion (Equation 3.222), three scalar equations:

$$\rho\, \mathrm{d}a_i = \rho\, \mathrm{d}b_i + \mathrm{d}\sigma_{ij,j}\,. \tag{3.252}$$

As decided earlier, we treat ρ as a constant. Therefore, we do not need the equation of conservation of mass. Thus, we have 15 scalar equations for 15 unknowns: (i) 3 incremental displacement components $\mathrm{d}u_i$, (ii) 6 incremental linear strain components $\mathrm{d}\varepsilon_{ij}$ and (iii) 6 incremental Cauchy stress components $\mathrm{d}\sigma_{ij}$. To solve these equations for the given material, the material properties have to be supplied: (i) density ρ, (ii) elastic properties λ and μ and (iii) the yield stress σ_Y and the hardening parameters K and n. Further, the incremental body force $\mathrm{d}b_i$ (per unit mass) also has to be specified.

Equations 3.250–3.252 are differential equations in spatial variables x_j and time t. Therefore, boundary and initial conditions are required for solving these equations.

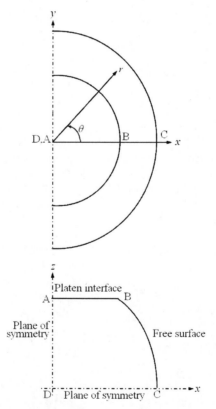

Figure 3.21. Domain for the updated Lagrangian formulation of forging of cylindrical block. It is the deformed configuration of quarter block at time t

Boundary Conditions

Typical boundary conditions are as follows. As before, we denote the boundary of the domain by S.

(i) On a part of the boundary (S_u), an incremental displacement vector $\mathbf{du}$ is specified. Thus,

$$du_i = du_i^* \quad \text{on } S_u,\tag{3.253}$$

where du_i^* represents the specified value. This is called the *kinematic* or *displacement* boundary condition.

(ii) On the remaining part of the boundary (S_t), an incremental stress vector $\mathbf{dt_n} = \mathbf{d\sigma}\hat{n}$ is specified,. Thus,

$$(dt_n)_i \equiv d\sigma_{ij}n_j = (dt_n^*)_i \quad \text{on } S_t,\tag{3.254}$$

where $(dt_n)_i^*$ represents the specified value. This is called the *stress* or *traction* boundary condition.

Note that the parts S_u and S_t have to satisfy the relations similar to the one given by Equation 3.228. Further, each of S_u and S_t may consist of several disjoint segments. In practice, the boundary conditions differ from those specified by Equations 3.253 and 3.254. Sometimes there are *mixed* boundary conditions. Further, on a boundary inclined to coordinate axes or on a boundary where friction is present, a combination of the components of $\mathbf{du}$ or $\mathbf{dt_n}$ is known.

As an illustration of the boundary conditions for a practical problem, now, we write the boundary conditions for the problem of Figure 3.21. Note that the boundaries AB and DC are plane surfaces of semi-circular shape. Further, the boundary BC represents a semi-cylindrical surface. Finally, the boundary AD is a rectangular plane surface. It would be convenient to use the cylindrical polar coordinates (r, θ, z) to write the boundary conditions.

Stress free boundary BC:

The boundary BC is a stress-free surface. On the stress-free surface, the incremental stress vector is zero at every point. Therefore, the boundary conditions at the boundary BC can be expressed as

$$dt_r = 0, \; dt_\theta = 0, \; dt_z = 0,\tag{3.255}$$

where dt_r, dt_θ and dt_z are the components of the incremental stress vector $\mathbf{dt_n}$ in cylindrical polar coordinates.

Plane of symmetry DC:

On the plane of symmetry, the normal component of the incremental displacement vector and both the shear components of the incremental stress vector are zero at every point. Therefore, the boundary conditions at the boundary DC can be written as

$$dt_r = 0, \ dt_\theta = 0, \ du_z = 0, \tag{3.256}$$

where du_z is the z-component of the incremental displacement vector. Note that these are mixed type of boundary conditions.

Plane of symmetry AD:

On the plane of symmetry, the normal component of the incremental displacement vector and both the shear components of the incremental stress vector are zero at every point. Therefore, the boundary conditions at the boundary AD can be written as

$$dt_r = 0, \ du_\theta = 0, \ dt_z = 0, \tag{3.257}$$

where du_θ is the θ-component of the incremental displacement vector. Note that these are mixed type of boundary conditions.

Platen interface AB:

At the interface, z-component of the incremental displacement vector must be equal to the incremental platen displacement.

As far as other two boundary conditions are concerned, we observe the following. Nearer to point A (center of the platen), the block material sticks to the platen while nearer to the free edge (point B), the block material slips relative to the platen in outward direction. We first discuss the boundary condition corresponding to the *slipping case*. Here, we assume that the platen does not exert any frictional (or shear) stress in θ-direction. Further, the frictional (or shear) stress exerted by the platen in r-direction is assumed to be governed by the *Coulomb's law*:

$$\left| t_r + dt_r \right| = f \left| t_z + dt_z \right| \text{ if } \left| t_r + dt_r \right| \geq f \left| t_z + dt_z \right|, \tag{3.258}$$

where f is the coefficient of friction and $t_r + dt_r$ and $t_z + dt_z$ are the components of the stress vector $t_n + dt_n$ along the directions r and z respectively. Note that the friction boundary condition has to be in terms of the total stress vector at time $t+dt$ and not in terms of the incremental stress vector dt_n. The material flow at the interface is in the positive r-direction. Therefore, the frictional stress will be in the opposite direction, *i.e.*, in the negative r-direction. Further, the normal stress exerted by the platen is always compressive, *i.e.*, in the negative *z*-direction. Therefore, both $t_r + dt_r$ and $t_z + dt_z$ are negative. Then, Equation 3.258 becomes

$$t_r + dt_r = f(t_z + dt_z) \quad \text{if } \left| t_r + dt_r \right| \geq f \left| t_z + dt_z \right|. \tag{3.259}$$

Next, we discuss the boundary condition corresponding to the *sticking case*. In this case, r and θ components of the incremental displacement vector must be zero.

Now, the boundary conditions at the boundary AB become

$$
\begin{aligned}
&t_r + dt_r - f(t_z + dt_z) = 0, \; dt_\theta = 0, \\
&\text{if } \left| t_r + dt_r \right| \geq f \left| t_z + dt_z \right| \text{ (Slipping)}, \\
&du_r = 0, \quad du_\theta = 0 \quad \text{if } \left| t_r + dt_r \right| < f \left| t_z + dt_z \right| \text{ (Sticking)}, \\
&du_z = du_z^*,
\end{aligned}
\tag{3.260}
$$

where du_z^* is the prescribed *incremental displacement of the platen*. Note that these boundary conditions involve a combination of $t_r + dt_r$ and $t_z + dt_z$. Here also, the shear stress at the platen interface is subject to a constraint that it can not exceed its maximum value.

Initial Conditions

The governing equation (Equation 3.252) contains incremental acceleration vector, which involves two time derivatives of the incremental displacement vector. Therefore, we need to specify the values of the incremental displacement vector du_i and the incremental velocity vector dv_i at the beginning of the increment, (*i.e.*, at the current time t) at every point of the current configuration. Thus, the initial conditions are

$$du_i = du_i^t, \; dv_i = dv_i^t \quad \text{at time } t, \tag{3.261}$$

where du_i^t and dv_i^t are the specified values at time t. For a *slow* forming process, the incremental acceleration term can be neglected from Equation 3.252. Then, we do *not* need the initial conditions.

Updating Scheme

After solving the incremental equations (Equations 3.250–3.252) along with the boundary and initial conditions, we get the incremental displacement vector du_i, the incremental linear strain tensor $d\varepsilon_{ij}$ and the incremental Cauchy stress tensor $d\sigma_{ij}$. Using du_i, we update the geometry to get the deformed configuration at time $t+dt$. Further, by adding du_i and $d\sigma_{ij}$ to u_i and σ_{ij}, we get the displacement vector and the Cauchy stress tensor at time of $t+dt$. This completes the analysis of the current increment. After this, we go for the next increment. For this increment, the deformed configuration at time $t+dt$ is used as the reference configuration. Therefore, we solve the incremental equations (Equations 3.250–3.252) over the deformed configuration at time $t+dt$ along with the boundary and initial conditions for this configuration. We continue this process till we achieve the desired deformation.

3.10 Eulerian Formulation for Machining Processes

As stated in Chapter 1, machining processes are difficult to model. Here, we present Eulerian formulation for the simplest machining process, namely orthogonal cutting. We choose a small region of the work-piece around the cutting edge as the domain of the problem, since the deformation is confined to this region only. This region is called the cutting zone. With this choice of control volume, it is possible to analyze the process using Eulerian formulation. To make the problem two-dimensional, we assume that the width of cut is large compared to the dimensions of the cutting zone.

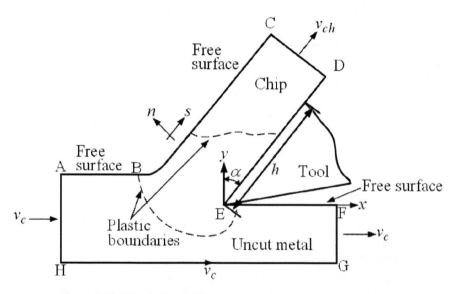

Figure 3.22. Domain for the Eulerian formulation of orthogonal cutting

Figure 3.22 shows the domain selected for the Eulerian formulation. It is a region in the cross-sectional plane of the work-piece perpendicular to the cutting edge. Point E is the projection of the cutting edge. The z-axis is along the cutting edge or the width of cut. The boundaries AB and EF are actually circular. But, since the cutting zone dimensions are small compared to the work-piece radius, they are taken to be straight. The angle α is equal to the rake angle of the cutting tool. The distance h is called the *tool-chip contact length*. It is given by [13]

$$h = \frac{f \sin \theta}{\sin \phi \cos (\theta + \alpha - \phi)}, \tag{3.262}$$

where ϕ is the shear angle (*i.e.*, the inclination of shear plane with the direction of cutting velocity), θ is the angle between the shear force and the resultant force and f is the feed. The boundaries AH, HG, FG and CD are placed sufficiently away from the cutting edge projection E so as to simplify the boundary conditions on these boundaries by taking advantage of the uniform velocity fields existing there. The figure also shows the possible *plastic boundaries*. These boundaries are not known *a priori* but have to be determined as a part of the solution.

Temperature rise in the cutting process is substantial. To incorporate the temperature rise in the analysis, one needs to solve the heat transfer equation governing the temperature field in conjunction with the usual three equations governing the deformation field: strain rate-velocity relations, stress-strain relations and equations of motion. For plastic deformation, these equations are coupled, and hence, difficult to solve. We can decouple this problem as follows. We first estimate the average temperature in the cutting zone either experimentally or by simple analytical methods. Then, we solve the governing equations of the deformation field by evaluating the material properties at the estimated average temperature of the cutting zone.

In the present formulation, it is assumed that the problem has been *decoupled*. We further assume that the elastic deformation is small. Therefore, we use the governing equations of the *rigid-plastic materials*. For the decoupled problem of the rigid-plastic materials, the velocity field v_i, the strain rate field $\dot{\varepsilon}_{ij}$, the deviatoric stress field σ'_{ij} and the hydrostatic stress field p in the control volume are governed by the following equations. For the sake of completeness, these equations have been reproduced below.

Governing Equations for Decoupled Problem of Rigid-Plastic Materials
(i) Strain rate-velocity relations (Equation 3.66), six scalar equations:

$$\dot{\varepsilon}_{ij} = \frac{1}{2}(v_{i,\,j} + v_{j,\,i}). \tag{3.263}$$

(ii) Rigid–plastic stress-strain rate relations (Equation 3.165), six scalar equations:

$$\sigma'_{ij} = \frac{2\sigma_{eq}}{3\dot{\varepsilon}_{eq}}\dot{\varepsilon}_{ij}. \tag{3.264a}$$

Since the strain rates are very high in the cutting process, the phenomenon of *viscoplasticity* must be accounted for. In viscoplastic behavior, equivalent stress σ_{eq} also depends on $\dot{\varepsilon}^p_{eq}$ besides being a function of ε^p_{eq}. Thus,

$$\sigma_{eq} = H(\varepsilon_{eq}^p, \dot{\varepsilon}_{eq}^p).$$
(3.264b)

(iii) Equations of motion (Equation 3.220), three scalar equations:

$$\rho\left(\frac{\partial v_i}{\partial t} + v_{i,j} v_j\right) = \rho b_i - p_{,i} + \sigma'_{ij,j}.$$
(3.265)

(iv) Incompressibility constraint (Equation 3.166), one scalar equation:

$$\dot{\varepsilon}_{kk} = 0.$$
(3.266)

Thus, we have 16 scalar equations for 16 unknowns: (i) 3 velocity components v_i, (ii) 6 strain rate components $\dot{\varepsilon}_{ij}$, (iii) 6 deviatoric stress components σ'_{ij} and (iv) 1 hydrostatic stress component p. To solve these equations for the given material, the material properties have to be supplied: (i) density ρ (which is treated as constant) and (ii) the hardening function H (a function of ε_{eq}^p as well as $\dot{\varepsilon}_{eq}^p$). These material properties are evaluated at the estimated average temperature of the cutting zone. Further, the body force b_i (per unit mass) also needs to be specified.

Equations 3.263, 3.265 and 3.266 are differential equations in spatial variables x_j and time t. Therefore, boundary and initial conditions are required for solving these equations. Further, like the governing equations of metal forming processes, these equations are also non-linear and therefore need to be solved by an iterative scheme.

Boundary Conditions
Typical boundary conditions for the Eulerian formulation have been described earlier. Now, we discuss the boundary conditions for the problem of Figure 3.22. We use the Cartesian coordinates to write the boundary conditions. Note that, since the problem is two-dimensional, only two boundary conditions are needed on each boundary instead of three.

Boundaries AH, HG and FG:

As stated earlier, the boundaries AH, HG and FG are chosen sufficiently away from the cutting edge projection E. Therefore, we can assume that the velocity vector has only x-component at these boundaries. Further, the velocity actually varies linearly from point H to point A and from point G to point F. But, since the distances AH and FG are very small compared to the work-piece radius, we assume the velocity to be uniform over these boundaries. Let v_c be the specified cutting velocity. Then, the boundary conditions at the boundaries AH, HG and FG become

$$v_x = v_c, \; v_y = 0, \tag{3.267}$$

where v_x and v_y are the Cartesian components of the velocity vector.

Boundary CD:

The boundary CD is also chosen sufficiently away from the cutting edge projection E. Therefore, we can assume that, at this boundary, the velocity vector is normal to the boundary and it is uniform over the whole boundary. Let v_{ch} be the *chip velocity*. The chip velocity can be calculated from the cutting velocity using the conservation of mass equation over the uncut depth and the chip thickness:

$$v_{ch} = v_c r , \tag{3.268}$$

where the cutting ratio r is given by Equation 1.19. Note that normal to the boundary CD makes an angle α with y-axis. Then the boundary conditions at the boundary CD become

$$v_x = v_{ch} \sin \alpha, \; v_y = v_{ch} \cos \alpha . \tag{3.269}$$

Stress free boundaries AB, BC and EF:

The boundaries AB, BC and EF are stress-free surfaces. On the stress-free surfaces, the stress vector is zero at every point. Therefore, the boundary conditions at the boundaries AB, BC and EF can be expressed as

$$t_x = 0, \; t_y = 0 , \tag{3.270}$$

where t_x and t_y are the Cartesian components of the stress vector $\boldsymbol{t_n}$. Sometimes an alternate set of boundary conditions is used on these boundaries. This set is as follows. Since the direction of the velocity vector at the boundaries AB and EF is always along x-axis, the boundary conditions (Equation 3.270) may be modified to specify v_y to be zero instead of t_y being zero. The boundary BC is inclined at an angle of α to y-axis. Normal component of the velocity vector and shear component of the stress vector on the boundary BC are

$$v_n = -v_x \cos \alpha + v_y \sin \alpha, \; t_s = t_x \sin \alpha + t_y \cos \alpha . \tag{3.271}$$

Therefore, on the boundary BC, we can specify v_n and t_s to be zero instead of t_x and t_y being zero. The modified boundary conditions are expected to give more accurate velocity field.

Tool-chip interface ED:

Velocity along the tool-chip interface ED can be approximated by the following relation [13]:

$$v_\xi = \frac{v_{ch}}{3}(1+8\frac{\xi}{h})^{1/2}, \quad \text{for} \quad \xi \le h,$$
$$= v_{ch}, \qquad\qquad \text{for} \quad \xi > h,$$

(3.272)

where ξ is the distance measured along the boundary from point E. Thus, the value of v_ξ varies from $v_{ch}/3$ at point E to v_{ch} when ξ is equal to h. Note that the boundary ED makes an angle α with y-axis. Then the boundary conditions at boundary ED become

$$v_x = v_\xi \sin\alpha, \; v_y = v_\xi \cos\alpha .$$

(3.273)

Initial Conditions

Note that the governing equation (Equation 3.265) involves the *first* (partial) time derivative of the velocity vector. Therefore, we need to specify the initial value of v_i at every point of the control volume. Thus, the initial condition is

$$v_i = v_i^0, \quad \text{at} \quad t = t_0,$$

(3.274)

where v_i^0 is the specified value at the initial time t_0. For a s*teady* process, the partial time derivative (*i.e.*, the unsteady part of the material time derivative) of the velocity vector is zero. Therefore, we do *not* need the initial condition.

Location of Plastic Boundaries

As in the case of Eulerian formulation of the metal forming processes, the plastic boundaries are located using a sufficiently small value of the cut-off on the equivalent plastic strain rate $\dot{\varepsilon}_{eq}^p$. Accuracy of the plastic boundaries depends on the choice of the cut-off value.

3.11 Summary

In this chapter, the classical theory of plasticity has been presented. First, the one-dimensional experimental observations on plasticity based on tension tests have been described. Then, two criteria for initial yielding of isotropic materials have been discussed : (i) Mises yield criterion and (ii) Tresca yield criterion. The Mises yield criterion is found to be in better agreement with experimental predictions on yielding, and therefore adopted in the rest of the book. Anisotropic yield criteria will be presented in the next chapter. Two common measures of plastic deformation, namely the incremental linear strain tensor and the strain rate tensor, have been developed next. The incremental linear strain tensor is used in the analysis of processes like forging, deep drawing, sheet bending *etc.* which are amenable to incremental formulation, called the updated Lagrangian formulation. On the other hand, strain rate measure is employed in the processes of rolling, drawing, extrusion *etc.* which are analyzed by the Eulerian formulation where the domain is a fixed region in space, known as the control volume. The incremental linear strain tensor is a valid measure of deformation only for small incremental deformation. Measures of deformation applicable for finite increment size will be discussed in the next chapter. Next, hardening behavior has been modeled by assuming that the hardening is isotropic. This assumption does not have much experimental support. However, in the absence of required experimental data, it is difficult to develop a better hardening model. Kinematic hardening will be discussed in the next chapter.

The plastic stress-strain relations, for isotropic materials, have been developed using the approach based on the postulate of plastic potential. Starting with the associated flow rule, the following two constitutive equations have been developed: (i) elastic-plastic incremental stress-strain relation for the updated Lagrangian formulation and (ii) elastic-plastic stress-strain rate relation for the Eulerian formulation. It has been assumed that the elastic and plastic parts of the deformation are additive. The validity of this assumption will be discussed in the next chapter. An unloading criterion has been presented next. Then, the concepts of objective stress rate and objective incremental stress measures have been discussed. A commonly used objective stress rate measure, namely the *Jaumann stress rate tensor,* has also been presented. The objective incremental stress tensor is taken to be the product of the Jaumann stress rate and the time increment. Other objective stress rate measures and objective incremental stress measures will be discussed in the next chapter. Next, the Eulerian and updated Lagrangian formulations for the metal forming processes have been presented. These formulations, along with the boundary and initial conditions, have been illustrated through the examples of wire drawing and forging of cylindrical block respectively. In the end, Eulerian formulation, along with the boundary and initial conditions, has been presented for the simplest machining process, namely orthogonal cutting.

3.12 References

[1] Johnson, W. and Mellor, P.B. (1972), Engineering Plasticity, von Nostrand Co. Ltd.
[2] Khan, A.S. and Huang, S. (1995), Continuum Theory of Plasticity, John Wiley and Sons Inc., New York.
[3] Chakrabarty, J. (1987), Theory of Plasticity, McGraw-Hill Book Company, New York.
[4] Hill, R. (1950), The Mathematical Theory of Plasticity, Oxford University, Press, Oxford.
[5] Lode, W. (1925), Versuche uber den Einfluss der mittleren Hauptspannung auf die Fliessgrenze, Zeitschrift für Angewandte Mathematik und Mechanik, Vol. 5, pp. 142.
[6] Taylor, G.I. and Quinney, H. (1931), The plastic distortion of metals, The Philosphical Transaction of the Royal Society of London, Vol. A230, pp. 323–362.

[7]   Malvern, L.E. (1969), Introduction to the Mechanics of a Continuous Medium, Prentice Hall Inc., Englewood Cliffs.

[8]   Naghdi, P.M., Essenberg, F. and Koff, W. (1958), An experimental study of initial and subsequent yield surfaces in plasticity, Transaction of ASME, Journal of Applied Mechanics, Vol. 25, pp. 201–209.

[9]   Philips, A. and Das, P.K. (1985), Yield surfaces and loading surfaces of aluminum and brass: An experimental investigation at room and elevated temperatures, International Journal of Plasticity, Vol. 1, pp. 89–109.

[10]   Lubahn, J.D. and Felgar, R.P. (1961), Plasticity and Creep of Metals, John Wiley and Sons Inc., New York.

[11]   Chadwick, P. (1976), Continuum Mechanics, Concise Theory and Problems, George Allen and Unwin Ltd., London.

[12]   Dunne, F. and Petrinic, N. (2005), Introduction to Computational Plasticity, Oxford University Press, Oxford.

[13]   Tay, A.O., Stevenson, M.G., Vahal Davis, G. and Oxley, P.L.B. (1976), A numerical method for calculating temperature distribution in machining from force and shear angle measurements, International Journal of Machine Tool Design & Research, Vol. 16, pp. 335–349.

4

Plasticity of Finite Deformation and Anisotropic Materials, and Modeling of Fracture and Friction

4.1 Introduction

In Chapter 3, we developed the *Eulerian formulation* of metal forming problems. In deriving the constitutive equation of this formulation, it is assumed that the elastic and plastic parts of the rate of deformation tensor are additive. This assumption is usually true for small rotation. Further, the objective stress rate measure used in the constitutive equation, namely the Jaumann stress rate tensor, also remains objective only for small rotation. Therefore, we need to modify this *constitutive equation for the case of finite deformation and rotation*. In Sections 4.2 and 4.3 of this chapter, we first discuss the kinematics of finite deformation, and then the constitutive equation for the case of finite deformation and rotation. Since, in this case, the elastic and plastic parts of the rate of deformation tensor are not additive, the elastic and plastic parts of the constitutive equation are expressed separately. The elastic constitutive equation is expressed in terms of the stress and logarithmic strain tensors and the plastic constitutive equation in terms of the deviatoric stress and the plastic part of the rate of deformation tensors. Since no stress rates are involved, an objective stress rate measure is not needed in this constitutive equation. Next, we discuss the *iterative solution procedure* to be adopted while using this constitutive equation.

In the last chapter, we also developed the *updated Lagrangian formulation* of metal forming problems. The formulation uses a measure of incremental plastic deformation that is valid only for small incremental deformation. Further, the incremental constitutive relation is derived on the basis of the assumption that the elastic and plastic parts of the incremental measure of deformation are additive. (Such an assumption is usually true for the case of small increment size only.) Additionally, the incremental objective stress measure used in the constitutive equation is obtained from the Jaumann stress rate tensor, which remains objective only for a small incremental rotation. Therefore, we need to develop a measure of finite incremental deformation and then modify the constitutive equation accordingly. In Section 4.4, starting from the kinematics of finite incremental deformation, we first develop some commonly used *measures of finite incremental deformation* like the incremental Green-Lagrange strain tensor and the incremental logarithmic strain tensor. It is shown that, for the case of incremental logarithmic strain tensor, the additive decomposition into the elastic and plastic parts still remains valid. Then, in Section 4.5, we derive the *constitutive equation for finite incremental deformation* that involves integration over the increment size. We also discuss a stress updating procedure that makes the incremental stress objective. Sometimes, for finite incremental deformation, the constitutive equation of the updated Lagrangian formulation is expressed in terms of the stress rate and rate of deformation tensors. The objective stress rate measures to be used in this constitutive relation must remain objective even for finite incremental rotation. We discuss some commonly used such measures like the Truesdell rate and the Green-Naghdi rate.

The Eulerian and updated Lagrangian formulations of Chapter 3 use the Mises criterion of initial yielding in developing the hardening relationship and the associated flow rule. However, after cold forming, the crystallographic directions of a metal gradually rotate towards a common axis thus creating anisotropy in the metal. When this metal is subjected to further forming processes without annealing, the hardening relationship and the associated flow rule need to be based on an anisotropic yield criterion. In Section 4.6, we discuss some *criteria for initial yielding of anisotropic materials*. Various approaches have emerged for developing anisotropic yield criteria. We follow the approach based on phenomenological observations. We first discuss the 1948 and 1979 anisotropic yield criteria of Hill. These criteria have certain drawbacks. Lately, anisotropic yield criteria are being developed by applying a tensor transformation (mostly linear) to isotropic yield criteria. We discuss two such criteria developed by Barlat and his co-workers: (i) a plane stress anisotropic yield criterion based on a single linear transformation and (ii) a three-dimensional anisotropic yield criterion based on two linear transformations. These transformations are applied to Hosford's isotropic yield criterion. A plane strain anisotropic yield criterion based on a modification of Hill's 1979 criterion is also presented.

Section 4.7 discusses the development of the constitutive equations corresponding to two of the anisotropic yield criteria: (i) the three-dimensional anisotropic yield criterion of Barlat and his co-workers and (ii) the plane strain anisotropic yield criterion based on a modification of Hill's 1979 criterion.

The Bauschinger effect is normally modeled by a rigid translation of the initial yield surface away from its center but without incorporating the change in its size or shape. This change in the yield surface is called *kinematic hardening*. In Section 4.8, two kinematics hardening models due to Prager and Ziegler are presented.

Section 4.9 is devoted to modeling of *ductile fracture*. It is, now, well established that the ductile fracture occurs mainly due to *micro-void nucleation, growth and finally coalescence* into a micro-crack. If the extent of void growth up to fracture is small, it is possible to ignore its effect on the constitutive equation. However, a realistic model for ductile fracture prediction must include void nucleation, void growth and a condition for void coalescence. We discuss three broad approaches which predict ductile fracture on the basis of void nucleation, growth and coalescence: (i) porous plasticity model of Berg and Gurson, (ii) void nucleation, growth and coalescence model (of Goods and Brown, Rice and Trace and Thomason) and (iii) continuum damage mechanics models of Lemaitre and Rousselier. In the absence of reliable quantitative models for incorporating the phenomena of void nucleation, growth and coalescence, many *phenomenological fracture criteria* have been used in metal forming processes. A few such criteria are presented at the end of this section.

In both the formulations of Chapter 3, the friction at the die-work interface has been modeled using the Coulomb's law or sticking friction model. However, sometimes other *friction models* like the friction factor model, Wanheim and Bay model *etc.* are also used. The Wanheim and Bay model is more general in the sense that it indicates a smooth transition from the Coulomb's law (applicable at lower forming loads) to the friction factor model (applicable at higher forming loads). Both these models are presented in Section 4.10. The last section, namely Section 4.11, summarizes the chapter.

4.2 Kinematics of Finite Deformation and Rotation

In Chapter 2, we discussed the kinematics of small deformation and rotation. Further, we assumed the deformation to be only elastic. In this section, we shall discuss the kinematics of finite as well as elastic-plastic deformation and rotation. Figure 4.1 shows the initial (undeformed) configuration of a body and its deformed configuration at time t. A material particle occupies the position P_0 in the initial configuration and the position P in the deformed configuration. The position vector x of the particle in the deformed configuration depends on its position x_0 in the initial configuration and time t. This functional dependence is called the *motion* of the particle and is represented mathematically as

$$x = x(x_0, t).$$
(4.1)

Consider a small line element $P_0 Q_0$ at point P_0 of the initial configuration which occupies the position PQ at point P of the deformed configuration. We denote the line segments $P_0 Q_0$ and PQ by dx_0 and dx respectively. The relation between dx and dx_0 is obtained by taking the differential of Equation 4.1 while holding t constant. Thus, we get

$$dx = F dx_0,$$
(4.2)

where

$$F = \frac{\partial x}{\partial x_0}.$$
(4.3)

The tensor F is called the *deformation gradient tensor*. For real motions, the neighborhood of point P_0 never deforms into a point. Thus, F is non-singular, *i.e.*, its determinant is non-zero. In index notation, Equation 4.3 can be written as

$$F_{ij} = \frac{\partial x_i}{\partial (x_0)_j} = x_{i,j}, \tag{4.4}$$

where the comma indicates the derivatives with respect to the components of x_0.

The *displacement vector* u of the particle also depends on its position vector x_0 in the initial configuration and time t:

$$u = u(x_0, t). \tag{4.5}$$

Taking the differential of above equation while holding t constant, we get the following expression for the *relative displacement vector* of point Q_0 with respect to point P_0:

$$du = (\nabla_0 u) dx_0, \tag{4.6}$$

where

$$\nabla_0 u = \frac{\partial u}{\partial x_0}, \tag{4.7}$$

is the *displacement gradient tensor* introduced earlier in Section 2.4. The subscript zero used for the symbol ∇_0 emphasizes the fact that the derivative is to be taken with respect to x_0.

From the geometry of Figure 4.1, we get the following relation between u, x and x_0:

$$u = x - x_0. \tag{4.8}$$

Taking the differential of the above equation while holding t constant, we obtain

$$du = dx - dx_0. \tag{4.9}$$

This is consistent with the geometry of Figure 4.1. Substituting Equations 4.2 and 4.6 in the above equation, we get the following relationship between F and $\nabla_0 u$:

$$F = 1 + \nabla_0 u, \tag{4.10}$$

where **1** is the unit tensor. In index notation, the above equation can be written as

$$F_{ij} = \delta_{ij} + \frac{\partial u_i}{\partial (x_0)_j} = \delta_{ij} + u_{i,j},$$

(4.11)

where the comma indicates the derivatives with respect to the components of x_0. Using Equation 4.10, the expression for the relative displacement vector (Equation 4.6) becomes

$$d\mathbf{u} = (\mathbf{F} - \mathbf{1})d\mathbf{x_0}.$$

(4.12)

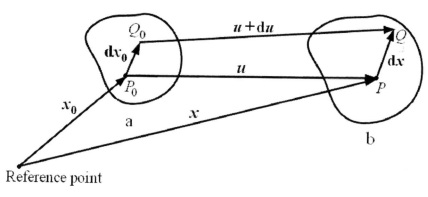

Figure 4.1. Kinematics of finite deformation. **a** Undeformed configuration. **b** Deformed configuration

As stated in Section 2.4, the displacement at a point consists of three parts: (i) displacement due to translation of a neighborhood of the point, (ii) displacement due to rotation of the neighborhood of the point and (iii) displacement due to deformation of the neighborhood of the point. However, if we consider only the relative displacement of a point with respect to the center of its neighborhood, then it contains only two parts: displacement due to the rotation and displacement due to the deformation. Equations 4.6 and 4.12 suggest that, we can consider either the displacement gradient tensor $\nabla_0 \mathbf{u}$ or the deformation gradient tensor $\mathbf{F}$ as a *measure of the deformation and rotation* at a point. In Chapter 2, we considered $\nabla_0 \mathbf{u}$ as the measure of the deformation and rotation at a point. Further, when both the deformation and rotation are small, we showed that the symmetric part of $\nabla_0 \mathbf{u}$ represents the deformation and the antisymmetric part represents the rotation. Thus, for the case of *small deformation and rotation*, the *decomposition* of $\nabla_0 \mathbf{u}$ into the deformation and rotation is *additive*. For the case of *finite deformation and rotation*, however, it is convenient to consider $\mathbf{F}$ as the measure of deformation and rotation at a point. The *decomposition* of $\mathbf{F}$ into the deformation and rotation is governed by the *polar decomposition theorem*, which states that a non-singular tensor can be decomposed into the product of a symmetric positive definite tensor and an orthogonal tensor. The proof of the theorem is given in standard books on continuum mechanics like Malvern [1], Jaunzemis [2] *etc.* We have defined symmetric as well as orthogonal tensors earlier. Now, we shall define a positive definite tensor. A *positive definite tensor A* is defined by the condition that

$$(Aa) \cdot a > 0 \quad \text{for all non-zero vectors } a,$$
$$(Aa) \cdot a = 0 \quad \text{only if } a \text{ is a zero vector.} \tag{4.13}$$

From the polar decomposition theorem, we get

$$F = RU = VR, \tag{4.14}$$

where U and V are the symmetric positive definite tensors, called as *right and left stretch tensors* respectively and R is the orthogonal tensor called the *rotation tensor*. The tensor U (or V) represents the deformation of a neighborhood of the point while the tensor R represents its rotation about the point. Thus, the *decomposition* of F into the deformation and rotation is *multiplicative* and not additive. To show that U (or V) and R represent respectively the deformation and rotation, we consider the principal values λ_i and the principal directions $\hat{e}_i$ of the tensor U. Note that $\hat{e}_i$ are the directions in the undeformed configuration. It can be shown [1, 2] that λ_i represents the ratio of the deformed length to the initial length of a small line element along the direction $\hat{e}_i$. Further, let $\hat{e}_i^d$ be the positions of the directions $\hat{e}_i$ in the deformed configuration. (It can be shown [1, 2] that $\hat{e}_i^d$ are the principal directions of V, its principal values being the same as λ_i). Then,

$$\hat{e}_i^d = R\hat{e}_i. \tag{4.15}$$

To obtain U from F, we use the symmetry of U, the orthogonality of R and the polar decomposition of F (Equation 4.14). Note that the operation of taking the transpose changes the order of tensors in the product. Thus, we get U as the square-root of the following tensor:

$$\begin{aligned}
U^2 &= U^T U, \\
&= U^T R^T RU, \\
&= (RU)^T (RU), \\
&= F^T F.
\end{aligned} \tag{4.16}$$

In index notation, it can be written as

$$U_{ij}^2 = F_{ik}^T F_{kj}. \tag{4.17}$$

Velocity v is the rate of change of the position vector with time. Using this definition and differentiating Equation 4.3 with respect to time, we get the time derivative of F:

$$\dot{F} = \frac{\partial v}{\partial x_0} \, . \tag{4.18}$$

Consider v as a function of x and t, *i.e.*, $v = v(x, t)$. Using the chain rule, the definition of the *velocity gradient tensor* ∇v (given in Section 3.4) and the definition of F (Equation 4.3), we obtain the following relation between F and ∇v :

$$\dot{F} = \frac{\partial v}{\partial x} \frac{\partial x}{\partial x_0} = (\nabla v)F \qquad \text{(holding } t \text{ constant)} \tag{4.19}$$

or

$$\nabla v = \dot{F}F^{-1} \, . \tag{4.20}$$

The symbol ∇ in Equations 4.19 and 4.20 does not have the subscript zero as it involves the derivative with respect to x.

Now, we decompose F into the elastic and plastic parts. For this purpose, we assume that the forces acting on the deformed configuration of Figure 4.1 have been removed and the body has been brought to the state of zero stress. In the process, the elastic deformation becomes zero and only the plastic deformation is left in the body. This configuration of zero stress and only the plastic deformation is shown in Figure 4.2c. Let the position of the particle in this configuration be P_p and the position of the small line element be $P_p Q_p$. Let the vector representation of this line element be dp. Note that we can obtain the elastic-plastic deformed configuration of Figure 4.2 b in two stages (i) We first obtain the intermediate configuration of Figure 4.2 c from the initial configuration (Figure 4.2a) by purely plastic deformation. We write the relationship between dp and dx_0 as

$$dp = F^p dx_0 , \tag{4.21}$$

where

$$F^p = \frac{\partial p}{\partial x_0} \, . \tag{4.22}$$

The superscript p in the symbol emphasizes the fact that the deformation up to this intermediate configuration is purely plastic. (ii) Next, we deform the intermediate configuration elastically to obtain the final elastic-plastic configuration. We write the relationship between $\mathbf{dx}$ and $\mathbf{dp}$ as

$$\mathbf{dx} = \boldsymbol{F}^e \mathbf{dp}, \tag{4.23}$$

where

$$\boldsymbol{F}^e = \frac{\partial \mathbf{x}}{\partial \mathbf{p}}. \tag{4.24}$$

The superscript e in the symbol emphasizes the fact that this deformation is only elastic. Eliminating $\mathbf{dp}$ from Equations 4.21 and 4.23 and comparing the result with Equation 4.2, we get the following *multiplicative decomposition* of the deformation gradient tensor *into the elastic and plastic parts*:

$$\boldsymbol{F} = \boldsymbol{F}^e \boldsymbol{F}^p. \tag{4.25}$$

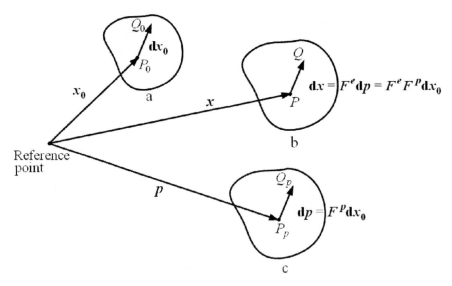

Figure 4.2. Multiplicative decomposition of deformation gradient $\boldsymbol{F}$. **a** Undeformed configuration. **b** Deformed configuration. **c** Configuration after unloading (zero stress, only plastic deformation)

Often the *plastic deformation* is *non-homogeneous*. In that case, the body develops the self-equilibrating *residual stresses* after unloading, and does not return to the state of zero stress. Then, a non-continuous mapping is required to obtain the intermediate configuration of Figure 4.2c that makes the deformation gradient tensors $\boldsymbol{F}^e$ and $\boldsymbol{F}^p$ as point functions of the particle position vectors. However, this causes no difficulty in the analysis [3]. Another point to be noted about this decomposition is that the *intermediate configuration cannot be uniquely determined* since the superposition of an arbitrary rotation or spin tensor leaves the state of stress unchanged. There are quite a few ways of removing this non-uniqueness. Here, *we assume that the plastic part of spin tensor is zero* [4]. To emphasize this fact, the plastic part of the deformation gradient tensor and all the related tensors are denoted by an overbar. Thus,

$$F = F^e \bar{F}^p . \tag{4.26}$$

Now, we use the decomposition of F (Equation 4.26) to obtain the expression for the strain rate tensor $\dot{\varepsilon}$ in terms of its elastic and plastic parts. As a first step, we substitute Equation 4.26 into the expression for the velocity gradient tensor (Equation 4.20) and use the product rule for the differentiation. Note that the operation of taking the inverse changes the order of tensors in the product. Further, we use the identity that the product of a tensor with its inverse is the unit tensor. Thus, we get

$$
\begin{aligned}
\nabla v &= \dot{F} F^{-1}, \\
&= (\dot{F}^e \bar{F}^p + F^e \dot{\bar{F}}^p)(\bar{F}^p)^{-1}(F^e)^{-1}, \\
&= \dot{F}^e \bar{F}^p (\bar{F}^p)^{-1}(F^e)^{-1} + F^e \dot{\bar{F}}^p (\bar{F}^p)^{-1}(F^e)^{-1}, \\
&= \dot{F}^e (F^e)^{-1} + F^e \dot{\bar{F}}^p (\bar{F}^p)^{-1}(F^e)^{-1}.
\end{aligned}
\tag{4.27}
$$

Next, we use Equation 4.20 to write the expressions for the velocity gradient tensors corresponding to the elastic and plastic parts:

$$
\begin{aligned}
(\nabla v)^e &= \dot{F}^e (F^e)^{-1}, \\
(\nabla \bar{v})^p &= \dot{\bar{F}}^p (\bar{F}^p)^{-1}.
\end{aligned}
\tag{4.28}
$$

Note that we have assumed the plastic part of spin tensor to be zero. Therefore, $(\nabla \bar{v})^p$ is equal to $\dot{\bar{\varepsilon}}^p$. Further, using Equations 3.65 and 3.70, the elastic part of the velocity gradient tensor $(\nabla v)^e$ can be decomposed into the symmetric ($\dot{\varepsilon}^e$) and antisymmetric ($\dot{\omega}^e$) parts. Then, Equation 4.28 becomes

$$
\begin{aligned}
\dot{F}^e (F^e)^{-1} &= \dot{\varepsilon}^e + \dot{\omega}^e, \\
\dot{\bar{F}}^p (\bar{F}^p)^{-1} &= \dot{\bar{\varepsilon}}^p.
\end{aligned}
\tag{4.29}
$$

Next, we substitute Equation 4.29 into Equation 4.27 for the velocity gradient tensor to obtain

$$\nabla v = (\dot{\varepsilon}^e + \dot{\omega}^e) + F^e \dot{\bar{\varepsilon}}^p (F^e)^{-1}. \tag{4.30}$$

Note that, using the polar decomposition theorem, the elastic part of the deformation gradient tensor can be decomposed as

$$F^e = R^e U^e . \tag{4.31}$$

We substitute the above decomposition to simplify Equations 4.30. Note that the operation of taking the inverse changes the order of tensors in the product. Further, we use the identity that the inverse of an orthogonal tensor is the same as its transpose. Then Equation 4.30 becomes

$$\nabla v = (\dot{\varepsilon}^e + \dot{\omega}^e) + R^e U^e \dot{\bar{\varepsilon}}^p (U^e)^{-1} (R^e)^{\mathrm{T}} . \tag{4.32}$$

It can be shown that the tensors U^e and $\dot{\bar{\varepsilon}}^p$ commute [5]. Using this commutative property and the identity that the product of a tensor with its inverse is the unit tensor, the above equation simplifies to

$$\nabla v = (\dot{\varepsilon}^e + \dot{\omega}^e) + R^e \dot{\bar{\varepsilon}}^p (R^e)^{\mathrm{T}} . \tag{4.33}$$

Finally, we take the symmetric part of the above equation. Note that the symmetric parts of ∇v, $\dot{\varepsilon}^e$ and $\dot{\omega}^e$ are respectively, $\dot{\varepsilon}$, $\dot{\varepsilon}^e$ and zero. Further, the third term on the right side is already symmetric. Thus, we get

$$\dot{\varepsilon} = \dot{\varepsilon}^e + R^e \dot{\bar{\varepsilon}}^p (R^e)^{\mathrm{T}} . \tag{4.34}$$

The above equation shows that

$$\dot{\varepsilon} \neq \dot{\varepsilon}^e + \dot{\bar{\varepsilon}}^p . \tag{4.35}$$

*It means the elastic and plastic parts of the strain rate tensor $\dot{\varepsilon}$ are **not** additive* as assumed earlier (Equation 3.154) while developing the elastic-plastic stress-strain rate relation for the Eulerian formulation.

Now, assume that the rotation is very small. Note that the plastic part of the deformation gradient has no rotation part since the plastic part of the spin tensor has been assumed to be zero. So, the above assumption means that the tensor R^e is small compared to the unit tensor. Thus,

$$R^e \cong 1 . \tag{4.36}$$

Substituting the above approximation into Equation 4.34, we get

$$\dot{\varepsilon} \cong \dot{\varepsilon}^e + \dot{\bar{\varepsilon}}^p . \tag{4.37}$$

Then, *the elastic and plastic parts of the strain rate tensor $\dot{\varepsilon}$ are additive.*

Now we shall discuss *two measures of finite deformation.* They would be useful in developing the elastic part of the constitutive equation for Eulerian formulation for the case of finite deformation and rotation. Both these measures are related to the right stretch tensor U, the part of F which represents the deformation of a neighborhood of the point. The first measure, called the *Green-Lagrange strain tensor*, is defined by the following relation:

$$e = \frac{1}{2}(U^2 - 1).$$

(4.38)

Using Equation 4.16, it can be expressed in terms of the deformation gradient tensor F:

$$e = \frac{1}{2}(F^T F - 1).$$

(4.39)

It can be expressed in terms of the displacement gradient tensor $\nabla_0 u$ using Equation 4.10:

$$e = \frac{1}{2}\left((1 + \nabla_0 u)^T(1 + \nabla_0 u) - 1\right),$$
$$= \frac{1}{2}\left(\nabla_0 u + (\nabla_0 u)^T + (\nabla_0 u)^T(\nabla_0 u)\right).$$

(4.40)

In terms of the index notation, the above expression can be written as

$$e_{ij} = \frac{1}{2}(u_{i,j} + u_{j,i} + u_{k,i}u_{k,j}),$$

(4.41)

where the comma indicates the derivatives with respect to the components of x_0. The above equations represent the strain-displacement relations, when we choose the Green-Lagrange strain tensor as the measure of finite deformation. These relations are *non-linear* unlike that for the linear strain tensor (Equation 2.147). Now, assume that the deformation is small. Mathematically, it means the tensor $\nabla_0 u$ is small compared to the unit tensor or the components $u_{i,j}$ are small compared to unity. Therefore, when the deformation is small, the non-linear term in Equations 4.40 and 4.41 becomes negligible and the expression for e reduces to that of ε. Thus, when the deformation is small, the Green-Lagrange strain tensor reduces to the linear strain tensor.

The physical interpretation of the components of e is as follows. The component e_{xx} represents *half the change in square length per unit square length* along the direction which was initially along the x-direction. The other two normal components e_{yy} and e_{zz} have similar interpretation. The component e_{xy} represents the product of λ_x, λ_y and *half the value of the sine function of change in angle* between the directions which were initially along the x and y directions. The quantity λ_x represents the ratio of the deformed length to the initial length of a small line element which was initially along the x-direction. The quantity λ_y has a similar interpretation. The physical interpretation of other two shear components e_{yz} and e_{zx} is similar. The sign convention for the components of e is similar to that of the components of the linear strain tensor. Just like the linear strain tensor, the tensor e has the principal values, principal directions, principal invariants and the hydrostatic and deviatoric parts. They are defined similarly.

The second measure of finite deformation is the *logarithmic strain* tensor. It is denoted by ε^L and is defined as follows. Note that the principal directions $\hat{e}_i$ of U remain orthogonal in the deformed configuration. Therefore, if we choose the coordinate directions along the $\hat{e}_i$, the shear associated with these directions is zero. To define the normal strains associated with these directions, we use the result that the principal values λ_i of U represent the ratio of the deformed length to the initial length of a small line element along $\hat{e}_i$. Then, the logarithm of λ_i can be used as a measure of the normal strain. Thus, with respect to the coordinate system of $\hat{e}_i$, we define the components of the logarithmic strain tensor ε^L as

$$\begin{aligned} \varepsilon^L_{ij} &= \ln\lambda_i \quad \text{if } i=j, \\ &= 0 \quad\quad\; \text{if } i \ne j. \end{aligned} \tag{4.42}$$

In the above equation, the symbol ln represents the natural logarithm, *i.e.*, the logarithm with respect to the base e. Note that the tensor U is a positive definite tensor. Therefore, all λ_i are positive and hence the components ε^L_{ij} are well-defined.

In tensor notation, Equation 4.42 is expressed as

$$\varepsilon^L = \ln U. \tag{4.43}$$

The symbol ln in the above equation means first the matrix of U is to be obtained in the coordinate system of its principal directions $\hat{e}_i$ so as to make its non-diagonal components zero and then the operation of taking the natural logarithm is to be carried out only on the diagonal components λ_i (which are positive). We can use the tensor transformation relation (Equation 2.56) to find the components of ε^L with respect to any other coordinate system.

The strain-displacement relations for the logarithmic strain tensor cannot be expressed by a single equation like that for the Green-Lagrange strain tensor (Equation 4.41). They are given by Equations 4.42, 4.17 and 4.11 along with the fact that λ_i are the principal values of the tensor U. These relations are also non-linear. It can be shown that when the deformation is small, the tensor ε^L reduces to the linear strain tensor ε.

The physical interpretation of the components of ε^L with respect to $\hat{e}_i$ as the coordinate system is obvious from Equation 4.42. The sign convention for the components of ε^L is similar to that of the components of the linear strain tensor. Just like the linear strain tensor, the logarithmic strain tensor has the principal directions (which are the same as those of U), the principal values (which are the natural logarithm of the principal values of U), the principal invariants and the hydrostatic and deviatoric parts.

We next discuss the elastic-plastic constitutive equation for Eulerian formulation for the case of finite deformation and rotation. This is done in the next subsection.

4.3 Constitutive Equation for Eulerian Formulation When the Rotation Is Not Small

First, we shall develop the *elastic part of the constitutive equation*. In the last chapter, we used the rate form of the elastic constitutive equation as it is found to be convenient for the Eulerian formulation. However, even in Eulerian formulation, it is possible to use the other form of the constitutive equation in which the stress is expressed in terms of the deformation, rather than the rate of deformation. But, if the elastic deformation is not small, we cannot use the linear strain tensor as a measure of the deformation. We have to use one of the two measures of finite deformation developed in the last section.

For the case of large deformation and rotation, the constitutive equation of elastic behavior is usually expressed as a relation between the Cauchy stress tensor σ and the deformation gradient tensor F:

$$\sigma = f(F). \tag{4.44}$$

As stated earlier, the Cauchy stress σ is an objective tensor. But the tensor F is not objective, as its mathematical representation F^* with respect to a moving frame is $Q(t)F$ and not $Q(t)FQ^T(t)$ [1]. The requirement that the response function f should be objective leads to the following form of Equation 4.44:

$$\sigma = R\,f(U)R^T, \tag{4.45}$$

where the tensors R and U are related to the tensor F through the polar decomposition theorem (Equation 4.14). Thus, for the case of large deformation and rotation, the stress depends on both the deformation (through U) as well as rotation (through R).

Now, we *assume* that

- The dependence of f on U is through the *logarithmic strain tensor* $\varepsilon^L = \ln U$. Further, this dependence is *linear*.
- The material is *isotropic*.

Then, the constitutive equation (Equation 4.45) becomes

$$\sigma = R(\lambda tr\varepsilon^L 1 + 2\mu\varepsilon^L)R^T , \qquad (4.46)$$

where the material constants λ and μ are the same as the Lame's constants of the linear constitutive relation for the case of small deformation (Equation 2.214). If we choose the Green-Lagrange strain tensor as the measure of large deformation, then the material constants in the constitutive equation (Equation 4.46) would be different.

In the above discussion, we have assumed the deformation to be purely elastic. For elasto-plastic deformation, we use the above constitutive equation only for the elastic part of the deformation. Let ε^{eL} be the logarithmic strain tensor and R^e be the rotation tensor corresponding to the elastic part of the deformation gradient F^e. Then, the constitutive equation for the elastic part can be stated as

$$\sigma = R^e(\lambda tr\varepsilon^{eL} 1 + 2\mu\varepsilon^{eL})(R^e)^T . \qquad (4.47)$$

Even for the elastic part, the change in volume is quite small. To simplify the solution methodology, we assume that the elastic deformation is also incompressible. Then, the hydrostatic part of stress $-p$ remains constitutively indeterminate. Further, the relation between the deviatoric part and the elastic deformation becomes

$$\sigma' = R^e(2\mu\varepsilon^{eL})(R^e)^T . \qquad (4.48)$$

This is the constitutive equation we use for the elastic part. Next, we develop the constitutive equation for the plastic part.

The tensor $\dot{\bar{\varepsilon}}^p$ is the *actual* strain rate tensor corresponding to the plastic part of the deformation gradient tensor (Equation 4.29). However, for developing the *plastic part of the constitutive equation*, we do not choose it as the deformation rate measure. Instead, we choose the *modified* strain rate tensor

$$\dot{\varepsilon}^p = R^e \dot{\bar{\varepsilon}}^p (R^e)^T \qquad (4.49)$$

as the deformation rate measure. Objectivity of this measure can be verified from Equation 4.34:

$$\dot{\varepsilon}^p = \dot{\varepsilon} - \dot{\varepsilon}^e , \qquad (4.50)$$

Since the tensors $\dot{\varepsilon}$ and $\dot{\varepsilon}^e$ are objective, the tensor $\dot{\varepsilon}^p$, which is their difference, also becomes objective. Now we write the constitutive equation for the plastic part using the following version of the associated flow rule (Equation 3.159):

$$\sigma' = \frac{2\sigma_{eq}}{3\dot{\varepsilon}^p_{eq}} \dot{\varepsilon}^p,$$
(4.51)

where the equivalent stress σ_{eq} is given by Equation 3.103:

$$\sigma_{eq} = H(\varepsilon^p_{eq}) \equiv \sigma_Y + K(\varepsilon^p_{eq})^n.$$
(4.52)

Here, σ_Y is the yield stress and K and n are the hardening parameters of the material. Note that the equivalent plastic strain rate $\dot{\varepsilon}^p_{eq}$ in Equation 4.51 is to be calculated from the modified strain rate tensor $\dot{\varepsilon}^p$ (Equation 4.49) and not the actual strain rate tensor $\bar{\dot{\varepsilon}}^p$. However, the argument of the hardening function H should be related to the actual strain rate tensor as this function is obtained from the results of tension test on the basis of equivalence between $\bar{\dot{\varepsilon}}^p_{eq}$ (equivalent plastic strain rate related to the actual strain rate tensor) and the plastic part of the axial strain rate in the tension test. This inconsistency can be resolved very easily by showing that $\dot{\varepsilon}^p_{eq}$ and $\bar{\dot{\varepsilon}}^p_{eq}$ are equal. To show it, we substitute the relation between $\dot{\varepsilon}^p$ not $\bar{\dot{\varepsilon}}^p$ (Equation 4.49) in the definition of $\dot{\varepsilon}^p_{eq}$ (Equation 3.156). Then we simplify the resulting equation by using the orthogonality of rotation tensor $\boldsymbol{R}^e$ and the identity at Equation 2.15 in δ. Thus we get

$$
\begin{aligned}
\dot{\varepsilon}^p_{eq} &= \left(\frac{2}{3}\dot{\varepsilon}^p_{ij}\dot{\varepsilon}^p_{ij}\right)^{1/2}, \\
&= \left(\frac{2}{3}R^e_{ik}\bar{\dot{\varepsilon}}^p_{kl}(R^{eT})_{lj}\, R^e_{im}\bar{\dot{\varepsilon}}^p_{mn}(R^{eT})_{nj}\right)^{1/2}, \\
&= \left(\frac{2}{3}\bar{\dot{\varepsilon}}^p_{kl}\bar{\dot{\varepsilon}}^p_{mn}(R^{eT})_{ki}R^e_{im}(R^{eT})_{lj}R^e_{jn}\right)^{1/2}, \\
&= \left(\frac{2}{3}\bar{\dot{\varepsilon}}^p_{kl}\bar{\dot{\varepsilon}}^p_{mn}\delta_{km}\delta_{nl}\right)^{1/2}, \\
&= \left(\frac{2}{3}\bar{\dot{\varepsilon}}^p_{mn}\bar{\dot{\varepsilon}}^p_{mn}\right), \\
&= \bar{\dot{\varepsilon}}^p_{eq}.
\end{aligned}
$$
(4.53)

Using Equation 4.50, the above constitutive equation (Equation 4.51) can be expressed in terms of the total strain rate tensor and its elastic part:

$$\sigma' = \frac{2\sigma_{eq}}{3\dot{\varepsilon}^p_{eq}}(\dot{\varepsilon} - \dot{\varepsilon}^e). \tag{4.54}$$

This is the constitutive equation we use for the plastic part.

No stress rate is involved in these two parts of the constitutive equation (Equations 4.48 and 4.54). Therefore, we do not need objective stress rate tensor. Further, we do not combine these two parts of the constitutive equation (Equations 4.48 and 4.54) into a single equation as was done earlier. Earlier they were combined on the basis of the assumption that the elastic and plastic parts of the strain rate tensor were additive. But this assumption is not true for the case of large rotation. Further, the elastic constitutive equation (Equation 4.48) is in terms of the deformation while the plastic part (Equation 4.54) involves the rate of deformation. Hence, they cannot be combined. How these equations are to be used in the solution procedure is described in the next subsection.

4.3.1 Solution Procedure

When the rotation is not small, the Eulerian formulation for the metal forming problems is similar to that of Subsection 3.9.3 (Equations 3.223–3.225) except that now the constitutive equation (Equation 3.224) is replaced by Equations 4.48 and 4.54. Since the deformation has been assumed to be incompressible, an additional equation, in the form of incompressibility constraint (Equation 3.249), is needed for determining the hydrostatic part. Further, it is convenient to express the equation of motion in the form of Equation 3.248 which separates the hydrostatic and deviatoric parts.

These equations need to be solved by iteration. In every iteration, the velocity, strain rate, hydrostatic stress and deviatoric stress fields are found by solving these equations without the elastic constitutive equation. Note that the plastic constitutive equation involves the elastic part ($\dot{\varepsilon}^e$) of the strain rate tensor. In the first iteration, an initial guess is used for it. Then, in subsequent iterations, it is estimated from the solution of the previous iteration. The procedure for estimating $\dot{\varepsilon}^e$ from the solution of the previous iteration is given below. The quantity $\dot{\varepsilon}^e$ is determined at certain discrete points along all the stream lines.

- First, using the velocity gradient tensor of the previous iteration, the deformation gradient tensor F is obtained by integrating, along the stream lines, the following differential equation (Equation 4.20):

$$\dot{F} = (\nabla v)F \tag{4.55}$$

This is done by using a suitable finite difference scheme for the time variable.

- Next, at a point on the stream line, the elastic part of the deformation gradient tensor is determined from Equation 4.26 using the F of the above step and the plastic part ($\overline{F}^p$) of the deformation gradient tensor at the point which the particle occupies at a *previous time instant* (on the stream line). This is called a *trial* elastic deformation gradient tensor and is denoted by F^{e*}. The *trial* right stretch tensor U^{e*} and the *trial* rotation tensor R^{e*} are found from the polar decomposition of F^{e*} (Equation 4.14). Further, the *trial* logarithmic strain tensor is found from U^{e*} using Equation 4.43.

- Next, at that point, the *trial* deviatoric stress tensor σ'^* is determined from the elastic constitutive equation (Equation 4.48). Then, the trial equivalent stress σ_{eq}^* is found. The actual value of the equivalent stress σ_{eq} at the point is calculated from the hardening relation of the previous iteration. If σ_{eq}^* is less than σ_{eq}, then the state of stress is elastic, otherwise it is plastic.

- If the state of stress is elastic at the point, then the plastic part of the strain rate tensor is zero and $\dot{\varepsilon}^e$ is equal to $\dot{\varepsilon}$ of the previous iteration.

- If the state of stress is plastic at that point, then the plastic part ($\dot{\overline{\varepsilon}}^p$) of the strain rate tensor is determined using the procedure given by Weber and Anand [6]. In this procedure, $\dot{\overline{\varepsilon}}^p$ is assumed to be constant over the time step. Then, a relationship is obtained between the values of $\overline{F}^p$ at the points which the particle is occupying now and at the previous time instant (on the stream line). This is done by integrating Equation 4.29. Combining this relationship with the *trial* elastic deformation gradient tensor F^{e*}, the polar decomposition theorem and the elastic constitutive equation, a relationship is developed between $\dot{\overline{\varepsilon}}^p$, σ'^*, σ_{eq}, the shear modulus μ and the time increment. The plastic part ($\dot{\overline{\varepsilon}}^p$) of the strain rate tensor is determined from this relation.

- It can be shown that the *trial* rotation tensor R^{e*} is exactly equal to the actual rotation tensor R^e.

- Finally, at that point, the elastic part ($\dot{\varepsilon}^e$) of the strain rate tensor is obtained from Equation 4.34 by substituting the above values of R^e and $\dot{\overline{\varepsilon}}^p$.

- This procedure also gives the value of $\overline{F}^p$ at that point. It will be needed in determining $\dot{\varepsilon}^e$ at the point which the particle will occupy at the next time instant.

4.4 Kinematics of Finite Incremental Deformation and Rotation

In Section 4.2, we discussed the kinematics of finite elastic-plastic deformation and rotation required for the Eulerian formulation. In this section, we shall discuss the kinematics of *finite elastic-plastic incremental deformation and rotation* which is needed in the updated Lagrangian formulation. Figure 4.3 shows two deformed configurations of a body: (i) one at the current time t (called as the *current* configuration) and (ii) the other at time $t + \Delta t$, *i.e.*, after the finite time increment Δt (called the *incremental* configuration). A material particle occupies the position $^t P$ in the current configuration and the position $^{t+\Delta t} P$ in the incremental configuration. The position vectors of the particle in the current and the incremental configurations are respectively $^t x$ and $^{t+\Delta t} x$. Consider a small line element $^t P^t Q$ at point $^t P$ of the current configuration which occupies the position $^{t+\Delta t} P^{t+\Delta t} Q$ at point $^{t+\Delta t} P$ of the incremental configuration. We denote the line segments $^t P^t Q$ and $^{t+\Delta t} P^{t+\Delta t} Q$ by $\mathbf{d}^t x$ and $\mathbf{d}^{t+\Delta t} x$ respectively. Similar to Equations 4.2 and 4.3, the relation between $\mathbf{d}^t x$ and $\mathbf{d}^{t+\Delta t} x$ is given by

$$\mathbf{d}^{t+\Delta t} x = (_t \Delta F) \mathbf{d}^t x, \tag{4.56}$$

where

$$_t \Delta F = \frac{\partial^{t+\Delta t} x}{\partial^t x}. \tag{4.57}$$

The partial derivative in Equation 4.57 is to be taken by holding the time constant. The tensor $_t \Delta F$ is called the *incremental deformation gradient tensor*. The left subscript t on the symbol Δ denotes that the increment is to be taken at time t. For real incremental deformation, $_t \Delta F$ is non-singular, *i.e.*, its determinant is non-zero. In index notation, Equation 4.57 can be written as

$$_t \Delta F_{ij} = \frac{\partial^{t+\Delta t} x_i}{\partial^t x_j}. \tag{4.58}$$

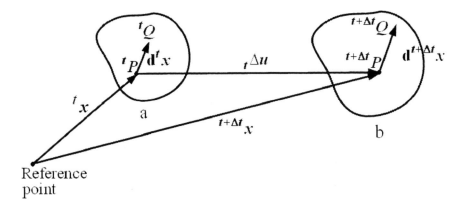

Figure 4.3. Kinematics of finite incremental deformation. **a** Deformed configuration at time t (current configuration). **b** Deformed configuration at time $t + \Delta t$ (incremental configuration)

Let the *incremental displacement vector* of the particle in the time increment Δt be $_t\Delta u$. From the geometry of Figure 4.3, we get the following relation between $_t\Delta u$, $^t x$ and $^{t+\Delta t} x$:

$$^{t+\Delta t}x = {}^t x + {}_t\Delta u. \tag{4.59}$$

Substituting the above expression in Equation 4.57, we get the following relationship between $_t\Delta F$ and the derivative of $_t\Delta u$ with respect to $^t x$:

$$_t\Delta F = 1 + \frac{\partial(_t\Delta u)}{\partial x^t}. \tag{4.60}$$

In index notation, the above equation can be written as

$$_t\Delta F_{ij} = \delta_{ij} + \frac{\partial(_t\Delta u)_i}{\partial^t x_j} = \delta_{ij} + {}_t\Delta u_{i,j}. \tag{4.61}$$

Here, the comma indicates the derivatives with respect to the components of $^t x$.

Similar to the decomposition of F into the rotation and deformation parts (Eq. 4.14), the tensor $_t\Delta F$ can also be decomposed using the polar decomposition theorem:

$$_t\Delta F = (_t\Delta R)(_t\Delta U) = (_t\Delta V)(_t\Delta R). \tag{4.62}$$

Here, $_t\Delta U$ and $_t\Delta V$ are the symmetric positive definite tensors, called the *incremental right and left stretch tensors* respectively and $_t\Delta R$ is the orthogonal tensor called the *incremental rotation tensor*. The tensor $_t\Delta U$ (or $_t\Delta V$) represents the incremental deformation of a neighborhood of the point while the tensor $_t\Delta R$ represents its incremental rotation about the point. Let $_t\Delta\lambda_i$ be the principal values and $^t\hat{e}_i$ be the principal directions of $_t\Delta U$. (Note that $^t\hat{e}_i$ are the directions in the current configuration.) Then $_t\Delta\lambda_i$ represents the ratio of the deformed length at time $t + \Delta t$ to the deformed length at time t of a small line element along the direction $^t\hat{e}_i$. Further, let $^{t+\Delta t}\hat{e}_i$ be the positions of the directions $^t\hat{e}_i$ in the incremental configuration. (It can be shown that, $^{t+\Delta t}\hat{e}_i$ are the principal directions of $_t\Delta V$, its principal values being the same as $_t\Delta\lambda_i$.) Then

$$^{t+\Delta t}\hat{e}_i = (_t\Delta R)^t\hat{e}_i. \tag{4.63}$$

Similar to Equations 4.16 and 4.17, we get the following relation between $_t\Delta U$ and $_t\Delta F$:

$$_t\Delta U^2 = (_t\Delta F^{\mathrm{T}})(_t\Delta F), \tag{4.64}$$

$$_t\Delta U_{ij}^2 = (_t\Delta F)_{ik}^{\mathrm{T}}(_t\Delta F)_{kj}. \tag{4.65}$$

Similar to the *decomposition* of F into the *elastic and plastic parts* (Equation 4.25), the tensor $_t\Delta F$ can also be decomposed as

$$_t\Delta F = (_t\Delta F^e)(_t\Delta F^p). \tag{4.66}$$

Like Equation 4.25, the above decomposition of $_t\Delta F$ is also not unique. To remove this non-uniqueness, *we assume that the incremental rotation tensor has no plastic part.* To emphasize this fact, the plastic part of the incremental deformation gradient and all the related tensors are denoted by an overbar. Thus,

$$_t\Delta F = (_t\Delta F^e)(_t\Delta\bar{F}^p). \tag{4.67}$$

Now, we decompose the elastic part of $_t\Delta F$ into the rotation and deformation parts using the polar decomposition theorem (Equation 4.62). Further, we have assumed that $_t\Delta R$ has no plastic part. Thus, we obtain

$$
{}_t\Delta F^e = ({}_t\Delta R^e)({}_t\Delta U^e),
$$
$$
{}_t\Delta\bar{F}^p = {}_t\Delta\bar{U}^p. \tag{4.68}
$$

Substituting the decompositions (Equation 4.62 and 4.68) in Equation 4.67 and equating the symmetric positive definite and orthogonal parts separately, we get

$$
{}_t\Delta R = {}_t\Delta R^e,
$$
$$
{}_t\Delta U = ({}_t\Delta U^e)({}_t\Delta\bar{U}^p). \tag{4.69}
$$

Thus, if we assume that the plastic part of the incremental rotation tensor is zero, it means the incremental right stretch tensor can be decomposed multiplicatively into the elastic and plastic parts.

In the last chapter, we discussed a measure of incremental deformation called the incremental linear strain tensor. It is not a valid deformation measure when the incremental deformation is large. Now, we present *two* commonly used *measures of finite incremental deformation*.

The first measure, called the *incremental Green-Lagrange strain* tensor, is defined by a relation similar to Equations 4.38 and 4.39:

$$
{}_t\Delta e = \frac{1}{2}\left(({}_t\Delta U)^2 - 1\right),
$$
$$
= \frac{1}{2}\left(({}_t\Delta F)^{\mathrm{T}}({}_t\Delta F) - 1\right). \tag{4.70}
$$

Using the index expression for the components of ${}_t\Delta F$ (Equation 4.61), we get the following expression for the components of ${}_t\Delta e$ in terms of the derivatives of the incremental displacement:

$$
{}_t\Delta e_{ij} = \frac{1}{2}\left({}_t\Delta u_{i,j} + {}_t\Delta u_{j,i} + ({}_t\Delta u)_{k,i}({}_t\Delta u)_{k,j}\right), \tag{4.71}
$$

where the comma indicates the derivatives with respect to the components of ${}^t x$. The above equations represent the incremental strain-displacement relations, when we choose the incremental Green-Lagrange strain tensor as the measure of finite incremental deformation. These relations are *non-linear* unlike that of the incremental linear strain tensor (Equation 3.55). Now, assume that the incremental deformation is small. Mathematically, it means the components ${}_t\Delta u_{i,j}$ are small compared to unity. Therefore, when the incremental deformation is small, the non-linear term in Equation 4.71 becomes negligible and the expression for ${}_t\Delta e$ reduces to that of $\mathbf{d}\varepsilon$. We, of course, need to change the notation for the incremental displacement vector from ${}_t\Delta u$ to $\mathbf{d}u$. Thus, when the incremental deformation is small, the incremental Green-Lagrange strain tensor reduces to the incremental linear strain tensor.

The physical interpretation of the components of $_t\Delta e$ is similar to that of the components of e. The sign convention for the components of $_t\Delta e$ is similar to that of the components of the linear strain tensor. Just like the linear strain tensor, the tensor $_t\Delta e$ has the principal values, principal directions, principal invariants and the hydrostatic and deviatoric parts. They are defined similarly.

The elastic and plastic parts of $_t\Delta e$ would be

$$
\begin{aligned}
_t\Delta e^e &= \frac{1}{2}\left(({}_t\Delta F^e)^{\mathrm{T}}({}_t\Delta F^e)-\mathbf{1}\right), \\
_t\Delta \bar{e}^p &= \frac{1}{2}\left(({}_t\Delta \bar{F}^p)^{\mathrm{T}}({}_t\Delta \bar{F}^p)-\mathbf{1}\right).
\end{aligned}
\tag{4.72}
$$

Now, we use the decomposition of $_t\Delta F$ (Equation 4.67) to obtain the expression for $_t\Delta e$ in terms of its elastic and plastic parts. First, we substitute Equation 4.67 into the definition of $_t\Delta e$ (Equation 4.70). Note that the operation of taking the transpose changes the order of tensors in the product. Then, we substitute for $({}_t\Delta F^e)^{\mathrm{T}}({}_t\Delta F^e)$ and $({}_t\Delta \bar{F}^p)^{\mathrm{T}}({}_t\Delta \bar{F}^p)$ using Equation 4.72. Thus, we get

$$
\begin{aligned}
_t\Delta e &= \frac{1}{2}\left(({}_t\Delta F)^{\mathrm{T}}({}_t\Delta F)-\mathbf{1}\right) \\
&= \frac{1}{2}\left(({}_t\Delta F^e \,_t\Delta \bar{F}^p)^{\mathrm{T}}({}_t\Delta F^e \,_t\Delta \bar{F}^p)-\mathbf{1}\right) \\
&= \frac{1}{2}\left(({}_t\Delta \bar{F}^p)^{\mathrm{T}}({}_t\Delta F^e)^{\mathrm{T}}({}_t\Delta F^e)({}_t\Delta \bar{F}^p)-\mathbf{1}\right) \\
&= \frac{1}{2}\left(({}_t\Delta \bar{F}^p)^{\mathrm{T}}(2\,_t\Delta e^e+\mathbf{1})({}_t\Delta \bar{F}^p)-\mathbf{1}\right) \\
&= ({}_t\Delta \bar{F}^p)^{\mathrm{T}}({}_t\Delta e^e)({}_t\Delta \bar{F}^p)+\,_t\Delta \bar{e}^p.
\end{aligned}
\tag{4.73}
$$

The above equation shows that

$$
_t\Delta e \neq \,_t\Delta e^e + \,_t\Delta \bar{e}^p.
\tag{4.74}
$$

It means the elastic and plastic parts of the incremental Green-Lagrange strain tensor $_t\Delta e$ are **not** additive.

From Equation 4.60, we can express $_t\Delta F$ as

$$_t\Delta F = 1 + \alpha,$$

$$\alpha = \frac{\partial(_t\Delta u)}{\partial x^t}. \tag{4.75}$$

Now, *assume that the incremental deformation is small*. As stated earlier, it means the components $_t\Delta u_{i,j}$ are small compared to unity. Then, Equation 4.75 implies that the tensor α is small compared to the unit tensor. We can write similar equations for the elastic and plastic parts of $_t\Delta F$:

$$_t\Delta F^e = 1 + \alpha^e,$$

$$_t\Delta \bar{F}^p = 1 + \bar{\alpha}^p. \tag{4.76}$$

Here, the tensors α^e and $\bar{\alpha}^p$ are small compared to the unit tensor. Substituting Equation 4.76 in Equation 4.72 for $_t\Delta e^e$ and neglecting higher order terms in α^e, we find that the tensor $_t\Delta e^e$ is of the order of α^e. Similarly, we find that the tensor $_t\Delta \bar{e}^p$ is of the order of $\bar{\alpha}^p$. Now, we substitute Equation 4.76 in Equation 4.73 for $_t\Delta e$ and neglect the higher order terms in $\bar{\alpha}^p$. Then we get

$$\begin{aligned}
_t\Delta e &= (_t\Delta\bar{F}^p)^T(_t\Delta e^e)(_t\Delta\bar{F}^p) + _t\Delta\bar{e}^p \\
&= (1 + (\bar{\alpha}^p)^T)(_t\Delta e^e)(1 + \bar{\alpha}^p) + _t\Delta\bar{e}^p \\
&= {}_t\Delta e^e + (\bar{\alpha}^p)^T(_t\Delta e^e) + (_t\Delta e^e)(\bar{\alpha}^p) + (\bar{\alpha}^p)^T(_t\Delta e^e)(\bar{\alpha}^p) + _t\Delta\bar{e}^p \\
&\cong {}_t\Delta e^e + _t\Delta\bar{e}^p.
\end{aligned} \tag{4.77}$$

As stated earlier, the tensor $_t\Delta e$ reduces to the incremental linear strain tensor $d\varepsilon$ when the deformation is small. Therefore, the elastic ($_t\Delta e^e$) and plastic ($_t\Delta\bar{e}^p$) parts of $_t\Delta e$ will also reduce to the elastic ($d\varepsilon^e$) and plastic ($d\varepsilon^p$) parts of $d\varepsilon$. Thus, for the case of small deformation, Equation 4.77 reduces to

$$d\varepsilon \cong d\varepsilon^e + d\varepsilon^p. \tag{4.78}$$

Equation 4.78 shows that the assumption used in developing the elastic-plastic incremental stress-strain relation for the case of updated Lagrangian formulation is valid for the case of small incremental deformation.

The second measure of finite incremental deformation is the *incremental logarithmic strain* tensor. It is denoted by $_t\Delta\varepsilon^L$ and defined by an equation similar to Equation 4.43:

$$_t\Delta\varepsilon^L = \ln(_t\Delta U). \tag{4.79}$$

The symbol ln in the above equation means first the matrix of $_t\Delta U$ is to be obtained in the coordinate system of its principal directions $^t\hat{e}_i$ so as to make its non-diagonal components zero and then the operation of taking the *natural* logarithm is to be carried out only on the diagonal components (which are the principal values of $_t\Delta U$). Note that the tensor $_t\Delta U$ is a positive definite symmetric tensor. Therefore, all its principal values are positive and hence the tensor $_t\Delta\varepsilon^L$ is well-defined. The components of $_t\Delta\varepsilon^L$ in the coordinate system of $^t\hat{e}_i$ are given by

$$_t\Delta\varepsilon_{ij}^L = \ln(_t\Delta\lambda_i) \qquad \text{if} \quad i = j,$$
$$= 0 \qquad \text{if} \quad i \neq j. \tag{4.80}$$

The strain-displacement relations for the incremental logarithmic strain tensor cannot be expressed by a single equation like that for the incremental Green-Lagrange strain tensor (Equation 4.71). They are given by Equations 4.80, 4.65 and 4.61 along with the fact that $_t\Delta\lambda_i$ are the principal values of the tensor $_t\Delta U$. These relations are also non-linear. It can be shown that when the deformation is small, the tensor $_t\Delta\varepsilon^L$ reduces to the incremental linear strain tensor.

The physical interpretation of the components of $_t\Delta\varepsilon^L$ with respect to $^t\hat{e}_i$ as the coordinate system is obvious from Equation 4.80. The sign convention for the components of $_t\Delta\varepsilon^L$ is similar to that of the components of the linear strain tensor. Just like the linear strain tensor, the logarithmic strain tensor has the principal directions (which are the same as those of $_t\Delta U$), the principal values (which are the natural logarithm of the principal values of $_t\Delta U$), the principal invariants and the hydrostatic and deviatoric parts.

The elastic and plastic parts of $_t\Delta\varepsilon^L$ can be defined as

$$_t\Delta\varepsilon^{eL} = \ln(_t\Delta U^e),$$
$$_t\Delta\bar{\varepsilon}^{pL} = \ln(_t\Delta\bar{U}^p). \tag{4.81}$$

Now, we use the decomposition of $_t\Delta U$ (Equation 4.69) to obtain the expression for $_t\Delta\varepsilon^L$ in terms of its elastic and plastic parts. First, we substitute Equation 4.69 into the definition of $_t\Delta\varepsilon^L$ (Equation 4.79). Then we use a property of the logarithmic function. Finally, we substitute for $\ln(_t\Delta U^e)$ and $\ln(_t\Delta\bar{U}^p)$ using Equations 4.81. Thus, we get

$$_t\Delta\varepsilon^L = \ln(_t\Delta U),$$
$$= \ln(_t\Delta U^e {_t}\Delta\bar{U}^P),$$
$$= \ln(_t\Delta U^e) + \ln(_t\Delta\bar{U}^P), \tag{4.82}$$
$$= {_t}\Delta\varepsilon^{eL} + {_t}\Delta\bar{\varepsilon}^P.$$

This shows that the *elastic and plastic parts* of the incremental logarithmic strain tensor $_t\Delta\varepsilon^L$ are *additive*.

4.5 Constitutive Equation for Updated Lagrangian Formulation for Finite Incremental Deformation and Rotation

In the last chapter, we derived the elastic-plastic incremental stress-strain relation needed for the updated Lagrangian formulation using a measure of small incremental deformation. In this section, we develop this relationship for the case of finite incremental deformation. We use the incremental logarithmic strain tensor $_t\Delta\varepsilon^L$ as the measure of finite incremental deformation. This allows us to use the elastic constants and the hardening function determined from the results of tension test on the variation of true stress *vs* logarithmic strain. If we choose the incremental Green-Lagrange strain tensor as the measure of finite incremental deformation, then the results of tension test need to be modified appropriately.

We *assume* that the elastic behavior remains linear even for the case of large incremental deformation. Then, the elastic incremental stress-strain relation developed for the case of small incremental deformation (Equation 3.150) remains valid for the finite deformation also. But, we need to change the measure of incremental deformation in Equation 3.150 from $d\varepsilon_{ij}^e$ to $_t\Delta\varepsilon_{ij}^{eL}$. We also need to change the notation of the incremental stress tensor from $d\sigma_{kl}$ to $_t\Delta\sigma_{kl}$. We further *assume* that the associated flow rule derived for the case of small incremental deformation remains valid when we replace the measure of incremental deformation from $d\varepsilon_{ij}^P$ to $_t\Delta\varepsilon_{ij}^{PL}$. As in the case of small incremental deformation, we use the Mises yield function as the plastic potential. Then, the plastic incremental stress-strain relation is given by Equation 3.143 where the measure of incremental deformation needs to be changed from $d\varepsilon_{ij}^P$ to $_t\Delta\varepsilon_{ij}^{PL}$ and the notation of incremental stress needs to be changed from $d\sigma_{kl}$ to $_t\Delta\sigma_{kl}$. Now, as shown in the last section, the tensor $_t\Delta\varepsilon_{ij}^L$ consists of the addition of its elastic $(_t\Delta\varepsilon_{ij}^{eL})$ and plastic $(_t\Delta\varepsilon_{ij}^{PL})$ parts. Therefore, as in the case of small incremental deformation, we add the elastic and plastic incremental stress-strain relations to get a single elastic-plastic constitutive equation in terms of $_t\Delta\varepsilon_{ij}^L$. Next, we invert this equation to express it in terms of $_t\Delta\sigma_{kl}$. Then, similar to the elastic-plastic incremental stress-strain relation of small incremental deformation (Equations 3.152 and 3.153), we get

$$_t\Delta\sigma_{ij} = {}^t C_{ijkl}^{EP}({}_t\Delta\varepsilon_{kl}^L),$$

(4.83)

where the *fourth order elastic-plastic tensor* $^t C_{ijkl}^{EP}$ is given by

$$^t C_{ijkl}^{EP} = 2\mu\left[\frac{v}{1-2v}\delta_{ij}\delta_{kl} + \delta_{ik}\delta_{jl} - \frac{9\mu}{2}\frac{{}^t\sigma_{ij}'\,{}^t\sigma_{kl}'}{({}^tH'+3\mu){}^t\sigma_{eq}^2}\right].$$

(4.84)

Here, the left superscript t is added to the symbols $^t C_{ijkl}^{EP}$, $^t\sigma_{ij}'$, $^t\sigma_{eq}$ and $^tH'$ to emphasize the fact that these quantities are to be evaluated at time t. Note that, when the increment size is large, these quantities vary continuously from time t to $t + \Delta t$. To take care of this variation, we *assume* that the above constitutive equation (Equation 4.83) can be modified as

$$_t\Delta\sigma_{ij} = \int_t^{t+\Delta t} {}^t C_{ijkl}^{EP}\,\mathrm{d}({}_t\Delta\varepsilon_{kl}^L).$$

(4.85)

The integration in the above equation is usually performed using the forward Euler integration scheme. Details of this scheme are discussed in Chapter 6. Further, it is convenient to use the coordinate system of $^t\hat{e}_i$ (the principal directions of $_t\Delta U$) to determine the incremental logarithmic strain. Therefore, the evaluation of the incremental stress using the above constitutive equation is also done in this coordinate system only.

In the constitutive equation (Equation 4.85), the response function must be *objective*. We can make the response function objective using the method of Section 4.3. However, a simpler method exists [7]. In this method, incremental rotation is incorporated while updating the stress tensor. Employing this method, we make the response function objective by using the following updating relation:

$$^{t+\Delta t}\sigma = ({}_t\Delta R)({}^t\sigma)({}_t\Delta R)^T + {}_t\Delta\sigma.$$

(4.86)

Thus, the updating procedure can be described as follows:

- Evaluate the incremental rotation tensor $_t\Delta R$.

  - Determine the rotated stress tensor: $({}_t\Delta R)({}^t\sigma)({}_t\Delta R)^T$. This is equivalent
    to rotating the element through $_t\Delta R$ so as to rotate the stress components

acting on the element. This quantity should be evaluated in coordinate system of $'\hat{e}_i$ as $_t\Delta\sigma$ is also to be evaluated in the same coordinate system. Then the addition becomes easy. For evaluating $(_t\Delta R)('\sigma)(_t\Delta R)^\mathrm{T}$ in the coordinate system of $'\hat{e}_i$, the stress components must be transformed from the reference coordinate system to this coordinate system. Further, the components of $_t\Delta R$ should be obtained in this coordinate system.

- Evaluate the incremental stress tensor $_t\Delta\sigma$ from the elastic-plastic incremental constitutive equation (Equation 4.85). This should be done in the coordinate system of $'\hat{e}_i$.

- Update the stresses by adding the incremental stress $_t\Delta\sigma$ to the rotated stress at time t: $(_t\Delta R)('\sigma)(_t\Delta R)^\mathrm{T}$. The addition should be done in the coordinate system of $'\hat{e}_i$.

- Transform the components of the updated stress $^{t+\Delta t}\sigma$ to the reference coordinate system.

There are other ways of making the response function objective. One such method [8] is the use of the increment of the second Piola-Kirchhoff stress tensor $^{t+\Delta t}_t S$ in the incremental elastic-plastic stress-strain relation. The tensor $^{t+\Delta t}_t S$ is defined by the relation

$$^{t+\Delta t}_t S = \left(\det(_t\Delta F)\right)(_t\Delta F)^{-1}(^{t+\Delta t}\sigma)(_t\Delta F)^{-\mathrm{T}}. \tag{4.87}$$

Since the tensor $^{t+\Delta t}_t S$ is work-conjugate to the incremental Green-Lagrange strain tensor $_t\Delta e$, in this case, $_t\Delta e$ is used as the measure of finite incremental deformation. Thus, the elastic-plastic incremental stress-strain relation is expressed as

$$_t\Delta S_{ij} = {}^tL^{EP}_{ijkl}(_t\Delta e_{kl}), \tag{4.88}$$

where $_t\Delta S$ is an increment of the second Piola-Kirchhoff stress tensor $^{t+\Delta t}_t S$. Note that, in the above equation, the expression for the fourth order elastic-plastic tensor $^tL^{EP}_{ijkl}$ cannot be the same as in Equation 4.84. To obtain the elastic constants and the hardening function appearing in $^tL^{EP}_{ijkl}$, appropriate modifications need to be made on the results of tension test. The expression for $^tL^{EP}_{ijkl}$ is given in [8] and its derivation is given in [9].

Another way of making the response function (of the elastic-plastic incremental stress-strain relation) objective is to write it in the rate form:

$$\overset{o}{\sigma}_{ij} = C^{EP}_{ijkl}\dot{\varepsilon}_{kl}, \tag{4.89}$$

where $\overset{o}{\sigma}$ is an objective stress tensor, $\dot{\varepsilon}$ is the strain rate tensor and $\boldsymbol{C}^{EP}$ is the fourth order tensor appearing in the elastic-plastic incremental stress-strain relation for the case of small deformation (Equation 3.153). Then, the incremental stress tensor $_t\Delta\sigma$ is obtained by integrating Equation 4.89:

$$_t\Delta\sigma_{ij} = \overset{o}{\sigma}_{ij}\, dt = \int\limits_{t}^{t+\Delta t} C^{EP}_{ijkl}\dot{\varepsilon}_{kl}\, dt = \int\limits_{t}^{t+\Delta t} C^{EP}_{ijkl}\, d\varepsilon_{kl}. \tag{4.90}$$

Here, $d\varepsilon_{kl}$ is the incremental linear strain tensor. The integration is usually carried out by the Euler forward integration scheme.

Quite a few choices are available for the objective stress rate tensor to be used in the constitutive equation (Equation 4.89). One choice is the Jaumann stress rate tensor (Equation 3.200), which uses the spin tensor $\dot{\omega}$ as the rate of rotation. However, in incremental formulation, use of the Jaumann stress rate is equivalent to assuming that the incremental rotation is small and is given by the product of $\dot{\omega}$ and the time increment. Thus, when the incremental rotation is large, the Jaumann stress rate is not a good choice. The other two commonly used objective stress rate measures are: (i) Truesdell Rate and (ii) Green-Naghdi Rate (also called the Green-McInnis Rate). These rates are defined by the following expressions:

(i) *Truesdell Rate*:

$$\dot{\sigma}_T = \dot{\sigma} - (\nabla v)\sigma - \sigma(\nabla v)^{\mathrm{T}} + \sigma\, tr(\nabla v). \tag{4.91}$$

(ii) *Green-Naghdi Rate*:

$$\dot{\sigma}_{GN} = \dot{\sigma} - (\dot{R}R^{\mathrm{T}})\sigma - \sigma(\dot{R}R^{\mathrm{T}})^{\mathrm{T}}. \tag{4.92}$$

In the above equations, ∇v is the velocity gradient tensor, σ is the Cauchy stress tensor, R is the rotation tensor, $\dot{\sigma}$ is the rate of Cauchy stress tensor and $\dot{R}$ is the rate of rotation tensor.

For finite deformation problems, use of different objective stress measures leads to different solutions, unless the constitutive equation is properly adjusted. To obtain physically meaningful solution, the choice of the objective stress measure should be based on the underlying physics of the finite elastic-plastic deformation. Unfortunately, the comparison of various objective stress measures is limited to just two cases: finite simple shear deformation problem of (i) linear hypoelastic materials and (ii) linear kinematically hardening materials. In both the cases, use of the Jaumann stress rate gives an oscillatory solution. In the case of hypoelastic materials, Prager [10] and Dienes [11] attributed the oscillations to improper use of the Jaumann stress rate tensor. Further, Prager obtained a non-oscillatory solution by adding some non-linear terms to the constitutive equation whereas Dienes obtained it by using the Green-Naghdi rate. In the case of kinematically hardening materials also, the oscillations were due to improper use of the Jaumann stress rate tensor [12]. Further, Lee *et al.* [12] and Dafalias [13] proposed new objective stress rate measures to obtain a non-oscillatory solution.

At present, the research on objective stress rate measures is still in progress. Therefore, there is no agreement about which is the best objective stress rate tensor to be used in elastic-plastic analysis. However, as far as elastic-plastic analysis is concerned, the Truesdell rate has a certain inconsistency. Note that, when the stress rate is zero, there is no plastic flow, and therefore no change in the yield surface. Since the yield surface depends on the invariants of σ, it means there is also no change in the invariants. However, when the Truesdell rate becomes zero, the invariants of σ do not necessarily remain constant [14].

4.6 Anisotropic Initial Yield Criteria

Microstructure of metals is polycrystalline in nature. In an annealed metal, the crystallographic directions are randomly oriented. At macroscopic level, this means there is no preferred direction. Thus, an annealed metal is isotropic at macroscopic level. However, when it is subjected to cold forming processes like drawing, extrusion, rolling *etc.*, the crystallographic directions gradually rotate towards a common axis, thus creating a preferred direction. Therefore, after cold forming, the metal usually becomes anisotropic in nature. When this metal is subjected to further forming processes without annealing, the yield criteria and the plastic stress-strain relations to be used for the analysis of these processes should incorporate the anisotropy. In this section, we shall discuss various anisotropic yield criteria. The corresponding plastic stress-strain relations will be discussed in next section. In this book, we shall consider the anisotropy of cold-rolled sheets only. These sheets possess a symmetry called orthotropy. It means there exist three mutually orthogonal planes of symmetry at every point, the intersections of which are called the principal axes of anisotropy. These axes are the rolling direction, the transverse direction (in the plane of sheet) and the direction normal to the sheet (*i.e.*, the thickness direction).

In cold-rolled sheets, plastic properties differ along the thickness direction (known as *normal* anisotropy) as well as vary with the orientation in the plane of the sheet (known as *planar* anisotropy). Both types of anisotropy play an important role in metal forming processes. For example, the planar anisotropy leads to formation of ears in cup drawing while the deep drawability of sheets depends strongly on the normal anisotropy.

A measure of anisotropy of sheets is the *strain rate ratio*, which can be defined as follows. In a rolled sheet, let x be the rolling direction, y be the transverse direction and z be the normal (or thickness) direction (Figure 4.4). Now, suppose a tensile test specimen is cut from this sheet such that its longitudinal axis x' makes an angle θ with the rolling direction. Let y' be the transverse axis of the specimen and z' be its normal axis (which of course coincides with z). Then, the strain rate ratio r_θ is defined as

$$r_\theta = \frac{\dot{\varepsilon}^P_{y'y'}}{\dot{\varepsilon}^P_{z'z'}}, \tag{4.93}$$

where $\dot{\varepsilon}^P_{y'y'}$ and $\dot{\varepsilon}^P_{z'z'}$ are the normal components of the plastic parts of the strain rate tensor along y' and z' directions respectively. In *planar* anisotropy, r_θ varies with θ. In many metals, r_θ decreases from $0°$ to $45°$ and then increases up to $90°$ and this pattern repeats in the other three quadrants. An *average strain rate ratio* $\bar{r}$ is defined as follows:

$$\bar{r} = \frac{1}{4}(r_0 + 2r_{45} + r_{90}), \tag{4.94}$$

where r_0 and r_{90} are the strain rate ratios measured by cutting the specimens along the rolling and transverse directions respectively and r_{45} along the direction at $45°$ to these axes. When the planar anisotropy is negligible, $\bar{r}$ is considered as a measure of the *normal* anisotropy.

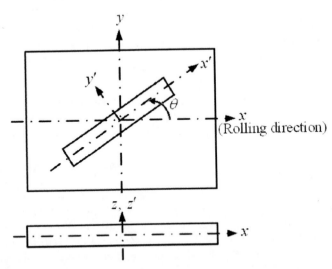

Figure 4.4. Sheet from which a tensile specimen is cut for finding strain rate ratio

Determination of the material constants appearing in various anisotropic yield criteria needs measurement of yield stresses along various directions using tension tests. In case of thin sheets, it is difficult to conduct a tension test along the thickness direction. Instead, a compression test along the thickness direction or *balanced bi-axial test* is conducted. In the balanced biaxial test, tensile stresses of equal magnitude (σ_b) are applied along the rolling and transverse directions on a square-shaped specimen. This state of stress is equal to the sum of tensile hydrostatic stress σ_b and the compressive stress of magnitude σ_b along the thickness direction. But the hydrostatic stress does not affect yielding. Thus, the balanced biaxial test is equivalent to the compression test in the thickness direction.

Various approaches have emerged for incorporating anisotropy in metal forming analysis. In the first approach, crystal plasticity models are used to simulate the anisotropy [15, 16]. The advantage of these models is that they can incorporate the crystallographic texture. However, use of such models in metal forming analysis is computationally time-consuming. In the second approach, strain rate potentials have been proposed to incorporate anisotropy in metal forming analysis [17, 18]. In the third approach, anisotropic yield criteria have been developed based on phenomenological observations. These criteria can describe complete anisotropy, unlike the crystal plasticity models which only account for the crystallographic texture. Further, these criteria can be easily adapted to different materials by changing the values of certain parameters. In this book, we shall discuss only the third approach.

Quite a few anisotropic yield functions have been proposed by Hill [19, 20], Hosford [21, 22] and others for orthotropic materials. These anisotropic yield functions do not completely represent the general state of anisotropy, even in plane stress conditions. Hill's 1948 criterion is simple to implement, but possesses a certain anomaly. Hill's 1979 criterion is free of this defect but does not contain the shear stress term. It is possible to develop an anisotropic yield criterion by applying an appropriate tensor transformation (corresponding to the symmetry group of materials) to an isotropic yield criterion. However, for a general transformation, it is usually difficult to satisfy the convexity condition. Therefore, mostly linear transformations have been used. Barlat and Lian [23] applied a linear transformation to modified Hosford [21] criterion to develop an anisotropic yield function for the plane stress conditions. Their yield surface matches with the yield surface calculated from the Bishop and Hill's [24] model for textured polycrystalline sheets. Later on Barlat *et al.* [25] and Karafillis and Boyce [26] extended this technique to the three-dimensional state of stress. These three-dimensional anisotropic yield functions were not able to capture the full anisotropy of aluminum sheets. Therefore, Barlat and his co-workers introduced two linear transformations first for the plane stress case [27] and then later for the three-dimensional case [28]. Further, Bron and Besson [29] incorporated two linear transformations in the Karafillis and Boyce criterion. In this book, we shall describe only the following criteria: (i) Hill's 1948 and 1979 criteria [19, 20], (ii) plane stress [23] and three-dimensional criteria [28] of Barlat and co-workers and (iii) a plane strain yield criterion based on a modification of Hill's 1979 yield criterion [20].

4.6.1 Hill's Anisotropic Yield Criteria

We choose the Cartesian coordinate system such that x-axis is along the rolling direction, y-axis along the transverse direction and z-axis along the normal direction. Thus, (x,y,z) are the principal axes of anisotropy. With respect to this coordinate system, Hill's 1948 criterion [19] can be stated as

$$F(\sigma_{ij}) \equiv \{f(\sigma_{yy} - \sigma_{zz})^2 + g(\sigma_{zz} - \sigma_{xx})^2 + h(\sigma_{xx} - \sigma_{yy})^2 \\ + 2l\sigma_{yz}^2 + 2m\sigma_{zx}^2 + 2n\sigma_{xy}^2\} - \sigma_0^2 = 0 \tag{4.95}$$

where σ_{ij} are the stress components with respect to the chosen coordinate system and σ_0 is a scaling factor. Note that this criterion contains six material constants: f, g, h, l, m and n. These parameters are determined by measuring the three uni-axial yield stresses and three shear yield stresses associated with the principal axes of anisotropy. If $f = g = h = (1/3)l = (1/3)m = (1/3)n = \sigma_0^2/(2\sigma_Y^2)$, then the above criterion reduces to the von Mises yield criterion.

Using the associated flow rule, one can find the strain rate ratio r_0 in terms of g and h. If we assume that there is no planar anisotropy, then r_0, r_{45} and r_{90} are all equal to the average strain rate ratio $\bar{r}$. This gives an expression for $\bar{r}$ in terms of g and h. By applying Equation 4.95 to the tension test along the rolling direction, we get an expression for σ_Y (uni-axial yield stress along the rolling direction) in terms of σ_0, g and h. Since, the biaxial test is equivalent to the compression test along the thickness direction, by applying Equation 4.95 to the compression test, we get an expression for σ_b (yield stress in balanced biaxial test) in terms of σ_0, f and g. Note that no planar anisotropy means f and g are equal. Eliminating f, g and h from the three expressions for $\bar{r}$, σ_Y and σ_b with the help of equality $f = g$, we get the following relation:

$$\frac{\sigma_b}{\sigma_Y} = \left(\frac{1 + \bar{r}}{2}\right)^{1/2}. \tag{4.96}$$

The detailed derivation of this equation (which was first presented by Hosford and Backofen) is given in Woodthorpe and Pearce [30]. According to Equation 4.96, σ_b should be less than σ_Y whenever $\bar{r}$ is less than unity. However, Woodthorp and Pearce [30] observed that σ_b was greater than σ_Y in some commercial aluminum alloy and steel sheets even when $\bar{r}$ was less than unity. This is the anomaly exhibited by Hill's 1948 criterion.

In 1979, Hill [20] proposed another anisotropic criterion for initial yielding which does not exhibit the above-mentioned anomaly. This criterion can be stated as

$$F(\sigma_{ij}) \equiv f|\sigma_2 - \sigma_3|^m + g|\sigma_3 - \sigma_1|^m + h|\sigma_1 - \sigma_2|^m + a|2\sigma_1 - \sigma_2 - \sigma_3|^m$$
$$+ b|2\sigma_2 - \sigma_3 - \sigma_1|^m + c|2\sigma_3 - \sigma_1 - \sigma_2|^m - \sigma_0^m = 0 \qquad , (4.97)$$

where σ_i are the principal stresses, σ_0 is a scaling factor and the material constants f, g, h, a, b and c are determined through experiments. The parameter m is assumed to be known depending on the crystal structure. Thus, $m=6$ for BCC (body-centered cubic) metals and $m=8$ for FCC (face-centered cubic) metals. Here, it is assumed that the principal directions of the stress tensor coincide with the principal axes of anisotropy. Thus, this criterion does not contain the shear stress terms. As a result, it is restricted to loading along the principal axes of orthotropy only. Further, this criterion does not always satisfy the convexity condition which is a requirement for every yield function.

4.6.2 Plane Stress Anisotropic Yield Criterion of Barlat and Lian

The starting point of the development of plane stress anisotropic yield criterion of Barlat and Lian [23] is the following *isotropic* yield criterion due to Hosford [21]:

$$f(\sigma_{ij}) \equiv |\sigma_2 - \sigma_3|^m + |\sigma_3 - \sigma_1|^m + |\sigma_1 - \sigma_2|^m - 2\sigma_Y^m = 0 , \qquad (4.98)$$

where σ_i are the principal stresses and σ_Y is the uni-axial yield stress. This criterion closely approximates the *isotropic* yield surface calculated from the Bishop and Hill's crystal plasticity model [24] for BCC metals for $m = 6$ and FCC metals for $m = 8$. (Further, this criterion reduces to von Mises criterion for $m = 2$.) For the *plane stress* condition, the above criterion can be expressed in terms of the stress components σ_{ij} with respect to a Cartesian coordinate system (x,y,z) as follows:

$$f(\sigma_{ij}) \equiv |K_1 + K_2|^m + |K_1 - K_2|^m + |2K_2|^m - 2\sigma_Y^m = 0 , \qquad (4.99)$$

where

$$K_1 = \frac{1}{2}(\sigma_{xx} + \sigma_{yy}),$$
$$K_2 = \left[\frac{1}{4}(\sigma_{xx} - \sigma_{yy})^2 + \sigma_{xy}^2\right]^{1/2} . \qquad (4.100)$$

are the stress invariants.

Barlat and Lian [23] extended the above criterion to the case of planar isotropy by multiplying the first three terms by the coefficients a, b and c which characterize the normal anisotropy:

$$f(\sigma_{ij}) \equiv a|K_1 + K_2|^m + b|K_1 - K_2|^m + c|2K_2|^m - 2\sigma_0^m = 0, \qquad (4.101)$$

where σ_0 is a scaling factor. However, it can be shown that only one of the three coefficients is independent. The requirement that the derivatives of f with σ_{ij} at $\sigma_{xx} = \sigma_{yy}$ and $\sigma_{xy} = 0$ in balanced bi-axial test be finite leads to the equality of the first two coefficients, i.e., $a = b$. Further, the choice of the in-plane yield stress as the scaling factor σ_0 gives the following relation: $a = 2 - c$. Therefore, we get

$$a = b = 2 - c. \qquad (4.102)$$

As stated earlier, this criterion does not describe the planar anisotropy of the plane stress case. To include the planar anisotropy, Barlat and Lian [23] made use of the observation that the tri-component plane stress yield surfaces of FCC sheet metals, for various textures, can be approximated by a dilatation of the normalised isotropic yield surfaces [31]. Based on this observation, they extended the above yield criterion to the case of planar anisotropy (of plane stress problems) by using the following *linear transformation* of the Cauchy stress components:

$$\sigma'_{xx} = \sigma_{xx}, \quad \sigma'_{yy} = h\sigma_{yy}, \quad \sigma'_{xy} = p\sigma_{xy}, \qquad (4.103)$$

where the constants h and p, like a, characterize the material anisotropy. Then, in the criterion (Equation 4.101), the expressions for K_1 and K_2 become

$$K_1 = \frac{1}{2}(\sigma'_{xx} + \sigma'_{yy}),$$

$$K_2 = \left[\frac{1}{4}(\sigma'_{xx} - \sigma'_{yy})^2 + (\sigma'_{xy})^2 \right]^{1/2}. \qquad (4.104)$$

The primed notation used here of stress components should not be confused with the components of the deviatoric part of the stress tensor. For $m = 2$, the yield criterion given by Equations 4.101–4.104 reduces to Hill's 1948 criterion.

In this criterion, there are three parameters which characterize the material anisotropy: a, h and p. These parameters are evaluated using the following expressions:

$$a = \dfrac{2\left(\dfrac{\sigma_0}{\tau_{s2}}\right)^m - 2\left(1+\dfrac{\sigma_0}{\sigma_{90}}\right)^m}{1+\left(\dfrac{\sigma_0}{\sigma_{90}}\right)^m - \left(1+\dfrac{\sigma_0}{\sigma_{90}}\right)^m},$$

$$h = \dfrac{\sigma_0}{\sigma_{90}},$$

$$p = \dfrac{\sigma_0}{\tau_{s1}}\left(\dfrac{2}{2a+2^m(2-a)}\right)^{1/m}. \tag{4.105}$$

Here, σ_{90} is the (uni-axial) yield stress in the transverse direction and τ_{s1} and τ_{s2} are the yield stresses corresponding to the following two situations respectively: (i) $\sigma_{xx} = \sigma_{yy} = 0, \sigma_{xy} = \tau_{s1}$ and (ii) $\sigma_{yy} = -\sigma_{xx} = \tau_{s2}, \sigma_{xy} = 0$. The scaling factor is chosen to be the (uni-axial) yield stress in the rolling direction. The material parameters a, h and p can also be determined from the measurements of the strain rate ratio r. The relations between (a, h) and (r_0, r_{90}) are given by

$$a = 2 - 2\left(\dfrac{r_0}{1+r_0}\dfrac{r_{90}}{1+r_{90}}\right)^{1/2}, \quad h = \left(\dfrac{r_0}{1+r_0}\dfrac{1+r_{90}}{r_{90}}\right)^{1/2}. \tag{4.106}$$

However, there is no analytical expression for p in terms of r. Therefore, the variation of p with r needs to be determined either numerically or graphically using the variations of a and h with r. Since the graphical (or numerical) method gives only an approximate variation of p with r, the two methods of calculating the parameters a, h and p do not give the identical set of values. The parameter m is chosen so as to match the predictions of this criterion with experimental results.

Barlat and Lian [23] have also shown that the yield function f given by Equations 4.101–4.104 is convex. Further, the yield locus predicted by the above criterion, for an FCC material containing 50% of grains having Gaussian distribution around the $\{1\bar{1}0\} <112>$ ideal orientation (brass texture) and 50% of randomly distributed grains, matches well with the one obtained from Bishop and Hill's model [24].

4.6.3 A Three-Dimensional Anisotropic Yield Criterion of Barlat and Co-workers

As stated earlier, Barlat et al. [25] extended the method of linear transformation to obtain a three-dimensional anisotropic yield criterion. However, this yield criterion was not able to capture the full anisotropy of aluminum sheets. Therefore, later, Barlat et al. [28] used two linear transformations for this purpose. In this subsection, first, we introduce *briefly* the three-dimensional anisotropic yield criterion with *one* linear transformation [25] and, later, we discuss in *detail* the criterion with *two* linear transformations [28].

The development of the plane stress anisotropic yield criterion (Equations 4.101–4.104) starts with the *isotropic* yield criterion of Hosford [21]. The anisotropy is introduced, first, through the three coefficients *a*, *b* and *c* (Equation 4.101) and later through the linear transformation (Equation 4.103) of the Cauchy stress components. In the three-dimensional case [25] also, the starting point is Hosford's [21] *isotropic* yield criterion but expressed in terms of the principal values σ_i' of the *stress deviator* σ' (*i.e.*, the deviatoric part of the Cauchy stress tensor):

$$f(\sigma_{ij}) \equiv \left|\sigma_2' - \sigma_3'\right|^m + \left|\sigma_3' - \sigma_1'\right|^m + \left|\sigma_1' - \sigma_2'\right|^m - 2\sigma_Y^m = 0. \qquad (4.107)$$

Barlat *et al.* [25] introduced anisotropy in the above criterion by expressing it in terms of the principal values S_i' of the *modified stress deviator* S' :

$$f(\sigma_{ij}) \equiv \left|S_2' - S_3'\right|^m + \left|S_3' - S_1'\right|^m + \left|S_1' - S_2'\right|^m - 2\sigma_0^m = 0, \qquad (4.108)$$

where S' is obtained from the *stress deviator* σ' by the following linear transformation:

$$S' = C\sigma'. \qquad (4.109)$$

Here, as before, σ_0 is a scaling factor. Further, the parameter *m* is chosen so as to match the predictions of this criterion with experimental results. By using the transformation

$$\sigma' = T\sigma, \qquad (4.110)$$

the tensor S' can be expressed in terms of the Cauchy stress tensor σ by the following relation:

$$S' = L\sigma, \quad L = CT. \qquad (4.111)$$

The tensors C, T and L are all fourth order tensors. Since, the tensors S', σ' and σ are all symmetric tensors, the tensors C, T and L can have at the most 36 independent components.

We express the components of the tensors S', σ' and σ with respect to a Cartesian coordinate system (x,y,z) as the following one-dimensional arrays:

$$\{S'\} = \begin{Bmatrix} S'_{xx} \\ S'_{yy} \\ S'_{zz} \\ S'_{xy} \\ S'_{yz} \\ S'_{zx} \end{Bmatrix}, \quad \{\sigma'\} = \begin{Bmatrix} \sigma'_{xx} \\ \sigma'_{yy} \\ \sigma'_{zz} \\ \sigma'_{xy} \\ \sigma'_{yz} \\ \sigma'_{zx} \end{Bmatrix}, \quad \{\sigma\} = \begin{Bmatrix} \sigma_{xx} \\ \sigma_{yy} \\ \sigma_{zz} \\ \sigma_{xy} \\ \sigma_{yz} \\ \sigma_{zx} \end{Bmatrix},$$
(4.112)

Then the array form of the linear transformation (Equations 4.109–4.111) can be expressed as

$$\{S'\} = [C]\{\sigma'\} = [C][T]\{\sigma\} = [L]\{\sigma\},$$
(4.113)

where $[C]$, $[T]$ and $[L]$ denote respectively the matrices of C, T and L containing the 36 independent components. The matrix $[T]$ is given by

$$[T] = \frac{1}{3}\begin{bmatrix} 2 & -1 & -1 & 0 & 0 & 0 \\ -1 & 2 & -1 & 0 & 0 & 0 \\ -1 & -1 & 2 & 0 & 0 & 0 \\ 0 & 0 & 0 & 3 & 0 & 0 \\ 0 & 0 & 0 & 0 & 3 & 0 \\ 0 & 0 & 0 & 0 & 0 & 3 \end{bmatrix}.$$
(4.114)

Barlat *et al.* [25] assumed that only the following nine components of the matrix $[C]$ are non-zero:

$$[C] = \begin{bmatrix} 0 & -c_{12} & -c_{13} & 0 & 0 & 0 \\ -c_{21} & 0 & -c_{23} & 0 & 0 & 0 \\ -c_{31} & -c_{32} & 0 & 0 & 0 & 0 \\ 0 & 0 & 0 & c_{44} & 0 & 0 \\ 0 & 0 & 0 & 0 & c_{55} & 0 \\ 0 & 0 & 0 & 0 & 0 & c_{66} \end{bmatrix},$$
(4.115)

(The corresponding $[L]$ matrix can be obtained by multiplying the matrices $[C]$ and $[T]$). Further, they assumed that the matrix $[C]$ is symmetric:

$$c_{12} = c_{21}, \quad c_{23} = c_{32}, \quad c_{31} = c_{13}.$$
(4.116)

Thus, in this anisotropic yield criterion, besides the parameter m, there are six more parameters $(c_{12}, c_{23}, c_{31}, c_{44}, c_{55}, c_{66})$ which characterize the material anisotropy. This anisotropic yield criterion has been labeled as Yld91 by Barlat *et al.* [25].

When all the anisotropic parameters are equal to 1, and *m* is equal to 2, Yld91 reduces to the von Mises yield criterion. However, for the plane stress problems, this criterion does not reduce to the 3-parameter anisotropic criterion of Barlat and Lian (Equations 4.101–4.104). Barlat *et al.* [25] determined the six anisotropic parameters of Yld91 for the aluminum alloys 2008-T4 and 2024-T3 using the three uni-axial yield stresses and three shear yield stresses associated with the three axes of anisotropy. As stated earlier, the parameter *m* is determined so as to match the predictions of this criterion with experimental results. The yield locus of 2008-T4 material predicted by this anisotropic yield criterion (with *m* = 11) matches well with the one obtained from Bishop and Hill's model [24]. Further, the predictions of the yield stress variation are also in good agreement with the experimental results for both the materials. But this criterion underpredicts the values of the strain rate ratio *r*.

To represent the anisotropy of aluminum sheets to a better degree of accuracy, Barlat *et al.* [28] used *two* linear transformations by introducing *two modified stress deviators*:

$$S' = C'\sigma', \quad S'' = C''\sigma'. \tag{4.117}$$

The yield criterion is expressed as

$$f(\sigma_{ij}) \equiv \left|S_1' - S_1''\right|^m + \left|S_1' - S_2''\right|^m + \left|S_1' - S_3''\right|^m + \left|S_2' - S_1''\right|^m + \left|S_2' - S_2''\right|^m + \left|S_2' - S_3''\right|^m$$
$$+ \left|S_3' - S_1''\right|^m + \left|S_3' - S_2''\right|^m + \left|S_3' - S_3''\right|^m - 4\sigma_0^m = 0 \tag{4.118}$$

where S_i' and S_i'' are the principal values of the *modified stress deviators* S' and S''. As before, σ_0 is a scaling factor, but *m* is taken to be 6 for BCC metals and 8 for FCC metals. Here also, out of the 36 independent components of C' and C'', only the following 9 components each are assumed to be non-zero.

$$[C'] = \begin{bmatrix} 0 & -c_{12}' & -c_{13}' & 0 & 0 & 0 \\ -c_{21}' & 0 & -c_{23}' & 0 & 0 & 0 \\ -c_{31}' & -c_{32}' & 0 & 0 & 0 & 0 \\ 0 & 0 & 0 & c_{44}' & 0 & 0 \\ 0 & 0 & 0 & 0 & c_{55}' & 0 \\ 0 & 0 & 0 & 0 & 0 & c_{66}' \end{bmatrix}, \tag{4.119}$$

$$[C''] = \begin{bmatrix} 0 & -c''_{12} & -c''_{13} & 0 & 0 & 0 \\ -c''_{21} & 0 & -c''_{23} & 0 & 0 & 0 \\ -c''_{31} & -c''_{32} & 0 & 0 & 0 & 0 \\ 0 & 0 & 0 & c''_{44} & 0 & 0 \\ 0 & 0 & 0 & 0 & c''_{55} & 0 \\ 0 & 0 & 0 & 0 & 0 & c''_{66} \end{bmatrix}. \tag{4.120}$$

(The corresponding $[L']$ and $[L'']$ matrices can be obtained by multiplying the matrices $[C']$ and $[C'']$ with $[T]$). However, the matrices $[C']$ and $[C'']$ are not assumed to be symmetric. Thus, in this anisotropic yield criterion, there are 18 parameters which characterize the material anisotropy. This anisotropic yield criterion has been labeled as Yld2004-18p by Barlat et al. [28]. When the matrices $[C']$ and $[C'']$ are identical and symmetric, this criterion reduces to Yld91. Further, when all the anisotropic parameters are equal to 1, and m is equal to 2, Yld2004-18p reduces to the von Mises yield criterion. Barlat et al. [28] have also shown that the yield function given by Equation 4.118 is convex.

Barlat et al. [28] determined the 18 anisotropic parameters of Yld2004-18p for the aluminum alloys 2090-T3 and 6111-T4 (mildly anisotropic) by minimising the following weighted error function:

$$E(c'_{ij}, c''_{ij}) = \sum_p w_p \left(\frac{\sigma^{pr}_p}{\sigma^{ex}_p} - 1 \right)^2 + \sum_q w_q \left(\frac{\sigma^{qr}_q}{\sigma^{ex}_q} - 1 \right)^2, \tag{4.121}$$

where w_p and w_q are the weight functions. The indices p and q denote respectively the number of experimental uni-axial, bi-axial or shear yield stresses and experimental strain rate ratios (r). Further, the superscripts pr and ex indicate whether it is a predicted value or an experimental value. The experimental data in the error function consists of the seven (uni-axial) yield stresses and seven strain rate ratios in the plane of sheet at every 15° angular increment: 0, 15, 30, 45, 60, 75 and 90. Further, it includes the yield stress and the strain rate ratio in the biaxial test. It also includes the following four out-of-plane properties : shear yield stresses in y-z and z-x planes and the (uni-axial) yield stresses along 45° direction in y-z and z-x planes. Polycrystalline simulations were performed to obtain the out-of-plane material properties.

Variations of the uni-axial yield stress and the strain rate ratio (with the angular position from the rolling direction) predicted by the Yld2004-18p anisotropic yield criterion match well with the experimental results for both the materials. (However, note that, the same experimental data was used in minimising the error function.) Further, the cup height profile (i.e., the earing profile) for 2090-T3 material with six ears predicted by this criterion is in good agreement with the experimental cup height profile [32].

Before ending this section, we shall briefly describe the other three-dimensional anisotropic yield criteria which are based on the method of linear transformation but are different from the criteria of Barlat *et al.* [25, 28]. Karafillis and Boyce [26] also used only a single linear transformation but used two functions (rather, a linear combination of them) of the principal values of the modified stress deviator. Bron and Benson [29] extended this criterion by using two linear transformations. Their criterion can be expressed as

$$f(\sigma_{ij}) \equiv [\alpha(\psi^1)^{m/b^1} + (1-\alpha)(\psi^2)^{m/b^2}]^{1/m} - \sigma_0 = 0,$$ (4.122)

where

$$\psi^1 = \frac{1}{2}\left(\left|S_2' - S_3'\right|^{b^1} + \left|S_3' - S_1'\right|^{b^1} + \left|S_1' - S_2'\right|^{b^1} \right),$$ (4.123)

and

$$\psi^2 = \frac{3^{b^2}}{2^{b^2}+2}\left(\left|S_1''\right|^{b^2} + \left|S_2''\right|^{b^2} + \left|S_3''\right|^{b^2} \right).$$ (4.124)

Here, σ_0 is a scaling factor. Further, S_i' and S_i'' are the principal values of the *modified stress deviators* S' and S'' defined by Equation 4.117. Like Barlat *et al.* [28], Bron and Benson [29] also assumed that the matrices $[C']$ and $[C'']$ have only nine non-zero components each (Equations 4.119 and 4.120), But, they further assumed that these matrices are symmetric and therefore have only six independent components. Thus, the anisotropic yield criterion of Bron and Benson [29] have 16 material parameters: m, b^1, b^2, α, 6 components of $[C']$ and 6 components of $[C'']$. The first four parameters (m, b^1, b^2 and α) do not influence the anisotropy but only cause the change of shape of the yield surface. Only the remaining 12 parameters represent the material anisotropy. When the matrices $[C']$ and $[C'']$ are identical and $m = b^1 = b^2$, this criterion reduces to that of Karafillis and Boyce [26]. Additionally, if $\alpha = 1$, it reduces to the Yld91 criterion of Barlat *et al.* [25]. Further, when all the non-zero elements of $[C']$ and $[C'']$ are unity, $\alpha = 1$ and $m = b^1 = 2$, it reduces to the Mises criterion. Bron and Benson [29] determined the 16 material parameters, for various aluminum alloys by minimising the error between the simulated and experimental results for the variation of load with the opening displacement for smooth and notched specimens in tension tests. They found that the criterion is accurate in describing the anisotropy of various aluminum alloys. They have also shown that this criterion satisfies the convexity condition.

Hu [33] also used the method of linear transformation to propose the following anisotropic yield criterion:

$$f(\sigma_{ij}) \equiv (J_2)^2 + (J_2^{(1)})^2 + (J_2^{(2)})^2 - \frac{1}{c^2} = 0, \tag{4.125}$$

where J_2, $J_2^{(1)}$ and $J_2^{(2)}$ are the second invariants of the deviatoric parts of the tensors σ, $\sigma^{(1)}$ and $\sigma^{(2)}$ and C is a constant. The tensors $\sigma^{(1)}$ and $\sigma^{(2)}$ are obtained from σ by the following linear transformations:

$$\frac{\sigma_{xx}^{(1)} - \sigma_{yy}^{(1)}}{\sigma_{xx} - \sigma_{yy}} = a_{11}, \quad \frac{\sigma_{yy}^{(1)} - \sigma_{zz}^{(1)}}{\sigma_{yy} - \sigma_{zz}} = a_{22}, \quad \frac{\sigma_{zz}^{(1)} - \sigma_{xx}^{(1)}}{\sigma_{zz} - \sigma_{xx}} = a_{33},$$

$$\frac{\sigma_{xy}^{(1)}}{\sigma_{xy}} = a_{12}, \quad \frac{\sigma_{yz}^{(1)}}{\sigma_{yz}} = a_{23}, \quad \frac{\sigma_{zx}^{(1)}}{\sigma_{zx}} = a_{31}. \tag{4.126}$$

$$\frac{\sigma_{xx}^{(2)} - \sigma_{yy}^{(2)}}{\sigma_{xx} - \sigma_{yy}} = b_{11}, \quad \frac{\sigma_{yy}^{(2)} - \sigma_{zz}^{(2)}}{\sigma_{yy} - \sigma_{zz}} = b_{22}, \quad \frac{\sigma_{zz}^{(2)} - \sigma_{xx}^{(2)}}{\sigma_{zz} - \sigma_{xx}} = b_{33},$$

$$\frac{\sigma_{xy}^{(2)}}{\sigma_{xy}} = b_{12}, \quad \frac{\sigma_{yz}^{(2)}}{\sigma_{yz}} = b_{23}, \quad \frac{\sigma_{zx}^{(2)}}{\sigma_{zx}} = b_{31}. \tag{4.127}$$

where a_{ij} and b_{ij} are the anisotropic coefficients to be determined from experiments.

If we assume that the two principal axes are in the two diagonal directions to the rolling axis, then the expansion of Equation 4.125 leads to the following expression:

$$\begin{aligned}
f(\sigma_{ij}) &= X_1(\sigma_{xx} - \sigma_{zz})^4 + X_2(\sigma_{xx} - \sigma_{zz})^3(\sigma_{yy} - \sigma_{zz}) + X_3(\sigma_{xx} - \sigma_{zz})^2(\sigma_{yy} - \sigma_{zz})^2 \\
&\quad + X_4(\sigma_{xx} - \sigma_{zz})(\sigma_{yy} - \sigma_{zz})^3 + X_5(\sigma_{yy} - \sigma_{zz})^4 + X_6(\sigma_{xy}^2 + \sigma_{yz}^2 + \sigma_{zx}^2) \\
&\quad \times [(\sigma_{xx} - \sigma_{zz})^2 + (\sigma_{yy} - \sigma_{zz})^2 - (\sigma_{xx} - \sigma_{zz})(\sigma_{yy} - \sigma_{zz})] \\
&\quad + X_7(\sigma_{xy}^2 + \sigma_{yz}^2 + \sigma_{zx}^2) - 1 = 0,
\end{aligned} \tag{4.128}$$

where the seven parameters X_i (depending on a_{ij} and b_{ij}) can be expressed in terms of the following seven measurable quantities: uni-axial yield stresses along the rolling direction (σ_0), at 45° to the rolling direction (σ_{45}) and at 90° to the rolling direction (σ_{90}), the yield stress in balanced bi-axial test (σ_b), strain rate ratios in tension tests along the rolling direction (R_0), at $45°$ to the rolling direction (R_{45}) and at $90°$ to the rolling direction (R_{90}):

$$X_1 = \frac{1}{\sigma_0^4}, \quad X_2 = -\frac{4R_0}{(1+R_0)\sigma_0^4},$$

$$X_3 = \frac{1}{\sigma_b^4} - \frac{1}{\sigma_0^4} + \frac{1}{\sigma_{90}^4} + \frac{4R_0}{(1+R_0)\sigma_0^4} + \frac{4R_{90}}{(1+R_{90})\sigma_{90}^4}, \quad X_4 = -\frac{4R_{90}}{(1+R_{90})\sigma_{90}^4},$$

$$X_5 = \frac{1}{\sigma_{90}^4}, \quad X_6 = \frac{16}{(1+R_{45})\sigma_{45}^4} - \frac{2}{\sigma_b^4}, \quad X_7 = \frac{1}{\sigma_b^4} + \frac{16R_{45}}{(1+R_{45})\sigma_{45}^4}.$$

$$(4.129)$$

Hu [33] has shown that the above anisotropic yield criterion satisfies the convexity requirement.

Hu [33] validated this criterion by comparing its predictions about the variation of the yield stress and strain rate ratio (with the angular position from the rolling direction) with experimental results for Y350 MPa high strength steel with two different coatings: (i) cold-rolled and (ii) hot dip galvanized. He also compared the yield surfaces predicted by his criterion with experimentally predicted yield surfaces for high strength steel and aluminum alloys. The agreement in all cases is reported to be good.

4.6.4 A Plane Strain Anisotropic Yield Criterion

When the width of the sheet is sufficiently large compared to the length of the arc of contact, the rolling process can be modeled as a plane strain problem in the x-z plane where x is the rolling direction and z is the thickness direction. In this section, we describe an anisotropic yield criterion for the plane strain rolling problem proposed by Dixit and Dixit [34]. The description here differs from the one in [34] in the following respect: whereas the y axis is taken as the thickness direction in [34], here it is assumed that z is the thickness direction. For plane strain problems in the x-z plane, the anisotropy is restricted to the x-z plane only and therefore, there is planar *isotropy* in the plane of the sheet, i.e., in the x-y plane. For this case, the coefficients f, g, a and b in the Hill's 1979 anisotropic criterion (Equation 4.97) satisfy the relation $f = g$ and $a = b$. We further assume that $f = 0$ and $a = 0$. Then, the criterion reduces to

$$f(\sigma_{ij}) \equiv c|\sigma_1 + \sigma_2 - 2\sigma_3|^m + h|\sigma_1 - \sigma_2|^m - \sigma_0^m = 0. \quad (4.130)$$

Note that in Equation 4.97, it is assumed that the loading is along the principal axes of the anisotropy and therefore, the principal directions of the stress tensor coincide with the principal axes of anisotropy (x,y,z). Thus, the principal stresses

$(\sigma_1, \sigma_2, \sigma_3)$ are equal to $(\sigma_{xx}, \sigma_{yy}, \sigma_{zz})$. Rewriting Equation 4.130 in terms of $(\sigma_{xx}, \sigma_{yy}, \sigma_{zz})$, we get

$$f(\sigma_{ij}) \equiv c \left| \sigma_{xx} + \sigma_{yy} - 2\sigma_{zz} \right|^m + h \left| \sigma_{xx} - \sigma_{yy} \right|^m - \sigma_0^m = 0 . \tag{4.131}$$

When the loading is not along the principal axes of anisotropy, one needs to include the shear stress terms in the above criterion. For the plane strain problems in the x-z plane, the only non-zero shear stress component is σ_{xz}. Thus, to get an anisotropic yield criterion for plane strain problems in the x-z plane, we add an extra term containing σ_{xz} to Equation 4.131. Thus, we get

$$f(\sigma_{ij}) \equiv c \left| \sigma_{xx} + \sigma_{yy} - 2\sigma_{zz} \right|^m + h \left| \sigma_{xx} - \sigma_{yy} \right|^m + N(\left| \sigma_{xz} \right|^m + \left| \sigma_{zx} \right|^m) - \sigma_0^m = 0 . \tag{4.132}$$

Here, the scaling factor σ_0 is taken to be equal to the uni-axial yield stress in the rolling direction and the coefficient N accounts for the presence of in-plane shear stress component σ_{xz}. Thus, there are four parameters which characterize the material anisotropy: c, h, N and m.

To determine the material parameters c, h, N and m in terms of measurable quantities like the uni-axial yield stresses and the strain rate ratios, we proceed as follows. We assume that $\sigma_{xx} \geq \sigma_{yy}$ and $\sigma_{xx} + \sigma_{yy} > 2\sigma_{zz}$. By applying Equation 4.132 to the tension test along the rolling direction $(i.e., \sigma_{xx} = \sigma_0, \sigma_{yy} = \sigma_{zz} = \sigma_{xz} = 0)$, we get

$$c + h = 1 . \tag{4.133}$$

Further, using the associated flow rule for $\dot{\varepsilon}_{yy}^P$ and $\dot{\varepsilon}_{zz}^P$ in the tension test, we find the strain rate ratio r_0 in terms of c and h. Since we have assumed that there is *planar isotropy*, the ratios r_0, r_{45} and r_{90} are all equal to the average strain rate ratio $\bar{r}$. Thus, we get

$$\bar{r} = \frac{1}{2} \left(\frac{h}{c} - 1 \right) . \tag{4.134}$$

From Equations 4.133 and 4.134, we get the following expressions for the material parameters c and h in terms of $\bar{r}$:

$$c = \frac{1}{2(\bar{r}+1)},$$

$$h = \frac{2\bar{r}+1}{2(\bar{r}+1)}.$$

(4.135)

Thus, the material parameters c and h are positive. Now, consider a coordinate system (x', y', z') such that $y' = y$ and the axes (x', z') are in the x-z plane. Further, the x' axis makes a counter-clockwise angle of 45°with the rolling direction. In the uni-axial state of stress along x' direction, the stress components with respect to (x', y', z') system at yielding are given by

$$\sigma_{x'x'} = \sigma_{45}, \quad \sigma_{z'z'} = 0, \quad \sigma_{x'z'} = 0, \quad \sigma_{y'y'} = 0, \qquad (4.136)$$

where σ_{45} is the uni-axial yield stress along the 45° direction to the rolling direction. Transforming these components to the (x,y,z) system, we get

$$\sigma_{xx} = \frac{1}{2}(\sigma_{x'x'} + \sigma_{z'z'} - 2\sigma_{x'z'}),$$

$$\sigma_{zz} = \frac{1}{2}(\sigma_{x'x'} + \sigma_{z'z'} + 2\sigma_{x'z'}),$$

$$\sigma_{xz} = \frac{1}{2}(\sigma_{x'x'} - \sigma_{z'z'}),$$

$$\sigma_{yy} = \sigma_{y'y'}.$$

(4.137)

Using Equation 4.136, the stress components with respect to the (x,y,z) system become

$$\sigma_{xx} = \frac{1}{2}\sigma_{45}, \quad \sigma_{zz} = \frac{1}{2}\sigma_{45}, \quad \sigma_{xz} = \frac{1}{2}\sigma_{45}, \quad \sigma_{yy} = 0. \qquad (4.138)$$

Substituting these values in the anisotropic yield criterion of Equation 4.132 and using Equation 4.133, we get the following expression for the material parameter N:

$$N = \frac{\sigma_0^m - (\sigma_{45}/2)^m}{2(\sigma_{45}/2)^m}. \qquad (4.139)$$

Finally, the parameter m is evaluated from the bi-axial test using the following expression:

$$\frac{\sigma_b}{\sigma_0} = \left(\frac{1+\bar{r}}{2^{m-1}}\right)^{1/m}.$$

(4.140)

Note that for $m = 2$, this expression reduces to Equation 4.96.

Dixit and Dixit [34] have also shown that this criterion satisfies the convexity condition if $m \geq 1$ and $N \geq 0$. They have employed this criterion in the analysis of anisotropic plane strain rolling [34].

4.7 Elastic-Plastic Incremental Stress-Strain and Stress-Strain Rate Relations for Anisotropic Materials

In this section, we discuss the procedure of obtaining the elastic-plastic incremental stress-strain and stress-strain rate relations for anisotropic materials. Various criteria for initial yielding of these materials have been discussed in the last section. However, not much is known about their behavior in subsequent yielding. Therefore, in the absence of enough data to model their hardening behavior, we assume that their hardening is *isotropic*. First, we develop the incremental stress-strain relation for the material which obeys the three-dimensional initial anisotropic yield criterion of Barlat *et al.* [28] (Section 4.6.3) and which hardens according to the strain-hardening hypothesis. Later, we develop the stress-strain rate relation for the material which obeys the plane strain anisotropic yield criterion of Section 4.6.4.

4.7.1 Elastic-Plastic Incremental Stress-Strain Relation for Anisotropic Materials

For this material, using Equation 4.118, the criterion for subsequent yielding can be expressed as

$$f(\sigma_{ij}, \varepsilon_{eq}^p) \equiv \sigma_{eq}(\sigma_{ij}) - H(\varepsilon_{eq}^p) = 0,$$

(4.141)

where the equivalent stress σ_{eq} is defined by

$$\sigma_{eq} = \left(\frac{\phi}{4}\right)^{1/m},$$

$$\phi = \left|S_1' - S_1''\right|^m + \left|S_1' - S_2''\right|^m + \left|S_1' - S_3''\right|^m + \left|S_2' - S_1''\right|^m + \left|S_2' - S_2''\right|^m + \left|S_2' - S_3''\right|^m$$
$$+ \left|S_3' - S_1''\right|^m + \left|S_3' - S_2''\right|^m + \left|S_3' - S_3''\right|^m.$$

(4.142)

Further, H is the hardening function (Equation 3.103) to be determined along the rolling direction and ε_{eq}^p is the equivalent plastic strain which is obtained from the integration of $d\varepsilon_{eq}^p$ along the deformation path:

$$\varepsilon_{eq}^p = \int d\varepsilon_{eq}^p . \tag{4.143}$$

The equivalent plastic strain increment $d\varepsilon_{eq}^p$ is defined in such a way that σ_{eq} and $d\varepsilon_{eq}^p$ are work-conjugate to each other. Thus,

$$d\varepsilon_{eq}^p = \frac{\sigma_{ij} d\varepsilon_{ij}^p}{\sigma_{eq}} . \tag{4.144}$$

Applying the associated flow rule to the yield function of Equation 4.141, we get the following expression for the plastic part of the incremental linear strain tensor:

$$d\varepsilon_{ij}^p = d\lambda \frac{\partial f}{\partial \sigma_{ij}} = d\lambda \frac{\partial \sigma_{eq}}{\partial \sigma_{ij}} , \tag{4.145}$$

where $d\lambda$ is a positive scalar.

To determine $d\lambda$, we substitute Equation 4.145 for $d\varepsilon_{ij}^p$ into Equation 4.144 to obtain

$$d\varepsilon_{eq}^p = \frac{\sigma_{ij} d\lambda \dfrac{\partial \sigma_{eq}}{\partial \sigma_{ij}}}{\sigma_{eq}} . \tag{4.146}$$

Since Equation 4.142 for σ_{eq} is a first degree homogeneous function of σ_{ij}, we get

$$\frac{\partial \sigma_{eq}}{\partial \sigma_{ij}} \sigma_{ij} = \sigma_{eq} . \tag{4.147}$$

Then, Equations 4.146 and 4.147 imply

$$d\lambda = d\varepsilon_{eq}^p . \tag{4.148}$$

From the hardening relation (Equation 3.103), we obtain

$$d\varepsilon_{eq}^p = \frac{d\sigma_{eq}}{H'},$$ (4.149)

where H' is the slope of the hardening curve. Eliminating $d\varepsilon_{eq}^p$ from Equations 4.148 and 4.149, we get

$$d\lambda = \frac{d\sigma_{eq}}{H'}.$$ (4.150)

After substituting the above equation for $d\lambda$, the associated flow rule (Equation 4.145) becomes

$$d\varepsilon_{ij}^p = \frac{d\sigma_{eq}}{H'} \frac{\partial\sigma_{eq}}{\partial\sigma_{ij}}.$$ (4.151)

To obtain the incremental stress-strain relation from the above equation, we express $d\sigma_{eq}$ in terms of $d\sigma_{kl}$ as follows:

$$d\varepsilon_{ij}^p = \frac{1}{H'} \frac{\partial\sigma_{eq}}{\partial\sigma_{ij}} \frac{\partial\sigma_{eq}}{\partial\sigma_{kl}} d\sigma_{kl}.$$ (4.152)

Now, define a fourth order tensor:

$$M_{ijkl} = \frac{1}{H'} A_{ij} A_{kl},$$ (4.153)

where

$$A_{ij} = \frac{\partial\sigma_{eq}}{\partial\sigma_{ij}}.$$ (4.154)

Then, the incremental plastic strain-stress relationship (Equation 4.152) becomes

$$d\varepsilon_{ij}^p = M_{ijkl} d\sigma_{kl}.$$ (4.155)

To determine the fourth order constitutive tensor M_{ijkl}, we need to obtain the expression for A_{ij}, i.e., the derivative of σ_{eq} with respect to σ_{ij}. For this purpose, first we define the following invariants of the modified stress deviators S' and S'':

$$H_1' = \frac{1}{3}trS', \quad H_2' = \frac{1}{3}\left\{\frac{1}{2}[tr(S'^2)-(trS')^2]\right\}, \quad H_3' = \frac{1}{2}\det S',$$

$$H_1'' = \frac{1}{3}trS'', \quad H_2'' = \frac{1}{3}\left\{\frac{1}{2}[tr(S''^2)-(trS'')^2]\right\}, \quad H_3'' = \frac{1}{2}\det S'',$$

$$(4.156)$$

Then, the principal values of S' and S'' satisfy the following characteristic equations:

$$S_i'^3 - 3H_1'S_i'^2 - 3H_2'S_i' - 2H_3' = 0, \qquad i = 1,2,3$$
$$S_i''^3 - 3H_1''S_i''^2 - 3H_2''S_i'' - 2H_3'' = 0, \qquad i = 1,2,3$$

$$(4.157)$$

As per the Cardan's solution of a cubic equation, S_i' and S_i'' are given by

$$S_1' = H_1' + 2(H_1'^2 + H_2')^{1/2}\cos\left(\frac{\theta'}{3}\right),$$

$$S_2' = H_1' + 2(H_1'^2 + H_2')^{1/2}\cos\left(\frac{\theta'+4\pi}{3}\right),$$

$$S_3' = H_1' + 2(H_1'^2 + H_2')^{1/2}\cos\left(\frac{\theta'+2\pi}{3}\right),$$

$$S_1'' = H_1'' + 2(H_1''^2 + H_2'')^{1/2}\cos\left(\frac{\theta''}{3}\right),$$

$$S_2'' = H_1'' + 2(H_1''^2 + H_2'')^{1/2}\cos\left(\frac{\theta''+4\pi}{3}\right),$$

$$S_3'' = H_1'' + 2(H_1''^2 + H_2'')^{1/2}\cos\left(\frac{\theta''+2\pi}{3}\right),$$

$$(4.158)$$

where

$$\theta' = \cos^{-1}\left(\frac{q'}{p'^{3/2}}\right), \quad p' = H_1'^2 + H_2', \quad q' = \frac{1}{2}(2H_1'^3 + 3H_1'H_2' + 2H_3'),$$

$$\theta'' = \cos^{-1}\left(\frac{q''}{p''^{3/2}}\right), \quad p'' = H_1''^2 + H_2'', \quad q'' = \frac{1}{2}(2H_1''^3 + 3H_1''H_2'' + 2H_3'').$$

$$(4.159)$$

Next, the chain rule is used to obtain the expression for A_{ij}, i.e., the derivative of σ_{eq} with respect to σ_{ij}. Thus, we get

$$A_{ij} \equiv \frac{\partial \sigma_{eq}}{\partial \sigma_{ij}} = \frac{\partial \sigma_{eq}}{\partial \phi} \left[\frac{\partial \phi}{\partial S'_p} \frac{\partial S'_p}{\partial H'_q} \frac{\partial H'_q}{\partial S'_{rs}} \frac{\partial S'_{rs}}{\partial \sigma_{ij}} + \frac{\partial \phi}{\partial S''_p} \frac{\partial S''_p}{\partial H''_q} \frac{\partial H''_q}{\partial S''_{rs}} \frac{\partial S''_{rs}}{\partial \sigma_{ij}} \right]. \quad (4.160)$$

In this expression, the derivatives $\partial \sigma_{eq} / \partial \phi$, $\partial \phi / \partial S'_p$ and $\partial \phi / \partial S''_p$ are obtained from the definitions of σ_{eq} and ϕ (Equation 4.142), the derivatives $\partial S'_p / \partial H'_q$ and $\partial S''_p / \partial H''_q$ are evaluated from Equation 4.158, the derivatives $\partial H'_q / \partial S'_{rs}$ and $\partial H''_q / \partial S''_{rs}$ are calculated from the definitions of the invariants (Equation 4.156), whereas the derivatives $\partial S'_{rs} / \partial \sigma_{ij}$ and $\partial S''_{rs} / \partial \sigma_{ij}$ are determined from the linear transformations at Equations 4.117 and 4.110.

In Section 3.6, while deriving the elastic-plastic incremental stress-strain relationship for isotropic materials, it was assumed that the elastic and plastic parts of the incremental linear strain tensor were additive. We make the same assumption while deriving the elastic-plastic incremental constitutive equation for anisotropic materials as well. Then we combine the incremental plastic strain-stress relationship (Equation 4.155) with the incremental elastic strain-stress relationship (Equation 3.150). Finally, the combined relationship is inverted to express the incremental stress in terms of the incremental strain.

4.7.2 Elastic-Plastic Stress-Strain Rate Relation for Anisotropic Materials

Now we develop the stress-strain rate relation for the material which obeys the plane strain anisotropic criterion (of Section 4.6.4) for initial yielding and which hardens according to the strain-hardening hypothesis. For this material, using Equation 4.132, the criterion for subsequent yielding can be expressed as

$$f(\sigma_{ij}, \varepsilon^p_{eq}) \equiv c|\sigma_{xx} + \sigma_{yy} - 2\sigma_{zz}|^m + h|\sigma_{xx} - \sigma_{yy}|^m$$
$$+ N(|\sigma_{xz}|^m + |\sigma_{zx}|^m) - H^m(\varepsilon^p_{eq}) = 0 . \quad (4.161)$$

where H is the hardening function (Equation 3.103) to be determined along the rolling direction. Note that we can define the equivalent stress σ_{eq} as

$$\sigma_{eq} = \left\{ c|\sigma_{xx} + \sigma_{yy} - 2\sigma_{zz}|^m + h|\sigma_{xx} - \sigma_{yy}|^m + N(|\sigma_{xz}|^m + |\sigma_{zx}|^m) \right\}^{1/m}, \quad (4.162)$$

the equivalent plastic strain increment $d\varepsilon^p_{eq}$ as the work-conjugate to σ_{eq} (Equation 4.144) and then the equivalent plastic strain ε^p_{eq} as the integration of $d\varepsilon^p_{eq}$ along the deformation path (Equation 4.143). However, here we assume that

ε_{eq}^{p} is the von Mises equivalent plastic strain defined by Equations 3.158 and 3.156.

The associated flow rule (Equation 4.145) can be expressed in terms of the plastic part of the strain rate tensor by using the relationship at Equation 3.155 between $d\varepsilon_{ij}^{p}$ and $\dot{\varepsilon}_{ij}^{p}$ and defining

$$\dot{\lambda} = \frac{d\lambda}{dt}.$$ (4.163)

Thus, we get

$$\dot{\varepsilon}_{ij}^{p} = \dot{\lambda}\frac{\partial f}{\partial \sigma_{ij}},$$ (4.164)

where now f is given by Equation 4.161. (Since $d\lambda$ is a positive scalar, $\dot{\lambda}$ is also a positive scalar.)

We assume that $\sigma_{xx} > \sigma_{yy}$ and $\sigma_{xx} + \sigma_{yy} > 2\sigma_{zz}$. Then we write the associated flow rule (Equation 4.164) in the component form by differentiating the yield function f (Equation 4.161) with respect to the non-zero stress components of σ:

$$\dot{\varepsilon}_{xx}^{p} = \dot{\lambda}[cm(\sigma_{xx} + \sigma_{yy} - 2\sigma_{zz})^{m-1} + hm(\sigma_{xx} - \sigma_{yy})^{m-1}],$$ (4.165a)

$$\dot{\varepsilon}_{yy}^{p} = \dot{\lambda}[cm(\sigma_{xx} + \sigma_{yy} - 2\sigma_{zz})^{m-1} - hm(\sigma_{xx} - \sigma_{yy})^{m-1}],$$ (4.165b)

$$\dot{\varepsilon}_{zz}^{p} = \dot{\lambda}[-2cm(\sigma_{xx} + \sigma_{yy} - 2\sigma_{zz})^{m-1}],$$ (4.165c)

$$\dot{\varepsilon}_{xz}^{p} = \dot{\lambda}[(sign\sigma_{xz})Nm|\sigma_{xz}|^{m-1}].$$ (4.165d)

These relations need to be expressed only in terms of σ_{xx}, σ_{zz} and σ_{xz}. To eliminate σ_{yy} from these relations, the plane strain condition $\dot{\varepsilon}_{yy}^{p} = 0$ is used. Substitution of this condition in Equation 4.165b leads to

$$\sigma_{yy} = \frac{1-d}{1+d}\sigma_{xx} + \frac{2d}{1+d}\sigma_{zz},$$ (4.166)

where

$$d = \left(\frac{c}{h}\right)^{1/(m-1)}.$$ (4.167)

Now, the following relations can be obtained from Equation 4.166:

$$\sigma_{xx} + \sigma_{yy} - 2\sigma_{zz} = \frac{2}{1+d}(\sigma_{xx} - \sigma_{zz}),$$

$$\sigma_{xx} - \sigma_{yy} = \frac{2d}{1+d}(\sigma_{xx} - \sigma_{zz}).$$

(4.168)

Using these relations, the associated flow rule (Equations 4.165a,c,d) becomes

$$\dot{\varepsilon}_{xx}^p = 2\dot{\lambda}cm\left(\frac{2}{1+d}\right)^{m-1}(\sigma_{xx} - \sigma_{zz})^{m-1},$$

(4.169a)

$$\dot{\varepsilon}_{zz}^p = -2\dot{\lambda}cm\left(\frac{2}{1+d}\right)^{m-1}(\sigma_{xx} - \sigma_{zz})^{m-1},$$

(4.169b)

$$\dot{\varepsilon}_{xz}^p = \dot{\lambda}[(sign\sigma_{xz})Nm|\sigma_{xz}|^{m-1}].$$

(4.169c)

Since $d\varepsilon_{eq}^p$ is not defined as work-conjugate to σ_{eq}, $d\lambda$ cannot be equal to $d\varepsilon_{eq}^p$. Therefore, we cannot obtain $\dot{\lambda}$ by this method. However, we can determine $\dot{\lambda}$ from the consistency condition. But here, we obtain $\dot{\lambda}$ by a different method. To determine $\dot{\lambda}$, we express the yield criterion (Equation 4.161) only in terms of σ_{xx}, σ_{zz} and σ_{xz}. This is done by substituting Equation 4.168 into the yield criterion (Equation 4.161):

$$c\left\{\frac{2}{1+d}(\sigma_{xx} - \sigma_{zz})\right\}^m + h\left\{\frac{2d}{1+d}(\sigma_{xx} - \sigma_{zz})\right\}^m + 2N|\sigma_{xz}|^m - H^m(\varepsilon_{eq}^p) = 0.$$

(4.170)

Next we express the yield criterion in terms of $\dot{\varepsilon}_{xx}^p$ and $\dot{\varepsilon}_{xz}^p$ by using the associated flow rule (Equations 4.169a,c):

$$c\left(\frac{|\dot{\varepsilon}_{xx}^p|}{2cm\dot{\lambda}}\right)^{m/(m-1)} + h\left(\frac{|\dot{\varepsilon}_{xx}^p|}{2hm\dot{\lambda}}\right)^{m/(m-1)} + 2N\left(\frac{|\dot{\varepsilon}_{xz}^p|}{Nm\dot{\lambda}}\right)^{m/(m-1)} - H^m(\varepsilon_{eq}^p) = 0. \quad (4.171)$$

(While deriving the above equation, c, h, m and N are taken to be positive. However, it has been shown in Section 4.6.4 that all the material parameters, i.e., c, h, m and N are positive.) From the above relation, the following expression for $\dot{\lambda}$ is obtained:

$$\lambda = \frac{\left[\left(\dfrac{1}{c^{1/(m-1)}}+\dfrac{1}{h^{1/(m-1)}}\right)\left(\dfrac{\left|\dot{\varepsilon}^p_{xx}\right|}{2m}\right)^{m/(m-1)}+\dfrac{2}{N^{1/(m-1)}}\left(\dfrac{\left|\dot{\varepsilon}^p_{xz}\right|}{m}\right)^{m/(m-1)}\right]^{(m-1)/m}}{H^{m-1}(\varepsilon^p_{eq})}.$$

$$(4.172)$$

The associated flow rule (Equations 4.169a–c) needs to be expressed in terms of the deviatoric stress components σ'_{ij}. Using Equation 4.166, we obtain the following expressions for the two diagonal components σ'_{xx} and σ'_{zz} :

$$\sigma'_{xx} = \frac{1+3d}{3(1+d)}(\sigma_{xx}-\sigma_{zz}),$$

$$\sigma'_{zz} = \frac{2}{3(1+d)}(\sigma_{zz}-\sigma_{xx}),$$

$$(4.173a)$$

The only non-zero diagonal component σ'_{xz} is of course equal to σ_{xz} :

$$\sigma'_{xz} = \sigma_{xz}.$$

$$(4.173b)$$

Eliminating σ_{xx}, σ_{zz} and σ_{xz} from Equations 4.169 and 4.173, we get

$$\dot{\varepsilon}^p_{xx} = \left(\frac{1}{k_1}\right)\sigma'_{xx},$$

$$\dot{\varepsilon}^p_{zz} = \left(\frac{1}{k_2}\right)\sigma'_{zz},$$

$$\dot{\varepsilon}^p_{xz} = \left(\frac{1}{k_3}\right)\sigma'_{xz},$$

$$(4.174)$$

where

$$k_1 = \frac{1+3d}{6}\left(\frac{\dot{\varepsilon}^p_{xx}}{2\lambda cm}\right)^{1/(m-1)}\frac{1}{\dot{\varepsilon}^p_{xx}},$$

$$k_2 = \frac{1}{3}\left(\frac{(-1)^m\dot{\varepsilon}^p_{zz}}{2\lambda cm}\right)^{1/(m-1)}\frac{1}{\dot{\varepsilon}^p_{zz}},$$

$$(4.175)$$

$$k_3 = \left(\frac{\left|\dot{\varepsilon}^p_{xz}\right|}{\lambda Nm}\right)^{1/(m-1)}\frac{1}{\left|\dot{\varepsilon}^p_{xz}\right|}.$$

If the material is rigid-plastic (*i.e.*, $\dot{\varepsilon}_{ij} = \dot{\varepsilon}_{ij}^{p}$), then Equation 4.174 is the complete constitutive equation. For elastic-plastic material, as before, we assume that the decomposition of $\dot{\varepsilon}_{ij}$ into the elastic and plastic parts is additive.

Therefore, to get the complete constitutive equation, we combine the elastic constitutive equation (Equation 3.162) with the above plastic constitutive equation (Equation 4.174) in an additive way.

4.8 Kinematic Hardening

The Bauschinger effect is modeled by a rigid translation of the initial yield surface away from its center. Further, the size as well as the shape of the initial yield surface also gets changed during subsequent yielding. Prager [35] was the first to model the Bauschinger effect by employing a rigid translation of the initial yield surface but *without* incorporating either *the change in size or the change in shape*. He used the word *kinematic hardening* to describe this model.

If the initial yield criterion is expressed by the yield function f:

$$f(\sigma_{ij}) = 0,\tag{4.176}$$

then the criterion for subsequent yielding, corresponding to a kinematic hardening, can be expressed as

$$f(\sigma_{ij} - d\alpha_{ij}) = 0,\tag{4.177}$$

where now σ_{ij} is the current stress tensor and $d\alpha_{ij}$ is the *incremental translation* of the *yield surface*. The tensor $d\alpha_{ij}$ is called the *incremental back stress*. If the material yields according to the *Mises* criterion (Equations 3.21 and 3.22), then the criterion for subsequent yielding, corresponding to the kinematic hardening (Equation 4.177) becomes

$$(\sigma_{ij}' - d\alpha_{ij})(\sigma_{ij}' - d\alpha_{ij}) - \frac{2}{3}\sigma_Y^2 = 0,\tag{4.178}$$

where σ_{ij}' is the deviatoric part of the *current* stress tensor and σ_Y is the *yield stress* in uni-axial tension.

Prager [35] assumed that the incremental translation of the yield surface takes place along the direction of the plastic part of the incremental linear strain tensor $d\varepsilon_{ij}^{p}$. This assumption can be expressed mathematically as

$$d\alpha_{ij} = cd\varepsilon_{ij}^{p},\tag{4.179}$$

where c is a material parameter representing the kinematic hardening. If c is a constant, then it is called *linear* kinematic hardening model. However, if c is taken as a function of the deformation history, then it is called *non-linear* kinematic hardening model.

Graphical representation of Prager's [35] model in the stress space (*i.e.*, a nine-dimensional space of the stress components σ_{ij}) is shown in Figure 4.5. Here, the point O represents the origin of the stress space and the points C_1 and C_2 denote the current and new centers of the yield surface. The vector OC_1 represents the translation of the current yield surface from the initial yield surface and the vector OP_1 represents the current stress tensor. Further, the vector C_1C_2, which represents the incremental translation $d\alpha_{ij}$ of the current yield surface, is parallel to the vector P_1Q representing $d\varepsilon_{ij}^p$. The vector P_1P_2 represents the incremental stress $d\sigma_{ij}$. For this model, the Mises criterion for subsequent yielding (Equation 4.178) becomes

$$(\sigma_{ij}' - cd\varepsilon_{ij}^p)(\sigma_{ij}' - cd\varepsilon_{ij}^p) - \frac{2}{3}\sigma_Y^2 = 0. \tag{4.180}$$

In uni-axial tension along the x-direction, the shear components of σ_{ij}, σ_{ij}' and $d\varepsilon_{ij}^p$ are zero. Further, σ_{yy} and σ_{zz} are also zero. Then, the non-zero components of σ_{ij}' become

$$\sigma_{xx}' = \frac{2}{3}\sigma_{xx}, \ \sigma_{yy}' = -\frac{1}{3}\sigma_{xx}, \ \sigma_{zz}' = -\frac{1}{3}\sigma_{xx}. \tag{4.181}$$

Since the volume remains constant during plastic deformation, we get the following relations between the non-zero normal components of $d\varepsilon_{ij}^p$:

$$d\varepsilon_{yy}^p = -\frac{1}{2}d\varepsilon_{xx}^p, \ d\varepsilon_{zz}^p = -\frac{1}{2}d\varepsilon_{xx}^p, \tag{4.182}$$

Substituting Equations 4.181 and 4.182 in Equation 4.180, we get the following relation between $d\sigma_{xx}$ and $d\varepsilon_{xx}^p$:

$$d\sigma_{xx} = \sigma_{xx} - \sigma_Y = \left(\frac{3c}{2}\right)d\varepsilon_{xx}^p. \tag{4.183}$$

Thus, for Prager's [35] kinematic hardening model, the material parameter c is 2/3 times the slope of the one-dimensional stress-strain curve.

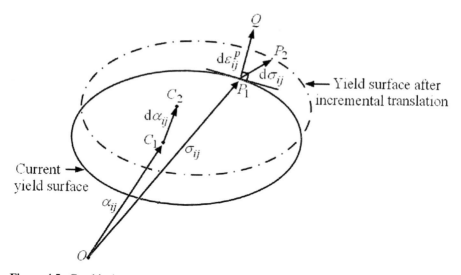

Figure 4.5. Graphical representation of Prager's kinematic hardening model in stress space

In uni-axial tension, the shear components of $d\varepsilon_{ij}^{p}$ are zero. Therefore, the shear components of $d\alpha_{ij}$ also become zero. Further, the normal components of $d\varepsilon_{ij}^{p}$ are related by Equation 4.182. Therefore, from Equation 4.179, we get the following expressions for the non-zero components of $d\alpha_{ij}$:

$$d\alpha_{xx} = (c)d\varepsilon_{xx}^{p}, \quad d\alpha_{yy} = \left(-\frac{c}{2}\right)d\varepsilon_{xx}^{p}, \quad d\alpha_{zz} = \left(-\frac{c}{2}\right)d\varepsilon_{xx}^{p}. \qquad (4.184)$$

The above equation shows that whereas in the x direction (*i.e.*, in the axial direction), the yield surface moves in the direction of $d\varepsilon_{xx}^{p}$, in the y and z directions (*i.e.*, in the transverse directions), it moves in the opposite direction of $d\varepsilon_{xx}^{p}$. Thus, there is (kinematic) hardening in the axial direction, but softening in the transverse directions.

As stated above, Prager's [35] model exhibits transverse softening in uni-axial tension. This transverse softening is not supported by experiments. Another drawback of Prager's [35] model is that the corresponding yield function takes two different forms in two- and three-dimensional problems [36, 37]. Ziegler [36] has reported that the yield surface of Prager's [35] model moves along the direction normal to the yield surface (*i.e.*, along $d\varepsilon_{ij}^{p}$) for three-dimensional problems as expected, but for two-dimensional problems it does not move along the same direction.

To remove these drawbacks, Ziegler [36] proposed a certain modification in Prager's model [35]. The modification can be represented by the following mathematical expression:

$$d\alpha_{ij} = (\sigma_{ij} - \alpha_{ij})d\mu, \tag{4.185}$$

where $d\mu$ is a material parameter representing the kinematic hardening. Graphical representation of Ziegler's [36] model in the stress space (*i.e.*, a nine-dimensional space of the stress components σ_{ij}) is shown in Figure 4.6. Here, the point O represents the origin of the stress space and the points C_1 and C_2 denote the current and new centers of the yield surface. Further, the vector C_1C_2, which represents the incremental translation $d\alpha_{ij}$ of the current yield surface is parallel to the vector P_1Q. But the vector P_1Q is not along the vector P_1R representing $d\varepsilon_{ij}^p$. As before, the vectors OC_1, OP_1 and P_1P_2 represent respectively α_{ij}, σ_{ij} and $d\sigma_{ij}$.

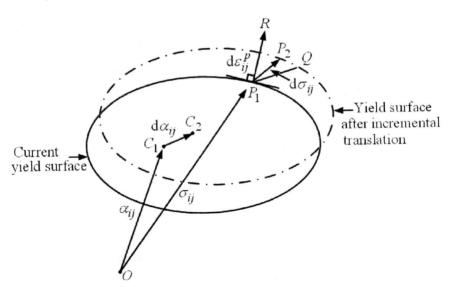

Figure 4.6. Graphical representation of Ziegler's kinematic hardening model in stress space

To determine $d\mu$, we use the condition that the vector QP_2 must be perpendicular to the vector P_1R representing $d\varepsilon_{ij}^p$ [38]. Thus,

$$(d\sigma_{ij} - d\alpha_{ij})d\varepsilon_{ij}^p = 0. \tag{4.186}$$

This condition, called the consistency condition, makes sure that the new stress point lies on the new yield surface. Multiplying both sides of Equation 4.185 by $d\varepsilon_{ij}^p$, eliminating $d\alpha_{ij}d\varepsilon_{ij}^p$ by using Equation 4.186 and changing the dummy indices of one side of the resulting equation from (i,j) to (k,l), we get the following expression for $d\mu$:

$$d\mu = \frac{d\sigma_{ij}d\varepsilon_{ij}^p}{(\sigma_{kl} - \alpha_{kl})d\varepsilon_{kl}^p}. \tag{4.187}$$

Both the Prager's [35] and Ziegler's [36] model coincide in tension and compression tests as well as in simple and pure shear tests.

The following non-linear modifications have been proposed to Prager's [35] and Ziegler's [36] model [39]:

$$d\alpha_{ij} = c_1 d\varepsilon_{ij}^p - \frac{c_2}{\sigma_Y} d\varepsilon_{eq}^p \alpha_{ij}, \tag{4.188}$$

and

$$d\alpha_{ij} = \frac{c_1}{\sigma_Y} d\varepsilon_{eq}^p (\sigma_{ij} - \alpha_{ij}) - \frac{c_2}{\sigma_Y} d\varepsilon_{eq}^p \alpha_{ij}. \tag{4.189}$$

Here, σ_Y is the yield stress, $d\varepsilon_{eq}^p$ is the equivalent plastic strain increment defined by Equation 3.93 and c_1 and c_2 are functions of the integrals of the second and third invariants of $d\varepsilon_{ij}^p$.

For these two models, the constitutive relation for $d\varepsilon_{ij}^p$ is obtained from the associated flow rule:

$$d\varepsilon_{ij}^p = d\lambda \frac{\partial f}{\partial \sigma_{ij}}, \tag{4.190}$$

where the positive scalar $d\lambda$ is determined from the consistency condition. Then Equation 4.190 is combined with the constitutive relation for $d\varepsilon_{ij}^e$ to obtain the elastic-plastic incremental stress-strain relation.

4.9 Modeling of Ductile Fracture

Ductile fracture is often a limiting factor in metal forming processes. Recent investigations have shown that a ductile fracture occurs mainly due to micro-void nucleation, growth and finally coalescence into a micro-crack. The void growth can also affect the macroscopic properties of the material. Therefore, the plastic stress-strain or stress-strain rate relations need to be changed appropriately to include the effect of void growth. However, if the extent of void growth up to fracture is small, then one can ignore the effect of void growth on the constitutive equations. However, a realistic model for the ductile fracture prediction must include void nucleation, void growth and a condition for void coalescence. Three broad approaches have emerged which try to predict ductile fracture on the basis of void nucleation, growth and coalescence. They are:

- Porous plasticity model of Berg and Gurson [40, 41]
- Void nucleation, growth and coalescence model (Goods and Brown, Rice and Tracy, and Thomason Model [42–44])
- Continuum damage mechanics model of Lemaitre [45] and Rousselier [46]

4.9.1 Porous Plasticity Model of Berg and Gurson

In this model, the material with voids is idealised as a porous material. Thus, its constitutive equation is derived from the plastic potential of a porous material. In a porous plastic material, the hydrostatic part of stress also influences yielding. As a result, the yield surface does not remain an infinitely long cylinder as shown in Figure 3.6, but is capped by elliptical surfaces. Based on Berg's [40] model of dilatational plasticity, Gurson [41] proposed a plastic potential for porous material. Starting from a unit spherical cell with a single void, he obtained the following expression for the plastic potential function for a porous material with randomly distributed voids of volume fraction v_f :

$$g(\sigma_{ij}, v_f) = \left(\frac{(\sigma_{eq})_p}{\sigma_{eq}} \right)^2 + 2q_1 v_f \cosh\left(\frac{3q_2(\sigma_h)_p}{2\sigma_{eq}} \right) - 1 - q_3(v_f)^2. \quad (4.191)$$

Here, $(\sigma_h)_p$ is the hydrostatic part of stress in the porous aggregate, $(\sigma_{eq})_p$ is the equivalent stress in the porous aggregate and (σ_{eq}) is the equivalent stress in the matrix. Gurson [41] assumed that $q_1 = q_2 = q_3 = 1$. (However, Tvergaard [47] assigned some other values to these constants ($q_1 = 1.5$, $q_2 = 1$, $q_3 = (q_1)^2$) to bring the predictions of the model into closer agreement with numerical analysis of a periodic array of voids.)

In this model, the rate of change of the void volume fraction $\dot{v}_f$ is considered as the sum of the void nucleation rate $(\dot{v}_f)_{nucleation}$ and the void growth rate $(\dot{v}_f)_{growth}$. Thus,

$$\dot{v}_f = (\dot{v}_f)_{nucleation} + (\dot{v}_f)_{growth}. \qquad (4.192)$$

In general, the void nucleation rate depends on the hydrostatic part of the plastic strain rate tensor $\dot{\varepsilon}_{ij}^p$ as well as equivalent plastic strain rate $\dot{\varepsilon}_{eq}^p$. Here, it is assumed to depend only on the equivalent plastic strain rate:

$$(\dot{v}_f)_{nucleation} = A\dot{\varepsilon}_{eq}^p, \qquad (4.193)$$

where A is a constant. Further, the void growth rate is related to the hydrostatic part of $\dot{\varepsilon}_{ij}^p$ by the following relation:

$$(\dot{v}_f)_{growth} = (1 - v_f)\dot{\varepsilon}_{kk}^p. \qquad (4.194)$$

The above relations can also be expressed in the increment form.

In this model, ductile fracture is regarded as the result of instability in the weak dilatational plastic flow field localised in a band called a shear band. Rudnicki and Rice [48] obtained the instability conditions without considering any imperfection in the material. Later, Tvergaard [47] and Yamamoto [49] incorporated material imperfection in the instability conditions. The fracture criterion is represented as a graph of critical localisation strain vs the critical void volume fraction with strain hardening exponent as a parameter.

In this model, the void coalescence by internal necking of inter-void matrix is considered as a secondary effect which develops only after the formation of the shear band. However, this is in contrast to the next model, which considers the internal necking of the inter-void matrix as the primary mechanism of ductile fracture.

Some researchers [50] have used the Berg-Gurson model without the instability conditions. They characterized the ductile fracture by a critical value of v_f which was determined experimentally.

4.9.2 Void Nucleation, Growth and Coalescence Model (Goods and Brown, Rice and Tracy and Thomason Model)

In this model, Thomason [44] combined the results of Goods and Brown [42] on void nucleation, those of Rice and Tracy [43] on void growth and his own on void coalescence to arrive at a fracture criterion in the form of a graph of fracture strain vs the hydrostatic part of stress. Thus, in this model, the effects of void nucleation and growth are incorporated not in the constitutive equations but in the fracture criterion itself. The void nucleation model of Goods and Brown, void growth model of Rice and Tracy and void coalescence model of Thomason are described below along with some additional models on void nucleation and void growth.

- Void nucleation models

A *de-cohesion* model of void nucleation is based on the condition that the void nucleation by de-cohesion of second phase particles occurs whenever the normal stress component on the particle/matrix interface reaches a critical value σ_c. Goods and Brown [42] used this condition to derive the following relation for the largest principal strain at void nucleation (called the *void nucleation strain* and denoted by ε_1^n) in terms of the hydrostatic part of stress σ_h:

$$\varepsilon_1^n = Kr(\sigma_c - \sigma_h)^2. \tag{4.195}$$

Here, r is the particle radius and K is a material constant depending on the volume fraction of the second phase particles. Experiments on the Fe-Fe$_3$C system confirm the linear relationship between ε_1^n and r for small spherical particles of radius less than 1 μm.

A *cracking* model of void nucleation was proposed by Gurland [51] based on his experimental observation on 1.05% C spherodised steel. As per this model, void nucleation (of cementite particles) takes place continuously at all strain levels depending on the size, shape and orientation with the maximum principal stress direction. This results in a linear relation between the void nucleation rate $(\dot{v}_f)_{nucleation}$ and the equivalent plastic strain rate $\dot{\varepsilon}_{eq}^p$ as in Equation 4.193.

It should be noted that the development of a very general model of void nucleation is a difficult task as a typical commercial alloy contains a wide range of particle types and particle morphologies which can result in a variety of void nucleation mechanisms operating simultaneously. Further, such a model may be too complicated to be amenable to analysis for making useful numerical predictions.

- Void growth model

Once the nucleation of a micro void takes place either by de-cohesion or by cracking of a second phase particle, the resulting stress-free surface of the void produces localised stress and strain concentrations in the surrounding plastically deforming matrix. With continuing plastic flow of the matrix, the void undergoes both the change in size and shape. If it is assumed that the inter-void distance is sufficiently large to prevent any interaction amongst them, then one can analyse the void growth phenomenon by considering a single void in an infinite medium.

Rice and Tracy [43] considered a single spherical void of initial radius R_0 in a remote uniform strain rate field $\dot{\varepsilon}_{ij}$ and remote stress field σ_{ij} in a rigid-plastic material. They derived the following expression for the rate of change of the radii of curvature ($\dot{R}_k$) in the principal strain rate directions:

$$\dot{R}_k = [(1+E)\dot{\varepsilon}_k + \dot{\varepsilon}_{eq}D]R_{mean} \qquad (k=1, 2, 3), \tag{4.196}$$

where

$E = \dfrac{2}{3}$ for linear hardening and for low values of σ_h for non-hardening,

$= 1$ for high values of σ_h for non-hardening,

$D = 0.75\dfrac{\sigma_h}{\sigma_{eq}}$ for linear hardening,

$= 0.558\sinh\left(\dfrac{3\sigma_h}{2\sigma_{eq}}\right) + 0.008\nu\cosh\left(\dfrac{3\sigma_h}{2\sigma_{eq}}\right)$ for non-hardening,

σ_{eq} = equivalent stress,

$R_{mean} = \dfrac{1}{3}(R_1 + R_2 + R_3)$,

$\dot{\varepsilon}_{eq}$ = equivalent strain rate,

$\dot{\varepsilon}_k$ = principal strain rates, $(k=1, 2, 3)$,

$\nu = -\dfrac{3\dot{\varepsilon}_2}{\dot{\varepsilon}_1 - \dot{\varepsilon}_3}$, Lode variable.

Note that an initial spherical void grows into an ellipsoidal void of the principal radii R_1, R_2 and R_3. Thomason [44] integrated the above equations assuming that the principal axes of the strain rate tensor remain fixed in direction throughout the deformation path and obtained the following expressions for the principal radii of the void:

$$R_1 = \left(A + \dfrac{3(1+\nu)}{2\sqrt{\nu^2+3}}B\right)R_0, \tag{4.197}$$

$$R_2 = \left(A - \dfrac{\nu}{\sqrt{\nu^2+3}}B\right)R_0, \tag{4.198}$$

$$R_3 = \left(A + \dfrac{(\nu-3)}{2\sqrt{\nu^2+3}}B\right)R_0, \tag{4.199}$$

where

$$A = \exp\left(\dfrac{2\sqrt{\nu^2+3}}{(3+\nu)}D\varepsilon_1^g\right), \tag{4.200}$$

$$B = \left(\frac{1+E}{D}\right)(A-1).$$ (4.201)

Here, ε_1^g is the integral of the largest principal strain rate. Thomason used these expressions (Equations 4.197–4.199) in his derivation of the condition for void coalescence.

For an array of void nucleating particles of diameter D_p and spacing d_p, setting the initial void radius as $D_p/2$ and integrating Equation 4.196 up to fracture, the fracture strain ε_f (for a non-hardening material) is obtained as

$$\varepsilon_f = \frac{\ln(d_p/D_p)}{0.28\exp(1.5\sigma_h/\sigma_{eq})}.$$ (4.202)

For a plane strain problem with cylindrical voids, similar analysis by McClintock [52] yields the following expression for the fracture strain:

$$\varepsilon_f = \frac{\ln(d_p/D_p)(1-n)}{\sinh[\sqrt{3}(1-n)(\sigma_1+\sigma_2)/2\sigma_{eq}]}.$$ (4.203)

Here, σ_i are the principal stresses and n is the hardening parameter of Equation 3.9.

Some research workers [53, 54] have used the above expressions of fracture strain to predict ductile fracture with the help of some experimentally determined parameters. A limitation of this approach is that it ignores the effects of void nucleation (as these expressions are derived from a pre-existing finite size crack) and void coalescence. Thus, usually ε_f is overestimated. However, these expressions do reveal the dependence of ε_f on the triaxiality (σ_h/σ_{eq}) of the stress state, hardening parameter n of the material and purity (d_p/D_p) of the material.

- Void coalescence condition

Thomason [44] modeled the void coalescence phenomenon as plastic instability due to necking of the inter-void matrix. According to him, the sufficient condition for plastic instability of the inter-void matrix is given by

$$\sigma_1 - \sigma_n \overline{A}_n = 0,$$ (4.204)

where $\overline{A}_n$ is the area fraction of the inter-void matrix perpendicular to the direction of the maximum principal stress σ_1 and σ_n is the plastic constraint stress. He considered a geometrically equivalent square prismatic void with the same principal dimensions as the ellipsoidal void and used the upper bound method to obtain the following expression for the plastic constraint stress:

$$\sigma_n = \sigma_{eq}\left(\frac{0.1}{(a/d_p)^2} + \frac{1.2}{[b/(b+d_p)]^{1/2}}\right),$$
(4.205)

where a and b are the void dimensions.

- Fracture criterion

Thomason [44] used Equations 4.197–4.199 for the void dimensions to express the void coalescence condition (Equations 4.204 and 4.205) in terms of the void growth strain ε_1^g and the hydrostatic part of stress σ_h. By superposing this condition onto the void nucleation relation (Equation 4.195), he obtained the fracture criterion as a graph of fracture strain ($\varepsilon_f = \varepsilon_1^n + \varepsilon_1^g$) vs the hydrostatic part of stress σ_h.

A limitation of this approach is the use of Equations 4.197–4.199 for the void growth. These expressions have been derived by assuming that the principal axes of the strain rate tensor remain fixed in direction throughout the deformation path. This is true only for the case of small deformation and rotation. As a result, this approach cannot be used when the deformation and/or rotation are large.

4.9.3 Continuum Damage Mechanics Models

Description of constitutive behavior of materials with micro-voids needs an additional variable, called the damage variable, which quantifies the intensity of micro-voids at a point. This variable can be defined as follows. On a plane with normal $\hat{n}$, the damage vector $\boldsymbol{D}$ at a point P is defined as

$$\boldsymbol{D} = D\hat{n},$$
(4.206)

where D is defined as the area void fraction around the point on that plane:

$$D = \lim_{\Delta A \to 0} \frac{\Delta A_v}{\Delta A}.$$
(4.207)

Here, ΔA is a small area around point P in the plane and ΔA_v is the area of void traces contained in ΔA. The set of $\boldsymbol{D}$s on all the planes passing through the point can be represented as a damage tensor. In this work, we consider the damage mechanics of only isotropic materials. In isotropic materials, the area void fraction has the same value on every plane passing through the point. Therefore, the damage tensor reduces to an isotropic tensor with D (given by Equation 4.207) as its diagonal components. As a result, for isotropic materials, the damage variable at a point can be considered as a scalar, defined by Equation 4.207, rather than a tensor. For isotropic materials, the damage is sometimes defined as a void volume fraction. However, in this work, we use the definition given by Equation 4.207.

According to the theory of continuum thermodynamics, when the temperature change is not significant, the constitutive equations for dissipative phenomenon like the plastic deformation with hardening and damage are derived from the plastic potential g expressed as a function of σ_{ij}, $-R$ and $-Y$ [45, 55]. Here, σ_{ij} is the Cauchy stress tensor and $-R$ and $-Y$ are the *dissipative parts of the thermodynamic forces* corresponding respectively to the variables $\dot{\varepsilon}_{eq}^p$ and $\dot{D}$.

(Note that, $\dot{\varepsilon}_{eq}^p$ is the equivalent plastic strain rate defined by Equation 3.156 and $\dot{D}$ is the time rate of change of D). Further, Y is interpreted as the rate of release of elastic energy with damage growth when the stress is held constant (*i.e.*, Y is the work-conjugate variable corresponding to D.) Thus, the expression for Y is obtained by taking the derivative of *the specific free energy* (or *the thermodynamic potential*) with respect to D while keeping σ_{ij} constant. When the temperature change is not significant, the expression for Y for the elastic-plastic material is given by [55]

$$Y = -\frac{\sigma_{eq}^2}{2E(1-D)^2}\left[\frac{2}{3}(1+v)+3(1-2v)\left(\frac{\sigma_h}{\sigma_{eq}}\right)^2\right]. \tag{4.208}$$

Here, E and v are the elastic constants of the material, σ_h is the hydrostatic part of the Cauchy stress tensor and σ_{eq} is the equivalent stress defined by Equation 3.23.

Additionally, it is assumed that it is possible to decompose the plastic potential g of the damaged material as [55]

$$g = g_1(\sigma_{ij}, -R, D) + g_D(-Y, D, \varepsilon_{eq}^p). \tag{4.209}$$

Here, the first part g_1 is the plastic potential associated with yielding and hardening of the material whereas the second part g_D is the plastic potential associated with damage of the material such that it reduces to zero whenever D is zero. R is the work-conjugate variable corresponding to ε_{eq}^p. For the first part, *i.e.*, for the plastic potential associated with yielding and hardening, we have chosen earlier ε_{eq}^p as the independent variable rather than $-R$. Thus, we can write g as

$$g = g_1(\sigma_{ij}, \varepsilon_{eq}^p, D) + g_D(-Y, D, \varepsilon_{eq}^p). \tag{4.210}$$

When D is zero, for the material yielding according to the von Mises criterion and hardening according to the strain hardening hypothesis, g_1 is taken to be the following yield function:

$$f(\sigma_{ij}, \varepsilon_{eq}^p) \equiv \sigma_{eq} - H(\varepsilon_{eq}^p),$$
(4.211)

where H is the hardening function defined by Equation 3.103. When D is not zero, g_1 is obtained from the *principle of strain equivalence*, which states that the deformation of a damaged material can be represented by the same constitutive relation as that of the virgin material if the Cauchy stress tensor σ is replaced by the *effective* Cauchy stress tensor $\overset{*}{\sigma}$:

$$\overset{*}{\sigma}_{ij} = \frac{\sigma_{ij}}{1-D}.$$
(4.212)

Therefore, when D is not zero, g_1 is taken as

$$g_1(\sigma_{ij}, \varepsilon_{eq}^p, D) = \overset{*}{\sigma}_{eq} - H(\varepsilon_{eq}^p),$$
(4.213)

where $\overset{*}{\sigma}_{eq}$ is the equivalent stress corresponding to the *effective* stress $\overset{*}{\sigma}$. Thus, $\overset{*}{\sigma}_{eq}$ is

$$\overset{*}{\sigma}_{eq} = \left(\frac{3}{2}\overset{*}{\sigma}_{ij}\overset{*}{\sigma}_{ij}\right)^{1/2}.$$
(4.214)

The associated flow rule now becomes

$$d\varepsilon_{eq}^p = d\lambda\frac{\partial g_1}{\partial\sigma_{ij}} = d\lambda\frac{\partial\overset{*}{\sigma}_{eq}}{\partial\sigma_{ij}}.$$
(4.215)

When the derivative of $\overset{*}{\sigma}_{eq}$ with respect to σ_{ij} is evaluated using Equations 4.214 and 4.212 and $d\lambda$ is obtained from Equation 3.138 by replacing f with g_1, we get the following expression for $d\varepsilon_{ij}^p$ for the damaged material:

$$d\varepsilon_{ij}^p = \left(\frac{1}{1-D}\right)\frac{9}{4}\frac{\sigma_{ij}'\sigma_{kl}'}{H'\sigma_{eq}^2}d\sigma_{kl}.$$
(4.216)

Note that this expression differs from Equation 3.143 for the undamaged material by a factor of $1/(1-D)$. Similarly, the expression for the elastic part $d\varepsilon_{ij}^e$ of the incremental linear strain tensor, for the damaged material, also differs by the same factor from the corresponding expression for the undamaged material (Equation 3.149). As a result, the fourth order elastic-plastic tensor C_{ijkl}^{EP} in the incremental stress-strain relation for the damaged material (for the updated Lagrangian formulation) differs from the corresponding relation for the undamaged material (Equation 3.153) by the factor of $(1-D)$:

$$C_{ijkl}^{EP} = 2\mu \left[\frac{\nu}{1-2\nu} \delta_{ij}\delta_{kl} + \delta_{ik}\delta_{jl} - \frac{9\mu}{2} \frac{\sigma_{ij}'\sigma_{kl}'}{(H'+3\mu)\sigma_{eq}^2} \right](1-D). \qquad (4.217)$$

For most metals, the value of D up to micro-crack initiation is quite small (of the order of 0.05) compared to 1. Therefore, the factor $(1-D)$ can be neglected in the expression for C_{ijkl}^{EP}. As far as the elastic-plastic stress-strain rate relation for the damaged material (for the Eulerian formulation) is concerned, it also differs from the corresponding relation for the undamaged material (Equations 3.163) by a factor containing $(1-D)$. Since D is small compared to 1 up to micro-crack initiation, we neglect this factor in the elastic-plastic stress-strain rate relation as well. Thus, even for the damaged material, we continue to use the constitutive equations of the undamaged material.

The damage growth law is obtained as the derivative of g_D with respect to $-Y$. Thus, we get

$$dD = d\lambda \frac{\partial g_D}{\partial(-Y)}. \qquad (4.218)$$

Unlike g_1, g_D is not well established in the literature. Therefore, experimental results on void measurement at different deformation levels are used to propose a damage growth law. Based on the experimental results of Le Roy et al. [56] on AISI 1090 steel, Dhar et al. [57] have proposed the following damage growth law:

$$dD = cd\varepsilon_{eq}^p + (a_1 + a_2 D)(-Y)d\varepsilon_{eq}^p. \qquad (4.219)$$

Here, the material constants c, a_1 and a_2 were determined by Dhar et al. [57] by fitting the above equation through the experiments results of Le Roy et al. [56]. They obtained the following values:

$$c = 1.898 \times 10^{-02}, \quad a_1 = 9.8 \times 10^{-04} \ (MPa)^{-1}, \quad a_2 = 1.84 \ (MPa)^{-1}. \ (4.220)$$

While getting these constants, they used Bridgeman's [58] relation to express the triaxiality $\left(\sigma_h / \sigma_{eq}\right)$ as a function of equivalent plastic strain.

The first term of Equation 4.219, which is independent of $-Y$, represents the damage evolution due to void nucleation. It states that the void nucleation takes place continuously at all strain levels. Thus, it is similar to the void nucleation model of Gurland [51]. The other terms of Equation 4.219 represent the evolution of damage due to void growth as proposed by others [59, 60]. The Lemaitre's [55] damage growth law does not contain the term corresponding to a_2, while in the model of Tai and Yang [59], the term corresponding to a_1 is missing. From the experimental results of Le Roy et al. [56], it is observed that the graph of area void fraction vs the equivalent plastic strain is linear at low strain level but it is highly non-linear at higher values of strain. Therefore, it seems to be wise to retain both the terms corresponding to void growth. One can add more non-linear terms like $a_3 D^2$, $a_4 D^3$ etc. if required.

Note that the constitutive equation of a damaged material contains the effect of void growth. The damage growth law contains the effects of both the void nucleation and void growth. However, the phenomenon of void coalescence has to be incorporated as an additional condition in terms of the continuum parameters. This condition, which serves as a fracture criterion, has to be based on an appropriate micro model. Thomason's [44] condition for void coalescence (Equations 4.204 and 4.205) is a good candidate for this purpose. However, while deriving this condition, Thomason [44] used the Rice and Tracy [43] expressions for the void dimensions and inter-void spacing that are valid only for the case of small strain and rotation. Dhar et al. [57] modified the Thomason's condition for void coalescence (Equations 4.204 and 4.205) by incorporating the finite strain expressions for the void dimensions and inter-void spacing. The modified version of Thomason's void coalescence condition, as proposed by Dhar et al. [57], is

$$\sigma_1 - \left[0.1 + \frac{1.2}{(1 - \exp(-\varepsilon_{eq}^p / 2))^{1/2}}\right] \exp(-\varepsilon_{eq}^p) \sigma_{eq} = 0. \tag{4.221}$$

- Critical damage criterion of Dhar et al. [57]

To find the critical value of damage parameter (D_c), Dhar et al. [57] applied the void coalescence condition (Equation 4.221) to AISI 1090 steel for various geometries and loading conditions by performing a large deformation elastic plastic finite element analysis. They observed that D_c is independent of geometry and loading and hence can be used as a material property for the prediction of micro-crack initiation in AISI 1090 steel. They found the value of D_c as 0.05 for this material.

At phenomenological level, ductile fracture is governed by both the equivalent plastic strain and the hydrostatic part of stress (or the triaxiality). The critical damage criterion incorporates both these parameters.

- Rousselier's model [46]

Another commonly used continuum damage mechanics model is due to Rousselier [46]. In this model, the plastic potential g is expressed as a function of the hydrostatic stress σ_h, the thermodynamic force Y (corresponding to the damage variable D) and the density ρ, besides the deviatoric stress σ'_{ij} and the equivalent plastic strain ε^p_{eq}. Thus, it is given by

$$g(\sigma'_{ij}, \sigma_h, \rho, Y, \varepsilon^p_{eq}) = \left[J_2 \left(\frac{\sigma'_{ij}}{\rho} \right) \right]^{1/2} + Yh \left(\frac{\sigma_h}{\rho} \right) - H(\varepsilon^p_{eq}). \qquad (4.222)$$

Here, the plastic potential g depends on σ'_{ij} / ρ through its second invariant J_2, on σ_h / ρ through the function h (which incorporates the dilatational effect of void growth) and on ε^p_{eq} through the hardening function H. The damage variable D is implicitly present in the plastic potential g through the following relation:

$$D = D(\rho). \qquad (4.223)$$

The mass conservation law is used to derive the expression for the function h, the thermodynamic force Y and the damage growth law.

Zheng et al. [61] used Rousselier's model to derive a macro damage parameter for prediction of crack initiation in metal forming processes.

4.9.4 Phenomenological Models

In the absence of reliable quantitative models for incorporating the phenomena of void nucleation, growth and coalescence, many empirical fracture criteria based on some phenomenological observations have been used in metal forming processes. Here, only a few such criteria are discussed.

- Freudenthal, Cockcroft and Latham and Oh criteria [62, 63, 54]

Freudenthal [62] postulated that the plastic work done per unit volume is a critical parameter in ductile fracture. Therefore, he proposed the following criterion for ductile fracture:

$$\int_0^{\varepsilon_f} \sigma_{eq} d\varepsilon^p_{eq} = C_1, \text{ a constant.} \qquad (4.224)$$

Here, ε_f stands for the equivalent plastic strain at fracture. The constant C_1 in this criterion as well as constants in subsequent criterion of this subsection can be determined from a tension test. For a tensile specimen, the change in neck geometry influences the fracture process. To take care of this change in geometry,

Cockcroft and Latham [63] modified the above criterion by incorporating the dimensionless stress concentration factor (σ_1 / σ_{eq}), where σ_1 is the maximum normal stress. Thus, their criterion can be stated as

$$\int_0^{\varepsilon_f} \sigma_{eq} \left(\frac{\sigma_1}{\sigma_{eq}} \right) d\varepsilon_{eq}^p = \text{another constant,} \tag{4.225}$$

or

$$\int_0^{\varepsilon_f} \sigma_1 d\varepsilon_{eq}^p = C_2 \text{, a constant.} \tag{4.226}$$

Oh *et al.* [54] modified the above criterion by replacing σ_1 with the dimensionless stress concentration factor (σ_1 / σ_{eq}). Thus, their criterion is

$$\int_0^{\varepsilon_f} \frac{\sigma_1}{\sigma_{eq}} d\varepsilon_{eq}^p = C_3 \text{, a constant.} \tag{4.227}$$

The main drawback of all the above three criteria is that the effect of hydrostatic stress is not incorporated in this model, even though it is known that hydrostatic stress also influences ductile fracture.

- Oyane's criterion [64]

Oyane's criterion [64] is based on a porous plasticity theory, where it is assumed that the dilatational stress-strain relation is given by

$$d\varepsilon_h = \frac{d\varepsilon_{eq}^p}{A} \left(\frac{\sigma_h}{\sigma_{eq}} + A_0 \right). \tag{4.228}$$

Here, $d\varepsilon_h$ is the hydrostatic part of the incremental linear strain tensor $d\varepsilon_{ij}$ and A and A_0 are material constants. Oyane [64] integrated the above equation over the total strain path to obtain

$$\int_0^{\varepsilon_f} \frac{A}{A_0} d\varepsilon_h = \int_0^{\varepsilon_f} \left(1 + \frac{\sigma_h}{A_0 \sigma_{eq}} \right) d\varepsilon_{eq}^p . \tag{4.229}$$

The left side of the above equation is assumed as a material constant. Then, Oyane's criterion [64] can be stated as

$$\int_0^{\varepsilon_f} \left(1 + \frac{\sigma_h}{A_0 \sigma_{eq}}\right) d\varepsilon_{eq}^p = C_4 \text{, a constant.} \tag{4.230}$$

This criterion apparently incorporates the effect of the hydrostatic stress and plastic strain, the two phenomenological parameters which govern the ductile fracture process. However, in deriving Equation 4.230, it is assumed that the ductile fracture is governed by the hydrostatic part of strain independently of its deviatoric part. This is not true for ductile fracture in pure shear.

- Criteria of Norris et al. [65] and Osakada and Mori [66]

Norris et al. [65] developed a ductile fracture criterion which incorporates both the hydrostatic stress and plastic strain, the two phenomenological parameters governing ductile fracture. Their criterion is based on experimental works and finite difference analyses of various test geometries. It can be stated as

$$\int_0^{\varepsilon_f} \frac{1}{(1 - c\sigma_h)} d\varepsilon_{eq}^p = C_5 \text{, a constant.} \tag{4.231}$$

Here, c is a material constant. Similarly, Osakada and Mori [66] have also proposed a criterion involving both the hydrostatic stress and plastic strain. Their criterion can be stated as

$$\int_0^{\varepsilon_f} \left\langle a + \varepsilon_{eq} + b\sigma_h \right\rangle d\varepsilon_{eq}^p = C_6 \text{, a constant.} \tag{4.232}$$

Here, a and b are material constants and the value of the diamond bracket is taken to be zero if it is negative.

The above two criteria have had limited success in predicting fracture in various metal forming processes, as they are based on observations of certain experiments rather than on microscopic observations on nucleation, growth and coalescence of micro-voids.

- Hydrostatic stress criterion of Reddy et al. [67]

If the die geometry is such that the hydrostatic stress is compressive throughout the deformation zone, then the micro-voids either do not nucleate or the existing micro-voids remain closed. Thus, there is no scope for the initiation of micro-voids. The work of Clift et al. [53] shows that the value of hydrostatic stress at fracture in extrusion and drawing processes is close to zero. Based on these observations, Reddy et al. [67] proposed the following criterion for the prediction of central burst in extrusion and drawing: "Whenever the hydrostatic stress at a point in the plastic deformation zone becomes zero, fracture initiates at that point leading to central burst". They named this criterion 'Hydrostatic stress criterion'.

Predictions of the hydrostatic stress criterion are in good agreement [67, 68] with experimental predictions of the central burst in extrusion and drawing. However, these predictions are conservative [68] compared with the predictions of the critical damage criterion of Dhar *et al.* [57]. This happens because the critical damage criterion of Dhar *et al.* [57] predicts the onset of micro-crack whereas the hydrostatic stress criterion probably predicts the defect (*i.e.*, the central burst) when the micro-crack has grown to a certain measurable size.

Reddy and his co-workers have applied the hydrostatic stress criterion to predict ductile fracture in plane strain rolling [69] and axisymmetric upsetting [70].

4.10 Friction Models

While analysing the metal forming processes, the friction at the die-work interface is often modeled either by the Coulomb's law (as in Equation 3.237 or Equation 3.258) or by the sticking friction model (as in Equation 3.260) or by the friction factor model. The Coulomb's law is subjected to the constraint that the maximum value of the tangential stress component at the interface can not exceed $\sigma_{eq}/\sqrt{3}$ for the Mises material where $\sigma_{eq} = H(\varepsilon_{eq}^p)$. Thus, the Coulomb's law can be stated as (Equation 3.237)

$$|t_s| = f|t_n| \quad \text{for} \quad f|t_n| \le \frac{\sigma_{eq}}{\sqrt{3}}, \tag{4.233}$$

$$|t_s| = \frac{\sigma_{eq}}{\sqrt{3}} \quad \text{for} \quad f|t_n| > \frac{\sigma_{eq}}{\sqrt{3}}, \tag{4.234}$$

where t_s and t_n are respectively the tangential and normal components of the stress vector t_n and f is the coefficient of friction. The tangential (or frictional) stress component acting on the work-piece is in the opposite direction to that of its motion relative to the die.

In his upper bound approach to the analysis of metal forming processes, Avitzur [71] used a friction factor to model the interface friction. In this model, the tangential stress component is expressed as a fraction of its maximum value. Thus,

$$|t_s| = m\frac{\sigma_{eq}}{\sqrt{3}}, \tag{4.235}$$

where the fraction m is called the friction factor. Further, it is assumed to be independent of the normal stress component t_n. Thus, in the friction factor model, the tangential stress component t_s is treated as constant (*i.e.*, independent of the normal stress component t_n). For a *frictionless* case, the value of t_s is zero, while for the *sticking friction* condition, its value is $\sigma_{eq}/\sqrt{3}$. Therefore, the value of m ranges from zero for a frictionless case to unity for the sticking friction condition.

4.10.1 Wanheim and Bay Friction Model

A general friction model has been developed by Wanheim and Bay [72]. Wanheim [73] used the slip-line field technique (for the plane strain case) to study the frictional behavior during metal forming processes and observed that the ratio of real contact area to apparent contact area increases with the normal stress component and approaches the value of unity asymptotically. This happens because very high normal stress is needed in the last phase. This behavior has also been confirmed experimentally. Using the slip-line solution for the ratio of real contact area to apparent contact area and the adhesion theory of friction, a plot of the variation of tangential (or frictional) stress with normal stress has been generated for different values of m [74]. This plot is shown in Figure 4.7.

The curves of Figure 4.7 can be approximated by the following analytical expressions [75]:

$$t_s = \frac{\sigma_Y}{\sqrt{3}}(1-\sqrt{1-m})\frac{t_n}{t_n'}, \quad \text{for} \quad t_n \le t_n' \, , \tag{4.236}$$

$$t_s = t_s' + \left(m\frac{\sigma_Y}{\sqrt{3}} - t_s' \right)\left(1-\exp\left[\frac{(t_n' - t_n)t_s'}{\left(m\frac{\sigma_Y}{\sqrt{3}} - t_s' \right)t_n'} \right] \right), \quad \text{for} \quad t_n > t_n', \tag{4.237}$$

where t_s' and t_n' are respectively the values of the tangential and normal stress components at the proportional limits. They are given by

$$\frac{t_s'}{\sigma_Y} = \frac{(1-\sqrt{1-m})}{\sqrt{3}}, \tag{4.238}$$

$$\frac{t_n'}{\sigma_Y} = \frac{1+\frac{\pi}{2}+\cos^{-1}m+\sqrt{1-m^2}}{\sqrt{3}(1+\sqrt{1-m})}. \tag{4.239}$$

The above expressions are actually for perfectly plastic materials. To make them applicable to strain hardening materials, σ_Y should be replaced by the equivalent stress σ_{eq}.

Figure 4.7 shows that the Wanheim and Bay friction model indicates a smooth transition from the Coulomb's law (applicable at lower forming loads) to the friction factor model (applicable at higher forming loads) with an additional transition range. At low values of forming loads, the slope of the curves in Figure 4.7 is proportional to the coefficient of friction f whereas, at higher forming loads, the tangential stress t_s asymptotically approaches a constant value of the friction factor m.

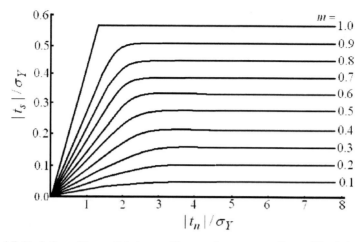

Figure 4.7. Variation of tangential stress with normal stress according to Wanheim and Bay friction model

Thus, for the Wanheim and Bay friction model, the friction boundary condition (Equations 4.233–4.235) is expressed as

$$|t_s| = f|t_n| \qquad \text{up to proportional limit } (i.e., \text{ for } t_n \leq t'_n), \qquad (4.240a)$$

$$|t_s| = f^*|t_n| \qquad \text{in the transition range}, \qquad (4.240b)$$

$$|t_s| = m\frac{\sigma_{eq}}{\sqrt{3}} \qquad \text{beyond the transition range}, \qquad (4.240c)$$

Thus, the coefficient of friction f is constant up to the proportional limit. In the transition range, the variable coefficient of friction f^* is found from Equation 4.237 by replacing σ_Y with σ_{eq}. Note that, f^* is a non-linear function of t_n and therefore, needs to be evaluated iteratively. Richelsen [76] has provided the following expression for f^*:

$$f^* = \frac{t'_s}{t'_n}\exp\left[\frac{(t'_n - t_n)t'_s}{\left(m\dfrac{\sigma_Y}{\sqrt{3}} - t'_s\right)t'_n}\right]. \qquad (4.241)$$

Beyond the transition range, the friction boundary condition is expressed in terms of the friction factor m. The equivalent coefficient of friction $\left(f_{eq} \right)$ is obtained by combining Equations 4.238 and 4.239 as follows:

$$f_{eq} = \frac{t'_s}{t'_n} = \frac{m}{1 + (\pi/2) + \cos^{-1} m + \sqrt{1-m}} \, . \tag{4.242}$$

4.11 Summary

In the first part of this chapter, the plasticity of finite deformation and anisotropic materials has been developed. In the remaining part, some models of ductile fracture and interface friction have been presented.

The kinematics of finite deformation and the corresponding constitutive equation for the Eulerian formulation have been discussed first. The elastic and plastic parts of the constitutive equation have been expressed separately. Since they do not involve stress rate, an objective stress measure is not needed in this constitutive equation. Next, starting from the kinematics of finite incremental deformation, some commonly used measures of finite incremental deformation have been developed. The corresponding constitutive equation, needed in the updated Lagrangian formulation, has been derived. This constitutive equation involves integration over the increment size. A stress updating procedure that makes the incremental stress tensor objective has also been discussed. Some criteria for initial yielding of anisotropic materials, based on phenomenological observations, have been discussed next. Here, the 1948 and 1979 anisotropic yield criteria of Hill have been presented. Since these criteria have certain drawbacks, two additional anisotropic yield criteria that do not have these drawbacks have been discussed: one for the plane stress case and the other for the three-dimensional case. These criteria have been developed by Barlat and his co-workers by applying linear tensor transformations to Hosford's isotropic yield criterion. A plane strain anisotropic yield criterion based on a modification of Hill's 1979 criterion has also been developed. Next, the constitutive equations corresponding to two of the above anisotropic yield criteria have been developed. At the end of this part of the chapter, two kinematic hardening models due to Prager and Ziegler are presented.

In the second part of this chapter, first, the modeling of ductile fracture has been discussed. Microscopic observations have shown that ductile fracture occurs mainly due to micro-void nucleation, growth and finally coalescence into a micro-crack. Three broad approaches which predict the ductile fracture on the basis of the above observation have been discussed: (i) porous plasticity model of Berg and Gurson, (ii) void nucleation, growth and coalescence model (of Goods and Brown, Rice and Tracy, and Thomason) and (iii) continuum damage mechanics models of Lemaitre and Rousselier. Some phenomenological fracture criteria have also been presented. Finally, the modeling of interface friction has been discussed by presenting some commonly used friction models like the Coulomb's law, the friction factor model and a more general Wanheim and Bay friction model.

4.12 References

[1]   Malvern, L.E. (1969), Introduction to the Mechanics of a Continuous Medium, Prentice-Hall Inc., Englewood Cliffs.

[2]   Jaunzemis, W. (1967), Continuum Mechanics, The Macmillan Company, New York.

[3]   Lee, E.H. (1981), Some comments on elastic-plastic analysis, International Journal of Solids & Structures, Vol. 17, pp. 859–872.

[4]   Boyce, M.C., Weber, G.G. and Parks, D.M. (1989), On the kinematics of finite strain plasticity, Journal of Mechanics & Physics of Solids, Vol. 37, pp. 647–665.

[5]   Segal, L.A. (1977), Mathematics Applied to Continuum Mechanics, Macmillan Publishing Co. Inc., New York.

[6]   Weber, G. and Anand, L. (1990), Finite deformation constitutive equations and a time integration procedure for isotropic, hyperelastic-viscoplastic solids, Computer Methods in Applied Mechanics and Engineering, Vol. 79, pp. 173–176.

[7]   Crisfield, M.A. (1997), Non-Linear Finite Element Analysis of Solids and Structures, John Wiley and Sons, Chichester, Vol. 2.

[8]   Kobayashi, S., Oh, S.I. and Altan, T. (1989), Metal Forming and the Finite-Element Method, Oxford university Press, Oxford.

[9]   Nagtegaal, J.C. and DeJong, J.E. (1981), Some computational aspects of elastic-plastic large strain analysis, International Journal of Numerical Methods for Engineering, Vol. 17, pp. 15–41.

[10]   Prager, W. (1961), Introduction to Mechanics of Continua, Ginn and Co., Boston

[11]   Dienes, J.K. (1979), On the analysis of rotation and stress rate in deforming bodies, Acta Mechanica, Vol. 32, pp. 217–232.

[12]   Lee, E.H., Mallet, R.L. and Wertheimer, T.B. (1983), Stress analysis of anisotropic hardening in finite-deformation plasticity, Transaction of ASME, Journal of Applied Mechanics, Vol. 50, pp. 554–560.

[13]   Dafalias, Y.F. (1983), A missing link in the macroscopic constitutive formulation of large plastic deformations: in Plasticity Today, ed. by Sawczuk, A. and Bianchi, G., pp. 135–151, International Symposium on Recent Trends and Results in Plasticity, Udine, Elsevier Applied Science Publishers, London.

[14]   Khan, A.S. and Huang, S. (1995), Continuum Theory of Plasticity, John Wiley and Sons Inc., New York.

[15]   Habraken, A.M. and Duchene, L. (2004), Anisotropic elasto-plastic finite element analysis using a stress-strain interpolation method based on a polycrystalline model, International Journal of Plasticity, Vol. 21, pp. 1525–1560.

[16]   Raabe, D. and Roters, F. (2004), Using texture components in crystal plasticity finite element simulations, International Journal of Plasticity, Vol. 21, pp. 339–361.

[17]   Barlat, F., Chung, K. and Richmond, O. (1993), Strain rate potential for metals and its application to minimum plastic work path calculations, International Journal of Plasticity,, Vol. 9, pp. 51–63.

[18]   Chung, K., Lee, S.Y., Barlat, F., Keum, Y.T. and Park, J.M. (1996), Finite element simulation of sheet forming based on a planar anisotropic strain-rate potential, International Journal of Plasticity,, Vol. 12, pp. 93–115.

[19]   Hill, R. (1948), A theory of the yielding and plastic flow of anisotropic metals, Proceedings of the Royal Society of London, Vol. A 193, pp. 281–297

[20] Hill, R. (1979), Theoretical plasticity of textured aggregates, Mathematical Proceedings of the Cambridge Philosophical Society, Vol. 85, pp. 179–191.

[21] Hosford, W.F. (1972), A generalized isotropic yield function, Transaction of ASME, Journal of Applied Mechanics, Vol. E39, pp. 607–609.

[22] Logan, R.W. and Hosford, W.F. (1980), Upper-bound anisotropic yield locus calculations assuming <111>-pencil glide, International Journal of Mechanical Sciences, Vol. 22, pp. 419–430.

[23] Barlat, F. and Lian, J. (1989), Plastic behavior and stretchability of sheet metals. Part I: A yield function for orthotropic sheets under plane stress conditions, International Journal of Plasticity, Vol. 5, pp. 51–66.

[24] Bishop, J.W.F. and Hill, R. (1951), A theory of the plastic distortion of a polycrystalline aggregate under combined stresses, Philosophical Magazine, Vol. 42, pp. 414–427 and A theoretical derivation of the plastic properties of a polycrystalline face-centered metal, Philosophical Magazine, Vol. 42, pp. 1298–1307.

[25] Barlat, F., Lege, D.J. and Brem, J.C. (1991), A six-component yield function for anisotropic materials, International Journal of Plasticity, Vol. 7, pp. 693–712.

[26] Karafillis, A.P. and Boyce, M.C. (1993), A general anisotropic yield criterion using bounds and a transformation weighting tensor, Journal of Mechanics & Physics of Solids, Vol. 41, pp. 1859–1886.

[27] Barlat, F., Brem, J.C., Yoon, J.W., Chung, K., Dick, R.E., Lege, D.J, Pourboghrat, F., Choi, S.H. and Chu, E. (2003), Plane stress yield function for aluminum alloy sheets, International Journal of Plasticity, Vol. 19, pp. 1297–1319.

[28] Barlat, F., Aretz, H., Yoon, J.W., Karabin, M.E., Brem, J.C. and Dick, R.E. (2005), Linear transformation-based anisotropic yield functions, International Journal of Plasticity, Vol. 21, pp. 1009–1039.

[29] Bron, F. and Besson, J. (2004), A yield function for anisotropic materials: Application to aluminum alloys, International Journal of Plasticity, Vol. 21, pp. 937–963.

[30] Woodthorpe, J. and Pearce, R. (1970), The anomalous behavior of aluminum sheet under balanced biaxial tension, International Journal of Mechanical Sciences, Vol. 12, pp. 341–347.

[31] Barlat, F. and Richmond, O. (1987), Prediction of tricomponent plane stress yield surfaces and associated flow and failure behavior of strongly textured FCC polycrystalline sheets, Material Science and Engineering, Vol. 95, pp. 15–29.

[32] Yoon, J.W., Barlat, F., Dick, R.E. and Karabin, M.E. (2006), Prediction of six or eight ears in a drawn cup based on a new anisotropic yield function, International Journal of Plasticity, Vol. 22, pp. 174–193.

[33] Hu, W. (2005), An orthotropic yield criterion in a 3-D general stress state, International Journal of Plasticity, Vol. 21, pp. 1771–1796.

[34] Dixit, U.S. and Dixit, P.M. (1997), Finite element analysis of flat rolling with inclusion of anisotropy, International Journal of Mechanical Sciences, Vol. 39, pp. 1237–1255.

[35] Prager, W. (1955), The theory of plasticity: A survey of recent achievements, Proceedings for the Institution of Mechanical Engineers, Vol. 169, pp. 41–57.

[36] Ziegler, H. (1959), A modification of Prager's hardening rule, Quarterly of Appied Mathematics, Vol. 17, pp. 55–65.

[37] Shield, R. and Ziegler, H. (1958), On Prager's hardening rule, Zeitschrift für Angewandte Mathematik und Physik , Vol. 9a, pp. 260–276.

[38] Chakrabarty, J. (1987), Theory of Plasticity, McGraw-Hill Book Co., New York.

[39] Rees, D.W.A. (2006), Basic Engineering Plasticity, Elsevier Ltd, Oxford.

[40] Berg, C.A. (1970), Plastic dilation and void interaction, Inelastic Behavior of Solids, Ed. By Kanninen, Adler, Rosenfield and Jaffe, pp. 171–210.

[41] Gurson, A.L. (1977), Continuum theory of ductile rapture by void nucleation and growth, Part I: Yield criteria and flow rules for porous ductile media, Transaction of ASME, Journal of Engineering Materials and Technology, Vol. 99, pp. 2–15.

[42] Goods S.H. and Brown L.M. (1979), The nucleation of cavities by plastic deformation, Acta Metallurgica, Vol 27, pp. 1–15.

[43] Rice, J.R. and Tracy D.M. (1969), On the ductile enlargement of voids in triaxial stress field, Journal of Mechanics & Physics of Solids, Vol. 17, pp. 201–217.

[44] Thomason, P.F. (1990), Ductile Fracture, Pergamon Press.

[45] Lemaitre, J. (1985), A continuous damage mechanics model for ductile fracture, Transaction of ASME, Journal of Engineering Materials and Technology, Vol. 107, pp. 83–89.

[46] Rousselier, G. (1987), Ductile fracture model and their potential in local approach of fracture, Nuclear Engineering and Design, Vol. 105. pp. 97–111.

[47] Tvergaard, V. (1981), Influence of voids on shear band instabilities under plane strain condition, International Journal of Fracture, Vol. 17, pp. 389–406.

[48] Rudnicki, J.W. and Rice, J.R. (1975), Conditions for the localization of deformation in pressure sensitive dilatant materials. Journal of Mechanics & Physics of Solids, Vol. 23, pp. 371–394.

[49] Yamamoto, H. (1978), Conditions for shear localization in the ductile fracture of void containing materials, International Journal of Fracture, Vol 14, pp. 347–365.

[50] Alberti, N., Barcellona, A. Cannizzaro, L. and Micari, F. (1994), Prediction of ductile fracture in metal forming processes: an approach based on the damage mechanics, Annals of CIRP, Vol. 43, pp. 207–210.

[51] Gurland, J. (1972), Observations on the fracture of cementite particles in spherodised 1.05 % C steel deformed at room temperature, Acta Metallurgica, Vol. 20, pp. 735–741.

[52] McClintock, F.A. (1968), A criterion for ductile fracture by the growth of holes, Transaction of ASME, Journal of Applied Mechanics, Vol. 90, pp. 363–371.

[53] Clift, S.E., Hartley, P., Sturgess, C.E.N. and Rowe, G.W. (1990), Fracture prediction in plastic deformation processes, International Journal of Mechanical Sciences, Vol. 32, pp. 1–17.

[54] Oh, S.I., Chen, C.C. and Kobayashi, S. (1979), Ductile fracture in axisymmetric extrusion and drawing, Transaction of ASME, Journal of Engineering for Industry, Vol. 101, pp. 36–44.

[55] Lemaitre, J. and Chaboche, J.L. (1990), Mechanics of Solid Materials, Cambridge University Press, Cambridge.

[56] Le Roy, G., Embury, J.D., Edward, G. and Ashby, M.F. (1981), A model of ductile fracture based on the nucleation and growth of voids, Acta Metallurgica, Vol. 29, pp. 1509–1522.

[57] Dhar, S., Sethuraman, R. and Dixit, P.M. (1996), A continuum damage mechanics model for void growth and micro-crack initiation, Engineering Fracture Mechanics, Vol. 53, pp. 917–928.

[58] Bridgeman, P.W. (1964), Studies in Large Plastic Flow and Fracture, Harvard University Press, Harward.

[59] Tai, W.H. and Yang, B.X. (1986), A new micro-void damage model for ductile fracture, Engineering Fracture Mechanics, Vol. 25, pp. 377–384.

[60] Jun, W.T. (1992), Unified CDM model and local criteria for ductile fracture-I, Engineering Fracture Mechanics, Vol. 42, pp. 177–183.

[61] Zheng, M., Luo, Z.J. and Zheng, X. (1992), A new damage model for ductile material, Engineering Fracture Mechanics, Vol. 41, pp. 103–110.

[62] Freudenthal, A.M. (1950), The Inelastic Behavior of Solids, John Wiley, New York.

[63] Cockcroft, M.G. and Latham, D.J. (1968), Ductility and workability of metals, Journal of the Institute of Metals, Vol. 96, pp. 33–39.

[64] Oyane, M. (1972), Criteria of ductile strain, Bulletin of JSME, Vol. 15, pp. 1507–1513.

[65] Norris, D.M., Reaugh, J.E., Moran, B. and Quinones, D.F. (1978), A plastic strain mean stress criterion for ductile fracture, Transaction of ASME, Journal of Engineering Materials and Technology, Vol. 100, pp. 279–286.

[66] Osakada, K. and Mori, K. (1978), Prediction of ductile fracture in cold forging, Annals of CIRP, Vol. 27, pp. 135–139.

[67] Reddy, N.V., Dixit, P.M. and Lal, G.K. (1996), Central bursting and optimal die profile for axisymmetric extrusion, Transaction of ASME, Journal of Manufacturing Science and Engineering, Vol. 118, pp. 579–584.

[68] Reddy, N.V., Dixit, P.M. and Lal, G.K. (2000), Ductile fracture criteria and its prediction in axisymmetric drawing, International Journal of Machine Tools & Manufacture, Vol. 40, pp. 495–111.

[69] Rajak, S.A. and Reddy, N.V. (2005), Prediction of internal defects in plane strain rolling, Journal of Materials Processing Technology, Vol. 159, pp. 409–417.

[70] Gupta, S., Reddy, N.V. and Dixit, P.M. (2003), Ductile fracture prediction in axisymmetric upsetting using continuum damage mechanics, Journal of Materials Processing Technology, Vol. 143, pp. 256–265.

[71] Avitzur, B. (1968), Metal Forming: Processes and Analysis, McGraw-Hill Book Co., New York.

[72] Wanheim, T. and Bay, N. (1978), A model for friction in metal forming processes, Annals of CIRP, Vol. 27, pp. 189–194.

[73] Wanheim, W. (1973), Friction at high normal pressure, Wear, Vol. 25, pp. 225–244.

[74] Wanheim, W., Bay, N. and Petersen, A.S. (1974), A theoretically determined model for friction in metal working processes, Wear, Vol. 28, pp. 251–258.

[75] Christensen, P., Everfelt, K. and Bay, N. (1986), Pressure distribution in plate rolling, Annals of CIRP, Vol. 35, pp. 141–146.

[76] Richelsen, A.B. (1991), Viscoplastic analysis of plane strain rolling using different friction models, International Journal of Mechanical Sciences, Vol. 33, pp. 761–774.

5

Finite Element Modeling of Metal Forming Processes Using Eulerian Formulation

5.1 Introduction

In Chapter 3 we discussed two methods of formulating a metal forming process—updated Lagrangian formualtion and Eulerian formulation. Eulerain formulation is convenient for processes like rolling, wire drawing, extrusion *etc.*, where there is a continuous flow of material and we can concentrate on a region in space for the analysis purposes. The fixed region in the space is called control volume. The material can be considered as a fluid passing through the control volume. Therefore, this formulation is also called flow formulation. In this formulation, we attempt to find the velocity and pressure (negative of hydrostatic stress) field throughout the region.

In this chapter, we will describe flow formulation using the finite element method (FEM). In the FEM, a region is discretized into a number of small elements called finite elements. The primary variables like displacements in solid mechanics problems are approximated using piecewise continuous functions that are continuous inside the elements. The parameters of the functions are adjusted to minimize the error in the solution. After the solution is obtained, it is post-processed to compute the desired secondary quantities like strain, stress *etc*. The following section provides a background of finite element method, assuming that the reader has not done a course on finite element method. The reader may also wish to read a textbook on FEM before going through this chapter [1, 2].

Flow formulation has been widely employed for rigid-plastic analysis of metal forming processes. It has also been employed for elasto-plastic analysis, particularly for finding out the residual stresses in metal forming. However, in elasto-plastic flow formulation, the researchers could achieve only limited success and the area is still open for research.

5.2 Background of Finite Element Method

The finite element method is a numerical method for solving differential and integral equations. In this method, the unknown variables to be determined are approximated by piecewise continuous functions. The coefficients of the functions are adjusted in such a manner that the error in the solution is minimized. Usually, the coefficients of the functions of a particular element are the values at certain points in the element called nodes. During the solution process, the differential equations get converted to algebraic equations or ordinary differential equations that can be solved by finite difference equations. The finite element method consists of the following steps: (i) Pre-processing; (ii) Developing elemental equations; (iii) Assembling equation; (iv) Applying boundary conditions; (v) Solving the system of equations; and (v) Post-processing. We shall describe each step in the following subsections.

5.2.1 Pre-processing

In this step, the domain is discretized into a number of small elements. The elements can be of different shapes and sizes. Of course, in one dimensional problems, the element has only one shape *i.e.,* a line. Figure 5.1a shows a line element with two nodal points at the corners of the element. With this element, one can approximate a primary variable u by the following interpolation function:

$$u = a + bx . \tag{5.1}$$

Then, the task of solution is just to find out the constants a and b. However, it is better to replace these constants by the unknown values of u at nodes. If the coordinates of two nodes are x_1 and x_2, then the values at the nodes are

$$u_1 = a + bx_1 , \tag{5.2}$$
$$u_2 = a + bx_2 . \tag{5.3}$$

Solving these two equations for a and b, and substituting these values in Equation 5.1, u can be written as

$$u = \frac{x_2 - x}{x_2 - x_1} u_1 + \frac{x - x_1}{x_2 - x_1} u_2 \equiv N_1 u_1 + N_2 u_2 , \tag{5.4}$$

where N_1 and N_2 are called shape functions, because they give an idea of the shape of the approximating function u. With this, the task of solution gets transformed into finding the nodal values of the primary variable u. We could have directly dealt with unknown constants a and b. However, replacing these by nodal variables offers two main advantages:

  (1) The nodal variables convey physical meaning, whereas the constants a and b do not represent any physical quantity.

(2) The continuity of function representing primary variable is ensured (at least at nodes) by expressing the interpolation function in terms of nodal variables, because from two adjacent elements the values at the nodes will be same. By proper choice of the shape of elements in two and three dimensions, it can be ensured that; between the two elements, the function is continuous not only at the nodes, but also at the entire interfacial boundary. If the continuity of derivatives is also required, the derivatives can also be included into nodal variables.

The two-noded element can provide only a linear approximation. If a quadratic approximation is required, a three-noded element as shown in Figure 5.1b can be used. It can easily be shown that the approximating polynomial u can be expressed in terms of the nodal variables as

$$u = \frac{(x-x_2)(x-x_3)}{(x_1-x_2)(x_1-x_3)}u_1 + \frac{(x-x_1)(x-x_3)}{(x_2-x_1)(x_2-x_3)}u_2 + \frac{(x-x_1)(x-x_2)}{(x_3-x_1)(x_3-x_2)}u_3,$$

$$\equiv N_1 u_1 + N_2 u_2 + N_3 u_3.$$

$$(5.5)$$

These shape functions are basically Lagrangian interpolation functions and are therefore called Lagrangian shape functions. The elements shown in Figures 5.1a, b are called one-dimensional Lagrangian elements. They ensure that the function is a continuous function of x throughout the domain, but do not guarantee the continuity of the first derivative of the function. The continuity of the function but not of its derivatives is called the continuity of zero order or C^0 continuity. The continuity of the function as well as its first derivative is called C^1 continuity. It can easily be verified that elements shown in Figure 5.1a, b possess C^0 continuity property but not C^1 continuity.

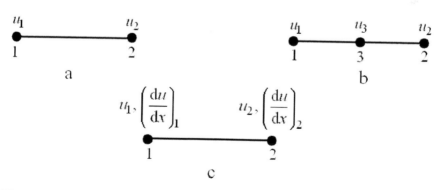

Figure 5.1. a A two-noded Lagrangian element. **b** A three-noded Lagrangian element. **c** A two-noded Hermitian element

A two noded element having nodal variables as the values of u and du/dx at the nodes can ensure C^1 continuity. One such element is shown in Figure 5.1c. It can be shown that the approximating function u can be expressed in terms of nodal variables as

$$u = N_1 u_1 + N_2 \left(\frac{du}{dx} \right)_1 + N_3 u_2 + N_4 \left(\frac{du}{dx} \right)_2, \tag{5.6}$$

where

$$N_1 = 1 - 3 \left(\frac{x - x_1}{x_2 - x_1} \right)^2 + 2 \left(\frac{x - x_1}{x_2 - x_1} \right)^3, \quad N_2 = (x - x_1) \left(1 - \frac{x - x_1}{x_2 - x_1} \right)^2,$$

$$N_3 = 3 \left(\frac{x - x_1}{x_2 - x_1} \right)^2 - 2 \left(\frac{x - x_1}{x_2 - x_1} \right)^3, \quad N_4 = (x - x_1) \left(\left(\frac{x - x_1}{x_2 - x_1} \right)^2 - \frac{x - x_1}{x_2 - x_1} \right). \tag{5.7}$$

These shape functions are called Hermitian shape function. They all are cubic functions; hence the approximating function in Equation 5.6 is a cubic polynomial.

It is convenient first to transform the physical coordinate in natural coordinate varying from -1 to $+1$ and express the interpolation function in natural coordinate. This practice offers the following advantages:

(1) The expressions for shape function are conveniently written and can be stored in the library of FEM software.
(2) It facilitates numerical integration, as the popular technique of numerical integration, the Gaussian-quadrature requires the function to lie in the range of -1 to $+1$.
(3) The primary variable of the problem can be expressed as one polynomial function of the natural coordinates and the physical coordinate can be expressed as another polynomial function of the natural coordinates. The result may be that the variable u becomes a non-polynomial function of the physical coordinate. For example, if u is expressed as a linear function of natural corrdinate ξ:

$$u = a + b\xi, \tag{5.8}$$

and x as a quadratic function of ξ, say $x = c\xi^2$, then

$$u = a + \frac{b}{\sqrt{c}} \sqrt{x}. \tag{5.9}$$

Usually, a linear relation between the physical and natural coordinates is assumed, i.e.,

$$x = \frac{x_1 + x_2}{2} + \frac{x_2 - x_1}{2} \xi. \tag{5.10}$$

In that case, in natural coordinates, Lagrangian shape functions of two noded element used in Equation 5.4 are given by

$$N_1 = \frac{(1-\xi)}{2}, \qquad N_2 = \frac{(1+\xi)}{2}. \tag{5.11}$$

It can be clearly seen that the shape function N_i has the value 1 at node i. Also, N_1 is 0 at node 2 and N_2 is 0 at node 1. The sum of the shape functions is constant and is equal to 1. For three-noded elements, the shape functions in the natural coordinates are

$$N_1 = \frac{1}{2}\xi(\xi-1), \quad N_2 = \frac{1}{2}\xi(\xi+1), \quad N_3 = 1-\xi^2. \tag{5.12}$$

Here, node 1 and node 2 are the corner nodes with natural coordinates -1 and $+1$ respectively and node 3 is the middle node with $\xi = 0$. It can be verified that a particular shape function N_i is 1 at the i-th node and 0 at other nodes. Also, the sum of the shape functions is 1. The Lagrangian shape functions satisfy the following properties:

Property 1: The shape function N_i is 1 at the i-th node.
Property 2: The shape function N_i is 0 at all other nodes.
Property 3: The sum of shape functions is 1.

In natural coordinates, the Hermitian shape functions for two noded elements (Equation 5.7) are given by

$$
\begin{aligned}
N_1 &= \frac{1}{4}\left(2-3\xi+\xi^3\right), & N_2 &= \frac{h}{8}\left(1-\xi-\xi^2+\xi^3\right), \\
N_3 &= \frac{1}{4}\left(2+3\xi-\xi^3\right), & N_4 &= -\frac{h}{8}\left(1+\xi-\xi^2-\xi^3\right),
\end{aligned}
\tag{5.13}
$$

where h is the length of the element.

In two dimensions, the simplest shape is a three-noded linear triangle, as shown in Figure 5.2a. In this case, the displacement components u_x and u_y can be approximated by first degree complete polynomials:

$$u_x = a_1 + b_1 x + c_1 y, \quad u_y = a_2 + b_2 x + c_2 y. \tag{5.14}$$

With this interpolation, planar strains are found as

$$\varepsilon_{xx} = \frac{\partial u_x}{\partial x} = b_1, \quad \varepsilon_{yy} = \frac{\partial u_y}{\partial y} = c_2, \quad \varepsilon_{xy} = \frac{1}{2}\left(\frac{\partial u_x}{\partial y}+\frac{\partial u_y}{\partial x}\right) = \frac{1}{2}(c_1+b_2). \tag{5.15}$$

We observe that these strains inside the element do not depend on x and y, but are constant. Hence, this element is called constant strain triangle (CST). As in the case of one-dimensional elements, the constants in the approximating function may be replaced by nodal values. Thus,

$$u_x = N_1 u_1 + N_2 u_2 + N_3 u_3, \quad u_y = N_1 v_1 + N_2 v_2 + N_3 v_3, \tag{5.16}$$

where

$$N_1 = \frac{(xy_2 - x_2 y) + (x_2 y_3 - x_3 y_2) + (x_3 y - x y_3)}{(x_1 y_2 - x_2 y_1) + (x_2 y_3 - x_3 y_2) + (x_3 y_1 - x_1 y_3)}, \tag{5.117a}$$

$$N_2 = \frac{(x_1 y - x y_1) + (x y_3 - x_3 y) + (x_3 y_1 - x_1 y_3)}{(x_1 y_2 - x_2 y_1) + (x_2 y_3 - x_3 y_2) + (x_3 y_1 - x_1 y_3)}, \tag{5.117b}$$

$$N_3 = \frac{(x_1 y_2 - x_2 y_1) + (x_2 y - x y_2) + (x y_1 - x_1 y)}{(x_1 y_2 - x_2 y_1) + (x_2 y_3 - x_3 y_2) + (x_3 y_1 - x_1 y_3)}. \tag{5.117c}$$

As in the case of one-dimensional shape functions, the three properties mentioned before are also being satisfied by these shape functions. Thus, the shape function N_i will be zero at all nodes except at node i (where its value is equal to 1) and sum of all the shape functions will be 1. Also, the shape functions are linear functions of x and y. Referring to Figure 5.2a, we may define the shape functions at a point P as

$$N_1 = \frac{\text{area of triangle P23}}{\text{area of triangle 123}}, \; N_2 = \frac{\text{area of triangle P13}}{\text{area of triangle 123}}, \; N_3 = \frac{\text{area of triangle P12}}{\text{area of triangle 123}}. \tag{5.18}$$

The coordinates of point P in Figure 5.2a are (x, y). We can transform these coordinates into natural coordinates so that their values lie between 0 and 1. It is a common practice to use three natural coordinates ξ_1, ξ_2 and ξ_3 for a triangle, out of which only two are independent as they satisfy the following relation:

$$\xi_1 + \xi_2 + \xi_3 = 1. \tag{5.19}$$

The coordinates of the three vertices of the triangle in the natural coordinates are taken as $(1, 0, 0)$, $(0, 1, 0)$ and $(0, 0, 1)$, respectively. The following relation is employed to map the natural coordinates into physical coordinates:

$$x = x_1 \xi_1 + x_2 \xi_2 + x_3 \xi_3, \quad y = y_1 \xi_1 + y_2 \xi_2 + y_3 \xi_3. \tag{5.20}$$

The natural coordinates of point P can be obtained by inverting Equations 5.19 and 5.20. Alternatively, they can also be obtained using the following relations:

$$\xi_1 = \frac{\text{area of triangle P23}}{\text{area of triangle 123}}, \quad \xi_2 = \frac{\text{area of triangle P13}}{\text{area of triangle 123}}, \quad \xi_3 = \frac{\text{area of triangle P12}}{\text{area of triangle 123}}.$$

$$(5.21)$$

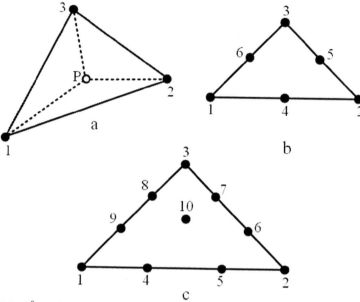

Figure 5.2. C^0 continuity triangular elements. **a** With three nodes. **b** With six nodes. **c** With ten nodes

In terms of natural coordinates, the shape functions of three noded linear triangular elements are given as

$$N_1 = \xi_1, \quad N_2 = \xi_2, \quad N_3 = \xi_3.$$

$$(5.22)$$

Figure 5.2b shows a six noded element. In terms of the natural coordinates, the shape functions for this element are

$$N_1 = \xi_1(2\xi_1 - 1), \quad N_2 = \xi_2(2\xi_2 - 1), \quad N_3 = \xi_3(2\xi_3 - 1),$$
$$N_4 = 4\xi_2\xi_3, \quad N_5 = 4\xi_2\xi_3, \quad N_6 = 4\xi_1\xi_3,$$

$$(5.23)$$

where the mid-side nodes 4, 5 and 6 have the natural coordinates $(1/2, 1/2, 0)$, $(0, 1/2, 1/2)$ and $(1/2, 0, 1/2)$, respectively. This element interpolates the displacement components u_x and u_y by second degree complete polynomials.

In general, physical coordinates can be expressed as a polynomial function of the natural coordinates. We can make use of the shape functions to express the relation between physical and natural coordinates. Thus, if the shape functions are expressed in natural coordinates, then

$$x = \sum_{i=1}^{n} N_i x_i, \quad y = \sum_{i=1}^{n} N_i y_i, \quad z = \sum_{i=1}^{n} N_i z_i, \tag{5.24}$$

where n is the number of nodal coordinates to approximate the geometry and x_i, y_i and z_i are the nodal coordinates. Note that Equation 5.24 is valid not only for triangular elements, but in general for any element. If the degrees of polynomials approximating the nodal coordinates and physical variable are the same, the FEM formulation is called iso-parametric formulation, for example that represented by Equations 5.16 and 5.20. If the primary variable is approximated by a higher degree polynomial than the geometry, then the FEM formulation is called sub-parametric formulation. If the primary variable is approximated by a lower degree polynomial than the geometry, then the FEM formulation is called a super-parametric formulation.

The nodes in a triangle need not be on the sides, they can also be inside. Figure 5.2c shows a 10 noded element, the 10-th node being the centroid. The approximation for this element is a complete cubic polynomial in x and y.

A four noded quadrilateral element is shown in Figure 5.3. It can be mapped into a square element having the natural coordinates of corner nodes as $(-1,-1)$, $(1,-1)$, $(1,1)$ and $(-1,1)$. Choosing bi-linear approximation $u = a + b\xi + c\eta + d\xi\eta$, the shape functions in natural coordinates are obtained as

$$N_1 = \frac{1}{4}(1-\xi)(1-\eta), \quad N_2 = \frac{1}{4}(1+\xi)(1-\eta),$$
$$N_3 = \frac{1}{4}(1+\xi)(1+\eta), \quad N_4 = \frac{1}{4}(1-\xi)(1+\eta). \tag{5.25}$$

With these shape functions, the approximation for the primary variable becomes

$$u = \sum_{i=1}^{4} N_i u_i, \tag{5.26}$$

and in iso-parametric formulation, the geometry can be approximated as

$$x = \sum_{i=1}^{4} N_i x_i, \quad y = \sum_{i=1}^{4} N_i y_i. \tag{5.27}$$

Note that u is not necessarily a bilinear function of the physical coordinates. In the special case of rectangular element, where $\xi = c_1 x + c_2$ and $\eta = c_3 y + c_4$, Equation 5.26 provides a bilinear relation in the physical coordinates as well.

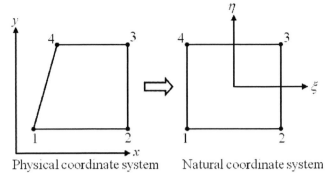

Physical coordinate system   Natural coordinate system

Figure 5.3. A four-noded quadrilateral element in the physical and natural coordinate systems

The above element is called four-noded (bilinear) Lagrangian element. The nine-noded Lagrangian element is shown in Figure 5.4a, where the approximation of primary variable in natural coordinates is biquadratic. The shape functions associated with the element are

$$N_1 = \frac{1}{4}\left(\xi^2 - \xi\right)\left(\eta^2 - \eta\right), \quad N_2 = \frac{1}{4}\left(\xi^2 + \xi\right)\left(\eta^2 - \eta\right),$$

$$N_3 = \frac{1}{4}\left(\xi^2 + \xi\right)\left(\eta^2 + \eta\right), \quad N_4 = \frac{1}{4}\left(\xi^2 - \xi\right)\left(\eta^2 + \eta\right),$$

$$N_5 = \frac{1}{4}\left(1 - \xi^2\right)\left(\eta^2 - \eta\right), \quad N_6 = \frac{1}{4}\left(\xi^2 + \xi\right)\left(1 - \eta^2\right), \tag{5.28}$$

$$N_7 = \frac{1}{4}\left(1 - \xi^2\right)\left(\eta^2 + \eta\right), \quad N_8 = \frac{1}{4}\left(\xi^2 - \xi\right)\left(1 - \eta^2\right),$$

$$N_9 = \left(1 - \xi^2\right)\left(1 - \eta^2\right).$$

A separate family of elements without internal nodes is used in FEM literature and has been named Serendipity elements. The eight-noded Serendipity element shown in Figure 5.4b has the following shape functions:

$$N_5 = \frac{1}{2}\left(1 - \xi^2\right)\left(1 - \eta\right), \qquad N_6 = \frac{1}{2}\left(1 + \xi\right)\left(1 - \eta^2\right),$$

$$N_7 = \frac{1}{2}\left(1 - \xi^2\right)\left(1 + \eta\right), \qquad N_8 = \frac{1}{2}\left(1 - \xi\right)\left(1 - \eta^2\right),$$

$$N_1 = \frac{1}{4}\left(1 - \xi\right)\left(1 - \eta\right) - \frac{N_5}{2} - \frac{N_8}{2}, \quad N_2 = \frac{1}{4}\left(1 + \xi\right)\left(1 - \eta\right) - \frac{N_5}{2} - \frac{N_6}{2}, \tag{5.29}$$

$$N_3 = \frac{1}{4}\left(1 + \xi\right)\left(1 + \eta\right) - \frac{N_6}{2} - \frac{N_7}{2}, \quad N_4 = \frac{1}{4}\left(1 - \xi\right)\left(1 + \eta\right) - \frac{N_7}{2} - \frac{N_8}{2}.$$

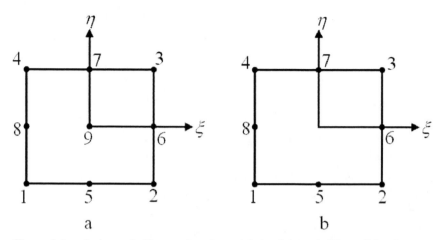

Figure 5.4. **a** A nine-noded Lagrangian element. **b** An eight-noded Serendipity element

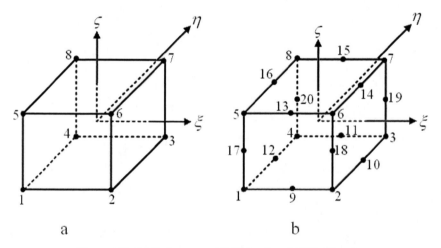

Figure 5.5. Brick elements. **a** With 8 nodes. **b** With 20 nodes

A three-dimensional brick element shown in Figure 5.5a consists of a cube of side 2. The origin of the natural coordinates system ξ-η-ζ is at the centroid of the cube. With tri-linear approximation for primary variables, the shape functions are given by

$$N_1 = \frac{1}{8}(1-\xi)(1-\eta)(1-\zeta), \qquad N_2 = \frac{1}{8}(1+\xi)(1-\eta)(1-\zeta),$$

$$N_3 = \frac{1}{8}(1+\xi)(1+\eta)(1-\zeta), \qquad N_4 = \frac{1}{8}(1-\xi)(1+\eta)(1-\zeta),$$

$$N_5 = \frac{1}{8}(1-\xi)(1-\eta)(1+\zeta), \qquad N_6 = \frac{1}{8}(1+\xi)(1-\eta)(1+\zeta),$$

$$N_7 = \frac{1}{8}(1+\xi)(1+\eta)(1+\zeta), \qquad N_8 = \frac{1}{8}(1-\xi)(1+\eta)(1+\zeta).$$

$$(5.30)$$

A 3-dimensional 20-noded brick element is shown in Figure 5.5b. The shape functions of this element are as follows:

(A) Mid-side nodes:

$$N_9 = \frac{(1-\xi^2)(1-\eta)(1-\zeta)}{4}, \quad N_{10} = \frac{(1-\eta^2)(1+\xi)(1-\zeta)}{4},$$

$$N_{11} = \frac{(1-\xi^2)(1+\eta)(1-\zeta)}{4}, \quad N_{12} = \frac{(1-\eta^2)(1-\xi)(1-\zeta)}{4},$$

$$N_{13} = \frac{(1-\xi^2)(1-\eta)(1+\zeta)}{4}, \quad N_{14} = \frac{(1-\eta^2)(1+\xi)(1+\zeta)}{4},$$

$$(5.31)$$

$$N_{15} = \frac{(1-\xi^2)(1+\eta)(1+\zeta)}{4}, \quad N_{16} = \frac{(1-\eta^2)(1-\xi)(1+\zeta)}{4},$$

$$N_{17} = \frac{(1-\zeta^2)(1-\xi)(1-\eta)}{4}, \quad N_{18} = \frac{(1-\zeta^2)(1+\xi)(1-\eta)}{4},$$

$$N_{19} = \frac{(1-\zeta^2)(1+\xi)(1+\eta)}{4}, \quad N_{20} = \frac{(1-\zeta^2)(1-\xi)(1+\eta)}{4}.$$

(B) Corner nodes:

$$N_1 = \frac{1}{8}(1-\xi)(1-\eta)(1-\zeta) - \frac{1}{2}(N_9 + N_{12} + N_{17}),$$

$$N_2 = \frac{1}{8}(1+\xi)(1-\eta)(1-\zeta) - \frac{1}{2}(N_9 + N_{10} + N_{18}),$$

$$N_3 = \frac{1}{8}(1+\xi)(1+\eta)(1-\zeta) - \frac{1}{2}(N_{10} + N_{11} + N_{19}),$$

$$N_4 = \frac{1}{8}(1-\xi)(1+\eta)(1-\zeta) - \frac{1}{2}(N_{11} + N_{12} + N_{20}),$$

$$(5.32)$$

$$N_5 = \frac{1}{8}(1-\xi)(1-\eta)(1+\zeta) - \frac{1}{2}(N_{13} + N_{16} + N_{17}),$$

$$N_6 = \frac{1}{8}(1+\xi)(1-\eta)(1+\zeta) - \frac{1}{2}(N_{13} + N_{14} + N_{18}),$$

$$N_7 = \frac{1}{8}(1+\xi)(1+\eta)(1+\zeta) - \frac{1}{2}(N_{14} + N_{15} + N_{19}),$$

$$N_8 = \frac{1}{8}(1-\xi)(1+\eta)(1+\zeta) - \frac{1}{2}(N_{15} + N_{16} + N_{20}).$$

It can be easily verified that all these shape functions are zero at all nodes except at one node, where they adopt a value 1. Also, the sum of the shape functions is 1.

Analogous to triangular element in 2-D, we have tetrahedral element (Figure 5.6a) in 3-D. It is common to use four natural coordinates ξ_1, ξ_2, ξ_3 and ξ_4 for a tetrahedron, out of which only three are independent as they also satisfy the following relation:

$$\xi_1 + \xi_2 + \xi_3 + \xi_4 = 1, \tag{5.33}$$

The coordinates of four vertices of the tetrahedron in natural coordinates are taken as (1, 0, 0, 0), (0, 1, 0, 0), (0, 0, 1, 0) and (0, 0, 0, 1) respectively. We can express the coordinates of a point, say P, in terms of natural coordinates ξ_1, ξ_2, ξ_3 and ξ_4 as

$$\xi_1 = \frac{\text{Volume of tetrahedron P234}}{\text{Volume of tetrahedron 1234}}, \quad \xi_2 = \frac{\text{Volume of tetrahedron P134}}{\text{Volume of tetrahedron 1234}},$$
$$\xi_3 = \frac{\text{Volume of tetrahedron P124}}{\text{Volume of tetrahedron 1234}}, \quad \xi_4 = \frac{\text{Volume of tetrahedron P123}}{\text{Volume of tetrahedron 1234}}. \tag{5.34}$$

Recall that the volume V of a tetrahedron is given by

$$V = \frac{1}{6} \begin{vmatrix} x_1 & y_1 & z_1 & 1 \\ x_2 & y_2 & z_2 & 1 \\ x_3 & y_3 & z_3 & 1 \\ x_4 & y_4 & z_4 & 1 \end{vmatrix}, \tag{5.35}$$

where the coordinates of vertex i are (x_i, y_i, z_i). In terms of natural coordinates, the shape functions of a four-noded tetrahedral element are

$$N_1 = \xi_1, \quad N_2 = \xi_2, \quad N_3 = \xi_3, \quad N_4 = \xi_4. \tag{5.36}$$

The shape functions of a 10-noded tetrahedral element (Figure 5.6b) are

(A) Mid-side nodes:

$$N_5 = 4\xi_1\xi_2, \quad N_6 = 4\xi_2\xi_3, \quad N_7 = 4\xi_1\xi_3,$$
$$N_8 = 4\xi_1\xi_4, \quad N_9 = 4\xi_2\xi_4, \quad N_{10} = 4\xi_3\xi_4; \tag{5.37}$$

(B) Corner nodes:

$$N_1 = \xi_1(2\xi_1 - 1), \quad N_2 = \xi_2(2\xi_2 - 1),$$
$$N_3 = \xi_3(2\xi_3 - 1), \quad N_4 = \xi_4(2\xi_4 - 1). \tag{5.38}$$

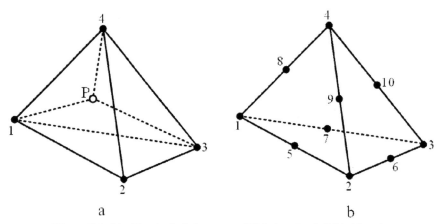

Figure 5.6. The Tetrahedral elements. **a** With 4 nodes. **b** With 10 nodes

Appropriate type of element has to be chosen as per the finite element formulation. The reader will get some idea after reading the following subsection. Usually, we carry out the analysis first with a coarse mesh. If the desired accuracy is not obtained, we either increase the number of elements by reducing the size (*h*-refinement) or increase the order of approximating polynomial (*p*-refinement). A combination of these two strategies of refinement called *hp*-refinement may also be employed.

5.2.2 Developing Elemental Equations

In this subsection, we discuss the method to convert a differential equation into a system of algebraic equations, solution of which provides nodal values of the primary variable. There are many ways to do this. In this book, we confine ourselves to the Galerkin method, which is one type of weighted residual method. In the weighted residual method, the approximating function is substituted in the differential equation to obtain the residual. Then to minimize the residual, its weighted integral set to zero. For example, consider the differential equation

$$L\phi + q = 0 , \tag{5.39}$$

where L is the differential operator, ϕ is the unknown primary variable and q is some function. Now, if we approximate ϕ by some function $\tilde{\phi}$ over an element, the residual R will be given by

$$R = L\tilde{\phi} + q . \tag{5.40}$$

(For the sake of convenience, henceforth, we shall drop symbol over-tilde for denoting the approximating function.) After multiplying this residual by a weight function w, we integrate it over the elemental domain D and equate it to zero, *i.e.*,

$$\int_D wR\,dD = \int_D w(L\phi + q)\,dD = 0. \tag{5.41}$$

Now, the following question arises. What should be the weight function? If interpolating function u^e has n unknown parameters to be determined, then we can make the residuals zero by taking n different independent weight functions. This provides n simultaneous equations, solution of which provides the values of the parameters of u. In finite element method, these parameters are usually (not always) the nodal values of the primary variables. Note that in Equation 5.41, ϕ should be chosen such that $L\phi$ is not zero at all points of the domain. Further, ϕ must be differentiable inside the element up to the order of differential equation. Also, if k is the order of the differential equation, then at the interface of two elements $(k-1)$-th derivative of ϕ should be continuous. Thus, there is a strong requirement of continuity of ϕ and its derivatives. Therefore, this type of formulation is called strong formulation and is not common in the literature of finite element method.

We can integrate the expression at Equation 5.41 by parts. In doing so, the order of derivative of ϕ will reduce and that of w will increase. We try to balance the order of derivative of w and ϕ. This way the differentiability requirement gets reduced. This type of formulation is called weak formulation. There are a number of different types of weak formulations. In the Galerkin method, the weight functions are the same as the shape functions of the interpolating functions. Instead of n separate weight functions, we can make a single weight function as a linear combination of the shape functions. Thus,

$$w = \sum_{i=1}^{n} w_i N_i , \tag{5.42}$$

where w_i are arbitrary constants called the nodal weights. The weighted residual corresponding to this weight function provides an expression, which is the sum of n sub-expressions multiplied by nodal weights. As the nodal weights are arbitrary and independent, the combined expression will be zero if and only if all sub-expressions become zero. Thus, we get n equations, the solution of which determines the interpolating function ϕ.

Example 5.1: The governing equation for the two-dimensional steady-state conduction with heat generation in an orthotropic material is given as

$$\frac{\partial}{\partial x}\left(k_x \frac{\partial T}{\partial x} \right) + \frac{\partial}{\partial y}\left(k_y \frac{\partial T}{\partial y} \right) + \dot{Q} = 0, \tag{5.43}$$

where T is the temperature, k_x and k_y are the thermal conductivities in the x and y directions respectively and $\dot{Q}$ is the rate of heat generation per unit volume. Obtain the FEM formulation for this problem.

Solution: Let the approximate temperature function in an element be T. Then residual in the element is given by

$$R = \frac{\partial}{\partial x}\left(k_x \frac{\partial T}{\partial x}\right) + \frac{\partial}{\partial y}\left(k_y \frac{\partial T}{\partial y}\right) + \dot{Q}.$$

(5.44)

We make the weighted integral of this expression over the elemental domain A to be zero, *i.e.*,

$$\int_A w \left\{ \frac{\partial}{\partial x}\left(k_x \frac{\partial T}{\partial x}\right) + \frac{\partial}{\partial y}\left(k_y \frac{\partial T}{\partial y}\right) + \dot{Q} \right\} dA = 0.$$

(5.45)

Now, we have to reduce the order of derivative of T at the cost of the order of derivative of w. For that, we can make use of the following theorem (which follows from the divergence theorem):

$$\int_A \phi_{,k}\, dA = \oint \phi n_k\, dB,$$

(5.46)

where ϕ is a scalar field and n_k is the component of unit normal on the boundary. Thus, making use of Equation 5.46, we can write

$$\int_A w \frac{\partial}{\partial x}\left(k_x \frac{\partial T}{\partial x}\right) dx\, dy = \int_A \frac{\partial}{\partial x}\left(w k_x \frac{\partial T}{\partial x}\right) dx dy - \int_A \frac{\partial w}{\partial x} k_x \frac{\partial T}{\partial x} dx dy,$$

$$= \oint w k_x \frac{\partial T}{\partial x} n_x dB - \int_A k_x \frac{\partial w}{\partial x} \frac{\partial T}{\partial x} dx dy.$$

(5.47)

Similarly,

$$\int_A w \frac{\partial}{\partial y}\left(k_y \frac{\partial T}{\partial y}\right) dx\, dy = \oint w k_y \frac{\partial T}{\partial y} n_y dB - \int_A k_y \frac{\partial w}{\partial y} \frac{\partial T}{\partial y} dx dy.$$

(5.48)

Thus, the weak form of the weighted integral of the residual becomes

$$-\int_A \left(k_x \frac{\partial w}{\partial x}\frac{\partial T}{\partial x} + k_y \frac{\partial w}{\partial y}\frac{\partial T}{\partial y} - \dot{Q}\right) dx dy + \oint w \left(k_x \frac{\partial T}{\partial x} n_x + k_y \frac{\partial T}{\partial y} n_y\right) dB.$$

(5.49)

Setting the weak form to zero and rearranging the terms, we get

$$\int_A \left(k_x \frac{\partial w}{\partial x}\frac{\partial T}{\partial x} + k_y \frac{\partial w}{\partial y}\frac{\partial T}{\partial y}\right) dx dy = \int_A w\dot{Q} dx dy + \oint w \left(k_x \frac{\partial T}{\partial x} n_x + k_y \frac{\partial T}{\partial y} n_y\right) dB.$$

(5.50)

Observing the weak form, we notice that the requirement is that the first derivative of temperature and weight function should exist inside the element. Also, the functions need to be C^0 continuous everywhere. Thus, a three-noded triangular or four-noded quadrilateral element is sufficient to model this problem.

Now, let

$$T = \{N\}^{\mathrm{T}} \{T\}^e \, , \tag{5.51}$$

where $\{N\}^{\mathrm{T}}$ is the row vector of the shape functions and $\{T\}^e$ is the vector containing the nodal temperatures. We approximate, the w in the same way, i.e.,

$$w = \{w\}^{e\mathrm{T}} \{N\} \, , \tag{5.52}$$

where $\{w\}^{e\mathrm{T}}$ is the row vector of nodal weights and $\{N\}$ is the column vectors of the shape functions. Substituting Equations 5.51 and 5.52 in Equation 5.50, the following expression is obtained:

$$\int_A \{w\}^{e\mathrm{T}} \left[k_x \{N_{,x}\}\{N_{,x}\}^{\mathrm{T}} + k_y \{N_{,y}\}\{N_{,y}\}^{\mathrm{T}} \right] \mathrm{d}x\,\mathrm{d}y \{T\}^e$$

$$= \int_A \{w\}^{e\mathrm{T}} \{N\} \dot{Q}\,\mathrm{d}x\,\mathrm{d}y + \oint \{w\}^{e\mathrm{T}} \{N\} \left(k_x \frac{\partial T}{\partial x} n_x + k_y \frac{\partial T}{\partial y} n_y \right) \mathrm{d}B. \tag{5.53}$$

As $\{w\}^{e\mathrm{T}}$ is an arbitrary row vector of weights, we can eliminate it to obtain

$$\int_A \left[k_x \{N_{,x}\}\{N_{,x}\}^{\mathrm{T}} + k_y \{N_{,y}\}\{N_{,y}\}^{\mathrm{T}} \right] \mathrm{d}x\,\mathrm{d}y \{T\}^e$$

$$= \int_A \{N\} \dot{Q}\,\mathrm{d}x\,\mathrm{d}y + \oint \{N\} \left(k_x \frac{\partial T}{\partial x} n_x + k_y \frac{\partial T}{\partial y} n_y \right) \mathrm{d}B. \tag{5.54}$$

The first term on the left hand side of the equality sign is called the element coefficient matrix:

$$[k]^e = \int_A \left[k_x \{N_{,x}\}\{N_{,x}\}^{\mathrm{T}} + k_y \{N_{,y}\}\{N_{,y}\}^{\mathrm{T}} \right] \mathrm{d}x\,\mathrm{d}y \, . \tag{5.55}$$

The two terms on the right hand side are together called the element right hand side vector. The first term on the right hand side provides the vector due to heat load:

$$\{f\}^e = \int_A \dot{Q} \{N\} \mathrm{d}x\,\mathrm{d}y \, . \tag{5.56a}$$

The second term on the right hand side of sign provides the vector due to heat flux across the boundaries:

$$\left\{ f_{\text{int}} \right\}^e = \oint \{N\} \left(k_x \frac{\partial T}{\partial x} n_x + k_y \frac{\partial T}{\partial y} n_y \right) dB . \tag{5.56b}$$

For a three-noded triangular element, $[k]^e$ will be of size 3×3 whereas $\{f\}^e$ and $\left\{ f_{\text{int}} \right\}^e$ will be of size 3×1.

In Equation 5.55, the derivatives of the shape function with respect to x and y need to be calculated. If the shape functions are expressed in natural coordinates, the chain rule of partial differentiation is applied to find out the derivatives. Applying the chain rule,

$$\begin{aligned}
\frac{\partial N_i}{\partial \xi} &= \frac{\partial N_i}{\partial x} \frac{\partial x}{\partial \xi} + \frac{\partial N_i}{\partial y} \frac{\partial y}{\partial \xi}, \\
\frac{\partial N_i}{\partial \eta} &= \frac{\partial N_i}{\partial x} \frac{\partial x}{\partial \eta} + \frac{\partial N_i}{\partial y} \frac{\partial y}{\partial \eta}.
\end{aligned} \tag{5.57}$$

Writing in the matrix form,

$$\begin{Bmatrix} \dfrac{\partial N_i}{\partial \xi} \\[2mm] \dfrac{\partial N_i}{\partial \eta} \end{Bmatrix} = \begin{bmatrix} \dfrac{\partial x}{\partial \xi} & \dfrac{\partial y}{\partial \xi} \\[2mm] \dfrac{\partial x}{\partial \eta} & \dfrac{\partial y}{\partial \eta} \end{bmatrix} \begin{Bmatrix} \dfrac{\partial N_i}{\partial x} \\[2mm] \dfrac{\partial N_i}{\partial y} \end{Bmatrix} = [J] \begin{Bmatrix} \dfrac{\partial N_i}{\partial x} \\[2mm] \dfrac{\partial N_i}{\partial y} \end{Bmatrix}, \tag{5.58}$$

where $[J]$ is called the Jacobian. Thus,

$$\begin{Bmatrix} \dfrac{\partial N_i}{\partial x} \\[2mm] \dfrac{\partial N_i}{\partial y} \end{Bmatrix} = [J]^{-1} \begin{Bmatrix} \dfrac{\partial N_i}{\partial \xi} \\[2mm] \dfrac{\partial N_i}{\partial \eta} \end{Bmatrix}. \tag{5.59}$$

If the geometry is approximated by n_g shape functions, then knowing the coordinates of n_g nodes the Jacobian matrix can be found as

$$[J] = \begin{bmatrix} \sum\limits_{i=1}^{n_g} \dfrac{\partial N_i^g}{\partial \xi} x_i & \sum\limits_{i=1}^{n_g} \dfrac{\partial N_i^g}{\partial \xi} y_i \\[3mm] \sum\limits_{i=1}^{n_g} \dfrac{\partial N_i^g}{\partial \eta} x_i & \sum\limits_{i=1}^{n_g} \dfrac{\partial N_i^g}{\partial \eta} y_i \end{bmatrix}, \tag{5.60}$$

where symbol N_i^g is used for the i-th shape function corresponding to geometry, which may be different than i-th shape function corresponding to the primary variable like temperature in the previous example.

Usually, the coefficient matrix and right hand side vector is evaluated using numerical integration. The commonly employed numerical integration procedure is Gauss-quadrature. In this procedure, the integration of a function is carried out by evaluating the function at certain points called Gauss points, multiplying them by suitable weights and adding all the terms. The table containing weights and Gauss-points is available for the case in which one-dimensional integration limits are from -1 to $+1$. Therefore, it is convenient to transform the element from the physical domain to the natural domain. Thus, the one-dimensional Gauss-quadrature formula is expressed as

$$\int\limits_{-1}^{+1} f(\xi)\, d\xi = \sum_{i=1}^{n} w_i f(\xi_i), \tag{5.61}$$

where n is the total number of Gauss-points. It can be shown that if we use n Gauss-points, then $(2n-1)$ degree polynomial can be integrated exactly. Thus, a three Gauss-point formula can integrate up to fifth degree polynomial exactly. Table 5.1 provides weights and coordinates of Gauss-points up to three Gauss-point formulae. The Gauss-points and weights for formulae with more number of Gauss-points are available in [3].

Table 5.1. Gauss-points and weights for one-dimensional integration between -1 and $+1$

Total number of Gauss-points	Coordinates of Gauss-points	Weights
1	0	2
2	$\pm 1/\sqrt{3}$	1
3	0	8/9
	$\pm \sqrt{3/5}$	5/9

If the integral $\int\limits_{x_1}^{x_2} f(x)\, dx$ is to be evaluated, using Equation 5.10 it can be transformed to

$$\int_{x_1}^{x_2} f(x)\,dx = \int_{-1}^{1} f\left(\frac{1}{2}(1-\xi)x_1 + \frac{1}{2}(1+\xi)x_2\right)\frac{x_2 - x_1}{2}\,d\xi. \tag{5.62}$$

Usually the shape functions and their derivatives are already expressed in term of natural coordinates. In that case, there is a need just to replace dx by $\dfrac{(x_2 - x_1)}{2}\,d\xi$ in the integral.

The one-dimensional Gauss-quadrature can be easily extended to a two-dimensional square domain in natural coordinates by successive application of one-dimensional Gauss formula. Thus,

$$\int_{-1}^{1}\int_{-1}^{1} f(\xi,\eta)\,d\xi\,d\eta = \int_{-1}^{1}\sum_{i=1}^{n_\xi} w_i f\left(\xi_i,\eta\right)d\eta = \sum_{j=1}^{n_\eta} w_j \sum_{i=1}^{n_\xi} w_i f\left(\xi_i,\eta_j\right), \tag{5.63}$$

where n_ξ and n_η are the Gauss-points in ξ and η directions respectively. The appropriate value of n_ξ is decided after treating η as constant and observing the degree of polynomial of the function in ξ-coordinates. If the degree of polynomial is m, for exact integration n_ξ should be equal to or more than $(m+1)/2$. The value of n_η can be decided similarly. The integration in physical domain has to be changed to natural coordinates by transformation of the coordinates. The term $dxdy$ in physical coordinates domain is changed to $|J|d\xi\,d\eta$ in natural coordinates domain, where $|J|$ is the determinant of the Jacobian matrix, called the Jacobian. The integration can be carried out similarly on three-dimensional domain. In this case,

$$\int_{-1}^{1}\int_{-1}^{1}\int_{-1}^{1} f(\xi,\eta,\zeta)\,d\xi\,d\eta\,d\zeta = \sum_{k=1}^{n_\zeta} w_k \sum_{j=1}^{n_\eta} w_j \sum_{i=1}^{n_\xi} w_i f\left(\xi_i,\eta_j,\zeta_k\right). \tag{5.64}$$

Successive application of one-dimensional Gauss-quadrature may also be applied to integrate over triangular domain. However, in this case, the limits of inside integration are not fixed, but keep changing as linear functions of the other coordinate. Thus, tedious calculations are involved and it is better to use tabulated values [1, 2]. The weights and Gauss points for triangular element in which the natural coordinates ξ_1, ξ_2 and ξ_3 change from 0 to 1 have been tabulated. In this case, the formula is

$$\int_A f\,dA = \frac{1}{2}\sum_{i=1}^{n} w_i |J_i| f_i, \tag{5.65}$$

where f_i is the value of the function at i-th Gauss-point, $|J_i|$ is the Jacobian at that point and w_i is the weight at that point. Table 5.2 provides the Gauss-points and weights for one, three and four point formula.

Table 5.2. Gauss-points and weights for a triangular domain

Total number of Gauss-points	Degree of polynomial for exact integration	Coordinates of Gauss-points (ξ_1, ξ_2, ξ_3)	Weights
1	1	(1/3, 1/3, 1/3)	1
3	2	(2/3, 1/6, 1/6) (1/6, 2/3, 1/6) (1/6, 1/6, 2/3)	1/3 1/3 1/3
4	3	(1/3, 1/3, 1/3) (0.6, 0.2, 0.2) (0.2, 0.6, 0.2) (0.2, 0.2, 0.6)	−0.562500 0.5208333 0.5208333 0.5208333

5.2.3 Assembly Procedure

It is not possible to obtain the nodal values of an element by just solving the elemental equations, because internal load vectors are undetermined due to unknown heat flux at the boundaries of the element. However, if we assemble elemental equations of all the elements together, the elements of the vectors given by Equation 5.56b will become zero except at the boundary of the domain, where boundary conditions are known. The simple way to assemble the equations is to express the elemental matrices and vectors in a global form and then add them. Supposing the total number of nodal values (degrees of freedom), which equals the primary variables per node times the number of node is N, it is possible to write for each element N equations in n-unknowns. Of course, some of these equations will be of the form zero equals to zero. For example, when a rod subjected to an axial load is discretized into three elements (Figure 5.7), the elemental equations of the second element are expressed as

$$\frac{AE}{h}\begin{bmatrix} 1 & -1 \\ -1 & 1 \end{bmatrix}\begin{Bmatrix} u_2 \\ u_3 \end{Bmatrix} = \begin{Bmatrix} qh/2 \\ qh/2 \end{Bmatrix} + \begin{Bmatrix} F_2^2 \\ F_3^2 \end{Bmatrix}, \tag{5.66}$$

where A and E are the cross-sectional area and Young's modulus of the element respectively and h is the length of the element. The degree of freedom at a node i is represented as u_i, the displacement of the node. The q is load intensity (load per unit length) in the element. The internal load at node i of element e is denoted by F_i^e. Then, the global form of Equation 5.66 is

$$\frac{AE}{h}\begin{bmatrix} 0 & 0 & 0 & 0 \\ 0 & 1 & -1 & 0 \\ 0 & -1 & 1 & 0 \\ 0 & 0 & 0 & 0 \end{bmatrix}\begin{Bmatrix} u_1 \\ u_2 \\ u_3 \\ u_4 \end{Bmatrix} = \begin{Bmatrix} 0 \\ qh/2 \\ qh/2 \\ 0 \end{Bmatrix} + \begin{Bmatrix} 0 \\ F_2^2 \\ F_3^2 \\ 0 \end{Bmatrix}. \tag{5.67}$$

Here, the first and the last equations are actually $0 = 0$. In this way, the elemental equations for a particular element i, can be expressed as

$$[K]^i \{\Delta\} = \{R\}^i + \{F\}^i, \tag{5.68}$$

where $\{\Delta\}$ is the global vector of nodal variables and $[K]^i$ is coefficient matrix of the element expressed in the global form and $\{R\}^i$ and $\{F\}^i$ together is the elemental right hand side vector expressed in the global form. All the elements of vector $\{R\}^i$ are known, whereas in general, the elements of vector $\{F\}^i$ contain the derivatives of primary variables and are unknown. Once the equations for all the elements have been expressed in the global form, they can be added to yield

$$\sum_{i=1}^{m} [K]^i \{\Delta\} = \sum_{i=1}^{m} \{R\}^i + \sum_{i=1}^{m} \{F\}^i \quad \text{or} \quad [K]\{\Delta\} = \{R\} + \{F\}, \tag{5.69}$$

where m is the number of elements. In the process of summation, elements of $\{F\}^i$ will add up to give 0 at all nodes except the boundary nodes. Thus, there is no need to calculate the vectors $\{F\}^i$ for the interior elements. The final assembled system of equations is

$$[K]\{\Delta\} = \{R\}, \tag{5.70}$$

where $[K]$ is called the global stiffness matrix and $\{R\}$ is the global right side vector that includes $\{R\}$ of Equation 5.69 as well as the non-zero terms of $\{F\}$.

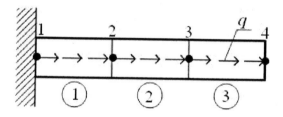

Figure 5.7. A rod subjected to axial load and discretized into three elements

Having explained the basic principle of assembly, we present a simple procedure of implementation. We need not actually rewrite the elemental equations in the global form and add them. All we have to do is to identify the place in the global coefficient matrix (and global right side vector) where a typical element of elemental coefficient matrix (and elemental right side vector) finds a place. If more than one element have a common location in the global coefficient matrix (or right side vector), they are simply added. To carry out the assembly in this manner, it must be known that a typical local node number corresponds to which global node number. This is easily found out from a connectivity matrix. Given an element number and local node number, the connectivity matrix provides the global node number. The row of the connectivity matrix corresponds to element number and column to local node number. For   example, connectivity matrix [C] of the mesh shown in Figure 5.8 is given by

$$[C] = \begin{bmatrix} 1 & 2 & 6 & 5 \\ 2 & 3 & 7 & 6 \\ 3 & 4 & 8 & 7 \\ 5 & 6 & 10 & 9 \\ 6 & 7 & 11 & 10 \\ 7 & 8 & 12 & 11 \\ 9 & 10 & 14 & 13 \\ 10 & 11 & 15 & 14 \\ 11 & 12 & 16 & 15 \end{bmatrix}.$$

Thus, i-th local node of element e corresponds to the global node $C(e, i)$. If the degree of freedom corresponding to each node is df, then corresponding to this node, the global primary variables are $(C(e, i) \times df - p)$, where $p = df-1, df-2, df-3,$ ....1, 0. The corresponding elemental primary variables are $df - p$. Thus, the assembly can be carried out in the following manner:

Step1: Initialize global stiffness matrix and right hand side vector.
Step2: Start from the first element. Put the $(df - p, df - p)$-th component of the elemental coefficient matrix into position (r, s) of the global coefficient matrix and the $(df - p)$-th component of the elemental right hand side vector into the r-th row of right hand side vector, where

$$r = C(e, i) \times df - p,$$
$$s = C(e, j) \times df - p. \tag{5.71}$$

Here, p varies from $df-1$ to 0. The i and j vary from 1 to number of nodes in the element.
Step3: Repeat Step 2 for all the elements.

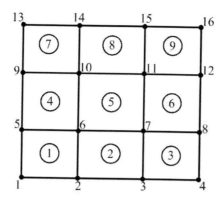

Figure 5.8. A finite element mesh of four-noded rectangular elements

5.2.4 Applying Boundary Conditions

Once the finite element equations have been assembled, they can be solved after the application of boundary conditions. There are two types of boundary conditions—essential (or geometric) and natural (or force). The essential boundary conditions prescribe the values of primary variables at the boundary, whereas the natural boundary conditions prescribe the gradients of the primary variables. Thus, in a steady-state heat conduction problem, essential boundary conditions prescribe the temperature at the boundary and the natural boundary conditions prescribe the heat flux on the boundary.

In the simplest way, an essential boundary condition is applied as follows. If the i-th degree of freedom is prescribed, the i-th equation of Equation 5.70 is replaced by

$$\Delta_i = \Delta^*, \tag{5.72}$$

where Δ^* is the prescribed value. Thus, the i-th row of global coefficient matrix $[K]$ is replaced by the row having the diagonal term as 1 and other terms zero. The i-th element of column vector $\{R\}$ is replaced by Δ^*. After the application of the boundary condition in this way, the coefficient matrix becomes unsymmetrical; even if the assembled coefficient matrix is symmetrical. To preserve the symmetry, all the elements of the i-th column of $[K]$ except the diagonal terms are made zero and the elements of the right hand side column vector are modified as

$$R_j = R_j - K_{ij} \times \Delta^*, \tag{5.73}$$

where $j = 1$ to n, but $j \neq n$. Of course, this needs a number of arithmetic operations.

A method that preserves the symmetry without the need of modifying other equations is to add a penalty number M to the i-th diagonal element of global stiffness matrix and make the right hand side equal to $M\Delta^*$, where M is a very large number. All other entries in the i-th row become insignificant in comparison to the diagonal term, which can be considered approximately equal to M. The i-th equation then becomes

$$M\Delta_i \approx M\Delta^*, \tag{5.74}$$

which enforces the required boundary condition.

Another method of applying the essential boundary condition is to eliminate the row and column of $[K]$ corresponding to prescribed boundary degree of freedom and modify the right hand side vector according to Equation 5.73. This reduces the size of the coefficient matrix and is advantageous in terms of storage requirement and computational time. However, the numbering of degrees of freedom gets changed and after the solution, it has to be restored to the original numbering.

Natural boundary conditions can be applied by evaluating the right hand side vector, such as given by Equation 5.56b at the boundary. With the value of derivatives known at the boundaries, the integral involved in the right hand side vector can be easily evaluated.

5.2.5 Solving the System of Equations

Gauss-elimination is the method in which $[K]$ is converted into upper triangular form by a number of row operations. At the k-th step of this method, the elements of the k-th column from $(k+1)$-th row to the n-th row are made zero by multiplying the elements of the k-th row by K_{ik}/K_{kk}, $i=\{k+1, k+2, \ldots,N)$ and subtracting them from rows $k+1, \ldots, n$ to produce zeros in position $(k+1, k), \ldots,(n, k)$. The entry K_{kk} is called the pivot and it should not be very small. Therefore, at the k-th step, it is better to interchange the rows to make the magnitude of the pivot large. This is called partial pivoting. In complete pivoting, both rows and columns are interchanged.

Gauss-elimination even with pivoting does not provide good results for ill-conditioned system of equations. Householder method [4] is a better method for solving the ill-conditioned system of equations. In this method, the upper triangular form is generated by successively multiplying the coefficient matrix by reflector matrices. This way, less round-off errors are propagated. There are a number of solvers that make use of the banded and sparse structure of the coefficient matrix.

5.2.6 Post-processing

Once the nodal values of the solution are found, they can be post-processed to provide the derivatives of the primary variables. Usually, the primary variables are more accurate at the nodes but not their derivatives. It can be shown that in one-dimension, if the primary variable is approximated by a linear approximation of two-noded element, the derivative of the primary variable is expected to be most accurate at the Gauss-point corresponding to one-Gauss point formula. If the primary variable is approximated by a quadratic approximation of three-noded element, the derivative of the primary variable is expected to be most accurate at Gauss-points corresponding to two-Gauss point formula. Extending this observation to two dimensions, for a four-noded quadrilateral element, the derivatives are expected to be accurate at the center of the element, corresponding to 1×1 Gauss-points. For eight and nine-noded element, the derivates are expected to be accurate at 2×2 Gauss-points *i.e.* at $\left(\pm 1/\sqrt{3}, \pm 1/\sqrt{3}\right)$. The derivatives can be extrapolated from Gauss-points to the nodes. If a node is shared between two or more elements, the nodal averaging may be employed, in which at a particular node, the values of derivatives obtained from various elements are averaged.

5.3 Formulation of Plane-Strain Metal Forming Processes

If the deformation takes place predominantly in one plane in a metal forming process, the process is called plane-strain metal forming process. Examples are drawing of a strip and flat rolling. Generally, if the width of the sheet is more than ten times the intial thickness, the rolling process can be modeled as a plane-strain process. We will explain the formulation of the plane-stain processes taking an example of cold flat rolling.

In the cold flat rolling process, the metal is plastically deformed by passing it between two counter rotating cylinders. The strip or sheet is drawn by means of the friction between the roll and work-material. Starting from the pioneering work of von Karman [5], a number of researchers have analyzed the rolling process. The process has been analyzed by slab method [6–10], slip-line method [11–12], visioplasticity (combination of experiments and analysis) [13], upper bound method [14–16], and finite element method [17–23]. Since rolling is a steady state process, Eulerian formulation has been used by many authors. Zienkiwicz *et al.* [17] considered the rolled material to be rigid-visco-plastic and incompressible. They simulated the friction at the roll-strip interface by introducing a thin layer of elements whose yield strength is assumed to depend on the coefficient of friction and mean (hydrostatic) stress. The neutral point is not modeled in this method. Mori *et al.* [18] have assumed that the rolled strip is made of a rigid-plastic, slightly compressible material. Using a constant frictional coefficient and rigid rolls, they reported a good agreement between the predicted and experimentally obtained front end shapes of rolled aluminum strips. In their formulation, the neutral point was determined by minimizing a certain functional. Li and Kobayashi [19] also considered the existence of rigid-plastic materials and rigid rolls but they modeled the neutral point using the velocity dependent frictional stress. The authors included a comparison of their predictions with the results of Al-Salehi *et al.* [24] and Shida and Awazuhara [25]. In acknowledging that some discrepancies do exist, they attributed them to their use of rigid rolls and some uncertainty in the modeling of interfacial friction. The authors also obtained non-steady state solution by simulating the deformation of the work piece in a step-by-step manner, updating the coordinates of material points and the material property after each step.

Hwu and Lenard [20] have studied the effects of roll deformation and the variation of the coefficient of friction in the roll gap. Prakash *et al.* [21] presented an FEM formulation in which the neutral point is found iteratively from the condition that the interfacial shear stress changes its sign at the neutral point. Dixit and Dixit [22] have incorporated roll deformation by using Hitchcock's formula in the mixed pressure-velocity formulation. In their formulation, the neutral point is found by minimizing the total rolling power. Chandra and Dixit [23] have carried out finite element analysis of temper rolling process in which the roll deformation is obtained by using a theory of elasticity solution.

5.3.1 Governing Equations and Boundary Conditions

The side view of a rolling process is shown in Figure 5.9, where the x-direction is the rolling direction and the y-direction is the thickness direction. Because of the symmetry, only the domain ABCDEF need be considered. We assume that the material is rigid-plastic. During plastic deformation of materials, there is no change in the volume. Thus, the conservation of mass provides, the following incompressibility constraint (Equation 3.249):

$$\dot{\varepsilon}_{ii} = \dot{\varepsilon}_{xx} + \dot{\varepsilon}_{yy} = 0 , \tag{5.75}$$

where the strain rate tensor $\dot{\varepsilon}_{ij}$ is defined as (Equation 3.246)

$$\dot{\varepsilon}_{ij} = \frac{1}{2}\left(v_{i,j} + v_{j,i}\right). \tag{5.76}$$

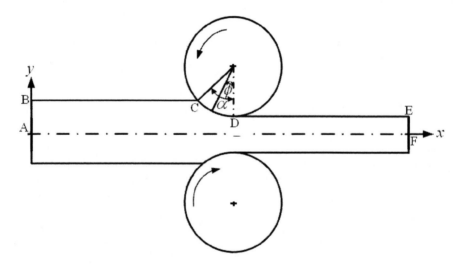

Figure 5.9. Plane-strain rolling process

The equation of motion (Equation 3.248) for the steady-state process, neglecting the body force, is

$$\rho v_j \frac{\partial v_i}{\partial x_j} - \frac{\partial \sigma_{ij}}{\partial x_j} = 0, \qquad (5.77)$$

For most of the practical rolling processes, the first term in Equation 5.77 may be neglected. Thus, the equation of motion reduce to equilibrium equation:

$$\frac{\partial \sigma_{xx}}{\partial x} + \frac{\partial \sigma_{xy}}{\partial y} = 0 \quad \text{or} \quad -\frac{\partial p}{\partial x} + \frac{\partial \sigma'_{xx}}{\partial x} + \frac{\partial \sigma'_{xy}}{\partial y} = 0,$$

$$\frac{\partial \sigma_{xy}}{\partial x} + \frac{\partial \sigma_{yy}}{\partial y} = 0 \quad \text{or} \quad -\frac{\partial p}{\partial y} + \frac{\partial \sigma'_{xy}}{\partial x} + \frac{\partial \sigma'_{yy}}{\partial y} = 0. \qquad (5.78)$$

The set of six equations given by Equation 5.75, Equation 5.76 and Equation 5.78 model the rolling process alongwith the boundary conditions. However, in this form, there are nine unknowns v_x, v_y, σ'_{xx}, σ'_{yy}, σ'_{xy}, $\dot{\varepsilon}_{xx}$, $\dot{\varepsilon}_{yy}$, $\dot{\varepsilon}_{xy}$, p and and only six equations are available. The additional equations are provided by the constitutive behavior of the material.

For an isotropic rigid-plastic metal, the constitutive equation is a relation between the deviatoric part σ'_{ij} and the strain-rate tensor $\dot{\varepsilon}_{ij}$ (Equation 3.247a):

$$\sigma'_{ij} = 2\eta_1 \dot{\varepsilon}_{ij}, \qquad (5.79)$$

where

$$\eta_1 = \frac{\sigma_{eq}}{3\dot{\varepsilon}_{eq}}. \qquad (5.80)$$

Here, the invariant $\dot{\varepsilon}_{eq}$ is called the equivalent strain-rate, defined by a relation similar to Equation 3.156 and the equivalent-stress σ_{eq} is given by the following hardening relation (Equation 3.247b):

$$\sigma_{eq} = \sigma_Y + K(\varepsilon_{eq})^n. \qquad (5.81)$$

After substituting the strain rate-velocity relation (Equation 5.76) and the material behavior in Equation 5.78, the stress components get eliminated. Instead, we get derivatives of v_1, v_2 and p. The values of three unknowns v_1, v_2 and p can be obtained by solving the set of three equations, Equation 5.75 and Equation 5.78 expressed in terms of the velocity components and pressure.

On any one part of the boundary, the boundary condition should be the following: either the velocity vector v or the traction vector t is specified, or one velocity and one traction component are specified. The following are the boundary conditions in the flat rolling process

On the entry boundary (AB), we should specify the following boundary conditions:

$$\text{either } v_x = U_1 \text{ or } t_x = t_b, \tag{5.82}$$

and

$$\text{either } v_y = 0 \text{ or } t_y = 0, \tag{5.83}$$

where U_1 is the inlet velocity of the strip, t_i are the components of traction vector and t_b is the back tension. In each of the Equations 5.82 and 5.83, the first boundary condition is the geometric (essential) and the second boundary condition is the natural boundary condition. One should not get the impression that on a particular boundary the essential boundary condition should be applied essentially. It is possible to choose among the essential and natural boundary conditions. Overall, some essential boundary conditions must be provided inorder to get a unique solution, which means at some places the velocity must be prescribed. Li and Kobayashi [19] have applied natural boundary conditions, whilst other researchers [21, 22] found it convenient to specify the essential boundary conditions. Provided that the surface AB is sufficiently away from the deformation zone, it is reasonably accurate to specify the velocity boundary condition on this surface. Specifying the essential boundary condition is expected to provide a more accurate solution in FEM compared to specifying the natural boundary condition. The reason is that, in the latter, the derivatives of primary variables are involved, which are less accurate compared to a primary variable. Also, the boundaries fall in the elastic zone, where the rigid-plastic analysis cannot determine the stresses and tractions accurately. Moreover, by specifying the essential boundary conditions, the number of equations to be solved can be reduced by eliminating the rows and columns corresponding to the specified boundary conditions.

In a similar way, the boundary conditions on the exit boundary (EF) can be specified:

$$\text{either } v_x = U_2 = U_1 / (1 - r) \text{ or } t_x = t_f \tag{5.84}$$

and

$$\text{either } v_y = 0 \text{ or } t_y = 0, \tag{5.85}$$

where r is the fractional reduction and t_f is the front tension. The same considerations as applicable to inlet boundary (AB) are applicable for exit boundary (EF).

At the top free surfaces (BC and DE), one can have the following boundary conditions:

$$t_x = 0, t_y = 0. \tag{5.86}$$

These boundary conditions allow v_y to adopt suitable values as *per* the deformation. However, it is observed that v_y is almost 0 at these surfaces except very near to entry point of roll-work interface. Before coming into contact with the roll, the sheet may bulge. Hence, v_y may not be 0 near the roll. Therefore, in place of the boundary conditions given by Equation 5.86, the following boundary conditions may be employed:

$t_x = 0$, $v_y = 0$ on BC and DE (except at two nodes near entry to roll-work interface),

$$(5.87)$$

$t_x = 0$, $t_y = 0$ on the two nodes near entry to roll-work interface . $\qquad$ (5.88)

The boundary conditions at the axis of symmetry (AF) are

$$t_x = 0, \ v_y = 0 . \qquad (5.89)$$

Here, because of symmetry, the traction t_x and velocity v_y must be zero.

At the roll strip interface (CD), assuming no movement of the work material normal to roll surface:

$$v_y + v_x \tan \phi = 0 \text{ on CD,} \qquad (5.90)$$

where ϕ is the angular position of the point as shown in Figure 5.9 at which the boundary condition is applied. The second boundary condition will be

$$|t_s| = c|t_n|, \qquad (5.91)$$

where t_s and t_n are the tangential and normal components of the stress vector t. In the case of Coulomb model, c is equal to friction coefficient f. In the case of Wanheim and Bay's model, c is equal to t_s/t_n where t_s and t_n are given by Equations 4.236 and 4.237 and thus a non-linear function of t_n.

5.3.2 Non-Dimensionalization

The non-dimensionalization is carried out using the following relationships:

$$\bar{x} = \frac{x}{(h_2/2)}, \ \bar{y} = \frac{y}{(h_2/2)}, \ \bar{v}_x = \frac{v_x}{U_2}, \ \bar{v}_y = \frac{v_y}{U_2}, \bar{p} = \frac{p}{(\sigma_Y/3)} . \qquad (5.92)$$

In the problem, there are only three independent dimensions—length, time and mass. Thus, three independent variables can be non-dimensionalized as per the convenience. Here, in the non-dimensionalized version, the final semi-thickness of the rolled sheet is taken as 1. Therefore, both the coordinates measuring the length dimension are non-dimensionalized with respect to $h_2/2$. The velocity components are non-dimensionalized with respect to the exit velocity U_2, which becomes 1 in the non-dimensionalized version. The dimension of velocity contains the dimensions of length and time. Thus, by carrying out the non-dimensionalization of coordinates and velocity components, we have non-dimensionalized time as well. The pressure is non-dimensionalized with respect to $(\sigma_y/3)$. As the dimension of pressure contains the dimensions of length, time and mass, by non-dimensionalizing it, we have non-dimensionalized mass also.

Let us obtain the relationship between the non-dimensional Levy-Mises coefficient $\bar{\eta}_1$ and dimensional η_1. Equation 5.80 expressed in non-dimensional form is

$$\bar{\eta}_1 = \frac{\bar{\sigma}_{eq}}{\left(3\bar{\dot{\varepsilon}}_{eq}\right)}. \tag{5.93}$$

As per our scheme of non-dimensionalization, the flow stress has to be non-dimensionalized with respect to $(\sigma_Y/3)$, and the equivalent strain-rate has to be non-dimesionalized with respect $U_2/(h_2/2)$, which has the dimension of strain-rate. Thus, Equation 5.93 can be written as

$$\bar{\eta}_1 = \frac{\sigma_{eq}}{3(\sigma_Y/3)\varepsilon_{eq}}\frac{U_2}{(h_2/2)} = \frac{\eta_1}{\eta_0}, \tag{5.94}$$

where η_0 is given by

$$\eta_0 = \frac{(\sigma_Y/3)}{2(U_2/h_2)}. \tag{5.95}$$

The governing equations and boundary conditions can be expressed in the non-dimensional form. The non-dimensional variables have been indicated by an over-bar. However, for the sake of convenience, henceforth we will show non-dimensional variables without an over-bar. Reader should treat the variables in the subsequent section as non-dimensional.

5.3.3 Weak Formulation

In this section, the mixed pressure-velocity formulation is explained. Let v_x, v_y, p be the functions that satisfy the essential boundary conditions exactly. Thus the weighted residual will become

$$\int_A \left[\left(\dot{\varepsilon}_{xx} + \dot{\varepsilon}_{yy} \right) w_p + \left(\frac{\partial \sigma_{xx}}{\partial x} + \frac{\partial \sigma_{xy}}{\partial y} \right) w_x + \left(\frac{\partial \sigma_{yx}}{\partial x} + \frac{\partial \sigma_{yy}}{\partial y} \right) w_y \right] dxdy = 0 \tag{5.96}$$

where A denotes domain of a typical area element. The above expression can be written in index notation as

$$\int_A \left[\dot{\varepsilon}_{ii} w_p + \sigma_{ij,j} w_i \right] dx dy = 0 . \tag{5.97}$$

The first part of the integral contains only the first order derivatives in velocity components. Therefore, there is no need to weaken this part. However, the second part contains second order derivatives of velocity components. This part can be reduced to weak form, in which the highest order of derivatives of velocity can be 1. Rewriting the second part, Equation 5.97 becomes

$$\int_A \dot{\varepsilon}_{ii} w_p dA + \int_A \left[\left(\sigma_{ij} w_i \right)_{,j} - \left(\sigma_{ij} w_{i,j} \right) \right] dA = 0 . \tag{5.98}$$

The application of the divergence theorem to Equation 5.98 yields

$$\int_A \dot{\varepsilon}_{ii} w_p dA + \int_\Gamma \sigma_{ij} n_j w_i d\Gamma - \int_A \left(\sigma_{ij} w_{i,j} \right) dA = 0 . \tag{5.99}$$

where 'Γ' may be considered only that part of the boundary where the tractions are specified. The reason is that, for the parts where the velocity components are specified, there cannot be any error and weighted residual for that part need not be considered. Substituting Cauchy's relation $\sigma_{ij} n_j = t_i$, Equation 5.99 can be expressed as

$$\int_A \dot{\varepsilon}_{ii} w_p dA + \int_\Gamma t_i w_i d\Gamma - \int_A \left(\sigma_{ij} w_{i,j} \right) dA = 0 , \tag{5.100}$$

or

$$\int_A \dot{\varepsilon}_{ii} w_p dA + \int_{\Gamma_1} t_x w_x d\Gamma_1 + \int_{\Gamma_2} t_y w_y d\Gamma_2 - \int_A \left(\sigma_{ij} w_{i,j} \right) dA = 0 , \tag{5.101}$$

where Γ_1 and Γ_2 are respectively those parts of the boundary where t_x and t_y are specified.

The last term in Equation 5.101 can be expressed as

$$\int_A \left(\sigma_{ij} w_{i,j} \right) dA = \int_A \sigma_{ij} \frac{1}{2} \left[\left(w_{i,j} + w_{j,i} \right) + \left(w_{i,j} - w_{j,i} \right) \right] dA. \tag{5.102}$$

Knowing that the scalar product of a symmetric and a skew-symmetric tensor is zero, Equation 5.102 simplifies to

$$\int_A \left(\sigma_{ij} w_{i,j} \right) dA = \int_A \sigma_{ij} \frac{1}{2} \left[w_{i,j} + w_{j,i} \right] dA. \tag{5.103}$$

Equation 5.103 can be written as

$$\int_A \left(\sigma_{ij} w_{i,j}\right) dA = \int_A \sigma_{ij} \dot{\varepsilon}_{ij}(w) dA \,, \tag{5.104}$$

where analogous to strain rate, $\dot{\varepsilon}_{ij}(w)$ can be called strain-rate weight. Thus, the weighted residual (Equation 5.101) becomes

$$\int_A \dot{\varepsilon}_{ii} w_p \; dA + \int_A \sigma_{ij} \dot{\varepsilon}_{ij}(w) \; dA - \int_{\Gamma_x} t_x w_x d\Gamma_1 - \int_{\Gamma_y} t_y w_y d\Gamma_2 = 0 \,. \tag{5.105}$$

Both the area integrals of Equation 5.105 contain the first order derivatives of velocity components. Thus, the weak form has been obtained for FEM modeling. The second term of Equation 5.105 can be simplified as

$$\begin{aligned} \sigma_{ij} \dot{\varepsilon}_{ij}(w) &= \left(-p\delta_{ij} + \sigma'_{ij}\right) \dot{\varepsilon}_{ij}(w), \\ &= -p\dot{\varepsilon}_{ii}(w) + 2\eta \dot{\varepsilon}_{ij}\dot{\varepsilon}_{ij}(w), \end{aligned} \tag{5.106}$$

which can be written in unabridged notation as

$$\sigma_{ij}\dot{\varepsilon}_{ij}(w) = -p\left[\dot{\varepsilon}_{xx}(w) + \dot{\varepsilon}_{yy}(w)\right] + 2\eta \left[\dot{\varepsilon}_{xx}\dot{\varepsilon}_{xx}(w) + 2\dot{\varepsilon}_{xy}\dot{\varepsilon}_{xy}(w) + \dot{\varepsilon}_{yy}\dot{\varepsilon}_{yy}(w)\right]. \tag{5.107}$$

For convenience, we write the weak form (Equation 5.105) as

$$\int_A I_1 \; dA + \int_A I_2 \; dA - \int_{\Gamma_x} I_3 d\Gamma - \int_{\Gamma_y} I_4 d\Gamma = 0. \tag{5.108}$$

where

$$\begin{aligned} I_1 &= -\left[\dot{\varepsilon}_{xx} + \dot{\varepsilon}_{yy}\right] w_p, \\ I_2 &= -p\left[\dot{\varepsilon}_{xx}(w) + \dot{\varepsilon}_{yy}(w)\right] + 2\eta_1 \left[\dot{\varepsilon}_{xx}\dot{\varepsilon}_{xx}(w) + 2\dot{\varepsilon}_{xy}\dot{\varepsilon}_{xy}(w) + \dot{\varepsilon}_{yy}\dot{\varepsilon}_{yy}(w)\right], \\ I_3 &= t_x w_x, \\ I_4 &= t_y w_y. \end{aligned} \tag{5.109}$$

5.3.4 Finite Element Formulation

The weak form of the governing equations (Equation 5.108) contains the first derivatives of the velocity and no derivatives of the pressure. Therefore, it is possible to choose C^0 continuity element for velocity and a discontinuous pressure field. Maniatty [26] has used discontinuous pressure field in his formulation. However, we describe the formulation which uses C^0 continuity element for both pressure and velocity field approximations. It is common to use the lower order approximation for pressure and higher order approximation for velocity. There are two reasons for doing this. The first reason concerns with the accuracy of secondary variables. As the computaion of stress requires a term with first derivative of the velocity and another term containing the pressure, in order to have the same order of accuracy in both the terms the velocity needs to be approximated by one higher order approximation. The second reason concerns the condition of the system of equations to be solved. If the ratio of pressure nodes to velocity nodes increases, the resulting system of equations becomes ill-conditioned. Using higher noded approximation for the velocities and lower noded for pressure makes the system of equations better-conditioned. We choose four-noded quadrilateral element for the pressure approximation and nine-noded quadrilateral element for velocity components. Thus there will be nine shape functions for the approximation of both components of velocity and four shape functions for pressure. Figure 5.10 shows a typical mesh consisting of 56 elements. Note that we have shown the local node numbering of element 1 separately for pressure and velocity. Actually it is the same element that interpolates pressure and velocity by different degrees of polynomials. The connectivity matrices for pressure and velocity can be prepared separately. The connectivity matrix for pressure will be a 56-by-4 matrix and for velocity a 56-by-9 matrix.

The approximation for v_x and v_y is

$$\begin{Bmatrix} v_x \\ v_y \end{Bmatrix} = \begin{bmatrix} N_1 & 0 & N_2 & 0 & N_3 & 0 & \ldots\ldots N_9 & 0 \\ 0 & N_1 & 0 & N_2 & 0 & N_3 & \ldots\ldots 0 & N_9 \end{bmatrix} \begin{Bmatrix} (v_x)_1^e \\ (v_y)_1^e \\ | \\ | \\ (v_x)_9^e \\ (v_y)_9^e \end{Bmatrix} = \begin{bmatrix} N^v \end{bmatrix} \{v\}^e$$

$$(5.110)$$

where $\{v\}^e$ is the vector containing the nodal velocities. The approximation for pressure is

$$p = \begin{bmatrix} N_1^p & N_2^p & N_3^p & N_4^p \end{bmatrix} \begin{Bmatrix} p_1^e \\ p_2^e \\ p_3^e \\ p_4^e \end{Bmatrix} = \left\{ N^p \right\}^T \{p\}^e . \qquad (5.111)$$

In Galerkin formulation, weight functions for velocity and pressure are approximated using the same shape functions as that of velocity and pressure respectively, *i.e.*,

$$\{w_v\} = \begin{Bmatrix} w_x \\ w_y \end{Bmatrix} = \begin{bmatrix} N^v \end{bmatrix} \{w_v\}^e . \tag{5.112}$$

$$w_p = \{w_p\}^{eT} \{N^p\} . \tag{5.113}$$

In order to model the curved boundary, it is better to approximate the geometry by nine noded shape functions. Thus, for geometry approximation, same shape functions can be considered as that of velocity variable. Therefore

$$x = \{N\}^T \{x\}^e \text{ and } y = \{N\}^T \{y\}^e , \tag{5.114}$$

where

$$\{N\}^T = \begin{bmatrix} N_1 & N_2 & \dots & N_9 \end{bmatrix}. \tag{5.115}$$

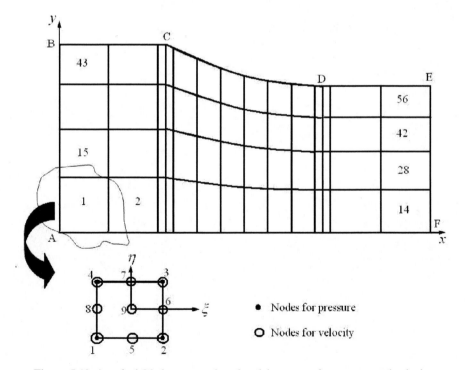

Figure 5.10. A typical 56 element mesh and nodal structure for pressure and velocity

To evaluate the integrals over the boundaries Γ_i, an approximation for the weight function over these boundaries is needed. The approximation consistent with the interior approximations is

$$
w_x = \{w_x^1 \ w_x^2 \ w_x^3\} \begin{Bmatrix} N_1^b \\ N_2^b \\ N_3^b \end{Bmatrix} = \{w_x\}^{bT} \{N\}^b ,
\tag{5.116}
$$

and

$$
w_y = \{w_y^1 \ w_y^2 \ w_y^3\} \begin{Bmatrix} N_1^b \\ N_2^b \\ N_3^b \end{Bmatrix} = \{w_y\}^{bT} \{N\}^b ,
\tag{5.117}
$$

where N_i^b are one-dimensional Lagrangian quadratic shape functions and $\{w_x\}^b$ and $\{w_y\}^b$ are the vectors of the nodal values of the velocity weight functions for a typical boundary element shown in Figure 5.11. Further, t_x and t_y on the boundary are approximated as

$$
t_x = \{N_1^b \ N_2^b \ N_3^b\} \begin{Bmatrix} (t_x)_1 \\ (t_x)_2 \\ (t_x)_3 \end{Bmatrix} = \{N\}^{bT} \{t_x\}^b ,
\tag{5.118}
$$

$$
t_y = \{N_1^b \ N_2^b \ N_3^b\} \begin{Bmatrix} (t_y)_1 \\ (t_y)_2 \\ (t_y)_3 \end{Bmatrix} = \{N\}^{bT} \{t_y\}^b ,
\tag{5.119}
$$

where $\{t_x\}^b$ and $\{t_y\}^b$ are the vectors of the nodal value of the traction. The expressions for N_i^b are given as

$$
N_1^b = \frac{1}{2}\left(\varsigma^2 - \varsigma\right), N_2^b = \frac{1}{2}\left(\varsigma^2 + \varsigma\right), \ N_3^b = \frac{1}{2}\left(1 - \varsigma^2\right),
\tag{5.120}
$$

where ς is the natural coordinate on the boundary.

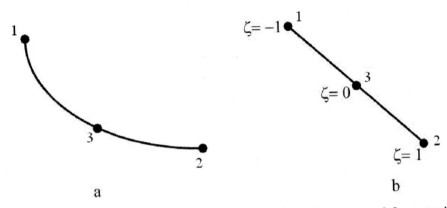

Figure 5.11. A typical boundary element. **a** In a physical coordinate system. **b** In a natural coordinate system

Equation 5.108 has to be expressed in matrix form to obtain the FEM equations. For that purpose, the following vectors are defined:

$$
\{\dot{\varepsilon}\} = \left\{ \begin{array}{c} \dot{\varepsilon}_{xx} \\ \dot{\varepsilon}_{yy} \\ \sqrt{2}\,\dot{\varepsilon}_{xy} \end{array} \right\} = \left\{ \begin{array}{c} \dfrac{\partial v_x}{\partial x} \\[2mm] \dfrac{\partial v_y}{\partial y} \\[2mm] \dfrac{1}{\sqrt{2}}\left(\dfrac{\partial v_x}{\partial y} + \dfrac{\partial v_y}{\partial x} \right) \end{array} \right\},
\tag{5.121}
$$

and

$$
\{\dot{\varepsilon}(w)\} = \left\{ \begin{array}{c} \dot{\varepsilon}_{xx}(w) \\ \dot{\varepsilon}_{yy}(w) \\ \sqrt{2}\,\dot{\varepsilon}_{xy}(w) \end{array} \right\} = \left\{ \begin{array}{c} \dfrac{\partial w_x}{\partial x} \\[2mm] \dfrac{\partial w_y}{\partial y} \\[2mm] \dfrac{1}{\sqrt{2}}\left(\dfrac{\partial w_x}{\partial y} + \dfrac{\partial w_y}{\partial x} \right) \end{array} \right\}.
\tag{5.122}
$$

Using the approximation for velocities and weights, the strain rate vectors $\{\dot{\varepsilon}\}$ and $\{\dot{\varepsilon}(w)\}$ become

$$
\{\dot{\varepsilon}\} = [B]\{v\}^e \text{ and } \{\dot{\varepsilon}(w)\} = [B]\{w_v\}^e,
\tag{5.123}
$$

where

$$[B] = \begin{bmatrix} \dfrac{\partial N_1}{\partial x} & 0 & \dfrac{\partial N_2}{\partial x} & 0 & ----- & \dfrac{\partial N_9}{\partial x} & 0 \\[2mm] 0 & \dfrac{\partial N_1}{\partial y} & 0 & \dfrac{\partial N_2}{\partial y} & ----- & 0 & \dfrac{\partial N_9}{\partial y} \\[2mm] \dfrac{1}{\sqrt{2}}\dfrac{\partial N_1}{\partial y} & \dfrac{1}{\sqrt{2}}\dfrac{\partial N_1}{\partial x} & \dfrac{1}{\sqrt{2}}\dfrac{\partial N_2}{\partial y} & \dfrac{1}{\sqrt{2}}\dfrac{\partial N_2}{\partial x} & ---- & \dfrac{1}{\sqrt{2}}\dfrac{\partial N_9}{\partial y} & \dfrac{1}{\sqrt{2}}\dfrac{\partial N_9}{\partial x} \end{bmatrix}.$$

$$(5.124)$$

The trace of the strain-rate vector is expressed as

$$\dot{\varepsilon}_{xx} + \dot{\varepsilon}_{yy} = \{1 \ \ 1 \ \ 0\}\{\dot{\varepsilon}\} = \{m\}^T [B]\{v\}^e, \qquad (5.125)$$

where

$$\{m\} = \begin{Bmatrix} 1 \\ 1 \\ 0 \end{Bmatrix}. \qquad (5.126)$$

Similarly,

$$\dot{\varepsilon}_{xx}(w) + \dot{\varepsilon}_{yy}(w) = \{m\}^T [B]\{w_v\}^e = \{w_v\}^{eT}[B]^T \{m\}. \qquad (5.127)$$

Substituting Equations 5.113, 5.116, 5.117, 5.118, 5.119, 5.123, 5.125, and 5.127, Equation 5.108 becomes

$$\int_A -\{w_p\}^{eT} \{N^P\}\{m\}^T [B]\{v\}^e \, dxdy$$

$$+ \int_A -\{w_v\}^{eT} [B]^T \{m\}\{N^P\}^T \{p\}^e \, dxdy$$

$$+ \int_A 2\eta_1 \{w_v\}^{eT} [B]^T [B]\{v\}^e \, dxdy$$

$$= \int_{\Gamma_x} \{w_x\}^{bT} \{N\}^b \{N\}^{bT} \{t_x\}^b \, d\Gamma + \int_{\Gamma_y} \{w_y\}^{bT} \{N\}^b \{N\}^{bT} \{t_y\}^b \, d\Gamma.$$

$$(5.128)$$

The above equations can be expressed as

$$\sum_{e=1}^{ne} \{w\}^{eT} [K]^e \{\delta\}^e = \sum_{b=1}^{nb_1} \{w_x\}^{bT} \{f_x\}^b + \sum_{b=1}^{nb_2} \{w_y\}^{bT} \{f_y\}^b, \qquad (5.129)$$

where ne is the number of area elements and nb_1, nb_2 are the number of boundary elements on Γ_x^b and Γ_y^b. Further

$$\{w\}^e = \left\{ \begin{array}{c} \{w_p\}^e \\ \{w_v\}^e \end{array} \right\}, \tag{5.130}$$

$$\{\delta\}^e = \left\{ \begin{array}{c} \{p\}^e \\ \{v\}^e \end{array} \right\}, \tag{5.131}$$

$$[K]^e = \begin{bmatrix} [0] & [K_{pv}]^e \\ [K_{vp}]^e & [K_{vv}]^e \end{bmatrix}, \tag{5.132}$$

$$
\begin{aligned}
[K_{pv}]^e &= \int_A -\{N^p\}\{m\}^T [B]dx_1 dx_2, \\
[K_{vp}]^e &= \int_A -[B]^T \{m\}\{N^p\}^T dx_1 dx_2 = [K_{pv}^e]^T, \\
[K_{vv}]^e &= \int_A 2\eta_1 [B]^T [B]dx_1 dx_2,
\end{aligned}
\tag{5.133}
$$

$$
\begin{aligned}
\{f_1\}^b &= \int_{\Gamma_x^b} \{N\}^b \{N\}^{bT} \{t_1\}^b d\Gamma, \\
\{f_2\}^b &= \int_{\Gamma_y^b} \{N\}^b \{N\}^{bT} \{t_2\}^b d\Gamma.
\end{aligned}
\tag{5.134}
$$

For the purpose of numerical evaluation, the area integrals are transformed to the natural coordinates (ξ, η) using the following transformation:

$$\int_{A^e} (\text{........})dxdy = \int_{-1}^{+1}\int_{-1}^{+1}(\text{.......})|J|d\xi \, d\eta, \tag{5.135}$$

where the Jacobian

$$|J| = \begin{vmatrix} \dfrac{\partial x}{\partial \xi} & \dfrac{\partial y}{\partial \xi} \\ \dfrac{\partial x}{\partial \eta} & \dfrac{\partial y}{\partial \eta} \end{vmatrix}, \tag{5.136}$$

is the determinant of the elemental Jacobian matrix. Similarly, the boundary integrals are transformed by the relation

$$\int_{\Gamma}(\ldots\ldots)\,d\Gamma = \int_{-1}^{+1}(\ldots\ldots\ldots)\,|\,J_b\,|\,d\zeta\;, \tag{5.137}$$

where $|\,J_b\,|$ is the Jacobian for boundary element and is given by

$$|\,J_b\,| = \sqrt{\left(\frac{dx}{d\zeta}\right)^2 + \left(\frac{dy}{d\zeta}\right)^2}\;. \tag{5.138}$$

Along the boundary, the coordinates $(x,\,y)$ are approximated using 1-D quadratic shape functions. All elemental matrices are evaluated using 3×3 Gauss-quadrature. Similarly the elemental vectors are evaluated using three point Gauss-quadrature. The assembled finite element equations can be written as

$$\{W\}^{\mathrm{T}}\,[K]\{\Delta\} = \{W\}^{\mathrm{T}}\,\{F\}\;, \tag{5.139}$$

where $\{W\},[K]\,,\{\Delta\}$ are the global vector of nodal values of weight function, global coefficient matrix and global vector of nodal values of pressure and velocity. $\{F\}$ is the global right hand side vector. Since weight functions are arbitrary, final FEM expression will be

$$[K]\{\Delta\} = \{F\}. \tag{5.140}$$

5.3.5 Application of Boundary Conditions

The friction condition at a typical node on the tool work interface (Figure 5.12) is given by Equation 5.91. Expressing t_s and t_n in the form of t_x and t_y as follows:

$$
\begin{aligned}
t_s &= t_x \cos\phi - t_y \sin\phi\,, \\
t_n &= -t_x \sin\phi - t_y \cos\phi.
\end{aligned}
\tag{5.141}
$$

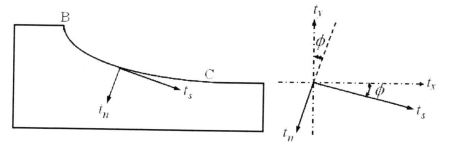

Figure 5.12. Shear and normal components of traction at work-tool interface

Thus, substituting above expressions of t_s and t_n into Equation 5.91, we get

$$(t_x \cos \phi - t_y \sin \phi) = ck_s (-t_x \sin \phi - t_y \cos \phi) , \qquad (5.142)$$

where k_s is 1 before the neutral point and -1 after the neutral point. The method to determine the neutral point is discussed in the next subsection. Rearranging Equation 5.142, we can write

$$t_x (\cos \phi + ck_s \sin \phi) - t_y (\sin \phi - ck_s \cos \phi) = 0 , \qquad (5.143)$$

or

$$t_x (1 + ck_s \tan \phi) - t_y (\tan \phi - ck_s) = 0 . \qquad (5.144)$$

The above expression for a boundary element of FEM can be written as

$$\begin{Bmatrix} (t_x)_1 \\ (t_x)_2 \\ (t_x)_3 \end{Bmatrix} (1 + ck_s \tan \phi) - \begin{Bmatrix} (t_y)_1 \\ (t_y)_2 \\ (t_y)_3 \end{Bmatrix} (\tan \phi - ck_s) = 0 . \qquad (5.145)$$

After pre-multiplying Equation 5.145 by

$$\int_{-1}^{+1} \{N\}^b \{N\}^{bT} \mid J_b \mid d\zeta , \qquad (5.146)$$

we get

$$\{f_x\}^b (1 + ck_s \tan \phi) - \{f_y\}^b (\tan \phi - ck_s) = \{0\} . \qquad (5.147)$$

Equation 5.147 holds good at all nodes of the element. At the middle node say 'k' (global node number), there is no contribution from the neighboring elements and therefore, in terms of the global right hand side vector, Equation 5.147 can be expressed as

$$\{F\}_{(d_p+2k-1)} (1 + ck_s \tan \phi) - \{F\}_{(d_p+2k)} (\tan \phi - ck_s) = 0 , \qquad (5.148)$$

where d_p is the total number of pressure nodes. At the end nodes, contributions from two elements have to be added and Equation 5.148 holds good for the end nodes too.

The velocity boundary condition at a node on the roll-strip interface is applied by replacing the (d_p+2k)-th equation by Equation 5.90. Procedure for applying the boundary conditions at the node 'k' is as follows:

- Replace (d_p+2k-1)-th row of global coefficient matrix $[K]$ by the following linear combination:
  $(1 + ck_s \tan \phi)$ times (d_p+2k-1)-th row of $[K] - (\tan \phi - ck_s)$ times (d_p+2k)-th row of $[K]$.
- Make (d_p+2k-1)-th row of global right hand vector $\{F\}$ zero.
- For applying the velocity boundary condition (Equation 5.90), set $(dp + 2k, dp + 2k - 1)$-th element of $[K]$ to $\tan \phi$ and $(dp + 2k, dp + 2k)$-th element of $[K]$ to 1.
- Make (d_p+2k)-th row of $\{F\}$ zero.

The essential boundary conditions at the other boundaries are applied following a procedure discussed in Section 5.2.4. For applying the natural boundary conditions of the form $t_i = 0$, we just have to make the corresponding element of global right hand side vector zero. After imposing the boundary conditions, the final matrix equation is solved iteratively by the Householder method, because the resulting matrix of the mixed formulation is ill-conditioned. Initially, a suitable position of neutral point is assumed, which is refined iteratively following the procedure discussed in subsequent section.

5.3.6 Estimation of Neutral Point

In order to apply the natural boundary condition on the roll-strip interface (Equation 5.142), the location of neutral point has to be known. The position of neutral point can be found iteratively. For example, Prakash et al. [21] found the neutral point iteratively from the condition that at the neutral point, interfacing shear stress changes sign. With a coarse mesh, the computation of shear stress may not be accurate. Therefore, in the formulation of Dixit and Dixit [22], the neutral point is found by minimizing total power with respect to the position of neutral point. The justification for this is as follows.

A generalized upper bound theorem [27] is expressed by

$$\int_{S_u} t_i v_i^* \, dS \leq \int_V \sigma_{ij}^* \dot{\varepsilon}_{ij}^* \, dV - \int_{S_F} t_i v_i^* \, dS, \tag{5.149}$$

where S_F is the part of surface where some or all of the traction components are specified and S_u is the remainder of the surface. Here, S_u is the interfacial boundary CD, while S_F includes all other boundaries. It is assumed that on inlet and outlet, instead of the velocity components, it is the traction component t_x which is specified with values equal to the back tension t_b and front tension t_f respectively. Further, v_i^* is any piecewise continuous velocity field defined over volume V with corresponding strain-rate field $\dot{\varepsilon}_{ij}^*$, and σ_{ij}^* is the plastic stress field related to $\dot{\varepsilon}_{ij}^*$

through the flow rule. Here, v_i^* satisfies the continuity equation, but need not be a kinematically admissible field, although in the formulation v_i^* is so chosen that on the interface CD, v_n^* also satisfies the velocity boundary conditions. Further, v_s^* can be expressed as the sum of roll velocity V_R and relative sliding velocity of strip $(v_s^* - V_R)$. Thus, Equation 5.149 becomes

$$\int_{CD} t_s V_R dS \leq \int_V \sigma_{ij}^* \dot{\varepsilon}_{ij} dV + \int_{AB} t_b v_1^* dS - \int_{EF} t_f v_1^* dS + \int_{CD} t_s (V_R - v_s^*) dS .$$

(5.150)

The integral on the left hand side of the inequality is the actual rolling power and the integral on the right side are plastic dissipation power, power due to tensions and friction power respectively. The powers on the right hand side of inequality are computed based on the assumed velocity field. The computations of plastic dissipation power and power due to tensions pose no problem, whilst for the computation of friction power, actual shear traction distribution is required which is not known beforehand. Avitzur [15] has assumed its magnitude to be $m\sigma_Y / \sqrt{3}$, where m is a friction factor. In the model of Dixit and Dixit [22], t_s is obtained from Equation 5.91, in which t_n is obtained from the solution of continuity and momentum equations, albeit at assumed location of neutral point. This t_n is found to be resonably close to the actual t_n except near the assumed neutral point. However, since the relative velocity between the roll and strip is very small near the assumed neutral point, the error in the evaluation of friction power is negligible. Thus, all the three components of power are computed for different assumed positions of neutral point and the position which minimizes the sum of these is treated as the correct position of neutral point.

The initial estimation of the neutral point is found by using the slab method formula [28]:

$$\tan^{-1} \sqrt{R' / (h_2 \alpha_n)} = \frac{1}{2} \sin^{-1} \sqrt{r} - \frac{1}{4a} \ln \left\{ \frac{h_1 (1 - t_f / 2k_2)}{h_2 (1 - t_b / 2k_1)} \right\}.$$

(5.151)

In Equation 5.151, $\alpha_n, h_1, h_2, k_1, k_2$ and R' are the angular position of neutral point, inlet height, exit height, yield shear stress at inlet, yield shear stress at exit and the deformed roll radius, respectively. The deformed roll radius can be obtained from Hitchcock's formula (Equation 1.5). The Hitchcock's formula provides highly inaccurate results for the rolling of thin and hard strips, especially at low reductions. Roychoudhari and Lenard [29] used Michell's [30] two-dimensional solution of the biharmonic equation to calculate the shape of deformed roll contour. The analysis employs the general expression for the stress function given by Michell, the coefficients of which are determined such that the boundary conditions on the roll surface and center are satisfied. The method has been seen to predict experimentally established stress field in a satisfactory manner. Mori et al.

[31] analyzed roll deformation by the boundary element method, while the strip was analyzed by FEM. Some authors have used the theory of elasticity solution considering the roll as an elastic half space [32–34]. The parameter a is given by

$$a = f\sqrt{\frac{R'}{h_2}},\tag{5.152}$$

where f is the Coulomb coefficient of friction. The yield shear stress is $1/\sqrt{3}$ times the yield stress in tension. Considering the deformation to be homogeneous, the equivalent strain at the outlet is approximated as $\dfrac{2}{\sqrt{3}}\ln(h_1/h_2)$. Then, from the hardening law, the approximated yield shear stress at exit can be found.

The minimization of power is carried out by an optimization process. Taking the assumed neutral point in the middle, an interval in which the neutral point may lie is assumed. Using the golden section search method [35], this interval is reduced. After the interval has been sufficiently reduced, the exact position of the neutral point is found by fitting a quadratic curve through two end points and one middle point of the interval. At each of the assumed positions of neutral point, a number of iterations are required to find out the corresponding velocity field. In order to accelerate the computational process, the methodology of Aitken and Steffenson may be employed [36]. According to this, let x_1, x_2 and x_3 denote the results of three consecutive iterations. Assume that they approach their limit x as a geometrical series *i.e.*,

$$\frac{x_1 - x}{x_2 - x} = \frac{x_2 - x}{x_3 - x}.\tag{5.153}$$

Then, the limit can be calculated as

$$x = \frac{x_1 x_2 - x_2^2}{x_1 - 2x_2 + x_3}.\tag{5.154}$$

Thus, for each assumed position of neutral point, only four to five iterations can be carried out. In total 20–25 are sufficient to give the position of neutral point and values of roll force and roll torque accurately. In order to compute accurate stress distribution, a higher number of iterations can be carried out with correct location of neutral point.

5.3.7 Formulation for Strain Hardening

The equivalent strain at a point is obtained by the time integration of the equivalent strain rate along the particle path. The first step in the determination of the equivalent strain field is the construction of particle paths, or flow lines. The slope of a flow line is given by

$$\frac{dy}{dx} = \frac{v_y}{v_x}. \qquad (5.155)$$

Given a point (x_i, y_i) on a flow line, the coordinates (x_{i+1}, y_{i+1}) of the adjacent point are found by using the relationship:

$$y_{i+1} = y_i + (x_{i+1} - x_i)\frac{v_y}{v_x}, \qquad (5.156)$$

where the adjacent point is chosen sufficiently small so that the path between two points may be approximated by a straight line. In this manner, the flow lines of various points on the inlet boundary can be found.

Now, along a flow line,

$$dt = \frac{dx}{v_x} = \frac{dy}{v_y}. \qquad (5.157)$$

Thus, Equation 3.158 can be written as

$$\varepsilon_{eq} = \int_0^L \frac{\dot{\varepsilon}_{eq}}{v_x} dx, \qquad (5.158)$$

where L is the length of the entire zone along x direction. The equivalent strain is zero at the first plastic boundary. However, this boundary is not known a priori; therefore, the equivalent strain is taken as zero at the inlet boundary. This is acceptable, since the strains it gives rise to at the first plastic boundary are quite small. Equation 5.158 can be integrated by Simpson's rule for obtaining the equivalent strain at various points along the flow lines. Three flow lines are taken in an element. Nine points in each element are selected and the equivalent strain values at these points are interpolated to obtain the equivalent strains at Gauss points. These values are then substituted in hardening law to update the flow stress for further iterations.

5.3.8 Modification of Pressure Field at Each Iteration

The finite element formulation does not require any pressure boundary condition to be satisfied. Therefore, the pressure values may be determined up to an additive constant. Moreover, the formulation must have provision to incorporate front and back tension. The effect of tensions is to reduce the roll pressure at each point approximately by the factor $t_b / 2k_1$ on the entry side (from inlet to neutral point) and by the factor $t_f / 2k_1$ on the exit side (from neutral point to outlet). The position of the neutral point is moved forward by the application of back tension and backward by the application of front tension.

When the front and back tensions (t_f and t_b) are present, it does not seem possible to incorporate them through imposition of traction boundary conditions as these boundaries fall in the rigid zone. The following approach may be adopted. It is assumed that:

(1) The velocity field and hence the strain rate field is not affected significantly in the presence of tensions. The tensions, therefore, influence the pressure (or hydrostatic stress) only. The implication of this assumption is that, if we solve a rolling problem without tensions in which the neutral point is approximately the same as in the presence of tensions, then keeping the same velocity field and modifying the pressure, we may obtain a solution in the presence of tensions. Since the hydrostatic stress does not participate in the work of deformation, the assumption also implies that the work expended per unit volume of the material is practically unaffected by the applied tensions [28].

(2) The effect of tensions is experienced uniformly across a cross-section of strip, so that one-dimensional equations of the slab method become applicable.

According to the slab method [28], the roll pressure with tensions (q) is related to one without tensions (q_0) by the relation

$$q_0 - q = t_f e^{2a\psi} \quad \text{on exit side.} \tag{5.159}$$

$$q_0 - q = t_b e^{2a(\psi_0 - \psi)} \quad \text{on entry side.} \tag{5.160}$$

The distribution of q_0 corresponding to both these equations is assumed to extend up to the neutral point with tensions. Here,

$$\psi = \tan^{-1}\left(\sqrt{\frac{R'}{h_2}}\,\phi\right), \tag{5.161}$$

and ψ_0 is the value of ψ when ϕ is equal to the angle of contact or angle of bite (α). Since the deviatoric part is unaffected by the tensions, $q - q_0$ will be equal to $p - p_0$ where p_0 is the pressure (negative of hydrostatic stress) without tensions and p is its value in their presence. Then, using Equations 5.159 and 5.160, we get the following relation between p and p_0:

$$p = p_0 - (t_f + p_{avo})e^{2a\psi} \quad \text{on exit side}, \tag{5.162}$$

$$p = p_0 - (t_b + p_{avi})e^{2a(\psi_0 - \psi)} \quad \text{on entry side}, \tag{5.163}$$

where p_{avi} and p_{avo} are the average pressures at the inlet and exit, respectively. These have been introduced to take care of the spurious pressure distribution.

5.3.9 Calculation of Secondary Variables

Once the solution of the problem is obtained in the form of nodal velocities and pressure, the secondary quantities are calculated. We describe the procedure to calculate secondary quantities:

(i) Roll torque (T)

It is preferable to calculate the roll torque from the relation

$$T = \frac{PR}{V_R},$$
(5.164)

where P is the total power. It consists of the following three parts:

(a) Power required for plastic deformation (P_p)

The power dissipated due to plastic deformation is given by

$$P_p = \int_A \sigma'_{ij}\dot{\varepsilon}_{ij}\, dx\, dy.$$
(5.165)

Substitution of Equation 5.79 leads to

$$P_p = \int_A \sigma_{eq}\dot{\varepsilon}_{eq}\, dx\, dy.$$
(5.166)

(b) Power required for overcoming friction at the roll-strip interface (P_f)

The power dissipated due to friction is given by

$$P_f = \int_0^l |t_s||\Delta v_s|\, ds,$$
(5.167)

where Δv_s is the relative velocity with respect to the velocity of neutral point along the roll-strip interface and l is the arc of contact.

(c) Power required due to tensions (P_t)

The power due to front and back tensions is

$$P_t = (t_b - t_f)\frac{h_1}{2}U_1.$$
(5.168)

The roll torque can also be computed by integrating the stresses along the roll-strip interface. However, the stresses are expected to be less accurate, first due to the pressure term contained in it and, second, due to errors in the computation of Levy-Mises coefficient and strain-rates. It would be interesting to compute the torque by the Bland and Ford formula [7] given by

$$T = \mu RR' \left(\int\limits_{\alpha_n}^{\alpha} t_n \mathrm{d}\phi - \int\limits_{0}^{\alpha_n} t_n \mathrm{d}\phi \right), \tag{5.169}$$

where α is the angle of contact. The value of torque computed by this formula may be compared with Equation 5.164. Less difference between two values ensures the accuracy of neutral point and t_n. A formula that is not dependent on the neutral angle is due to Hill used by Alexander [10] in his code. It is given by

$$T = R'^2 \int\limits_{0}^{\alpha} t_n \, \phi \, \mathrm{d}\phi + \frac{1}{2} R \left(t_b h_1 - t_f h_2 \right) - F_r (R' - R) \frac{1}{2} \alpha, \tag{5.170}$$

where F_r is the roll force.

(ii) Roll pressure or interfacial normal stress (t_n)

While calculating t_n, first the stresses σ_{ij} are calculated at 2×2 Gauss points and then they are extrapolated to various points on the roll-strip interface after which t_n is calculated from the expression

$$t_n = \boldsymbol{t}.\hat{\boldsymbol{n}} , \tag{5.171}$$

where $\boldsymbol{t}$ is the traction and $\hat{\boldsymbol{n}}$ unit outward normal to the interface.

(iii) Roll force (F_r)

The roll force is the vertical component of the resultant of interfacial stresses (Figure 5.12) which is given by

$$F_r = - \int\limits_{0}^{l} \left(t_n \cos\phi + t_s \sin\phi \right) \mathrm{d}s . \tag{5.172}$$

5.3.10 Some Numerical Aspects

The resulting system of equations (Equation 5.140) is highly ill-conditioned in this formulation. Therefore, it is advisable to use a solver that can handle ill-conditioning to some extent. The Householder method [4] is one such method. For the first iteration, the non-dimensional value of Levy-Mises coefficient η_1 can be taken about 100. For each assumed position of neutral point, five iterations are found sufficient for convergence of the power. Once the correct position of the neutral point is found by minimizing the power, the iterations are continued until the nodal velocity and pressure values converge within 0.1% between the successive iterations. Convergence of the deformed roll profile is attained simultaneously. The lengths of the inlet and exit zones are taken about three times the semi-exit thickness.

It is better to obtain the roll torque by computing the total power, although Hill's formula also provides reasonably accurate results. The interfacial shear traction can be found directly from stress values or by first calculating the normal traction and using a friction law. The latter procedure seems to be more accurate, because the normal traction is larger by an order of 10 in comparison to shear traction. Thus, the percentage error in the computation of the normal stress is expected to be lower.

5.3.11 Typical Results and Discussion

A number of researchers have obtained results for roll force and roll torque using the finite element method. Most of the researchers have compared their results with the experimental results of Shida and Awazuhara [25] and/or Al-Salehi et al. [24]. Shida and Awazuhara [25] have conducted experiments on cold flat rolling of steels. Although there is a good agreement between the experimental and FEM results of roll force, the roll torques are underestimated [19, 21]. The inclusion of roll deformation in the model [22] brings the FEM predicted roll torque values closer to experimental values, but the difference is significant. This may be due to uncertainties in the estimation of material parameters and friction. The experimental data in [24] show a large amount of scatter, indicating the presence of statistical variation in the parameters. Al-Salehi et al. [24] have conducted experiments on rolling of aluminum and copper. In most of the cases, there is a good matching between the experimental and FEM results. Figure 5.13 shows the predicted and experimental roll pressure distributions for a typical case. It is seen that there is a good qualitative matching between the two results. However, there is some difference in the magnitudes. It was reported in [24] that the roll force calculated from the normal pressure distribution is 20% higher than that measured from load cell. Thus, considering experimental uncertainties, the agreement between the predicted and experimental pressure distribution is good.

Finite element analysis can bring out detailed information about the stress, strain and strain-rate distributions. For a typical case of rolling of steel, the equivalent strain-rate distribution is shown in Figure 5.14. Corresponding equivalent strain distribution is shown in Figure 5.15. It is observed from FEM results that equivalent strain-rate and strain distributions do not depend on the material. They are mainly a function of initial thickness, roll radius, reduction and friction. Figure 5.14 shows that along the roll-strip interface, the equivalent strain-rate varies with multiple peaks. From Figure 5.15, it is seen that the equivalent strain increases continuously from the first plastic boundary (at which yielding occurs) at the inlet to the second plastic boundary (where unloading occurs) near to the exit. The plastic boundaries have been marked on the basis of 3% of maximum equivalent strain-rate, i.e., at the plastic boundaries the equivalent strain-rate is 3% of the maximum equivalent strain-rate.

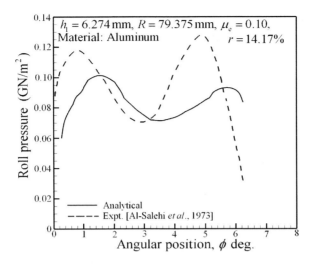

Figure 5.13. Comparison of analytical and experimental roll pressure distributions

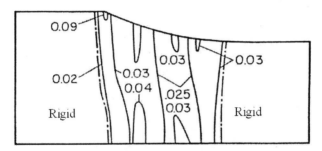

Figure 5.14. Equivalent strain-rate contours for steel (r=24%, R/h_1=65, equivalent Coulomb coefficient = 0.08)

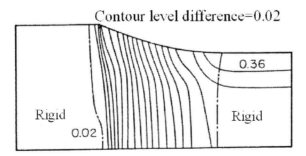

Figure 5.15. Equivalent strain contours for steel (r=24%, R/h_1=65, equivalent Coulomb coefficient = 0.08)

In the cold flat rolling process for metals, roll pressure and roll torque increase with reduction, R/h_1 and friction. The effect of both front and back tension is to decrease the roll force. Roll force is decreased more effectively by back tension than by front tension. Roll torque decreases with front tension but increases with back tension. When an equal amount of front and back tensions are present, roll torque decreases only slightly. The front tension causes the neutral point to move towards the inlet while the back tension moves it towards the exit. When the front and back tensions are of equal magnitude, the position of neutral point is not changed appreciably.

5.4 Formulation of Axisymmetric Metal Forming Processes

The examples of axisymmetric forming processes are wire drawing and extrusion. Wistreich [37] measured die-pressure in wire drawing by using the split die technique. Majors [38] measured the die pressure by measuring the hoop-strain in the die. Cook and Wistreich [39] have described the method for measuring the die-pressure by photo-elastic method. Thomoson et al. [40] employed visioplasticity method, which is a combination of experiments and analysis. Hoffman and Sachs [41] proposed the slab method for the analysis of the wire drawing process. The solution obtained by them is valid only for small die angle, as the redundant work is neglected in the formulation. Siebel [42] introduced a theory of wire drawing in which he assumed that the effects of homogeneous deformation, friction, and non-useful deformation were additive and has given an equation for drawing force. Avitzur [43] has proposed an extra term to account for redundant power in the drawing stress expression of Hoffmann and Sachs [41]. Avitzur [44, 45] has applied the upper bound theorem to the problem of wire drawing. He divided the wire in three zones in each of which velocity field was assumed to be continuous. At the interface, however, the tangential component of velocity was discontinuous. Avitzur [46] has also applied upper bound technique to strain hardening material considering linear hardening. He also analyzed central burst defects in extruded and drawn products [47]. Some researchers have analyzed the wire drawing process by using FEM. Chen et al. [48] obtained the steady-state deformation characteristics in extrusion and drawing as functions of material properties, die work interface friction, die angle and reduction. Using the elasto-plastic FEM model, Chevalier [49] studied the influence of geometrical parameters and friction condition on the quality of the final wire. Dixit and Dixit [50] employed rigid-plastic FEM model for the analysis of wire drwaing process. Gifford et al. [51] studied micro-hardness distribution in wire drawing by using commercial FEM package DEFORM-2D^{TM}.

When the deformation is axisymmetric, it is convenient to use cylindrical polar cordinates. In this case, the velocity v has only two components:

$$v = v_r \hat{i}_r + v_z \hat{i}_z .$$

(5.173)

Further, the velocity components are independent of the coordinate θ. The components of stress tensor σ and strain-rate tensor $\dot{\varepsilon}$ are also independent of θ. They are given by

$$[\sigma] = \begin{bmatrix} \sigma_{rr} & 0 & \sigma_{rz} \\ 0 & \sigma_{\theta\theta} & 0 \\ \sigma_{rz} & 0 & \sigma_{zz} \end{bmatrix}, \quad [\dot{\varepsilon}] = \begin{bmatrix} \dot{\varepsilon}_{rr} & 0 & \dot{\varepsilon}_{rz} \\ 0 & \dot{\varepsilon}_{\theta\theta} & 0 \\ \dot{\varepsilon}_{rz} & 0 & \dot{\varepsilon}_{zz} \end{bmatrix}, \tag{5.174}$$

The strain-rate velocity relations for axisymmetric problem are given by

$$\dot{\varepsilon}_{rr} = \frac{\partial u_r}{\partial r}, \quad \dot{\varepsilon}_{\theta\theta} = \frac{u_r}{r}, \quad \dot{\varepsilon}_{zz} = \frac{\partial u_z}{\partial z},$$
$$\dot{\varepsilon}_{rz} = \frac{1}{2}\left(\frac{\partial u_r}{\partial z} + \frac{\partial u_z}{\partial r} \right). \tag{5.175}$$

The incompressibility constraint and steady-state equations of motion without body force for axisymmetric problems are written as

$$\dot{\varepsilon}_{rr} + \dot{\varepsilon}_{\theta\theta} + \dot{\varepsilon}_{zz} = 0, \tag{5.176}$$

$$\rho\left(v_r \frac{\partial v_r}{\partial r} + v_z \frac{\partial v_r}{\partial z} \right) - \left(\frac{1}{r}\frac{\partial(r\sigma_{rr})}{\partial r} + \frac{\partial\sigma_{rz}}{\partial z} - \frac{\partial\sigma_{\theta\theta}}{r} \right) = 0, \tag{5.177}$$

$$\rho\left(v_r \frac{\partial v_z}{\partial r} + v_z \frac{\partial v_z}{\partial z} \right) - \left(\frac{1}{r}\frac{\partial(r\sigma_{rz})}{\partial r} + \frac{\partial\sigma_{zz}}{\partial z} \right) = 0. \tag{5.178}$$

In axisymmetric problems, the boundaries are such that the unit normal vector n has only two components, $i.e.$,

$$n = n_r \hat{i}_r + n_z \hat{i}_z. \tag{5.179}$$

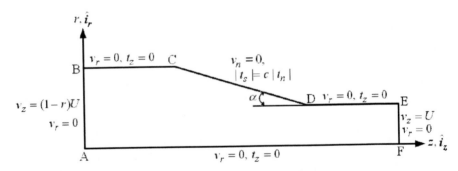

Figure 5.16. The domain and boundary conditions for a wire drawing process

For a wire drawing process (domain for analysis shown in Figure 5.16), the boundary conditions are as follows:

(1) Entry and exit boundaries (AB and EF):

The control volume is so selected that its entry and exit boundaries are sufficiently far away from either side of the deformation zone. If the drawing force F_d and back tension F_b are known, then they can be assumed to be uniformly distributed over AB and EF. Since the forces are in the z-direction, the r-component of the stress vector is zero. On the other hand, if the drawing speed U is specified, we know the z-component of the velocity at every point of EF. The r-component of this velocity is obviously zero at both EF and AB. The z-component of the velocity at AB can be found from the continuity equation:

$$(v_z)_{AB} A_{AB} = (v_z)_{EF} A_{EF} .$$
(5.180)

The ratio A_{EF}/A_{AB} is equal to $(1-r)$, where r is the fractional reduction. Thus,

$$(v_z)_{AB} = (1-r)(v_z)_{EF}.$$
(5.181)

Therefore, the following boundary conditions can be specified:

$$t_z = \frac{F_b}{\pi r_1^2}, \qquad t_r = 0 \text{ on AB},$$
(5.182)

$$t_z = \frac{F_d}{\pi r_2^2}, \qquad t_r = 0 \text{ on EF},$$
(5.183)

or

$$v_z = (1-r)U, \quad v_r = 0 \text{ on AB},$$
(5.184)

$$v_z = U, \quad v_r = 0 \text{ on EF.} \tag{5.185}$$

Dixit and Dixit [50] incorporated essential boundary conditions on AB and EF. Chen *et al.* [48] specified v_z and t_r on these boundaries. By running a computer code with different boundary conditions, it was observed that there is no significant difference in the results. However, the boundary conditions incorporated by Dixit and Dixit provide faster convergence.

(2) The top free surfaces (BC and DE):
Since these boundaries are free surfaces, they will be traction-free. Further, the flow of the material is along the boundary. Therefore, the r-component of velocity is zero. Thus, the boundary conditions are

$$t_z = 0 \text{ and either } v_r = 0 \text{ or } t_r = 0. \tag{5.186}$$

Strictly speaking, before coming into contact with the die, the wire bulges or converges; it may even bulge or converge in sequence. Therefore, very near to the inlet side of the die, the condition $v_r = 0$ is not valid. However, it is observed that even when we modify our program by incorporating the boundary condition $t_z = 0$ and $t_r = 0$ on the part of boundary which is close to the die, there is no appreciable change in the results. Bulge or convergence noticed is so small that for all practical purposes, it may be neglected.

(3) The axis of symmetry (AF):
Like in the case of plane-strain problems, here we use the following boundary condition:

$$t_z = 0 \text{ and } v_r = 0. \tag{5.187}$$

(4) The wire-die interface (CD):
The component of velocity in the direction normal to the die-wire interface at any point on the interface is zero. Thus,

$$v_n = v_r + v_z \tan \alpha = 0 , \tag{5.188}$$

where α is the die semi-angle. The second boundary condition will be

$$|t_s| = c|t_n|, \tag{5.189}$$

where t_s and t_n are the tangential and normal components of the stress vector t and c is a function of t_n which can be evaluated from Equations 4.236 and 4.237.

The finite element formulation does not require any pressure boundary condition to be satisfied. However, we may have a spurious pressure distribution in the solution. This constant can be determined from the condition that at the inlet boundary pressure values should be equal to one-third of back tension.

The non-dimensionalization of the governing equations is carried out in a manner similar to plane-strain problems. For moderate speed wire drawing, the inertial terms are ignored. The weighted residual form is given by

$$\int_A \left[\left(\dot{\varepsilon}_{rr} + \dot{\varepsilon}_{\theta\theta} + \dot{\varepsilon}_{zz} \right) w_p + \left(\frac{1}{r}\frac{\partial(r\sigma_{rr})}{\partial r} + \frac{\partial \sigma_{rz}}{\partial z} - \frac{\partial \sigma_{\theta\theta}}{r} \right) w_r + \left(\frac{1}{r}\frac{\partial(r\sigma_{rz})}{\partial r} + \frac{\partial \sigma_{zz}}{\partial z} \right) w_z \right] 2\pi r\, dr\, dz,$$
(5.190)

where w_p, w_r and w_z are weight functions which satisfy the homogeneous version of the essential boundary conditions and 'A' represents the area of the domain. Substituting the constitutive equation and then integrating the second and third parts of Equation 5.190, we get

$$\int_A I_1 2\pi r\, dr\, dz + \int_A I_2 2\pi r\, dr\, dz - \int_{\Gamma_r} I_3 2\pi r\, dr\, dz - \int_{\Gamma_z} I_4 2\pi r\, dr\, dz = 0, \quad (5.191)$$

where

$$
\begin{aligned}
I_1 &= -w_p \left(\dot{\varepsilon}_{rr} + \dot{\varepsilon}_{\theta\theta} + \dot{\varepsilon}_{zz} \right), \\
I_2 &= -p \left(\dot{\varepsilon}_{rr}(w) + \dot{\varepsilon}_{\theta\theta}(w) + \dot{\varepsilon}_{zz}(w) \right), \\
&\quad + 2\eta \left(\dot{\varepsilon}_{rr}\dot{\varepsilon}_{rr}(w) + \dot{\varepsilon}_{\theta\theta}\dot{\varepsilon}_{\theta\theta}(w) + \dot{\varepsilon}_{zz}\dot{\varepsilon}_{zz}(w) + 2\dot{\varepsilon}_{rz}\dot{\varepsilon}_{rz}(w) \right), \quad (5.192)\\
I_3 &= w_r t_r, \\
I_4 &= w_z t_z,
\end{aligned}
$$

where Γ_r and Γ_z are the boundaries where t_r and t_z are specified.

The shape functions for velocity and pressure approximations are same as in plane-strain problem. The strain-rate vector is defined as

$$
\{\dot{\varepsilon}\} = \left\{ \begin{array}{c} \dot{\varepsilon}_{rr} \\ \dot{\varepsilon}_{zz} \\ \sqrt{2}\,\dot{\varepsilon}_{rz} \\ \dot{\varepsilon}_{\theta\theta} \end{array} \right\} = \left\{ \begin{array}{c} \dfrac{\partial v_r}{\partial r} \\[2mm] \dfrac{\partial v_z}{\partial z} \\[2mm] \dfrac{1}{\sqrt{2}}\left(\dfrac{\partial v_r}{\partial z} + \dfrac{\partial v_z}{\partial r} \right) \\[2mm] \dfrac{v_r}{r} \end{array} \right\}.
\qquad (5.193)
$$

Substituting the approximation for velocities, we get

$$\{\dot{\varepsilon}\} = [B]\{v\}^e, \qquad (5.194)$$

where

$$[B] = \begin{bmatrix} \dfrac{\partial N_1}{\partial r} & 0 & \dfrac{\partial N_2}{\partial r} & 0 & \cdots & \dfrac{\partial N_9}{\partial r} & 0 \\[2ex] 0 & \dfrac{\partial N_1}{\partial z} & 0 & \dfrac{\partial N_2}{\partial z} & \cdots & 0 & \dfrac{\partial N_9}{\partial z} \\[2ex] \dfrac{1}{\sqrt{2}}\dfrac{\partial N_1}{\partial z} & \dfrac{1}{\sqrt{2}}\dfrac{\partial N_1}{\partial r} & \dfrac{1}{\sqrt{2}}\dfrac{\partial N_2}{\partial z} & \dfrac{1}{\sqrt{2}}\dfrac{\partial N_2}{\partial r} & \cdots & \dfrac{1}{\sqrt{2}}\dfrac{\partial N_2}{\partial z} & \dfrac{1}{\sqrt{2}}\dfrac{\partial N_2}{\partial r} \\[2ex] \dfrac{N_1}{r} & 0 & \dfrac{N_2}{r} & 0 & \cdots & \dfrac{N_9}{r} & 0 \end{bmatrix}.$$

$$(5.195)$$

Similarly,

$$\{\dot{\varepsilon}(w)\} = [B]\{w_v\}^e .$$

$$(5.196)$$

We note that

$$\dot{\varepsilon}_{rr} + \dot{\varepsilon}_{\theta\theta} + \dot{\varepsilon}_{zz} = \{1\ 1\ 0\ 1\}\{\dot{\varepsilon}\} = \{m\}^T [B]\{v\}^e ,$$

$$(5.197)$$

where

$$\{m\} = \begin{Bmatrix} 1 \\ 1 \\ 0 \\ 1 \end{Bmatrix}.$$

$$(5.198)$$

Similarly,

$$\dot{\varepsilon}_{rr}(w) + \dot{\varepsilon}_{\theta\theta}(w) + \dot{\varepsilon}_{zz}(w) = \{m\}^T [B]\{v\}^e .$$

$$(5.199)$$

Substitution of Equations 5.194, 5.196, 5.197 and 5.199 into Equation 5.191 leads to the following finite element equations in the local variables:

$$\sum_{i=1}^{ne} \{w\}^{eT} [k]^e \{\delta\}^e = \sum_{b=1}^{nb_1} \{w_r\}^{bT} \{f_r\}^b + \sum_{b=1}^{nb_2} \{w_z\}^{bT} \{f_z\}^b ,$$

$$(5.200)$$

where

$$\{w\}^e = \left\{ \begin{array}{c} \{w_p\}^e \\ \{w_v\}^e \end{array} \right\}, \tag{5.201}$$

$$\{\delta\}^e = \left\{ \begin{array}{c} \{p\}^e \\ \{v\}^e \end{array} \right\}, \tag{5.202}$$

$$[k]^e = \begin{bmatrix} [0] & [k_{pv}]^e \\ [k_{vp}]^e & [k_{vv}]^e \end{bmatrix}, \tag{5.203}$$

$$[k_{pv}]^e = \int_{A^e} -\left\{N^p\right\}\{m\}^{\mathrm{T}}[B]2\pi r\,dr\,dz,$$

$$[k_{vv}]^e = \int_{A^e} 2\eta_1[B]^{\mathrm{T}}[B]2\pi r\,dr\,dz, \tag{5.204}$$

$$[k_{vp}]^e = \int_{A^e} -[B]^{\mathrm{T}}\{m\}\{N^p\}2\pi r\,dr\,dz,$$

$$\left\{f_r\right\}^b = \int_{\Gamma_r^b} \{N\}^b\{N\}^{b\mathrm{T}}\{t_r\}^b\,2\pi r\,d\Gamma,$$

$$\left\{f_z\right\}^b = \int_{\Gamma_z^b} \{N\}^b\{N\}^{b\mathrm{T}}\{t_z\}^b\,2\pi r\,d\Gamma, \tag{5.205}$$

The remaining details of finite element formulation are similar to that of plane strain problem and will not be repeated here.

Figure 5.17 compares FEM results with Wistriech's results [37]. A good agreement is obtained between the two results. A look at Figure 5.17 reveals that there is an optimum die angle providing minimum drawing stress. The optimum die angle is different for different reductions and friction. With too small a cone angle, the length of contact between the wire and die is high, causing significantly high frictional losses. With too large a cone angle, the redundant work becomes a predominant factor. Therefore, there exists an optimum die angle. It is observed that, for a given friction coefficient, the optimum die angle increases with increasing reduction. The optimum cone angle increases with increasing coefficient of friction. It is seen that generally the optimum die angle is not dependent on the material. The variation of drawing stress is more significant at small die angle in comparison to large die angles. If the coefficient of friction is not known properly, it is better to design the die according to maximum expected coefficient of friction.

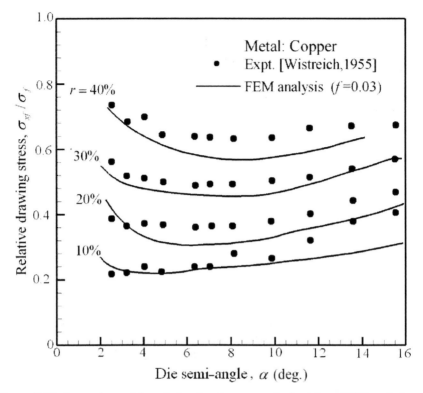

Figure 5.17. Comparison of the relative drawing stress obtained from FEM analysis and from Wistriech's experiments. With permission from Dixit and Dixit [50]. Copyright 1995 Elsevier

Figure 5.18 shows equivalent strain-rate contours for a low reduction and high die angle. In this case, the plastic region near to the axis of symmetry is very narrow and the strain rate near to the die surface becomes very high compared to that near to the axis of symmetry. It is observed that as the die angle is increased, at a particular angle the plastic zone disappears at the axis of symmetry. In this case, near to the axis of symmetry the zone of undeformed material and the zone of fully deformed product have a common boundary. Since the zone of fully-deformed material moves faster than the zone of undeformed product, both zones separate and form a central burst. Thus, using FEM analysis, the central burst defect can be studied by studying the strain-rate contours. Figure 5.19 shows the corresponding equivalent strain contour. It is observed that equivalent strain across the drawn wire cross-section is non-uniform. The non-uniformity is more pronounced near the wire-die interface. The strain and strain-rate distribution are also independent of the material. The friction mildly influences the strain and strain-rate distributions.

Metal: Steel

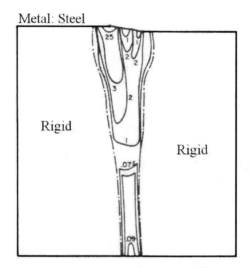

Figure 5.18. Equivalent strain-rate contours (r=5%, die semi-angle=10^0, Coulomb's coefficient = 0.03). With permission from Dixit and Dixit [50]. Copyright 1995 Elsevier

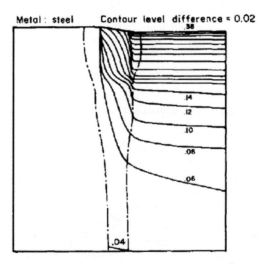

Figure 5.19. Equivalent strain contours (r=5%, die semi-angle=10^0, Coulomb's coefficient = 0.03). With permission from Dixit and Dixit [50]. Copyright 1995 Elsevier

As a result of parametric study, it is observed that for a fixed coefficient of friction and die angle, as the percentage reduction decreases, die pressure increases. It is also seen that die pressure is not uniform along the length of the die. The separation force increases with increasing reduction in spite of the decrease in die pressure. It is observed that friction has very little influence on the die pressure. It is observed that back tension decreases the die pressure but increases the drawing stress.

5.5 Formulation of Three-Dimensional Metal Forming Processes

The flow formulation procedure for three-dimensional metal forming problem is similar to the two-dimensional problem. The equations described in Section 5.3.1 are valid, except that the range of indices in the tensor equation is 1 to 3 instead of 1 to 2. The weak form is

$$\int_A \dot{\varepsilon}_{ii} w_p \, dA + \int_A \sigma_{ij} \dot{\varepsilon}_{ij}(w) \, dA - \int_{\Gamma_i} t_i w_i d\Gamma_i = 0, \tag{5.206}$$

where indices i and j range from 1 to 3. The second term in Equation 5.206 can be simplified as

$$\sigma_{ij} \dot{\varepsilon}_{ij}(w) = -p \dot{\varepsilon}_{ii}(w) + 2\eta \dot{\varepsilon}_{ij} \dot{\varepsilon}_{ij}(w). \tag{5.207}$$

The finite element formulation may be carried out as explained in Section 5.3.4. One can choose the brick element with 20 nodes for interpolating the three velocity components and eight corner nodes for interpolating the pressure. Compared to two-dimensional problems, the total degrees of freedom for a three-dimensional problem are enormous, increasing the memory and computational time requirement.

While analyzing the problem of 3-D symmetric rolling, it is enough to consider one fourth of the domain. At the surfaces of the symmetry, the normal component of velocity and tangential components of traction are zero. At the inlet and outlet surfaces, the velocity boundary conditions can be prescribed. At the top free surface, the components of traction can be put zero. Alternatively, away from the roll surface, one can make two velocity components zero, leaving only the longitudinal component of velocity. At the roll work interface, the normal component of velocity is zero. The other boundary conditions are provided by friction boundary conditions. For more details, reader can refer to [52, 53].

5.6 Incorporation of Anisotropy

The finite element formulation of anisotropic material is similar to the isotropic formulation, as the governing equations and the boundary conditions are the same, the only difference being in the constitutive equations. Section 4.7.2 provides the constitutive equations for plane strain condition. For a rigid-plastic anisotropic material, Equation 4.174 provides

$$\sigma'_{xx} = k_1 \dot{\varepsilon}_{xx}; \quad \sigma'_{yy} = k_2 \dot{\varepsilon}_{yy}; \quad \sigma'_{xy} = k_3 \dot{\varepsilon}_{xy}. \tag{5.208}$$

It can easily be verified that finite element formulation will provide element stiffness matrix similar to Equation 5.132, in which only the sub-matrix $[K^e_{vv}]$ is different from the isotropic case. The expression for sub-matrix $[K^e_{vv}]$ is

$$[K_{vv}]^e = \int_A [B_a]^T [B_a] dxdy , \qquad (5.209)$$

where

$$[B_a] = \begin{bmatrix} k_1 \dfrac{\partial N_1}{\partial x} & 0 & k_1 \dfrac{\partial N_2}{\partial x} & 0 & ----- & k_1 \dfrac{\partial N_9}{\partial x} & 0 \\[2ex] 0 & k_2 \dfrac{\partial N_1}{\partial y} & 0 & k_2 \dfrac{\partial N_2}{\partial y} & ---- & 0 & k_2 \dfrac{\partial N_9}{\partial y} \\[2ex] \dfrac{k_3}{\sqrt{2}} \dfrac{\partial N_1}{\partial y} & \dfrac{k_3}{\sqrt{2}} \dfrac{\partial N_1}{\partial x} & \dfrac{k_3}{\sqrt{2}} \dfrac{\partial N_2}{\partial y} & \dfrac{k_3}{\sqrt{2}} \dfrac{\partial N_2}{\partial x} & ---- & \dfrac{k_3}{\sqrt{2}} \dfrac{\partial N_9}{\partial y} & \dfrac{k_3}{\sqrt{2}} \dfrac{\partial N_9}{\partial x} \end{bmatrix}.$$

$$(5.210)$$

There have been few works in the anisotropic modeling of steady-state metal forming processes. Dixit and Dixit [54] have modeled the steady-state plane strain rolling process for anisotropic material. It was observed by them that, if the average of the flow stresses in thickness and longitudinal direction is used, an isotropic model provides almost the same results for roll torque and roll force as an anisotropic model. It is also observed that the deformation pattern is sensitive to exponent m in Equation 4.130. However, for fixed value of exponent m, the deformation pattern does not depend on average strain rate ratio $\bar{r}$. It is to be noted that Dixit and Dixit [54] presented their results for the hypothetical material parameters.

Figures 5.20 and 5.21 show that, for a particular value of m, roll force and roll torque increase with increasing value of average strain rate ratio $\bar{r}$. For a particular $\bar{r}$ value, roll torque decreases as m value is increased from 1.5 to 2.0, decrease in torque being more pronounced for higher $\bar{r}$ values. (For example, for $\bar{r}=2.5$, torque calculated on the basis of $m=1.5$ is 29% more than that calculated on the basis of $m=2$, i.e., Hill's quadratic criterion.) The parameter N, which is associated with shear stress term, also influences the results. However, the influence diminishes at higher value of R/h_1 ratio. Since in the cold flat rolling R/h_1 is more than 50, the determination of N is not so important for analyzing rolling problem. For the isotropic case N is equal to 1.5.

It is seen that for $m=2$, roll pressure and shear stress distributions remain the same with changing $\bar{r}$; only their magnitudes increase with increasing $\bar{r}$. However, change in m changes the distribution pattern. Also, it is seen that for $m=1.5$, the $\bar{r}$ does influence the distribution pattern for roll pressure and shear stress, although the position of neutral point is not affected significantly. At $m=1.5$, for certain values of $\bar{r}$, the multiple pressure peaks are obtained.

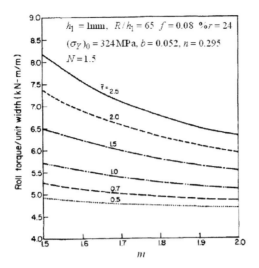

Figure 5.20. Effect of $\bar{r}$ and m values on the roll torque. With permission from Dixit and Dixit [54]. Copyright 1997 Elsevier

A study of strain and strain-rate contours indicates that anisotropy does not affect the distribution pattern of contours of the equivalent strain and equivalen strain rate. No influence of $\bar{r}$ is observed on magnitudes of these quantities. However, with decrease in m, there is a slight decrease in the values of equivalent strain and equivalent strain-rate. In the isotropic case, the deviatoric stresses along thickness and longitudinal direction are equal in magnitude (but opposite in sign) in the plastic deformation zone. Hence, for the isotropic case, the contours of deviatoric strains along the thickness and logitudinal directions look similar in the plastic deformation zone. This is not true for the anisotropic case.

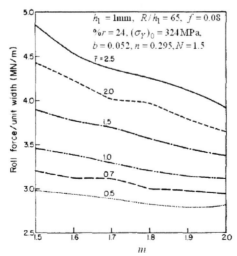

Figure 5.21. Effect of $\bar{r}$ and m values on the roll force. With permission from Dixit and Dixit [54]. Copyright 1997 Elsevier

5.7 Elasto-Plastic Formulation

There have been some attempts to solve the elasto-plastic problem in Eulerian reference frame. Elasto-plastic formulation is useful for predicting the plastic boundaries, accurate stress field and residual stresses. Dixit and Dixit [55] have reviewed three different types of formulation. In the mixed formulation, velocity, pressure and deviatoric stresses are treated as the primary variables. Neglecting hydrostatic part, the following constitutive equation is used (Equation 3.224a–c):

$$\dot{\varepsilon}_{ij} = \frac{\alpha}{2\eta_1}\sigma'_{ij} + \frac{1}{2\mu}\overset{o}{\sigma}'_{ij} , \tag{5.211}$$

where

$$\alpha = \begin{cases} 0 & \text{for elastic region} \\ 1 & \text{for plastic region} \end{cases}, \tag{5.212}$$

$$\overset{o}{\sigma}'_{ij} = v_k \frac{\partial \sigma'_{ij}}{\partial x_k} - \sigma'_{ik}\dot{\omega}_{jk} - \dot{\omega}_{ik}\sigma'_{kj} , \tag{5.213}$$

$$\dot{\omega}_{ij} = \frac{1}{2}\left(\frac{\partial v_i}{\partial x_j} - \frac{\partial v_j}{\partial x_i} \right) . \tag{5.214}$$

Here, μ is the shear modulus, η_1 is the Levy-Mises coefficient given by Equation 5.80 and $\overset{o}{\sigma}'_{ij}$ is the Jaumann rate of deviatoric stress tensor (Equation 3.224e).

For a plane strain problem, the weighted residual form of the governing equations is given by

$$\int_A [I_1 + I_2 + I_3 + I_4 + I_5 + I_6]\,dA = 0 , \tag{5.215}$$

where

$$I_1 = \left(\dot{\varepsilon}_{xx} + \dot{\varepsilon}_{yy} \right)w_p , \tag{5.216}$$

$$I_2 = \left(-\frac{\partial p}{\partial x} + \frac{\partial \sigma'_{xx}}{\partial x} + \frac{\partial \sigma'_{xy}}{\partial y} \right)w_{v_x} , \tag{5.217}$$

$$I_3 = \left(-\frac{\partial p}{\partial y} + \frac{\partial \sigma'_{yy}}{\partial y} + \frac{\partial \sigma'_{xy}}{\partial x} \right) w_{v_y}, \tag{5.218}$$

$$I_4 = \left(\dot{\varepsilon}_{xx} - \frac{\alpha}{2\eta_1} \sigma'_{xx} - \frac{1}{2\mu} \overset{o}{\sigma}'_{xx} \right) w_{\sigma_{xx}}, \tag{5.219}$$

$$I_5 = \left(\dot{\varepsilon}_{yy} - \frac{\alpha}{2\eta_1} \sigma'_{yy} - \frac{1}{2\mu} \overset{o}{\sigma}'_{yy} \right) w_{\sigma_{yy}}, \tag{5.220}$$

$$I_6 = \left(\dot{\varepsilon}_{xy} - \frac{\alpha}{2\eta_1} \sigma'_{xy} - \frac{1}{2\mu} \overset{o}{\sigma}'_{xy} \right) w_{\sigma_{xy}}. \tag{5.221}$$

where w_p, w_{v_x}, w_{v_y}, $w_{\sigma_{xx}}$, $w_{\sigma_{yy}}$, and $w_{\sigma_{xy}}$ are the weight functions which satisfy the homogeneous versions of the essential boundary conditions and A represents the area of the domain. Integrating the second and third parts of Equation 5.215, substituting the finite element approximations for velocity, pressure and deviatoric stress into it and assembling leads to a matrix equation of the following form:

$$\begin{bmatrix} [0] & [K_{pv}] & [0] \\ [K_{vp}] & [K_{vv}] & [K_{v\sigma}] \\ [0] & [K_{\sigma v}] & [K_{\sigma\sigma}(v_i)] \end{bmatrix} \begin{Bmatrix} \{P\} \\ \{V\} \\ \{\Sigma'\} \end{Bmatrix} = \begin{Bmatrix} \{0\} \\ \{F\} \\ \{0\} \end{Bmatrix}. \tag{5.222}$$

Here $\{P\}$, $\{V\}$ and $\{\Sigma'\}$ contain the nodal values of pressure, velocity components and deviatoric stress components, respectively. The vector $\{F\}$ contains the nodal forces corresponding to specified tractions on the boundary. These boundary terms arise from the integration by parts of the terms I_2 and I_3.

The system of equations given by Equation 5.222 is highly ill-conditioned. The decoupled system of equation is given by

$$\begin{bmatrix} [0] & [K_{pv}] \\ [K_{vp}] & [K_{vv}] \end{bmatrix} \begin{Bmatrix} \{P\} \\ \{V\} \end{Bmatrix} = \begin{Bmatrix} \{0\} \\ \{F(v_i, \sigma'_{ij})\} \end{Bmatrix}, \tag{5.223}$$

$$\left[K_{\sigma\sigma}(v_i) \right] \{\Sigma'\} = -\left[K_{\sigma v} \right] \{V\}.$$

Here all the submatrices except $\left[K_{vv} \right]$ are the same as those given by Equation 5.222. The submatrix $\left[K_{vv} \right]$ arises when the constitutive equation is substituted in the equilibrium equation by treating the whole domain as plastic. The right hand side vector contains the terms from the Jaumann stress rate term of the constitutive equation.

It was observed that this type of formulation does not provide proper values of stresses in the inlet and exit zone. This is owing to the assumption of the whole zone being plastic and restrictions on the value of Levy-Mises coefficient η_1 in the inlet and exit zones.

In rate formulation, the equilibrium equation is represented not in terms of Cauchy stress tensor σ_{ij}, but in terms of the time rate (or material derivative) of the first Piola-Kirchoff stress tensor:

$$\dot{T}_{ij} = v_k \frac{\partial \sigma_{ij}}{\partial x_k} - \frac{\partial v_i}{\partial x_k} \sigma_{kj} + \frac{\partial v_k}{\partial x_k} \sigma_{ij} . \qquad (5.224)$$

The equilibrium equation is

$$\frac{\partial \dot{T}_{ij}}{\partial x_j} = 0 . \qquad (5.225)$$

The rate formulation proposed by Thompson and Yu [56] converges rapidly for purely elastic behavior, but its rate of convergence deteriorates as the plastic zone increases in size. Therefore, this formulation may be applied only in the exit elastic zone to find the residual stresses. The constitutive equation becomes

$$\overset{o}{\sigma}_{ij} \equiv v_k \frac{\partial \sigma_{ij}}{\partial x_k} - \sigma_{ik} \dot{\omega}_{jk} - \dot{\omega}_{ik} \sigma_{kj} = C_{ijkl} \dot{\varepsilon}_{kl} , \qquad (5.226)$$

where

$$C_{ijkl} = 2\mu \left[\delta_{ik}\delta_{jl} + \frac{v}{(1-2v)} \delta_{ij}\delta_{kl} \right] . \qquad (5.227)$$

Applying finite element procedure, we obtain the following equations:

$$\left[K_v(\sigma_{ij}) \right]\{V\} = \{F_v\}, \qquad (5.228)$$

$$\left[K_\sigma(v_i) \right]\{\Sigma\} = \{F_\sigma(v_i)\} . \qquad (5.229)$$

Here, $\{V\}$ and $\{\Sigma\}$ contain the nodal values of velocity and stress components, respectively. The vector $\{F_v\}$ contains the nodal forces corresponding to the boundary terms arising from integration by parts of the equilibrium equation. The vector $\{F_\sigma(v_i)\}$ contains the terms associated with $C_{ijkl}\dot{\varepsilon}_{kl}$. The accuracy of this method depends upon the accuracy of prescribed velocities and stresses at the end of the plastic zone, and the shape of the second plastic boundary.

Maniatty *et al.* [57] has developed elasto-plastic Eulerian finite element formulation for large deformation and large rotation. The fundamental concepts of this formulation are explained in Section 4.2 and 4.3 and will not be repeated here. For an elasto-plastic material the flow rule is written as (Equation 4.54)

$$\sigma'_{ij} = 2\eta_1 \left(\dot{\varepsilon}_{ij} - \dot{\varepsilon}^e_{ij} \right). \tag{5.230}$$

This is substituted in the equilibrium equation and finite element formulation of equilibrium equations and incompressibility constraint is carried out using mixed pressure-velocity formulation. The following system of equations is obtained:

$$\begin{bmatrix} [0] & \left[K_{pv} \right] \\ \left[K_{vp} \right] & \left[K_{vv} \right] \end{bmatrix} \begin{Bmatrix} \{P\} \\ \{V\} \end{Bmatrix} = \begin{Bmatrix} \{0\} \\ \left\{ F(\dot{\varepsilon}^e) \right\} \end{Bmatrix}, \tag{5.231}$$

where the submatrices $\left[K_{pv} \right]$, $\left[K_{vp} \right]$ and $\left[K_{vv} \right]$ are given by Equation 5.133. The right side vector $\left\{ F(\dot{\varepsilon}^e) \right\}$ contains the nodal forces corresponding to the boundary terms arising from integration by parts of the equilibrium equation and an additional term associated with $\sigma'_{ij} = 2\eta_1 \dot{\varepsilon}^e_{ij}$. The system of Equation 5.231 is solved iteratively to find the pressure and velocity fields. At each iteration, the elastic strain rate tensor has to be determined to compute the right side vector. The procedure for its determination has been described in Section 4.3.1. Because of numerical difficulties, the reliable values of residual stresses could not be obtained in [55] using this method.

In [55], a simplified method is proposed to find out the longitudinal residual stress. Let l be the length of an infinitesimal fiber which was along x direction in the undeformed configuration and whose undeformed length was unity (Figure 5.22). Let l_p be the part of l which is due to plastic deformation. Then l_p is given by

$$l_p = \sqrt{\overline{F}^{p\mathrm{T}} \overline{F}^p \, \hat{i}_x . \hat{i}_x}, \tag{5.232}$$

where $\hat{i}_x$ is the unit vector along the x direction and $\overline{F}^p$ is the plastic part of the deformation gradient tensor (Equation 4.26) at the point of unloading. Because of inhomogeneity of the plastic deformation, l_p will be different for different flowlines. This implies that, during unloading, the elastic strains will not drop to zero, as it will result in incompatible deformation. Therefore, every fiber must retain a part of the elastic strain which will ensure that all the fibers have same length after unloading. Let l_f be the common length. Then assuming the elastic strains to be small, the longitudinal elastic strain retained by fibers can be expressed as

$$\varepsilon_{xx} = \frac{l_f - l_p}{l_p}.$$
(5.233)

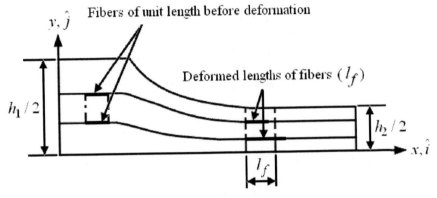

Figure showing top: "Fibers of unit length before deformation"

Figure 5.22. Figure showing non-uniform stretching of two fibers

In rolled strip, the top and bottom surfaces are stress-free and strip thickness is very small. Therefore, it may be assumed that the residual stress in the thickness direction is zero. Thus, the elastic stress-strain relations become

$$\frac{1}{E}\left(\sigma_{xx} - \nu\sigma_{zz}\right) = \varepsilon_{xx},$$
(5.234)

$$\frac{1}{E}\left(\sigma_{zz} - \nu\sigma_{xx}\right) = \varepsilon_{zz} = 0.$$
(5.235)

Eliminating σ_{zz} from the above two equations and substituting Equation 5.233, the following expression for the longitudinal residual stress is obtained:

$$\sigma_{xx} = \frac{E}{(1-\nu^2)}\frac{\left(l_f - l_p\right)}{l_p}.$$
(5.236)

The l_f can be determined from the requirement that the net longitudinal force should be zero. Using symmetry this condition can be stated as

$$\int_0^{h_2/2} \sigma_{xx}\,dy = 0.$$
(5.237)

This leads to

$$l_f = \frac{h_2}{2 \int_0^{h_2/2} \frac{dy}{l_p}} .$$
(5.238)

Thus, in this approach, we need to calculate l_p at unloading for various flowlines (varying with y). Then, Equation 5.238 can compute l_f and Equation 5.236 longitudinal residual stresses.

Figure 5.23 shows that residual stress is greater near the top fiber. Further, the stress gradients are also large. This behavior is consistent with the condition of high friction.

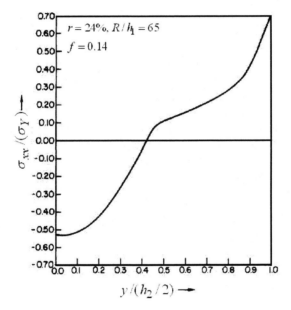

Figure 5.23. Distribution of longitudinal residual stress across the thickness (simplified approach). With permission from Dixit and Dixit [55]. Copyright 1997 Elsevier

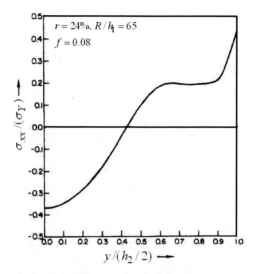

Figure 5.24. Distribution of longitudinal residual stress across the thickness for low friction. With permission from Dixit and Dixit [55]. Copyright 1997 Elsevier

Figure 5.24 depicts residual stress distribution for a reduced coefficient of friction. With reduced friction, deformation becomes more homogeneous thereby reducing the value of residual stress in the vicinity of top fibers. Figure 5.25 shows the effect of R/h_1 on the residual stress distribution. Process parameters are the same as those in Figure 5.24 except for R/h_1 which is now more. Increasing R/h_1 makes deformation more homogeneous. Therefore, the residual stress values are less now compared to those in Figure 5.24.

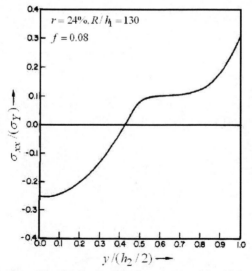

Figure 5.25. Distribution of longitudinal residual stress across the thickness for high R/h_1. With permission from Dixit and Dixit [55]. Copyright 1997 Elsevier

5.8 Summary

In this chapter, Eulerian (flow) formulation of metal forming problems is presented. First, a background of finite element method is presented. Then, the rigid-plastic plane strain formulation is explained with the example of a rolling problem. The formulation is explained in detail along with the mention of numerical difficulties. With the example of a wire-drawing process, rigid-plastic axisymmetric problem is explained. A brief note has been provided for the formulation of 3-D metal forming processes using flow formulation. Then, the method to incorporate anisotropy has been explained. The example has been provided from a plane-strain rolling problem. Finally, the elasto-plastic formulation and method for estimating residual stresses have been explained.

5.9 References

[1]   Reddy, J.N. (1993), An Introduction to the Finite Element Method, second edition, McGraw-Hill, New York.

[2]   Cook, R.D., Malkus, D.S. and Plesha, M.E. (1989), Concepts and Applications of Finite Element Analysis, third edition, John Wiley & Sons, New York.

[3]   Stoud, A.H. and Secrest, D. (1966), Gaussian Quadrature Formulas, Prentice-Hall, Englewood Cillfs, NJ.

[4]   Bathe, K.J. (1996), Finite Element Procedures, Prentice-Hall of India, New Delhi.

[5]   Karaman, T. von (1925), On the theory of rolling, Zeitschrift für Angewandte Mathematik und Mechanik, Vol. 5, pp. 139–141.

[6]   Orowan, E. (1943), The calculation of roll pressure in hot and cold flat rolling, Proceedings of the Institution of Mechanical Engineers, Vol. 150, pp. 140–167.

[7]   Bland, D.R. and Ford, H. (1948), The calculation of roll force and torque in cold strip rolling with tensions, Proceedings of the Institution of Mechanical Engineers, Vol. 159, pp. 144–153.

[8]   Bland, D.R. and Ford, H. (1952), An approximate treatment of the elastic compression of the strip in cold rolling, Journal of Iron and Steel Institute, Vol. 171, pp. 245–249.

[9]   Bland, D.R. and Sims, R.B. (1953), A note on the theory of rolling with tensions, Proceedings of the Institution of Mechanical Engineers, Vol. 167, pp. 371–374.

[10]   Alexander, J.M. (1972), On the theory of rolling, Proceedings of Royal Society of London, Vol. 326 A, pp. 535–563.

[11]   Firbank, T.C. and Lancastar, P.R. (1965), A suggested slip-line field for cold rolling with slipping friction, International Journal of Mechanical Sciences, Vol. 7, pp. 847–852.

[12]   Collins, I.F. (1969), Slip-line field solutions for compression and rolling with slipping friction, International Journal of Mechanical Sciences,, Vol. 11, pp. 971–978.

[13]   Thomson, P.F. and Brown, J.H. (1982), A study of deformation during cold rolling using visioplasticity, International Journal of Mechanical Sciences, Vol. 24, pp. 559–576.

[14]   Johnson, W. and Kudo, H. (1960), The use of upper-bound solutions for the determination of temperature distributions in fast hot rolling and axi-symmetric extrusion processes, International Journal of Mechanical Sciences, Vol. 1, pp. 175–191.

[15]   Avitzur, B. (1964), An upper bound approach to cold strip rolling, Transaction of ASME, Ser. B 86, pp. 31–48.

[16]   Martins, P.A.F. and Barata Marques, M.J.M. (1999), Upper bound analysis of plane strain rolling using a flow function and the weighted residual method, International Journal for Numerical Methods in Engineering, Vol. 44, pp. 1671–1683.

[17]   Zienkiwicz, O.C., Jain, P.C. and Onate, E. (1978), Flow of solids during forming and extrusion: some aspects of numerical solutions, International Journal of Solids and Structures, Vol. 14, pp. 15–38.

[18]   Mori, K., Osakada, K. and Oda, T. (1982), Simulation of plane strain rolling by rigid-plastic finite element method, International Journal of Mechanical Sciences, Vol. 24, pp. 519–527.

[19]   Li, G.J. and Kobayashi, S. (1982), Rigid-plastic finite element analysis of plane-strain rolling, Transaction of ASME, Journal of Engineering for Industry, Vol. 104, pp. 55–64.

[20]   Hwu, Y.G. and Lenard, J.G. (1988), A finite element study of flat rolling, ASME Journal of Engineering Materials and Technolgy, Vol. 110, pp. 22–27.

[21]   Prakash, R.S., Dixit, P.M. and Lal, G.K. (1995), Steady-state plane-strain cold rolling of a strain-hardening material, Journal of Materials Processing Technology, Vol. 52, pp. 338–358.

[22]   Dixit, U.S. and Dixit, P.M., 1995, An analysis of the steady-state wire drawing of strain-hardening materials. Journal of Materials Processing Technology, vol. 47, pp. 201–229.

[23]   Chandra, S. and Dixit, U.S., 2004, A rigid-plastic finite element analysis of temper rolling process, Journal of Materials Processing Technology, Vol. 152, pp. 9–16.

[24]   Al-Salehi, F.A.R., Firbank, T. C. and Lancaster P.R., 1973, An experimental determination of the roll pressure distributions in cold rolling, International Journal of Mechanical Sciences, Vol. 15, pp. 693–700.

[25]   Shida, S. and Awazuhara, H., 1973, Rolling load and torque in cold rolling, Journal of the Japan Society for Technology of Plasticity. Vol.14, pp. 267–278.

[26]   Maniatty, A.M. (1994), Predicting residual stresses in steady-state forming processes, Computing Systems in Engineering, Vol. 5, pp. 171–177.

[27]   Collins, I.F. (1969), The upper bound theorem for rigid/plastic solids generalized to include Coulomb friction, Journal of Mechanics and Physics of Solids, Vol. 17, pp. 323–338.

[28]   Chakrabarty, J. (1998), Theory of Plasticity, McGraw-Hill Book Company, Singapore.

[29]   Roychoudhari, R. and Lenard, J.G., 1984, A mathematical model of cold rolling-experimental substantiation, Proceedings of International Confrence on Technology of Plasticity Tokyo, pp. 1138–1145.

[30]   Michell, J.H., 1900, Some elementary distribution of stresses in three dimensions, Proceedings of London Mathematical Society, Vol. 32, pp. 23–35.

[31]   Mori, K., Nakadoi, K., Fukuda, M., 1986, Coupled analysis of steady state forming processes with elastic tools, NUMIFORM, pp. 237–242, Balkema, Rotterdam.

[32]   Fleck, N.A., Johnson, K.L., Mear, M.E. and Jhang, L.C. (1992), Cold rolling of foil, Proceedings of the Institution of Mechanical Engineers, Vol. 206, pp.119–131.

[33]   Le, H.R. and Sutcliffe, M.P.F. (2001), A robust model for rolling of thin strip and foil, International Journal of Mechanical Sciences, Vol. 43, pp. 1405–1419.

[34]   Kumar, D. and Dixit, U.S. (2006), A slab method study of strain hardening and friction effects in cold foil rolling process, Vol. 171, Journal of Materials Processing Technology, pp. 331–340.

[35]   Deb, K., 1995, Optimization for Engineering Design: Algorithms and Examples, Prentice-Hall India, New Delhi.

[36]   Zyczkowski, M., 1981, Combined loadings in the theory of plasticity, pp. 288–289, PWN-Polish Scientific, Warsaw.

[37]   Wistreich, J.G., 1955, Investigation of the mechanics of wire drawing, Proceedings of the Institution of Mechanical Engineers, Vol. 169, pp. 654–665.

[38]   Majors, H. (1956), Studies in cold-drawing, Transaction of ASME, Vol. 78, pp. 79–87.

[39]   Cook, P.M. and Wistreich, J.G. (1952), Measurement of die pressure in wire drawing by photo-elastic methods, Journal of Applied Physics, Vol. 3, pp. 159–165.

[40]   Thomsen, E.G., Yang, C.T. and Bierbower, T.B. (1959), An experimental investigation of the mechanics of plastic deformation of metals, University of California (Berkeley), Publication in Engineering, Vol. 5, pp. 89–144.

[41]   Hoffman, O. and Sachs, G. (1953), Introduction to the theory of plasticity for engineers, McGraw-Hill Book Company, New York.

[42]   Siebel, E. (1947), Derderzeitige Stand der Erkenntnisse über die Mechanischen Vorgange beim Drahtziehen, Stahl Und Eisen, Vol. 66–67, pp. 171–180.

[43]   Avitzur, B. (1968), Metal Forming: Process and Analysis, McGraw-Hill Book Company, New York.

[44]   Avitzur, B. (1963) Analysis of wire drawing and extrusion through conical dies of small cone angle. Transaction of ASME, Journal of Engineering for Industry, Vol. 85, pp. 89–96.

[45]   Avitzur, B. (1964), Analysis of wire Drawing and extrusion through conical dies of large cone angle, Transaction of ASME, Journal of Engineering for Industry, Vol. 86, pp. 305–316.

[46]   Avitzur, B. (1967), Strain-hardening and strain-rate effects in plastic flow through conical converging dies, Transaction of ASME, Journal of Engineering for Industry, Vol. 89, pp. 556–562.

[47]   Avitzur, B. (1968), Analysis of central bursting defects in extrusion and wire drawing, Transaction of ASME, Journal of Engineering for Industry, Vol. 90, pp. 79–91.

[48]   Chen, C.C., Oh S.I. and Kobayashi, S., 1979, Ductile fractures in axisymmetric extrusion and drawing; Part1: Deformation Mechanics of Extrusion and Drawing, Part2: Workability in extrusion and drawing, Transaction of ASME, Journal of Engineering for Industry, Vol. 101, pp. 23–44.

[49]   Chevalier, L. (1992), Prediction of defects in metal forming: application to wire drawing, Journal of Materials Processing Technology, Vol. 32, pp. 145–153.

[50]   Dixit, U.S. and Dixit, P.M. (1995), An analysis of the steady-state wire drawing of strain-hardening materials. Journal of Materials Processing Technology, vol. 47, pp. 201–229.

[51]   Gifford, R.B., Bandar, A.R., Coulter, J.P. and Misiolek, W.Z. (2004), The analysis and control of micro-hardness distribution during wire drawing, Transaction of ASME, Journal of Manufacturing Science and Engineering, Vol. 126, No. 2, pp. 247–254.

[52]   Zhengyi, J., Shangwu, X., Xianghua, L., Guodong, W. and Qiang, Z. (1998),   3-D rigid–plastic FEM analysis of the rolling of a strip with local residual deformation, Journal of Materials Processing Technology, Vol. 79, pp. 109–112.

[53]   Jiang, Z.Y. and Tieu, A.K. (2001), Modeling of the rolling processes by a 3-D rigid plastic/visco-plastic finite element method with shifted ICCG method, Computers & Structures, Vol. 79, pp. 2727–2740.

[54]   Dixit, U.S. and Dixit, P.M. (1997), Finite-element analysis of flat rolling with inclusion of anisotropy, International Journal of Mechanical Sciences, Vol. 39, pp. 1237–1255.

[55]   Dixit, U.S. and Dixit, P.M. (1997), A study on residual stresses in rolling, International Journal of Machine Tools & Manufacture, Vol. 37, pp. 837–853.

[56]   Thompson, E.G. and Yu, S. (1990), A flow formulation for rate equilibrium equations, International Journal for Numerical Methods in Engineering, Vol. 30, pp. 1619–1632.

[57]   Maniatty, A.M., Dawson, P.R. and Weber, G.G. (1991),   An Eulerian elasto-viscoplastic formulation for steady-state forming processes, International Journal of Mechanical Sciences, Vol. 33, pp. 361–377.

6

Finite Element Modeling of Metal Forming Processes Using Updated Lagrangian Formulation

6.1 Introduction

In Chapter 5, we discussed the details of finite element method as well as the finite element modeling of metal forming processes using Eulerian formulation. In this chapter, we extend the finite element technique to the updated Lagrangian formulation. In Eulerian formulation, the domain is a fixed region in space (called control volume). However, in a Lagrangian formulation, the domain consists of a set of material particles that changes its shape continuously with the deformation. The updated Lagrangian formulation is an incremental method in which the domain is updated incrementally. Further, the measure of deformation used in Eulerian formulation is the rate of deformation tensor and the constitutive equation is expressed in terms of the stress and rate of deformation tensors. On the other hand, in updated Lagrangian formulation, the measure of deformation is an incremental strain tensor and the constitutive equation is expressed in terms of the incremental stress and incremental strain tensors. We shall discuss how finite element modeling needs to be modified in the light of these changes in the governing equations. Like that of the Eulerian formulation, the governing equations of the updated Lagrangian formulation also are non-linear and need an iterative scheme to obtain a solution. But the iterative scheme we adopt here is different from that of the previous chapter. However, like in the previous chapter, here also we adopt the Galerkin formulation for developing finite element equations.

In Section 6.2 of this chapter, we develop the three-dimensional finite element model corresponding to the governing equations of the updated Lagrangian formulation. We use the incremental strain-displacement and stress-strain relations of Chapter 4. Thus, we use the incremental logarithmic strain tensor as the measure of finite incremental deformation and the integral constitutive equation to account for the change in the elastic-plastic constitutive tensor during the increment.

Further, we use the updating procedure of Section 4.5 to make the incremental stress tensor objective. We assume that the *process is slow*. Then, the acceleration term in the equation of motion becomes negligible, reducing it to the equilibrium equation. The incremental equilibrium equation is not found to be convenient for the finite element model. Instead, the equilibrium equation in the deformed configuration at time $t + \Delta t$ is used. To develop the finite element model, first, this equilibrium equation is converted into an integral form using the *weighted residual method*. Then the *Galerkin* finite element technique is used to convert this integral form into a set of algebraic equations. Since these algebraic equations are non-linear, an iterative scheme is needed to solve these equations. The *Newton-Raphson iterative scheme*, which is used in this book, is described next. Further, the Euler forward integration scheme, which is commonly used for the integration of constitutive equation, is also discussed. Finally, some divergence handling methods like the under-relaxation method, line search method, increment cutting method *etc.* are presented at the end of the section.

In Section 6.3, the finite element model developed for the updated Lagrangian formulation is applied to an *axisymmetric* problem of *open die cold forging of a cylindrical block of an isotropic material*. First, the boundary conditions are described. The platens are assumed rigid. The friction at the interface is modeled by *sticking friction* and *Coulomb's law*. The process is controlled by the movement of the platens. Since this is a displacement control problem, the Newton-Raphson technique does not always converge. An alternate iterative scheme, called *arc length technique*, is used. The model is verified by comparing the predicted forging load variation (with reduction) with experimental results available in the literature. Then the contact pressure distribution, deformed configuration, equivalent strain field, and equivalent stress field are presented for a typical set of process variables. Next, residual stresses are obtained after removal of the platens. A parametric study of the residual stresses is carried out with respect to the three process variables, namely, height-to-diameter ratio, reduction and friction coefficient. Finally, some studies on fracture prediction are presented using critical damage and hydrostatic stress criteria.

In the next section (*i.e.*, in Section 6.4), the finite element model of a *three-dimensional* problem of *deep drawing of a cylindrical cup of an anisotropic material* is developed. First, the boundary conditions are described. The punch, die and blank-holder are assumed rigid. *Friction* at the *punch* is assumed to be of the *sticking* type and that at the *die* is modeled by *Coulomb's law*. The *blank-holder force* is assumed to be equally distributed over the area and is applied incrementally up to a certain number of increments. Penetration of the cup with the punch and die is avoided. The friction and penetration boundary conditions are applied iteratively. Some of these boundary conditions vary with the punch movement. Because of the variation of the material properties in the circumferential direction, the problem cannot be treated as axisymmetric. However, because of the orthotropy of the sheet and the symmetry of the geometry and boundary conditions about the two axes, only a quarter of the sheet is selected as the domain. The model is verified by comparing the prediction of cup height variation with experimental results available in the literature. Then, for a typical set of process variables, the punch force variation, deformed configuration and thickness strain variation are presented. Because of anisotropy, the cup develops ears. However, in isotropic materials there is no earing. A parametric study is carried out with respect to the three process variables, namely the die profile radius, sheet thickness and material properties. Finally, the optimization of the initial shape of the sheet is carried out to minimize the earing. In the end, the whole chapter is summarized in Section 6.5.

6.2 Application of Finite Element Method to Updated Lagrangian Formulation

6.2.1 Governing Equations

The governing equations of the updated Lagrangian formulation, for the case of small incremental deformation, have been presented in Section 3.9 (Equations 3.250–3.352). However, now, we assume that the *incremental deformation is finite* and we choose the incremental logarithmic strain tensor $\Delta \varepsilon_{ij}^L$ as the measure of incremental deformation. Then, the *incremental strain displacement relation* is given not by Equation 3.250, but by the following relation between $\Delta \varepsilon_{ij}^L$ and the incremental displacement vector $_t\Delta u_i$ (Equations 4.61, 4.65 and 4.80):

$$
\begin{aligned}
_t\Delta\varepsilon_{ij}^L &= \ln(_t\Delta\lambda_i) &&\text{if } i = j \\
&= 0 &&\text{if } i \neq j
\end{aligned}
\tag{6.1a}
$$

$$
_t\Delta U_{ij}^2 = (_t\Delta F)_{ik}^{\mathrm{T}} (_t\Delta F)_{kj},
\tag{6.1b}
$$

$$
_t\Delta F_{ij} = \delta_{ij} + {}_t\Delta u_{i,j},
\tag{6.1c}
$$

along with the fact that $_t\Delta\lambda_i$ are the principal values of the incremental right stretch tensor $_t\Delta U$. Here, ln denotes the natural logarithm and the comma in Equation 6.1c indicates the derivatives with respect to the components of the position vector $^t x$ at time t.

Further, for the case of finite incremental deformation, the incremental stress-strain relation before yielding and after unloading remains the same (Equations 3.251e, f) except that $d\varepsilon_{kl}$ and $d\sigma_{ij}$ are replaced respectively by $_t\Delta\varepsilon_{kl}^L$ and $_t\Delta\sigma_{ij}$. But, the *incremental stress-strain relation after yielding* (Equations 3.251a, b) gets replaced by Equations 4.84 and 4.85:

$$_t\Delta\sigma_{ij} = \int\limits_{t}^{t+\Delta t} {}^tC_{ijkl}^{EP}\mathrm{d}(_t\Delta\varepsilon_{kl}^L),$$ (6.2a)

$${}^tC_{ijkl}^{EP} = 2\mu\left[\frac{v}{1-2v}\delta_{ij}\delta_{kl} + \delta_{ik}\delta_{jl} - \frac{9\mu}{2}\frac{{}^t\sigma_{ij}'\,{}^t\sigma_{kl}'}{({}^tH'+3\mu){}^t\sigma_{eq}^2}\right].$$ (6.2b)

Here, ${}^t\sigma_{ij}'$ is the deviatoric part of the stress tensor at time t, ${}^t\sigma_{eq}$ is the equivalent stress (defined by Equation 3.23) at time t and ${}^tH'$ is the derivative (at time t, with respect to equivalent plastic strain) of the hardening function H defined by

$${}^t\sigma_{eq} = H({}^t\varepsilon_{eq}^p) \equiv \sigma_Y + K({}^t\varepsilon_{eq}^p)^n.$$ (6.2c)

Here, σ_Y is the yield stress, K and n are the hardening parameters and ${}^t\varepsilon_{eq}^p$ is the equivalent plastic strain (defined by Equations 3.93 and 3.97) at time t. Further, v is the Poisson's ratio which is related to the Lame's constants λ and μ by

$$v = \frac{\lambda}{2(\lambda+\mu)}.$$ (6.2d)

For the case of finite incremental deformation, the incremental stress $_t\Delta\sigma_{ij}$ is not expressed as the product of the Jaumann stress rate and the time increment (Equations 3.251g–i). Instead, it is made *objective* by employing the following updating procedure:

$$^{t+\Delta t}\sigma = (_t\Delta R)({}^t\sigma)(_t\Delta R)^{\mathrm{T}} + {}_t\Delta\sigma,$$ (6.3)

where $_t\Delta R$ is the (finite) incremental rotation tensor at time t obtained from the polar decomposition of the incremental deformation gradient tensor $_t\Delta F$.

The incremental equation of motion (Equation 3.252) is not found to be convenient for the finite element model. Instead, the equation of motion in the deformed configuration at time $t+\Delta t$ is used. As stated earlier, we *assume* that the process is slow. Then, the *acceleration* term in the equation of motion becomes *negligible* reducing it to the equilibrium equation. We further assume that the *body forces* are also *negligible*. Then, the *equilibrium equation in the deformed configuration at time* $t+\Delta t$ is obtained from Equation 3.215 by removing the acceleration and body force terms and placing the left superscript on the notation of stress and the position vector. Thus, we get

$$\frac{\partial^{t+\Delta t}\sigma_{ij}}{\partial^{t+\Delta t}x_j} = 0,$$

(6.4)

where $^{t+\Delta t}\sigma_{ij}$ is the stress tensor and $^{t+\Delta t}x_j$ is the position vector both at time $t + \Delta t$.

As before, we assume the following boundary conditions: (i) on $^{t+\Delta t}S_u$ part of the domain boundary, the incremental displacement vector $^t\Delta u_i$ is specified and (ii) on $^{t+\Delta t}S_t$ part of the domain boundary, the incremental stress vector $^{t+\Delta t}\Delta t_i$ is specified. (The subscript n of the incremental stress vector is dropped henceforth.)

6.2.2 Integral Form of Equilibrium Equation

To convert the above equilibrium equation (Equation 6.4) into an integral form, we use the *weighted residual method*. We *choose* a vector function $_t\Delta u_i$ (of the coordinates ^tx_j) which only satisfies the incremental *displacement boundary condition* on $^{t+\Delta t}S_u$ but *otherwise* is *arbitrary*. As stated in Chapter 5, this boundary condition is called as the *essential boundary condition* in finite element formulation. From this $_t\Delta u_i$, one can find first the incremental logarithmic strain tensor $\Delta\varepsilon_{ij}^L$ using the incremental stress-strain relation (Equation 6.1) and then the incremental stress tensor $_t\Delta\sigma_{ij}$ using the incremental stress-strain relation (Equation 6.2). Further, one can find the stress tensor $^{t+\Delta t}\sigma_{ij}$ at time $t + \Delta t$ using the updating procedure of Equation 6.3. Since $_t\Delta u_i$ is an arbitrary function, the incremental stress $_t\Delta\sigma_{ij}$ obtained from it may not satisfy the boundary condition on $^{t+\Delta t}S_t$ or the updated stress $^{t+\Delta t}\sigma_{ij}$ obtained from it may not satisfy the equilibrium equation (Equation 6.4). In this case, the quantity $\partial^{t+\Delta t}\sigma_{ij}/\partial^{t+\Delta t}x_j$ is called as the *residue* of the differential equation (Equation 6.4) corresponding to the given choice of $_t\Delta u_i$. Thus, the chosen $_t\Delta u_i$ may not be an exact solution of the governing equations of the updated Lagrangian formulation. However, it can be made an approximate solution of the governing equations, if the following integral, called *weighted residual*, is made zero:

$$\int_{t+\Delta t_V} \frac{\partial^{t+\Delta t}\sigma_{ij}}{\partial^{t+\Delta t}x_j} w_i d^{t+\Delta t}V = 0,$$

(6.5)

where $^{t+\Delta t}V$ is the domain volume at time $t+\Delta t$ and w_i is a vector function of the coordinates $^{t}x_j$, called the *weight* function. The weight function is assumed to be completely *arbitrary* except that it is required to be *zero on the part* $^{t+\Delta t}S_u$ of the boundary.

To reduce the continuity requirement of the approximate solution, we carry out the following steps. First, we modify Equation 6.5 by using a vector identity:

$$\int_{t+\Delta t V} \frac{\partial}{\partial^{t+\Delta t}x_j}(^{t+\Delta t}\sigma_{ij}w_i)d^{t+\Delta t}V - \int_{t+\Delta t V} {}^{t+\Delta t}\sigma_{ij}\frac{\partial w_i}{\partial^{t+\Delta t}x_j}d^{t+\Delta t}V = 0. \quad (6.6)$$

Next, we convert the volume integral of the first term to a surface integral using the divergence theorem:

$$\int_V \frac{\partial A_{ij}}{\partial x_j}dV = \int_S A_{ij}n_j dS, \quad (6.7)$$

for any tensor function A_{ij}, of the coordinates x_j, defined over the domain volume V. Here, S is the domain boundary and n_j is the unit outward normal to S. This surface integral can be split into two parts: (i) integral over $^{t+\Delta t}S_u$ and (ii) integral over $^{t+\Delta t}S_t$. Note that the integral over $^{t+\Delta t}S_u$ is zero as the weight function w_i is zero there. The integral over $^{t+\Delta t}S_t$ can be expressed in terms of the stress vector $^{t+\Delta t}t_i$ by using the Cauchy's relation (Equation 2.64). We denote the modified surface integral over the boundary $^{t+\Delta t}S_t$ as $^{t+\Delta t}R$:

$$^{t+\Delta t}R = \int_{t+\Delta t S_t} {}^{t+\Delta t}t_i w_i d^{t+\Delta t}S. \quad (6.8)$$

We modify the second term of Equation 6.6 by using the symmetry of the stress tensor. Then, it can be expressed in terms of the symmetric part of $\partial w_i / \partial^{t+\Delta t}x_j$:

$$^{t+\Delta t}\varepsilon_{ij}(w) = \frac{1}{2}\left(\frac{\partial w_i}{\partial^{t+\Delta t}x_j} + \frac{\partial w_j}{\partial^{t+\Delta t}x_i}\right). \quad (6.9)$$

With these modifications, the weighted residual (Equation 6.5) becomes

$$\int_{t+\Delta t V} {}^{t+\Delta t}\sigma_{ij}{}^{t+\Delta t}\varepsilon_{ij}(w)d^{t+\Delta t}V = {}^{t+\Delta t}R. \quad (6.10)$$

This is called a *weak form* of the equilibrium equation (Equation 6.4).

6.2.3 Finite Element Formulation

First, we express Equation 6.10 in an array form. For this purpose, we define the following column arrays or vectors:

$$
{}^{t+\Delta t}\{t\} = \left\{ \begin{array}{c} {}^{t+\Delta t}t_x \\ {}^{t+\Delta t}t_y \\ {}^{t+\Delta t}t_z \end{array} \right\}, \qquad \{w\} = \left\{ \begin{array}{c} w_x \\ w_y \\ w_z \end{array} \right\},
$$

$$
{}^{t+\Delta t}\{\sigma\} = \left\{ \begin{array}{c} {}^{t+\Delta t}\sigma_{xx} \\ {}^{t+\Delta t}\sigma_{yy} \\ {}^{t+\Delta t}\sigma_{zz} \\ {}^{t+\Delta t}\sigma_{xy} \\ {}^{t+\Delta t}\sigma_{yz} \\ {}^{t+\Delta t}\sigma_{zx} \end{array} \right\}, \qquad {}^{t+\Delta t}\{\varepsilon(w)\} = \left\{ \begin{array}{c} {}^{t+\Delta t}\varepsilon_{xx}(w) \\ {}^{t+\Delta t}\varepsilon_{yy}(w) \\ {}^{t+\Delta t}\varepsilon_{zz}(w) \\ 2\,{}^{t+\Delta t}\varepsilon_{xy}(w) \\ 2\,{}^{t+\Delta t}\varepsilon_{yz}(w) \\ 2\,{}^{t+\Delta t}\varepsilon_{zx}(w) \end{array} \right\}. \qquad (6.11)
$$

Then, the integral equilibrium equation (Equation 6.10) becomes

$$
\int_{{}^{t+\Delta t}V} {}^{t+\Delta t}\{\varepsilon(w)\}^{\mathrm{T}}\, {}^{t+\Delta t}\{\sigma\}\, \mathrm{d}\,{}^{t+\Delta t}V = \int_{{}^{t+\Delta t}S_t} \{w\}^{\mathrm{T}}\, {}^{t+\Delta t}\{t\}\, \mathrm{d}\,{}^{t+\Delta t}S, \qquad (6.12)
$$

where the right superscript T denotes the transpose of the array.

Next, we discretize the domain into a number of *n-noded brick elements* (Figure 5.5). Over a typical element, we choose the following approximation for the incremental displacement $_t\Delta u_i$:

$$
_t\{\Delta u\} \equiv \left\{ \begin{array}{c} _t\Delta u_x \\ _t\Delta u_y \\ _t\Delta u_z \end{array} \right\} = {}^t[\phi]_t\,\{\Delta u\}^e, \qquad (6.13)
$$

where the *elemental incremental displacement vector* $_t\{\Delta u\}^e$ is given by

$$
_t\{\Delta u\}^{e\mathrm{T}} = \left\{ _t\Delta u_x^1, \;\; _t\Delta u_y^1, \;\; _t\Delta u_z^1, \;\; ..., \;\; _t\Delta u_x^n, \;\; _t\Delta u_y^n, \;\; _t\Delta u_z^n \right\}. \qquad (6.14)
$$

Here, the quantities $_t\Delta u_x^j$, $_t\Delta u_y^j$ and $_t\Delta u_z^j$ stand for the unknown incremental displacements at node j of element e in x, y and z directions respectively. The matrix ${}^t[\phi]$ is defined by

$$
{}^{t}[\phi] = \begin{bmatrix} {}^{t}\{\phi_1\}^{\mathrm{T}} \\ {}^{t}\{\phi_2\}^{\mathrm{T}} \\ {}^{t}\{\phi_3\}^{\mathrm{T}} \end{bmatrix},
\tag{6.15}
$$

where

$$
\begin{aligned}
{}^{t}\{\phi_1\}^{\mathrm{T}} &= \{{}^{t}N_1, \quad 0, \quad 0, \quad {}^{t}N_2, \quad 0, \quad 0, \quad {}^{t}N_3, \quad ......\}, \\
{}^{t}\{\phi_2\}^{\mathrm{T}} &= \{\, 0, \quad {}^{t}N_1, \quad 0, \quad 0, \quad {}^{t}N_2, \quad 0, \quad 0, \quad ......\}, \\
{}^{t}\{\phi_3\}^{\mathrm{T}} &= \{\, 0, \quad 0, \quad {}^{t}N_1, \quad 0, \quad 0, \quad {}^{t}N_2, \quad 0, \quad ......\}.
\end{aligned}
\tag{6.16}
$$

The ${}^{t}N_j$, which are functions of $({}^{t}x, {}^{t}y, {}^{t}z)$, are called *shape functions* and the matrix ${}^{t}[\phi]$ is called the *shape function matrix*. For the present problem, n is equal to 8. Then, ${}^{t}N_j$ become the tri-linear shape functions.

From the chosen approximation for ${}^{t}\Delta u_i$ (Equation 6.13), one can first calculate the incremental logarithmic strain $\Delta \varepsilon_{ij}^{L}$ using the incremental strain-displacement relation (Equation 6.1) and then the incremental stress ${}_{t}\Delta\sigma_{ij}$ using the incremental stress-strain relation (Equation 6.2). Then, one can determine ${}^{t+\Delta t}\sigma_{ij}$ using the updating procedure of Equation 6.3. Note that, because of the non-linear strain displacement relation (Equation 6.1), ${}^{t+\Delta t}\sigma_{ij}$ would be a non-linear function of ${}^{t}\Delta u_i$.

In the Galerkin finite element formulation, the functions chosen to construct the weight function are the same as those used in approximating the primary variable. Therefore, over a typical element, we express w_i as

$$
\{w\} \equiv \begin{Bmatrix} w_x \\ w_y \\ w_z \end{Bmatrix} = {}^{t+\Delta t}[\phi]\{w\}^{e},
\tag{6.17}
$$

where the *elemental weight vector* $\{w\}^{e}$ is given by

$$
\{w\}^{\mathrm{T}} = \left\{ w_x^1, \quad w_y^1, \quad w_z^1, \quad ...., \quad w_x^n, \quad w_y^n, \quad w_z^n \right\}.
\tag{6.18}
$$

Here, the quantities w_x^j, w_y^j and w_z^j denote the known but arbitrary values of the weight function components in x, y and z directions at node i of element e. Further, the superscript of the matrix $^{t+\Delta t}[\phi]$ is $t + \Delta t$, and not t, as the matrix needs to be a function of the coordinates at time $t + \Delta t$: $(^{t+\Delta t}x, ^{t+\Delta t}y, ^{t+\Delta t}z)$. This is because, the expression at Equation 6.17 is to be used in the integral over the domain at time $t + \Delta t$.

Differentiating the expression at Equation 6.17 for w_i with respect to the coordinates $^{t+\Delta t}x_j$, we get the following expression for $^{t+\Delta t}\{\varepsilon(w)\}$:

$$^{t+\Delta t}\{\varepsilon(w)\} = {}^{t+\Delta t}[B_L]\{w\}^e , \tag{6.19}$$

where the matrix $^{t+\Delta t}[B_L]$ is given by

$$^{t+\Delta t}[B_L] = \begin{bmatrix} ^{t+\Delta t}\{\phi_1\}_{,x}^{T} \\ ^{t+\Delta t}\{\phi_2\}_{,y}^{T} \\ ^{t+\Delta t}\{\phi_3\}_{,z}^{T} \\ ^{t+\Delta t}\{\phi_1\}_{,y}^{T} + {}^{t+\Delta t}\{\phi_2\}_{,x}^{T} \\ ^{t+\Delta t}\{\phi_2\}_{,z}^{T} + {}^{t+\Delta t}\{\phi_3\}_{,y}^{T} \\ ^{t+\Delta t}\{\phi_3\}_{,x}^{T} + {}^{t+\Delta t}\{\phi_1\}_{,z}^{T} \end{bmatrix} . \tag{6.20}$$

Here, the comma denotes the derivatives with respect to the coordinates $^{t+\Delta t}x_j$.

When the domain is discretized into volume elements, the boundary $^{t+\Delta t}S_t$ gets automatically divided into area elements. In the present discretization, the volume element is an eight-noded brick element. Therefore, the area element becomes a four-noded square element. The expression for $\{w\}$ over a typical area element, which is consistent with its expression over the volume element (Equation 6.17), is given by

$$\{w\} \equiv \begin{Bmatrix} w_x \\ w_y \\ w_z \end{Bmatrix} = {}^{t+\Delta t}[\phi]^b \{w\}^b , \tag{6.21}$$

where the matrix

$$^{t}[\phi]^b = \begin{bmatrix} ^{t}\{\phi_1\}^{bT} \\ ^{t}\{\phi_2\}^{bT} \\ ^{t}\{\phi_3\}^{bT} \end{bmatrix} , \tag{6.22}$$

now contains the *two-dimensional* bi-linear shape functions

$$
{}^{t}\{\phi_1\}^{bT} = \{{}^{t}N_1^b, \quad 0, \quad 0, \quad {}^{t}N_2^b, \quad 0, \quad 0, \quad {}^{t}N_3^b, \quad ......\},
$$
$$
{}^{t}\{\phi_2\}^{bT} = \{\quad 0, \quad {}^{t}N_1^b, \quad 0, \quad 0, \quad {}^{t}N_2^b, \quad 0, \quad 0, \quad ......\}, \tag{6.23}
$$
$$
{}^{t}\{\phi_3\}^{bT} = \{\quad 0, \quad 0, \quad {}^{t}N_1^b, \quad 0, \quad 0, \quad {}^{t}N_2^b, \quad 0, \quad ......\}.
$$

The *elemental weight vector* $\{w\}^b$ *for a typical area element* is given by

$$
\{w\}^{bT} = \left\{ w_x^{b1}, \quad w_y^{b1}, \quad w_z^{b1}, \quad ...., \quad w_x^{bn_b}, \quad w_y^{bn_b}, \quad w_z^{bn_b} \right\}, \tag{6.24}
$$

where the quantities w_x^{bj}, w_y^{bj} and w_z^{bj} stand for the known but arbitrary values of the weight function components in x, y and z directions at node j of area element b and n_b is the number of nodes per area element. In the present discretization, this number is 4.

Let the total number of volume elements be N_e and the total number of area elements on the boundary $^{t+\Delta t}S_t$ be N_b. Then, substitution of Equations 6.19 and 6.21 in the integral equilibrium equation (Equation 6.12) leads to

$$
\sum_{e=1}^{N_e} \{w\}^{eT} \, {}^{t+\Delta t}\{f\}_{in}^e = \sum_{b=1}^{N_b} \{w\}^{bT} \, {}^{t+\Delta t}\{f\}_{ex}^b, \tag{6.25}
$$

where

$$
{}^{t+\Delta t}\{f\}_{in}^e = \int_{{}^{t+\Delta t}V^e} {}^{t+\Delta t}[B_L]^T \, {}^{t+\Delta t}\{\sigma\} \, d\,{}^{t+\Delta t}V, \tag{6.26}
$$

and

$$
{}^{t+\Delta t}\{f\}_{ex}^b = \int_{{}^{t+\Delta t}S_t^b} {}^{t+\Delta t}[\phi]^{bT} \, {}^{t+\Delta t}\{t\} \, d\,{}^{t+\Delta t}S, \tag{6.27}
$$

are respectively the *elemental internal force vector* and *elemental external force vector*, both at time $t + \Delta t$. Here, $^{t+\Delta t}V^e$ and $^{t+\Delta t}S_t^b$ are respectively the domains of the typical volume element and typical area element.

Using the assembly procedure of Chapter 5, the elemental discretized equilibrium equation (Equation 6.25) becomes

$$
\{W\}^{T\,t+\Delta t}\{F\}_{in} = \{W\}^{T\,t+\Delta t}\{F\}_{ex}. \tag{6.28}
$$

Here, $^{t+\Delta t}\{F\}_{in}$ and $^{t+\Delta t}\{F\}_{ex}$ are respectively the *global internal force vector* and *global external force vector*, both at time $t+\Delta t$. Further, the vector $\{W\}$, called *the global weight vector*, contains the known but arbitrary values of the weight function components in x, y and z directions at all the nodes of the domain. Equation 6.28 is true for any arbitrary weight function vector $\{W\}$. Therefore, we get

$$^{t+\Delta t}\{F\}_{in} - {}^{t+\Delta t}\{F\}_{ex} = \{0\}. \tag{6.29}$$

This is a discretized form of equilibrium equation (Equation 6.4). As stated earlier, $^{t+\Delta t}\sigma_{ij}$ depends non-linearly on $^{t}\Delta u_i$. Therefore, $^{t+\Delta t}\{F\}_{in}$ becomes a non-linear function of the unknown nodal values of the incremental displacement vector. Thus, Equation 6.29 is a set of *non-linear algebraic equations*. We need an *iterative scheme* to solve Equation 6.29.

We use the *Newton-Raphson iterative scheme* for this purpose. To derive the iterative equations, we proceed as follows. We denote the *global displacement vector* by the symbol $_t\{\Delta U\}$. Let $_t\{\Delta U\}^{(i-1)}$ be the global displacement vector obtained in $(i-1)$-th iteration. Then, to find $_t\{\Delta U\}^{(i)}$ in i-th iteration, we expand $^{t+\Delta t}\{F\}_{in}$ in *Taylor's series* and retain only the first order terms. Thus, we get

$$^{t+\Delta t}\{F\}_{in}^{(i)} = {}^{t+\Delta t}\{F\}_{in}^{(i-1)} + \frac{\partial^{t+\Delta t}\{F\}_{in}}{\partial_t\{\Delta U\}}\bigg|_{t\{\Delta U\}=_t\{\Delta U\}^{(i-1)}} \{\Delta\} + O(\|\Delta\|^2), \tag{6.30}$$

where $\|\Delta\|$ is the norm of the vector $\{\Delta\}$:

$$\{\Delta\} =_t\{\Delta U\}^{(i)} -_t\{\Delta U\}^{(i-1)} \tag{6.31}$$

and $^{t+\Delta t}\{F\}_{in}^{(i-1)}$ and $^{t+\Delta t}\{F\}_{in}^{(i)}$ are respectively the global internal force vectors obtained from the iterative global displacement vectors $_t\{\Delta U\}^{(i-1)}$ and $_t\{\Delta U\}^{(i)}$. This means, using $_t\{\Delta U\}^{(i-1)}$, we first update the domain to $^{t+\Delta t}V^{(i-1)}$. Then, we obtain the corresponding shape function matrix $^{t+\Delta t}[\phi]^{(i-1)}$ by considering the shape functions $^{t+\Delta t}N_j^{(i-1)}$ as functions of the coordinates $^{t+\Delta t}x_j^{(i-1)}$. Let $_t\{\Delta u\}^{e(i-1)}$ be the elemental incremental displacement vector corresponding to the typical element e. Then, from the approximation

$$_t\{\Delta u\} = {}^{t+\Delta t}[\phi]^{(i-1)}{}_t\{\Delta u\}^{e(i-1)}, \tag{6.32}$$

we calculate the corresponding incremental logarithmic strain using the incremental strain-displacement relation (Equation 6.1) and the corresponding incremental stress using the incremental stress-strain relation (Equation 6.2). Next, we update the stress to $^{t+\Delta t}\sigma_{ij}^{(i-1)}$ using Equation 6.3 and calculate $^{t+\Delta t}[B_L]^{(i-1)}$ using Equation 6.20. Finally, we determine the corresponding elemental internal force vector $^{t+\Delta t}\{f\}_{in}^{e(i-1)}$ from the integral

$$^{t+\Delta t}\{f\}_{in}^{e(i-1)} = \int_{^{t+\Delta t}V^{e(i-1)}} {}^{t+\Delta t}[B_L]^{(i-1)T} \, {}^{t+\Delta t}\{\sigma\}^{(i-1)} \mathrm{d}^{t+\Delta t}V, \tag{6.33}$$

(where $^{t+\Delta t}V^{e(i-1)}$ is the updated domain corresponding to the typical element e) for all the elements and assemble them to obtain $^{t+\Delta t}\{F\}_{in}^{(i-1)}$. The vector $^{t+\Delta t}\{F\}_{in}^{(i)}$ corresponding to i-th iteration is obtained similarly from $_t\{\Delta U\}^{(i)}$.

6.2.4 Evaluation of the Derivative

It is quite tedious to evaluate the derivative $\partial^{t+\Delta t}\{F\}_{in}/\partial_t\{\Delta U\}$ when the incremental strain-displacement relation is given by Equation 6.1 or the incremental stress-strain relation is given by Equation 6.2a or the stress is updated by Equation 6.3. Therefore, we illustrate the determination of the derivative for a simpler case. Here, we use the following:
(i) The incremental linear strain tensor (Equation 3.55)

$$_t\Delta\varepsilon_{ij} = \frac{1}{2}(_t\Delta u_{i,j} + {}_t\Delta u_{j,i}), \tag{6.34}$$

as the measure of incremental deformation. The notation d(.) of Equation 3.55 for the incremental quantities has been changed to $_t\Delta(.)$ for the sake of consistency and the comma denotes the derivatives with respect to the components of the position vector $^t x$ at time t.
(ii) An increment ($_t\Delta S_{ij}$) of the second Piola-Kirchoff stress tensor (Equation 4.87)

$$^{t+\Delta t}_t S_{ij} = \det(_t\Delta F)(_t\Delta F)_{ik}^{-1}(^{t+\Delta t}\sigma_{kl})(_t\Delta F)_{lj}^{-T}, \tag{6.35}$$

as the measure of objective incremental stress tensor.
(iii) The tensor $^t C_{ijkl}^{EP}$ (Equations 6.2b–6.2d) as the constitutive tensor in the incremental stress-strain relation

$$_t\Delta S_{ij} = {}^t C_{ijkl}^{EP}(_t\Delta\varepsilon_{kl}). \tag{6.36}$$

6.2.4.1 Relation Between Internal Force Vectors at Times t and $t + \Delta t$

To evaluate the derivative, we first relate the elemental internal force vectors at times t and $t + \Delta t$. We start with the expression (Equation 6.26) of the elemental internal force vector $^{t+\Delta t}\{f\}_{in}^{e}$. Pre-multiplying it with $\{w\}^{eT}$ and using Equation 6.19, we get

$$\{w\}^{eT}\,{}^{t+\Delta t}\{f\}_{in}^{e} = \int_{t+\Delta t_{V}e} {}^{t+\Delta t}\{\varepsilon(w)\}^{T}\,{}^{t+\Delta t}\{\sigma\}\mathrm{d}^{t+\Delta t}V \,. \tag{6.37}$$

Using the definition at Equation 6.11 of the arrays $^{t+\Delta t}\{\varepsilon(w)\}$ and $^{t+\Delta t}\{\sigma\}$, the right side of the above equation is expressed in index notation as

$$\{w\}^{eT}\,{}^{t+\Delta t}\{f\}_{in}^{e} = \int_{t+\Delta t_{V}e} {}^{t+\Delta t}\sigma_{ij}\,{}^{t+\Delta t}\varepsilon_{ij}(w)\mathrm{d}^{t+\Delta t}V \,. \tag{6.38}$$

Further, using Equation 6.9 and symmetry of the stress tensor, the above equation is modified as

$$\{w\}^{eT}\,{}^{t+\Delta t}\{f\}_{in}^{e} = \int_{t+\Delta t_{V}e} {}^{t+\Delta t}\sigma_{ij}\frac{\partial w_{i}}{\partial^{t+\Delta t}x_{j}}\mathrm{d}^{t+\Delta t}V \,. \tag{6.39}$$

Adding the weighted equilibrium equation (Equation 6.5) to the above equation, we get

$$\{w\}^{eT}\,{}^{t+\Delta t}\{f\}_{in}^{e} = \int_{t+\Delta t_{V}e} \left({}^{t+\Delta t}\sigma_{ij}\frac{\partial w_{i}}{\partial^{t+\Delta t}x_{j}} + \frac{\partial^{t+\Delta t}\sigma_{ij}}{\partial^{t+\Delta t}x_{j}}w_{i} \right)\mathrm{d}^{t+\Delta t}V \,, \tag{6.40}$$

which is further modified to

$$\{w\}^{eT}\,{}^{t+\Delta t}\{f\}_{in}^{e} = \int_{t+\Delta t_{V}e} \frac{\partial}{\partial^{t+\Delta t}x_{j}}\left({}^{t+\Delta t}\sigma_{ij}w_{i} \right)\mathrm{d}^{t+\Delta t}V \,. \tag{6.41}$$

Using the divergence theorem (Equation 6.7), the volume integral on the right side of the above equation is converted to a surface integral

$$\{w\}^{eT}\,{}^{t+\Delta t}\{f\}_{in}^{e} = \int_{t+\Delta t_{S}e} \left({}^{t+\Delta t}\sigma_{ij}w_{i}\,{}^{t+\Delta t}n_{j} \right)\mathrm{d}^{t+\Delta t}S \,, \tag{6.42}$$

where $^{t+\Delta t}n_j$ is the unit outward normal to the element boundary $^{t+\Delta t}S^e$. Then, the surface integral is expressed in terms of the stress vector $^{t+\Delta t}t_i$ using the Cauchy's relation (Equation 2.64)

$$\{w\}^{eT}\,^{t+\Delta t}\{f\}^e_{in} = \int_{^{t+\Delta t}S^e} (^{t+\Delta t}t_i w_i) d^{t+\Delta t}S. \tag{6.43}$$

Note that, the domain boundary $^{t+\Delta t}S$ at time $t+\Delta t$ is not known. Therefore, we transform the above integral to the known domain boundary ^tS at time t. For this purpose, we use the transformation [1]

$$^{t+\Delta t}t_i d^{t+\Delta t}S = {}^{t+\Delta t}_{t}T^T_{ij} \, {}^t n_j d^tS, \tag{6.44}$$

where $^{t+\Delta t}_{t}T_{ij}$ is the *first Piola-Kirchoff stress tensor* and $^t n_j$ is the unit outward normal to ^tS. Eliminating $^{t+\Delta t}t_i d^{t+\Delta t}S$ from Equations 6.43 and 6.44, we get

$$\{w\}^{eT}\,^{t+\Delta t}\{f\}^e_{in} = \int_{^tS^e} (^{t+\Delta t}_{t}T^T_{ij} w_i \, {}^t n_j) d^tS, \tag{6.45}$$

where $^tS^e$ is the boundary of the typical element e at time t. Again using the divergence theorem (Equation 6.7), we convert this surface integral to the volume integral over the known domain at time t

$$\{w\}^{eT}\,^{t+\Delta t}\{f\}^e_{in} = \int_{^tV^e} \frac{\partial}{\partial^t x_j} \left(^{t+\Delta t}_{t}T^T_{ij} w_i \right) d^tV. \tag{6.46}$$

Here, $^tV^e$ is the volume of the typical element e at time t. The equilibrium equation (Equation 6.4) in terms of the first Piola-Kirchoff stress tensor becomes [1]

$$\frac{\partial\,^{t+\Delta t}_{t}T^T_{ij}}{\partial^t x_j} = 0. \tag{6.47}$$

Expanding the partial derivative on the right side of Equation 6.46 and using Equation 6.47, we get

$$\{w\}^{eT}\,^{t+\Delta t}\{f\}^e_{in} = \int_{^tV^e} {}^{t+\Delta t}_{t}T^T_{ij} \frac{\partial w_i}{\partial^t x_j} d^tV = \int_{^tV^e} {}^{t+\Delta t}_{t}T^T_{kj} \frac{\partial w_k}{\partial^t x_j} d^tV. \tag{6.48}$$

Using the relation between the first and second Piola-Kirchoff stress tensors [1]

$$^{t+\Delta t}T_{kj}^{T} = (_t\Delta F)_{ki} \, {}_{t}^{t+\Delta t}S_{ij} = \left(\delta_{ki} + {}_t\Delta u_{k,i}\right)_t^{t+\Delta t} S_{ij} , \qquad (6.49)$$

Equation 6.48 can be expressed in terms of the second Piola-Kirchoff stress tensor $^{t+\Delta t}_{t}S_{ij}$:

$$\{w\}^{eT \, t+\Delta t}\{f\}_{in}^{e} = \int_{{}_t V^e} \left(\delta_{ki} + {}_t\Delta u_{k,i}\right) {}_{t}^{t+\Delta t}S_{ij} \frac{\partial w_k}{\partial^t x_j} d^t V , \qquad (6.50)$$

where $_t\Delta F_{ki}$ is the incremental deformation gradient tensor (Equation 6.1c). Equation 6.35 implies that, at time t, $^{t+\Delta t}_{t}S_{ij}$ reduces to $^t\sigma_{ij}$ since $(_t\Delta F)$ becomes a unit tensor with the value of its determinant being unity. Therefore, we decompose $^{t+\Delta t}_{t}S_{ij}$ as

$$^{t+\Delta t}_{t} S_{ij} = {}^t\sigma_{ij} + {}_t\Delta S_{ij} , \qquad (6.51)$$

where $_t\Delta S_{ij}$ is the increment of $^{t+\Delta t}_{t}S_{ij}$. Further, using the incremental stress-strain relation (Equation 6.36), the above equation becomes

$$^{t+\Delta t}_{t} S_{ij} = {}^t\sigma_{ij} + {}^t C_{ijmn}^{EP}(_t\Delta\varepsilon_{mn}) . \qquad (6.52)$$

Substituting Equation 6.52 in Equation 6.50, we get

$$\{w\}^{eT \, t+\Delta t}\{f\}_{in}^{e} = \int_{{}_t V^e} \left(\delta_{ki} + {}_t\Delta u_{k,i}\right)\left({}^t\sigma_{ij} + {}^t C_{ijmn}^{EP}(_t\Delta\varepsilon_{mn})\right)\frac{\partial w_k}{\partial^t x_j} d^t V . \qquad (6.53)$$

Expanding the parentheses in Equation 6.53, we obtain

$$\{w\}^{eT \, t+\Delta t}\{f\}_{in}^{e}$$
$$= \int_{{}_t V^e} \left({}^t\sigma_{kj} + {}^t\sigma_{ij}(_t\Delta u_{k,i}) + {}^t C_{kjmn}^{EP}(_t\Delta\varepsilon_{mn}) + {}^t C_{ijmn}^{EP}(_t\Delta\varepsilon_{mn})(_t\Delta u_{k,i})\right)\frac{\partial w_k}{\partial^t x_j} d^t V . \qquad (6.54)$$

The last term in the parenthesis is of second order in the norm of the incremental displacement vector $_t\Delta u$. Therefore, we write the above equation as

$$\{w\}^{eT\ t+\Delta t}\{f\}_{in}^e = \int_{{}^t V^e} \left({}^t\sigma_{kj} + {}^t\sigma_{ij}({}_t\Delta u_{k,i}) + {}^t C_{kjmn}^{EP}({}_t\Delta\varepsilon_{mn}) \right) \frac{\partial w_k}{\partial\, {}^t x_j} \mathrm{d}\, {}^t V + O(\|{}_t\Delta u\|^2),$$

(6.55)

where $\|{}_t\Delta u\|$ is the norm of ${}_t\Delta u$. Similar to Equation 6.9, we define the symmetric part of $\partial w_k / \partial\, {}^t x_j$ as

$${}^t\varepsilon_{kj}(w) = \frac{1}{2}\left(\frac{\partial w_k}{\partial\, {}^t x_j} + \frac{\partial w_j}{\partial\, {}^t x_k} \right).$$

(6.56)

Since ${}^t\sigma_{kj}$ and ${}^t C_{kjmn}^{EP}({}_t\Delta\varepsilon_{mn})$ are symmetric in the indices k and j, using the above definition of ${}^t\varepsilon_{kj}(w)$, Equation 6.55 becomes

$$\{w\}^{eT\ t+\Delta t}\{f\}_{in}^e$$

$$= \int_{{}^t V^e} \left({}^t\sigma_{kj}\, {}^t\varepsilon_{kj}(w) + {}^t C_{kjmn}^{EP}({}_t\Delta\varepsilon_{mn})\, {}^t\varepsilon_{kj}(w) + {}^t\sigma_{ij} \frac{\partial\, {}_t\Delta u_k}{\partial\, {}^t x_i} \frac{\partial w_k}{\partial\, {}^t x_j} \right) \mathrm{d}\, {}^t V + O(\|{}_t\Delta u\|^2).$$

(6.57)

Now, we express the above equation in an array form. For this purpose, we define the following vectors:

$${}^t\{\sigma\} = \begin{Bmatrix} {}^t\sigma_{xx} \\ {}^t\sigma_{yy} \\ {}^t\sigma_{zz} \\ {}^t\sigma_{xy} \\ {}^t\sigma_{yz} \\ {}^t\sigma_{zx} \end{Bmatrix}, \qquad {}_t\{\Delta S\} = \begin{Bmatrix} {}_t\Delta S_{xx} \\ {}_t\Delta S_{yy} \\ {}_t\Delta S_{zz} \\ {}_t\Delta S_{xy} \\ {}_t\Delta S_{yz} \\ {}_t\Delta S_{zx} \end{Bmatrix},$$

$$
_t\{\Delta\varepsilon\} = \begin{Bmatrix} _t\Delta\varepsilon_{xx} \\ _t\Delta\varepsilon_{yy} \\ _t\Delta\varepsilon_{zz} \\ 2\,_t\Delta\varepsilon_{xy} \\ 2\,_t\Delta\varepsilon_{yz} \\ 2\,_t\Delta\varepsilon_{zx} \end{Bmatrix}, \qquad
{}^t\{\varepsilon(w)\} = \begin{Bmatrix} {}^t\varepsilon_{xx}(w) \\ {}^t\varepsilon_{yy}(w) \\ {}^t\varepsilon_{zz}(w) \\ 2\,{}^t\varepsilon_{xy}(w) \\ 2\,{}^t\varepsilon_{yz}(w) \\ 2\,{}^t\varepsilon_{zx}(w) \end{Bmatrix},
$$

$$
{}^t\{\nabla(\Delta u)\}^{\mathrm{T}} = \big\{ {}_t\Delta u_{x,x}, \quad {}_t\Delta u_{x,y}, \quad {}_t\Delta u_{x,z}, \quad {}_t\Delta u_{y,x}, \quad -, \quad -, \quad -, \quad -, \quad {}_t\Delta u_{z,z} \big\},
$$

$$
{}^t\{\nabla w\}^{\mathrm{T}} = \big\{ w_{x,x}, \quad w_{x,y}, \quad w_{x,z}, \quad w_{y,x}, \quad -, \quad -, \quad -, \quad -, \quad w_{z,z} \big\}.
$$

$$(6.58)$$

Then, the array form of the incremental stress-strain relation (Equation 6.36) becomes

$$
_t\{\Delta S\} = {}^t[C^{EP}]_t\{\Delta\varepsilon\}, \tag{6.59}
$$

where ${}^t[C^{EP}]$ is the 6×6 array form of the fourth[th] order tensor ${}^tC^{EP}_{ijkl}$. Using the definition of various arrays (Equations 6.58 and 6.59), Equation 6.57 becomes

$$
\{w\}^{eT}\,{}^{t+\Delta t}\{f\}^e_{in}
$$

$$
= \int_{{}^t V^e} \Big({}^t\{\varepsilon(w)\}^{\mathrm{T}}\,{}^t\{\sigma\} + {}^t\{\varepsilon(w)\}^{\mathrm{T}}\,{}^t[C^{EP}]_t\{\Delta\varepsilon\} + {}^t\{\nabla w\}^{\mathrm{T}}\,{}^t[\Sigma]\,{}^t\{\nabla(\Delta u)\} \Big) d\,{}^tV + O(\|_t\Delta u\|^2),
$$

$$(6.60)$$

where the matrix ${}^t[\Sigma]$ is given by

$$
{}^t[\Sigma] = \begin{bmatrix} {}^t[\sigma] & [0] & [0] \\ [0] & {}^t[\sigma] & [0] \\ [0] & [0] & {}^t[\sigma] \end{bmatrix},
$$

$$(6.61)$$

$$
{}^t[\sigma] = \begin{bmatrix} {}^t\sigma_{xx} & {}^t\sigma_{xy} & {}^t\sigma_{zx} \\ {}^t\sigma_{xy} & {}^t\sigma_{yy} & {}^t\sigma_{yz} \\ {}^t\sigma_{zx} & {}^t\sigma_{yz} & {}^t\sigma_{zz} \end{bmatrix}.
$$

Now, we substitute the approximation for the incremental displacement $_t\Delta u_i$ (Equation 6.13) and the corresponding expression for the weight function w_i (Equation 6.17). However, now, the superscript of the matrix $[\phi]$ has to be t, and not $t + \Delta t$, as the weight function expression is to be used in the integral over the domain at time t

$$\{w\} = {}^t[\phi]\{w\}^e . \tag{6.62}$$

Differentiating Equation 6.13 for $_t\Delta u_i$ and Equation 6.62 for w_i with respect to the coordinates ${}^tx_j$, we get the following expressions for $_t\{\Delta\varepsilon\}$, ${}^t\{\varepsilon(w)\}$, ${}^t\{\nabla(\Delta u)\}$ and ${}^t\{\nabla w\}$:

$$_t\{\Delta\varepsilon\} = {}^t[B_L]_t\{\Delta u\}^e, \quad {}^t\{\varepsilon(w)\} = {}^t[B_L]\{w\}^e , \tag{6.63}$$

where

$$
{}^t[B_L] = \begin{bmatrix}
{}^t\{\phi_1\}_{,x}^{\mathrm{T}} \\
{}^t\{\phi_2\}_{,y}^{\mathrm{T}} \\
{}^t\{\phi_3\}_{,z}^{\mathrm{T}} \\
{}^t\{\phi_1\}_{,y}^{\mathrm{T}} + {}^t\{\phi_2\}_{,x}^{\mathrm{T}} \\
{}^t\{\phi_2\}_{,z}^{\mathrm{T}} + {}^t\{\phi_3\}_{,y}^{\mathrm{T}} \\
{}^t\{\phi_3\}_{,x}^{\mathrm{T}} + {}^t\{\phi_1\}_{,z}^{\mathrm{T}}
\end{bmatrix} , \tag{6.64}
$$

and

$$
{}^t\{\nabla(\Delta u)\} = {}^t[B_{NL}]_t\{\Delta u\}^e, \quad {}^t\{\nabla w\} = {}^t[B_{NL}]\{w\}^e , \tag{6.65}
$$

where

$$
{}^t[B_{NL}]^{\mathrm{T}} = \left[{}^t\{\phi_1\}_{,x}, {}^t\{\phi_1\}_{,y}, {}^t\{\phi_1\}_{,z}, {}^t\{\phi_2\}_{,x}, -, -, -, {}^t\{\phi_3\}_{,z} \right] . \tag{6.66}
$$

Substitution of Equations 6.63 and 6.65 into Equation 6.60 leads to

$$
\{w\}^{e\mathrm{T}} {}^{t+\Delta t}\{f\}_{in}^e = \{w\}^{e\mathrm{T}} {}^t\{f\}_{in}^e + \{w\}^{e\mathrm{T}} {}^t[k]^e {}_t\{\Delta u\}^e + O\left(\left\| {}_t\Delta u^e \right\|^2 \right), \tag{6.67}
$$

where

$${}^t\{f\}_{in}^e = \int\limits_{{}^tV^e} {}^t[B_L]^{\mathrm{T}} {}^t\{\sigma\}\mathrm{d}^tV,$$
(6.68)

and

$${}^t[k]^e = {}^t[k_L]^e + {}^t[k_{NL}]^e,$$

$${}^t[k_L]^e = \int\limits_{{}^tV^e} {}^t[B_L]^{\mathrm{T}} {}^t[C^{EP}]^t[B_L]\mathrm{d}^tV,$$
(6.69)

$${}^t[k_{NL}]^e = \int\limits_{{}^tV^e} {}^t[B_{NL}]^{\mathrm{T}} {}^t[\Sigma]^t[B_{NL}]\mathrm{d}^tV.$$

Here, $\left\|{}_t\Delta u^e\right\|$ is the norm of the elemental incremental displacement vector ${}_t\{\Delta u\}^e$. Further, similar to Equation 6.26, ${}^t\{f\}_{in}^e$ is called the elemental internal force vector at time t.

6.2.4.2 Relation Between Internal Force Vectors of Two Different Iterations

Instead of starting from ${}^{t+\Delta t}\{f\}_{in}^e$ (Equation 6.26), if we start from its estimate corresponding to $(i-1)$-th iteration, *i.e.*, from

$${}^{t+\Delta t}\{f\}_{in}^{e(i)} = \int\limits_{{}^{t+\Delta t}V^{e(i)}} {}^{t+\Delta t}[B_L]^{(i)\mathrm{T}} {}^{t+\Delta t}\{\sigma\}^{(i)}\mathrm{d}^{t+\Delta t}V^{(i)},$$
(6.70)

and then transform its right side to the configuration corresponding to $(i-1)^{\mathrm{th}}$ iteration, we get the following expression:

$$\{w\}^{e\mathrm{T}} {}^{t+\Delta t}\{f\}_{in}^{e(i)}$$

$$= \{w\}^{e\mathrm{T}} {}^{t+\Delta t}\{f\}_{in}^{e(i-1)} + \{w\}^{e\mathrm{T}} {}^{t+\Delta t}[k]^{e(i-1)}\left({}_t\{\Delta u\}^{e(i)} - {}_t\{\Delta u\}^{e(i-1)}\right) + O\left(\left\|\delta^e\right\|^2\right),$$
(6.71)

where

$${}^{t+\Delta t}\{f\}_{in}^{e(i-1)} = \int\limits_{{}^{t+\Delta t}V^{e(i-1)}} {}^{t+\Delta t}[B_L]^{(i-1)\mathrm{T}} {}^{t+\Delta t}\{\sigma\}^{(i-1)}\mathrm{d}^{t+\Delta t}V^{(i-1)},$$
(6.72)

$$^{t+\Delta t}[k]^{e(i-1)} = {}^{t+\Delta t}[k_L]^{e(i-1)} + {}^{t+\Delta t}[k_{NL}]^{e(i-1)},$$

$${}^{t+\Delta t}[k_L]^{e(i-1)} = \int_{{}^{t+\Delta t}V^{e(i-1)}} {}^{t+\Delta t}[B_L]^{(i-1)\mathrm{T}} \, {}^{t+\Delta t}[C^{EP}]^{(i-1)} \, {}^{t+\Delta t}[B_L]^{(i-1)} \mathrm{d}^{t+\Delta t}V^{(i-1)},$$

$${}^{t+\Delta t}[k_{NL}]^{e(i-1)} = \int_{{}^{t+\Delta t}V^{e(i-1)}} {}^{t+\Delta t}[B_{NL}]^{(i-1)\mathrm{T}} \, {}^{t+\Delta t}[\Sigma]^{(i-1)} \, {}^{t+\Delta t}[B_{NL}]^{(i-1)} \mathrm{d}^{t+\Delta t}V^{(i-1)},$$

$$(6.73)$$

and $\left\| \delta^e \right\|$ is the norm of the vector $\{\delta\}^e$:

$$\{\delta\}^e = {}_t\{\Delta u\}^{e(i)} - {}_t\{\Delta u\}^{e(i-1)}. \tag{6.74}$$

Equation 6.71 involves the difference of ${}_t\{\Delta u\}^{e(i)}$ and ${}_t\{\Delta u\}^{e(i-1)}$ as that is the displacement vector from the configuration corresponding to $(i-1)$-th iteration to i-th iteration.

Assembling Equation 6.71 over all the elements (N_e) of the domain, we get

$$\{W\}^{\mathrm{T} \, t+\Delta t}\{F\}_{in}^{(i)}$$
$$= \{W\}^{\mathrm{T} \, t+\Delta t}\{F\}_{in}^{(i-1)} + \{W\}^{\mathrm{T} \, t+\Delta t}[K]^{(i-1)}\{\Delta\} + O(\|\Delta\|^2). \tag{6.75}$$

where ${}^{t+\Delta t}[K]^{(i-1)}$ is the global version of ${}^{t+\Delta t}[k]^{e(i-1)}$ and $\|\Delta\|$ is the norm of the vector $\{\Delta\}$:

$$\{\Delta\} = {}_t\{\Delta U\}^{(i)} - {}_t\{\Delta U\}^{(i-1)}. \tag{6.76}$$

Since the global weight vector $\{W\}$ is arbitrary, Equation 6.75 implies

$$^{t+\Delta t}\{F\}_{in}^{(i)} = {}^{t+\Delta t}\{F\}_{in}^{(i-1)} + {}^{t+\Delta t}[K]^{(i-1)}\{\Delta\} + O(\|\Delta\|^2). \tag{6.77}$$

6.2.4.3 Determination of the Derivative
Comparing Equations 6.30 and 6.77, we get

$$\left. \frac{\partial^{t+\Delta t}\{F\}_{in}}{\partial_t\{\Delta U\}} \right|_{{}_t\{\Delta U\} = {}_t\{\Delta U\}^{(i-1)}} = {}^{t+\Delta t}[K]^{(i-1)}, \tag{6.78}$$

and, therefore, Equation 6.30 becomes

$$^{t+\Delta t}\{F\}_{in}^{(i)} = {}^{t+\Delta t}\{F\}_{in}^{(i-1)} + {}^{t+\Delta t}[K]^{(i-1)}({}_t\{\Delta U\}^{(i)} - {}_t\{\Delta U\}^{(i-1)}) + O(\|\Delta\|^2), \tag{6.79}$$

where $^{t+\Delta t}[K]^{(i-1)}$ is the global assembly of the elemental matrices $^{t+\Delta t}[k]^{e(i-1)}$ given by Equation 6.73.

The matrix $^{t+\Delta t}[K]^{(i-1)}$ is called the *tangent stiffness matrix*. Note that the tangent stiffness matrix depends on the choice of incremental strain, objective incremental stress and incremental stress-strain relation. Crisfield [2] and Bathe [3] have used the virtual work expression to obtain the same expression for $^{t+\Delta t}[k]^{e(i-1)}$ as given by Equation 6.73.

The expression for the derivative $\partial^{t+\Delta t}\{F\}_{in}/\partial_t\{\Delta U\}$, as given by Equations 6.73 and 6.78, has been obtained by choosing $_t\Delta\varepsilon_{kl}$ as the incremental strain, $_t\Delta S_{ij}$ as the objective incremental stress and $^tC_{ijkl}^{EP}$ as the constitutive tensor relating the two. However, we wish to develop the finite element formulation for finite increment size for which we plan to use $\Delta\varepsilon_{ij}^L$ as the incremental strain, Equation 6.2a as the incremental stress-strain relation and Equation 6.3 as the updating scheme to make the incremental stress objective. For this combination, it is quite tedious to obtain the expression for the derivative $\partial^{t+\Delta t}\{F\}_{in}/\partial_t\{\Delta U\}$. Therefore, we use the expressions at Equations 6.73 and 6.78 for this derivative. Note that in the proposed iterative scheme, the derivative $\partial^{t+\Delta t}\{F\}_{in}/\partial_t\{\Delta U\}$ is to be used as the coefficient matrix of the algebraic equations to find the corrections to the initial guess. As long as the corrections lead to a converged solution, the choice of the coefficient matrix in the algebraic equations does not matter. Therefore, using a coefficient matrix different from the exact derivative $\partial^{t+\Delta t}\{F\}_{in}/\partial_t\{\Delta U\}$ does not affect the final solution as long as the iterative scheme converges.

6.2.5 Iterative Scheme

We now *assume* that the global external force vector $^{t+\Delta t}\{F\}_{ex}$ is independent of $_t\{\Delta U\}$. Then, for $_t\{\Delta U\}^{(i)}$ to be the solution of Equation 6.29, the vector $^{t+\Delta t}\{F\}_{in}^{(i)}$ as given by Equation 6.79 must satisfy Equation 6.29. Therefore, substituting Equation 6.79 in Equation 6.29 and neglecting the second order terms, we get

$$^{t+\Delta t}[K]^{(i-1)}\left(_t\{\Delta U\}^{(i)} - _t\{\Delta U\}^{(i-1)}\right) = {}^{t+\Delta t}\{R\}^{(i-1)}, \quad \text{for } i=2,3....., \quad (6.80)$$

where

$$^{t+\Delta t}\{R\}^{(i-1)} = {}^{t+\Delta t}\{F\}_{ex} - {}^{t+\Delta t}\{F\}_{in}^{(i-1)} \quad \text{for } i=2,3...., \quad (6.81)$$

is the difference between the global external force vector and global internal force vector in iteration $(i-1)$, and is called the *(global) unbalanced force vector*. Equations 6.80 and 6.81) are the iterative equations to be satisfied by $_t\{\Delta U\}^{(i)}$.

To start the iterations, we need an initial guess for the incremental displacement vector. To find a good initial guess, we use the expression at Equation 6.67 for the elemental internal force vector $^{t+\Delta t}\{f\}_{in}^e$ involving the integrals over the known domain at time t. Substituting this expression, after neglecting the second order terms, in the elemental discretized equilibrium equation (Equation 6.25), we get

$$\sum_{e=1}^{N_e} \{w\}^{eT}\,{}^t\{f\}_{in}^e + \{w\}^{eT}\,{}^t[k]^e\,{}_t\{\Delta u\}^e = \sum_{b=1}^{N_b} \{w\}^{bT}\,{}^{t+\Delta t}\{f\}_{ex}^b. \tag{6.82}$$

Using the assembly procedure of Chapter 5, the above equation becomes

$$\{W\}^{T}\,{}^t\{F\}_{in} + \{W\}^{T}\,{}^t[K]\,{}_t\{\Delta U\} = \{W\}^{T}\,{}^{t+\Delta t}\{F\}_{ex}, \tag{6.83}$$

where $^t[K]$ is the global assembly of the elemental matrices $^t[k]^e$ (Equation 6.69). Since the global weight vector $\{W\}$ is arbitrary, Equation 6.83 implies

$${}^t\{F\}_{in} + {}^t[K]\,{}_t\{\Delta U\} = {}^{t+\Delta t}\{F\}_{ex}. \tag{6.84}$$

Decomposing the global external force vector $^{t+\Delta t}\{F\}_{ex}$ as the sum of $^t\{F\}_{ex}$ and its increment during the time Δt

$${}^{t+\Delta t}\{F\}_{ex} = {}^t\{F\}_{ex} + {}_t\{\Delta F\}_{ex}, \tag{6.85}$$

Equation (6.84) becomes

$${}^t\{F\}_{in} + {}^t[K]_t\{\Delta U\} = {}^t\{F\}_{ex} + {}_t\{\Delta F\}_{ex}. \tag{6.86}$$

The configuration at time t is an equilibrium configuration. Therefore, using the discretized equilibrium equation (Equation 6.29) at time t, Equation 6.86 becomes

$${}^t[K]_t\{\Delta U\} = {}_t\{\Delta F\}_{ex}. \tag{6.87}$$

Solution of this equation is used as the initial guess for the iterative scheme and it is denoted by $_t\{\Delta U\}^{(1)}$. The corresponding coefficient matrix and the right side vector are denoted respectively by $^{t+\Delta t}[K]^{(0)}$ and $^{t+\Delta t}\{R\}^{(0)}$.

The Newton-Raphson iterative scheme can, now, be expressed as

$$'[K]_t \{\Delta U\}^{(1)} =_t \{\Delta F\}_{ex} , \tag{6.88}$$

$$^{t+\Delta t}[K]^{(i-1)} \left(_t\{\Delta U\}^{(i)} - _t\{\Delta U\}^{(i-1)} \right) = ^{t+\Delta t}\{R\}^{(i-1)}, \quad \text{for } i=2,3..., \tag{6.89}$$

$$^{t+\Delta t}\{R\}^{(i-1)} = ^{t+\Delta t}\{F\}_{ex} - ^{t+\Delta t}\{F\}_{in}^{(i-1)} \quad \text{for } i=2,3.... \tag{6.90}$$

Here, $'[K]$ and $^{t+\Delta t}[K]^{(i-1)}$, called the *global coefficient matrices*, are the global assemblies of the matrices $'[k]^e$ and $^{t+\Delta t}[k]^{e(i-1)}$ respectively. These matrices are called the *elemental coefficient matrices* and are given by Equations 6.69 and 6.73 respectively. The global external force vector $^{t+\Delta t}\{F\}_{ex}$ at time $t + \Delta t$ is the global assembly of the elemental external force vectors $^{t+\Delta t}\{f\}_{ex}^b$, given by Equation 6.27. Its increment $_t\{\Delta F\}_{ex}$ is obtained as the global assembly of $_t\{\Delta f\}_{ex}^b$, which is found from Equation 6.27 by replacing the stress vector array $^{t+\Delta t}\{t\}$ at time $t + \Delta t$ by the incremental stress vector array $_t\{\Delta t\}$. The global internal force vector $^{t+\Delta t}\{F\}_{in}^{(i-1)}$ is obtained as the global assembly of the elemental internal force vectors $^{t+\Delta t}\{f\}_{in}^{e(i-1)}$, given by Equation 6.72. Integration involved in the evaluation of the elemental quantities $'[k]^e$, $^{t+\Delta t}[k]^{e(i-1)}$, $_t\{\Delta f\}_{ex}^b$, $^{t+\Delta t}\{f\}_{ex}^b$ and $^{t+\Delta t}\{f\}_{in}^{e(i-1)}$ is done numerically using Gauss numerical integration scheme.

The iterations are continued untill the right side vector of Equation 6.89, namely $^{t+\Delta t}\{R\}^{(i)}$, becomes very small compared to the external force vector $^{t+\Delta t}\{F\}_{ex}$. Thus, the convergence criterion can be stated as

$$\frac{\left\| ^{t+\Delta t}\{R\}^{(i)} \right\|}{\left\| ^{t+\Delta t}\{F\}_{ex} \right\|} \le \varepsilon_c , \tag{6.91}$$

where $\| \{a\} \|$ represents the norm of the array $\{a\}$ defined as

$$\| \{a\} \| = \sum_i a_i^2 , \tag{6.92}$$

where the sum is to be taken over all the components. In Equation 6.91, ε_c is a specified small number, called as the convergence tolerance.

While deriving the iterative equations, it is assumed that the global external force vector $^{t+\Delta t}\{F\}_{ex}$ is independent of the incremental displacement vector

$_t\{\Delta U\}$. However, $^{t+\Delta t}\{F\}_{ex}$ is the global assembly of $^{t+\Delta t}\{f\}_{ex}^b$ involving integration over the boundary $^{t+\Delta t}S_t^b$. Since $^{t+\Delta t}S_t^b$ depends on $_t\{\Delta U\}$, $^{t+\Delta t}\{F\}_{ex}$ also depends on it. Therefore, it needs to be updated in each iteration. In that case, the global unbalanced force vector $^{t+\Delta t}\{R\}^{(i-1)}$ becomes

$$^{t+\Delta t}\{R\}^{(i-1)} = {}^{t+\Delta t}\{F\}_{ex}^{(i-1)} - {}^{t+\Delta t}\{F\}_{in}^{(i-1)} \quad \text{for } i=2,3\dots . \tag{6.93}$$

Further, the denominator of Equation 6.91 is to be changed to $^{t+\Delta t}\{F\}_{ex}^{(i)}$.

Updation of coefficient matrix in each iteration takes a lot of computational time. Therefore, sometimes, it is kept constant in each iteration. Thus, $^t[K]$ is used as the coefficient matrix instead of $^{t+\Delta t}[K]^{(i-1)}$ in every iteration. This version of the iterative scheme is called the *modified Newton-Raphson iterative scheme*. This scheme usually needs more iterations per increment than the full Newton-Raphson scheme. However, numerical experiments show that the computational time saved in not updating the coefficient matrix is usually larger than the time required for carrying out more iterations. Therefore, overall computational time per increment is less when the modified version of the Newton-Raphson scheme is used.

6.2.6 Determination of Stresses

Evaluation of the stress components (at the Gauss points of the elements) is done by the following stepwise procedure:

- Calculation of the incremental deformation gradient tensor $_t\Delta F$ from Equation 6.1c.
- Decomposition of $_t\Delta F$ into the incremental rotation tensor $_t\Delta R$ and the incremental right stretch tensor $_t\Delta U$ using the polar decomposition theorem (Equation 4.62).
- Determination of the principal values $_t\Delta\lambda_i$ and principal directions $^t\hat{e}_i$ of the incremental right stretch tensor $_t\Delta U$.
- Calculation of the components of the incremental logarithmic strain $_t\Delta\varepsilon^L$ in the coordinate system of $^t\hat{e}_i$ using Equation 6.1a. These components need to be transformed to the fixed frame using the tensor transformation law (Equation 2.56).
- Determination of the incremental (Cauchy) stress $_t\Delta\sigma$ using the constitutive equation (Equation 6.2). The integration in Equation 6.2 is performed using the Euler forward integration scheme which is described in the next subsection .

- Calculation of the (Cauchy) stress $^{t+\Delta t}\sigma$ at time $t+\Delta t$ using the updating scheme of Equation 6.3. This updating procedure makes the constitutive equation objective.

6.2.6.1 Integration of the Constitutive Equation

Different techniques exist for the integration of the constitutive equation (Equation 6.2a) [2]. A simple but robust technique is the Euler forward integration scheme which is described below.

Suppose that the incremental logarithmic strain $_t\Delta\varepsilon^L$ and the incremental rotation tensor $_t\Delta R$ at a Gauss point have been calculated and the state of the Gauss point at time t (elastic or plastic) is known.

(A) If the state of the Gauss point at time t is *elastic*:

1. First, the stress increment $_t\Delta\sigma$ is determined assuming elastic behavior:

$$_t\Delta\sigma = C^E(_t\Delta\varepsilon^L) \tag{6.94}$$

where C^E is given by Equation 3.251f.

2. Next, the (Cauchy) stress $^{t+\Delta t}\sigma$ at time $t+\Delta t$ is calculated using the updating scheme (Equation 6.3).

3. Next, the equivalent stresses $^t\sigma_{eq}$ and $^{t+\Delta t}\sigma_{eq}$ are determined using Equation 3.23.

4. If $^{t+\Delta t}\sigma_{eq} \leq \sigma_Y$, then the assumption of elastic behavior is correct and $^{t+\Delta t}\sigma$ calculated in step 2 is the correct (Cauchy) stress at the Gauss point at time $t+\Delta t$.

5. If $^{t+\Delta t}\sigma_{eq} > \sigma_Y$, a transition from elastic to plastic has occurred. Therefore, the state of the Gauss point is changed from elastic to plastic. Further, the elastic part of the increment is calculated as

$$fraction = \frac{\sigma_Y - {}^t\sigma_{eq}}{{}^{t+\Delta t}\sigma_{eq} - {}^t\sigma_{eq}}. \tag{6.95}$$

Thus, the incremental strain corresponding to the elastic part of the increment is equal to $fraction(_t\Delta\varepsilon^L)$ and that corresponding to the plastic part is $(1-fraction)(_t\Delta\varepsilon^L)$.

- First, the (Cauchy) stress corresponding to the elastic part of the increment is calculated from

$$^{t+\Delta t}\sigma^{(0)} = (_t\Delta R)(^t\sigma)(_t\Delta R)^T + C^E(fraction(_t\Delta\varepsilon^L)). \tag{6.96}$$

- Next, the incremental strain corresponding to the plastic part is divided into n sub-increments

$$_t d\varepsilon^L = \frac{(1 - fraction)_t \Delta\varepsilon^L}{n}.$$

(6.97)

- Then, the equivalent plastic strain $(^{t+\Delta t}\varepsilon_{eq}^p)^{(i)}$ at the end of i-th sub-increment is calculated from Equations 3.93 and 3.97 except that the integration in Equation 3.97 is replaced by the sum. Thus,

$$(^{t+\Delta t}\varepsilon_{eq}^p)^{(i)} = \sum_i (_t d\varepsilon_{eq}^{pL}), \quad \text{for } i = 1,...,n-1,$$

(6.98a)

$$_t d\varepsilon_{eq}^{pL} = \left(\frac{2}{3} (_t d\varepsilon_{ij}^{pL})(_t d\varepsilon_{ij}^{pL}) \right)^{1/2},$$

(6.98b)

where $_t d\varepsilon^{pL}$ is the plastic part of the $_t d\varepsilon^L$.

- Then, the elastic-plastic constitutive matrix $^{t+\Delta t}C^{EP(i)}$ for i-th sub-increment is calculated from Equation 6.2b:

$$^{t+\Delta t}C_{ijkl}^{EP(i)}$$

$$= 2\mu \left[\frac{v}{1-2v} \delta_{ij}\delta_{kl} + \delta_{ik}\delta_{jl} - \frac{9\mu}{2} \frac{(^{t+\Delta t}\sigma_{ij}'^{(i)})(^{t+\Delta t}\sigma_{kl}'^{(i)})}{(^{t+\Delta t}H'^{(i)} + 3\mu)(^{t+\Delta t}\sigma_{eq}^{(i)})^2} \right] \quad \text{for } i = 1,..,n-1,$$

(6.99)

where $^{t+\Delta t}\sigma'^{(i)}$ is the deviatoric part of $^{t+\Delta t}\sigma^{(i)}$, i.e., the (Cauchy) stress at the end of i-th sub-increment, which is calculated from

$$^{t+\Delta t}\sigma^{(i)} = (_t\Delta R)(^t\sigma)(_t\Delta R)^T + \sum_{j=0}^{i-1} (^{t+\Delta t}C^{EP(j)})(_t d\varepsilon^L) \quad \text{for } i = 1,..,n-1,$$

(6.100)

and $^{t+\Delta t}H'^{(i)}$ is the derivative of the hardening function defined by Equation 6.2c:

$$H(^{t+\Delta t}\varepsilon_{eq}^p) \equiv \sigma_Y + K(^{t+\Delta t}\varepsilon_{eq}^p)^n,$$

(6.101)

evaluated at $({}^{t+\Delta t}\varepsilon_{eq}^{p})^{(i)}$. Note that ${}^{t+\Delta t}C^{EP(0)}$ is calculated from ${}^{t+\Delta t}\sigma^{(0)}$ given by Equation 6.96 with ${}^{t+\Delta t}H'^{(0)}$ being taken as zero.

▪ Then, the (Cauchy) stress at the end of increment, *i.e.*, at time $t+\Delta t$ is calculated from

$$
{}^{t+\Delta t}\sigma = ({}_{t}\Delta R)({}^{t}\sigma)({}_{t}\Delta R)^{\mathrm{T}} + {}_{t}\Delta\sigma , \tag{6.102a}
$$

$$
{}_{t}\Delta\sigma = \sum_{i=0}^{n-1}({}^{t+\Delta t}C^{EP(i)})({}_{t}\mathrm{d}\varepsilon^{L}) . \tag{6.102b}
$$

(B) If the state of the Gauss point at time t is *plastic*, the sub-increment method described in step 5 above is used to calculate ${}^{t+\Delta t}\sigma$ with the fraction being set as zero.

(C) If the state of the Gauss point at time t is *plastic* and the stress increment ${}_{t}\Delta\sigma$ is such that it satisfies the following *unloading* criterion (Equation 3.212)

$$
({}^{t}\sigma'_{ij})({}_{t}\Delta\sigma_{ij}) < 0 , \tag{6.103}
$$

then

▪ The stress increment ${}_{t}\Delta\sigma$ is recalculated using the *elastic* constitutive equation (Equation 6.94).
▪ The (Cauchy) stress ${}^{t+\Delta t}\sigma$ at time $t+\Delta t$ is recalculated using the updating scheme (Equation 6.3).
▪ The state of the Gauss point is changed from *plastic to elastic*. Note that the yield stress corresponding to this elastic state would be ${}^{t}\sigma_{eq}$ and not σ_{Y}.

6.2.7 Divergence Handling Techniques

The Newton-Raphson iterative scheme diverges in some cases. In that case, the following simple but fairly effective techniques can be used to overcome the divergence. In all these techniques, if the divergence occurs in the iteration number i, then that iteration is repeated by reducing the norm of the unbalanced force vector ${}^{t+\Delta t}\{R\}^{(i-1)}$ corresponding to the solution of the previous iteration, *i.e.*, the iteration number $(i-1)$. Note that the solution of the $(i-1)$-th iteration is the iterative correction vector ${}_{t}\{\delta u\}^{(i-1)}$ defined by

$$
{}_{t}\{\delta u\}^{(i-1)} = \left({}_{t}\{\Delta U\}^{(i-1)} - {}_{t}\{\Delta U\}^{(i-2)}\right) . \tag{6.104}
$$

Thus, in a divergence handling technique, the iterative correction vector $_t\{\delta u\}^{(i-1)}$ is changed so as to reduce the norm of the unbalanced force vector $^{t+\Delta t}\{R\}^{(i-1)}$.

Under-Relaxation

In this technique, the iterative correction vector is directly scaled down by a chosen factor α_u $(0 < \alpha_u < 1)$. So, for repeating the iteration i, instead of the actual iterative correction vector $_t\{\delta u\}^{(i-1)}$ of the $(i-1)$-th iteration, its scaled version $(\alpha_u)_t\{\delta u\}^{(i-1)}$ is used.

Line Search

In this technique also, the iterative correction vector is scaled by a factor, but the scaling factor α_l is determined so as to minimize the norm of the unbalanced force vector $^{t+\Delta t}\{R\}^{(i-1)}$. Full line search technique, in which the norm of $^{t+\Delta t}\{R\}^{(i-1)}$ is minimized, is computationally expensive. Therefore, it is modified so as to reduce the computational time. In this modified version, the norm of $^{t+\Delta t}\{R\}^{(i-1)}$ is evaluated at n discrete values of α_l lying between $(\alpha_l)_{\min}$ and $(\alpha_l)_{\max}$. The value of α_l which leads to the lowest value of the norm of $^{t+\Delta t}\{R\}^{(i-1)}$ is chosen as the scaling factor. Note that α_l does not necessarily have to be less than 1.

Modification of the Coefficient Matrix

In this technique, the iterative correction vector is scaled down by increasing the size of the coefficient matrix $^t[K]$. This is done by adding a small percent of the elastic stiffness matrix to it.

Increment Cutting

If all the above techniques fail, then the divergence is overcome by cutting the size of the increment. In this case, the whole increment, and not just the iteration, is repeated unlike the earlier three techniques. The increment size is cut either by cutting the specified force if it is a force control problem or by cutting the specified displacement if it is a displacement control problem or by cutting both if both the force and displacement are specified at different parts of the boundary. This cut in the increment size automatically scales down the iterative displacement vector thereby reducing the norm of the unbalanced force vector. This happens in all the iterations of the repeated increment.

6.3 Modeling of Axisymmetric Open Die Forging by Updated Lagrangian Finite Element Method

In forging, the material is deformed plastically between two or more dies so as to give it the desired shape and size. In open-die forging, height of a solid work-piece is reduced by compressing it between two flat dies, also called platens. In closed die forging, the work-piece is given a desired shape by using shaped dies. In this book, only the open die cold forging would be considered as it is the simplest of all the forging processes. Open die forging problems appear to be similar to those of the simple compression. However, there are significant differences between the states of deformation and stress in the two cases. In the case of open die forging, the friction at the die–work-piece interface makes the deformation as well as the stress distribution non-uniform.

The forging process has been analyzed by the slab method [4], the slip-line method [5, 6] and the upper bound method [7, 8] mostly for determining the forging load and the contact pressure distribution. However, for determining the detailed deformation and stress fields, the finite element method (FEM) has been used.

There is a large body of literature on the application of FEM to forging problems. The early applications of FEM to forging problems were based on the incremental method proposed by Lee and Kobayashi [9]. The method uses the elastic-plastic stress-strain matrix based on the Prandtl-Reuss equations. Even though the stress-strain matrix and the geometry are updated after every increment, only the linearized incremental equations are used. The interfacial friction is modeled by the friction factor where the change in the direction of the shear stress is incorporated by introducing a velocity-dependent coefficient. This method has been applied to solid cylinder upsetting [10], ring compression [11] and for predicting defects [12]. Whereas Lee and Kobayashi [9] used the velocity as the primary unknown, Hartley et al. [13, 14] proposed an incremental method with the incremental displacement as the primary unknown. They also used only the linearized incremental equations and updated the stress-strain matrix and geometry after every increment. However, in their formulation, the friction factor is incorporated by the beta stiffness method [13]. Shima et al. [15] used the rigid-plastic constitutive equation based on the plasticity theory of porous metals and the Coulomb friction model to study the upsetting of a circular cylinder and validated their results by conducting experiments.

The finite element formulations involving the co-rotational strain measures and the objective stress measures have been discussed by Kobayashi et al. [16], Rowe et al. [17] and Hartley et al. [18]. The finite element formulation proposed by Bathe et al. [19], which involves solving non-linear incremental equations by an iterative scheme like the Newton-Raphson method, has been applied (with or without elastic effects) to axisymmetric forging problems by Dadras and Thomas [20] and Carter and Lee [21].

Some of the typical latest references on the application of FEM to axisymmetric forging process are discussed in this paragraph. All of them employ non-linear incremental equations. Further, in most of the references, the interfacial friction has been modeled by the friction factor rather than by Coulomb's law. Michel and Boyer [22] have carried out the elasto-viscoplastic finite element analysis of a cold upsetting process and validated it by residual stress measurements using the hole-drilling method. They have calculated and measured the residual stress variations on the flat end of the cylinder. They have used the friction factor model to represent the interfacial friction. Zhao et al. [23] have used forward and backward finite element simulations to design the preform shapes in forging processes. Even though the example considered is that of plane strain forging, the method is equally applicable to axisymmetric forging. Joun et al. [24] have proposed a finite element simulation technique for the forging process with a spring-attached die. The strategy of spring-attached die controls the metal flow lines in such a fashion that it results in the prevention of defects and the improvement of product quality. Kim *et al.* [25] have applied rigid-viscoplastic FEM to cold axisymmetric forging of aluminum alloy to study its ductile fracture. The Cockcroft and Latham criterion has been used for predicting the fracture. Yang *et al.* [26] have developed an intelligent system for the complete design methodology of the forging process by integrating FEM with expert systems and computer-aided design (CAD) interface modules. Gupta *et al.* [27] have predicted fractures in axisymmetric forging processes using the hydrostatic stress criterion as well as the critical value of the damage parameter. The damage is evaluated using the continuum damage mechanics theory. Mungi *et al.* [28] have carried out the parametric study of residual stresses in axisymmetric forging. Both the papers use the updated Lagrangian finite element formulation where non-linear incremental equations are solved by the Newton-Raphson technique. Incremental logarithmic strain has been used as the measure of incremental deformation and the stress is updated in a material frame to make the incremental stress objective.

6.3.1 Domain and Boundary Conditions

We assume that both the platens move with the same velocity but in the opposite direction. Therefore, due to *symmetry*, only the *upper half* of the cylindrical work-piece needs to be considered. Further, because of axisymmetry, only a typical *r-z plane* needs to be analyzed. Figure 6.1 shows the domain of the problem. It is the initial configuration of an *r-z plane* of the upper half of the cylindrical work-piece. Note that this domain is two-dimensional and undeformed and therefore looks different from the three-dimensional deformed domain of the same problem shown in Figure 3.21. The platens are assumed rigid and the friction at the interface is modeled by *sticking friction* and *Coulomb's law*.

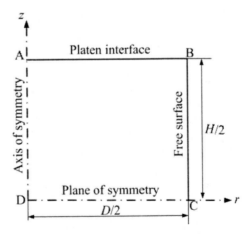

Figure 6.1. Domain of the problem: the initial configuration of a quarter of the cylindrical block

Note that an axisymmetric problem is a two-dimensional problem. Therefore, only two boundary conditions are needed on each boundary. The boundary conditions of the problem are the same as Equations 3.255–3.260 except that the prefix of the incremental quantities is changed from d(.) to $_t\Delta(.)$ and the notations $t_r + dt_r$ and $t_z + dt_z$ are changed to $^{t+\Delta t}t_r$ and $^{t+\Delta t}t_z$. Further, the boundary conditions do not involve any conditions on $_t\Delta u_\theta$ and $_t\Delta t_\theta$ (i.e., the θ-components of the incremental displacement vector and incremental stress vector respectively). For axisymmetric problems, both these quantities are identically zero everywhere in the domain. Thus, the boundary conditions of the problem are as follows. Whether the condition is essential (i.e., on the primary vaiable $_t\Delta u$) or natural (i.e., on the secondary variable $_t\Delta t$) is also mentioned against each condition. (The subscript n of the incremental stress vector $_t\Delta t_n$ is dropped henceforth.)

Free boundary BC:
The boundary BC is a free surface. On a free surface, both r and z components of the incremental stress vector $_t\Delta t$ are zero at every point. Therefore, the boundary conditions on the boundary BC are

$$_t\Delta t_r = 0, \quad _t\Delta t_z = 0 \quad \text{(natural)}, \tag{6.105}$$

Plane of symmetry DC:
On a plane of symmetry, the normal component of the incremental displacement vector $_t\Delta u$ and the shear component of the incremental stress vector $_t\Delta t$ are zero at every point. Since the boundary DC is perpendicular to z-axis, the boundary conditions on this boundary are:

$$_t\Delta t_r = 0, \quad \text{(natural)}, \tag{6.106a}$$

$$_t\Delta u_z = 0, \quad \text{(essential)}. \tag{6.106b}$$

Note that these are a mixed type of boundary condition.
Axis of symmetry AD:
On the axis of symmetry, the component of the incremental displacement vector $_t\Delta u$ normal to the axis and the component of the incremental stress vector $_t\Delta t$ along the axis are zero at every point. Since the axis AD is along z-axis, the boundary conditions on this boundary are

$$_t\Delta u_r = 0, \quad \text{(essential)}, \tag{6.107a}$$

$$_t\Delta t_z = 0, \quad \text{(natural)}. \tag{6.107b}$$

Note that these are also a mixed type of boundary condition.

Platen interface AB:

At the interface, z-component of the incremental displacement vector $_t\Delta u$ must be equal to the incremental platen displacement.

As far as the second boundary condition is concerned, we observe the following. Nearer to point A (center of the platen), the block material sticks to the platen while nearer to the free edge (point B), the block material slips relative to the platen in the outward direction. We first discuss the boundary condition corresponding to the *slipping case*. Here, we assume that the frictional (or shear) stress exerted by the platen in r-direction is governed by *Coulomb's law*:

$$\left|^{t+\Delta t}t_r\right| = f\left|^{t+\Delta t}t_z\right| \quad \text{if} \quad \left|^{t+\Delta t}t_r\right| \geq f\left|^{t+\Delta t}t_z\right|, \tag{6.108}$$

where f is the coefficient of friction and $^{t+\Delta t}t_r$ and $^{t+\Delta t}t_z$ are the components of the stress vector $^{t+\Delta t}t$ (at time $t+\Delta t$) along the directions r and z respectively. Note that the friction boundary condition has to be in terms of the stress vector at time $t+\Delta t$ and not in terms of the incremental stress vector $_t\Delta t$. The material flow at the interface is in the positive r-direction. Therefore, the frictional stress will be in the opposite direction, *i.e.*, in the negative r-direction. Further, the normal stress exerted by the platen is always compressive, *i.e.*, in the negative z-direction. Therefore, both $^{t+\Delta t}t_r$ and $^{t+\Delta t}t_z$ are negative. Then Equation 6.108 becomes

$$^{t+\Delta t}t_r = f(^{t+\Delta t}t_z) \quad \text{if} \quad \left|^{t+\Delta t}t_r\right| \geq f\left|^{t+\Delta t}t_z\right|. \tag{6.109}$$

Next, we discuss the boundary condition corresponding to the *sticking case*. In this case, r component of the incremental displacement vector must be zero.

Now, the boundary conditions on the boundary AB become

$$^{t+\Delta t}t_r - f(^{t+\Delta t}t_z) = 0 \quad \text{if} \quad \left|^{t+\Delta t}t_r\right| \geq f\left|^{t+\Delta t}t_z\right| \text{ (slipping)}, \tag{6.110a}$$

(natural)

$$_t\Delta u_r = 0, \quad \text{if} \quad \left|^{t+\Delta t}t_r\right| < f\left|^{t+\Delta t}t_z\right| \text{ (sticking)}, \tag{6.110b}$$

(essential)

$$_t\Delta u_z = \Delta^* \tag{6.111}$$

(essential).

where Δ^* is the prescribed *incremental displacement of the platen*. Note that these boundary conditions involve a combination of $^{t+\Delta t}t_r$ and $^{t+\Delta t}t_z$. Here also, the shear stress at the platen interface is subject to the constraint that it can not exceed its maximum value.

Since the nodal forces are proportional to the components of the stress vector $^{t+\Delta t}t$, the two conditions of Equation 6.110 can be expressed in terms of the r and z components of the global force vector at the interface nodes. In fact, this is the form which is convenient for the finite element formulation. In terms of the nodal forces, Equation 6.110 takes the following form. At node l on the interface,

$$^{t+\Delta t}F_{lr} - f\,^{t+\Delta t}F_{lz} = 0 \ \text{ if } \left|^{t+\Delta t}F_{lr}\right| \ge \left|^{t+\Delta t}F_{lz}\right| \ \text{ (slipping)}, \qquad (6.112a)$$

$$_t\Delta u_r = 0 \ \text{ if } \left|^{t+\Delta t}F_{lr}\right| < \left|^{t+\Delta t}F_{lz}\right| \ \text{ (sticking)}, \qquad (6.112b)$$

where $^{t+\Delta t}F_{lr}$ and $^{t+\Delta t}F_{lz}$ are the components of the global force vector $^{t+\Delta t}\{F\}$ indicating the nodal forces in the r and z direction at node l.

6.3.2 Cylindrical Arc Length Method for Displacement Control Problems

The boundary condition (Equation 6.111) indicates that the forging problem is a displacement control problem. Since there is no specified external force in displacement control problems, the denominator in the convergence criterion (Equation 6.91) does not exist. For such problems, one can try to achieve the convergence by making the unbalanced force vector $^{t+\Delta t}\{R\}^{(i)}$ small in an absolute sense. But, this slows down the rate of convergence to a considerable extent. To accelerate the rate of convergence, one can use an *arc length method* in conjunction with the modified Newton-Raphson method. In the arc length method, in every increment, the unknown nodal reaction vector at the boundary (where the incremental displacement vector is specified) is expressed as a linear combination of *known nodal force vectors* and *unknown coefficients*. Then, the problem is solved iteratively as a force control problem using the known nodal force vectors as the specified incremental force vectors. In every iteration, the unknown coefficients in the linear combination are found from the specified incremental displacement vector at the boundary.

Note that the nodal reaction vector also changes during iterations. We treat $^{t+\Delta t}\{F\}_{ex}$ of Equation 6.90 as the nodal reaction vector at time $t + \Delta t$, and define the nodal reaction vector of i-th iteration $^{t+\Delta t}\{F\}_{ex}^{(i)}$ as follows:

$$
\begin{aligned}
&^{t+\Delta t}\{F\}_{ex}^{(1)} = {}^{t}\{F\}_{ex} + {}_t\{\Delta F\}_{ex}^{(1)}, \\
&^{t+\Delta t}\{F\}_{ex}^{(i)} = {}^{t+\Delta t}\{F\}_{ex}^{(i-1)} + {}_t\{\Delta F\}_{ex}^{(i)} \quad \text{for } i = 1,2,......,
\end{aligned}
\qquad (6.113)
$$

where $^t\{F\}_{ex}$ is the known nodal reaction vector at the beginning of the increment and $_t\{\Delta F\}_{ex}^{(i)}$ is the iterative incremental nodal reaction vector of i-th iteration. Then, the iterative equations (Equation 6.88–6.90) get modified to

$$^t[K]_t\{\Delta U\}^{(1)} = _t\{\Delta F\}_{ex}^{(1)}, \tag{6.114}$$

$$^t[K]_t\{\delta u\}^{(i)} = _t\{\Delta F\}_{ex}^{(i)} + {}^{t+\Delta t}\{R\}^{(i-1)} \quad \text{for } i=2,3,......, \tag{6.115}$$

$$^{t+\Delta t}\{R\}^{(i-1)} = {}^{t+\Delta t}\{F\}_{ex}^{(i-1)} - {}^{t+\Delta t}\{F\}_{in}^{(i-1)}, \quad \text{for } i=2,3,....... \tag{6.116}$$

Here, as stated earlier, $_t\{\Delta U\}^{(1)}$ is the initial guess of the incremental displacement vector and $_t\{\delta u\}^{(i)}$ is the iterative correction vector of the i-th iteration defined by Equation 6.104. Further, this being the *modified* Newton-Raphson iterative scheme, the coefficient matrix is kept constant during the iterations at $^t[K]$, *i.e.*, its value at the beginning of the increment.

For structural analysis problems, the arc length method was first proposed by Riks [29]. The general description of the arc length method is given in Crisfield's book [2]. Here, we use the version of the cylindrical arc length method proposed by Batoz and Dhatt [30]. This version is for the case of proportional loading. However, in the forging problem, the loading is non-proportional. Therefore, the original method needs to be appropriately modified to take care of the non-proportional loading. The modified method is described in this subsection.

Let m be the number of nodes on the boundary where the incremental displacement vector is specified. (In the forging problem, this boundary is the platen interface.) Then, the iterative nodal reaction vector $_t\{\Delta F\}_{ex}^{(i)}$ is expressed as

$$_t\{\Delta F\}_{ex}^{(i)} = \sum_{k=1}^{m} (_t\Delta\lambda)_k^{(i)} \{P\}_k, \tag{6.117}$$

where the vectors $\{P\}_k$ are known nodal force vectors, called the *basic load vectors*, and $(_t\Delta\lambda)_k^{(i)}$ are the unknown coefficients. Note that whereas the vectors $\{P\}_k$ are the same in every iteration of each increment, the $(_t\Delta\lambda)_k^{(i)}$ change with the iteration as well as the increment. The vector $\{P\}_k$ is chosen such that all but one of its components are zero, the non-zero component being unity along the direction of the specified displacement at node k. (In the forging problem, this direction is the negative z-direction.)

Since Equations 6.114 and 6.115 are linear in $_t\{\Delta U\}^{(1)}$ and $_t\{\delta u\}^{(i)}$, their solutions are decomposed as

$$_t\{\Delta U\}^{(1)} = \sum_{k=1}^{m} (_t\Delta\lambda)_k^{(1)} \, _t\{\Delta U\}^{kI},$$

$$_t\{\delta u\}^{(i)} = \sum_{k=1}^{m} (_t\Delta\lambda)_k^{(i)} \, _t\{\Delta U\}^{kI} + _t\{\delta u\}^{(i)II} \quad \text{for } i = 2,3,....,$$

(6.118)

where the known vectors $_t\{\Delta U\}^{kI}$ and $_t\{\delta u\}^{(i)II}$ are obtained as the solutions of the problems

$$^t[K]_t \, _t\{\Delta U\}^{kI} = \{P\}_k \qquad \text{for } k = 1,2,....m,$$

$$^t[K]_t \, _t\{\delta u\}^{(i)II} = {}^{t+\Delta t}\{R\}^{(i-1)} \qquad \text{for } i = 2,3,.....$$

(6.119)

Note that the vector $_t\{\Delta U\}^{kI}$ is the contribution to the incremental displacement vector due to the basic load vector $\{P\}_k$ and it is the same in every increment. On the other hand, the vector $_t\{\delta u\}^{(i)II}$ is the contribution to the iterative correction vector due to the unbalanced force vector ${}^{t+\Delta t}\{R\}^{(i-1)}$ and it varies with the iteration as well as the increment. However, it is zero in the first iteration of every increment.

To obtain the complete solution of Equations 6.114 and 6.115, one needs to determine the unknown coefficients $(_t\Delta\lambda)_k^{(i)}$ corresponding to k-th basic load vector and i-th iteration for $k = 1,2,...,m$ and $i = 1,2,...$ They are determined from the specified incremental displacement vector at the boundary. For concreteness, assume that the z-component of the incremental displacement vector at each node is specified and the specified value is $_t\Delta u^*$. (This is actually the case for the forging problem.) We distribute $_t\Delta u^*$ in various iterations as follows. Let $_t\Delta U_{lz}^{(1)}$ denote the component of the initial guess of the incremental displacement vector $_t\{\Delta U\}^{(1)}$ which represents the z-component at node l of the boundary. Further, let $_t\delta u_{lz}^{(i)}$ denote the component of the iterative correction vector $_t\{\delta u\}^{(i)}$ which also represents the z-component at node l of the boundary. We assume that for $l = 1, 2, ...,m$, the entire $_t\Delta u^*$ is equal to $_t\Delta U_{lz}^{(1)}$, the z-component of the initial guess of the incremental displacement vector at node l. Then, in subsequent iterations, $_t\delta u_{lz}^{(i)}$, i.e., the z-component of the iterative correction at node l, is zero for $l = 1, 2, ...,m$. Thus,

$$_t\Delta U_{lz}^{(1)} = {}_t\Delta u^*,$$

$$_t\delta u_{lz}^{(i)} = 0 \text{ for } i = 2,3,\ldots\ldots$$

(6.120)

Substituting these values in Equation 6.118, we get the following equations for $\left({}_t\Delta\lambda\right)_k^{(i)}$:

$$\sum_{k=1}^{m} \left({}_t\Delta\lambda\right)_k^{(1)} {}_t\Delta U_{lz}^{kl} = {}_t\Delta u^*,$$

$$\sum_{k=1}^{m} \left({}_t\Delta\lambda\right)_k^{(i)} {}_t\Delta U_{lz}^{kl} + {}_t\delta u_{lz}^{(i)ll} = 0 \qquad \text{for } i = 2,3,\ldots\ldots$$

(6.121)

where $_t\Delta U_{lz}^{kl}$ and $_t\delta u_{lz}^{(i)ll}$ are the z-components of the vectors $_t\{\Delta U\}^{kl}$ and $_t\{\delta u\}^{(i)ll}$ at node l. By solving Equation 6.121, we get the iterative values of $\left({}_t\Delta\lambda\right)_k^{(i)}$ for $i = 1,2,\ldots$ Then, the iterative solution is obtained by substituting $\left({}_t\Delta\lambda\right)_k^{(i)}$ in Equation 6.118 and the iterative nodal reaction by Equation 6.117.

6.3.3 Friction Algorithm

The friction boundary conditions of Equation 6.112 can be applied only in an iterative fashion. Thus, besides the Newton-Raphson iterations, there is an additional set of iterations in the solution of the forging problem. First, for the specified incremental platen displacement, the friction iterations are carried out to determine the status (sticking or slipping) of the interface nodes. Then the Newton-Raphson iterations are carried out to minimize the unbalanced force vector. The stepwise algorithm for the friction iterations can be described as follows.

(i) *First Friction Iteration*:

- In the first iteration, all the interface nodes are assumed to be in sticking condition. Thus, the sticking boundary condition given by Equation 6.112b is applied to all the interface nodes.
- Next, the finite element equation (Equation 6.114) is solved to find $_t\{\Delta U\}^{(1)}$, i.e., the initial guess to the incremental displacement vector.
- Next, $_t\{\Delta F\}_{ex}^{(1)}$, i.e., the iterative nodal reaction vector of the first iteration, is found from Equation 6.117 and the nodal reaction vector $^{t+\Delta t}\{F\}_{ex}^{(1)}$ is updated using Equation 6.113.
- At the end of first friction iteration, the slipping nodes at the interface are identified using the following condition. Node l at the interface slips if

$$\left| \left(^{t+\Delta t}F_{ex}^{(1)}\right)_{lr} \right| > f \left| \left(^{t+\Delta t}F_{ex}^{(1)}\right)_{lz} \right|,$$

(6.122)

where $(^{t+\Delta t}F_{ex}^{(1)})_{lr}$ and $(^{t+\Delta t}F_{ex}^{(1)})_{lz}$ are the r and z components of the vector $^{t+\Delta t}\{F\}_{ex}^{(1)}$ at node l.

(ii) *Second Friction Iteration*:

- Now, the slipping boundary condition (Equation 6.112a) is applied to all the slipping nodes. Note that this condition is to be applied to the components of the nodal reaction vector $^{t+\Delta t}\{F\}_{ex}^{(1)}$. In terms of r and z components at a slipping node l, the condition at Equation 6.112a becomes

$$(^{t+\Delta r}F_{ex}^{(1)})_{lr} - f(^{t+\Delta t}F_{ex}^{(1)})_{lz} = 0.$$ (6.123)

We assume that similar condition is satisfied by the components of $_l\{\Delta F\}_{ex}^{(1)}$, *i.e.*, the iterative nodal reaction vector of the first iteration

$$(_l\Delta F_{ex}^{(1)})_{lr} - f(_l\Delta F_{ex}^{(1)})_{lz} = 0.$$ (6.124)

Since, r and z components of $_l\{\Delta F\}_{ex}^{(1)}$ at node l occupy respectively the $(2l-1)$-th and $(2l)$-th rows, Equation 6.124 means that the right sides of $(2l-1)$-th and $(2l)$-th equations in the set of Equation 6.114 are not known but they are related by this condition. To solve the set at Equation 6.114, the right sides of both these equations should be known. Equation 6.124 implies that if we replace either of these equations by an equation consisting of $(2l-1)$-th equation of the set minus f times $(2l)$-th equation, then the right side of that equation is zero and therefore known. Thus, we apply the condition at Equation 6.124 by performing the following operations on the coefficient matrix $^t[K]$ and the right side vector $_l\{\Delta F\}_{ex}^{(1)}$. We replace the $(2l-1)$-th row of $^t[K]$ by a combination of $(2l-1)$-th row minus f times $(2l)$-th row and the corresponding row of $_l\{\Delta F\}_{ex}^{(1)}$ by zero. We make the right side of $(2l)$-th equation known by replacing it with Equation 6.111.

- Then, the finite element equation (Equation 6.114) is solved again to find the new initial guess to the incremental displacement vector.
- Next, the new iterative nodal reaction vector of the first iteration is found from Equation 6.117 and the nodal reaction vector is updated afresh using Equation 6.113.
- Finally, a check is made to find whether any additional nodes are slipping by using Equation 6.122.

(iii) *Further Friction Iterations*:
- The second friction iteration is repeated until the condition

$$\left|(^{t+\Delta t}F_{ex}^{(1)})_{lr}\right| \le f\left|(^{t+\Delta t}F_{ex}^{(1)})_{lz}\right|,\qquad(6.125)$$

is satisfied at all the nodes.

6.3.4 Convergence Study and Evaluation of Secondary Variables

Eight-noded isoparametric brick elements are used to discretize the domain. A non-uniform mesh is used by placing smaller elements in the vicinity of the edge of the interface (*i.e.*, point B of Figure 6.1) as this happens to be a high gradient region. Convergence studies are carried out to select proper mesh and increment sizes. It is observed that good convergence is achieved at 10×10 mesh with an increment size of 0.05 mm.

From the incremental displacement vector, the Cauchy stress tensor (at the Gauss points of the elements) is updated using the procedure described in Section 6.2.6. Then the equivalent stress is calculated using Equation 3.23. The procedure for obtaining the contact pressure is as follows. First, the Cauchy stress tensor is extrapolated from the Gauss points to the corresponding points on the interface *AB* of Figure 6.1. Then the stress vector at these points is calculated using the Cauchy's relation (Equation 2.64). Finally, the contact pressure at these points is obtained as the normal component of the stress vector. The forging load is calculated either by integrating the contact pressure or by adding the z-components of the reactions at the interface nodes.

The equivalent plastic strain increment is calculated using Equation 3.93 and the equivalent plastic strain is obtained by replacing the integral of Equation 3.97 by a sum. Thus, the equivalent plastic strain is determined as the sum of the equivalent plastic strain increments. The deformed mesh is obtained by updating the nodal coordinates.

After the desired reduction is achieved, the reactions (on the interface *AB*) are reduced to zero to obtain the (equivalent) residual stress distribution. Thus, the boundary conditions at the interface, during unloading, are different to those given by Equations 6.110 and 6.111. The unloading is carried out in one increment only as, now, the material behavior is elastic.

The results reported in Subsections 6.3.5–6.3.7 are mostly from [28].

6.3.5 Validation of the Finite Element Formulation

A static, large deformation, elastic plastic, finite element (FE) code is developed based on the formulation described in Section 6.2. The code is validated by comparing the variation of forging load with percentage reduction with experimental results of Shima *et al.* [15]. The material properties of the work-piece, the friction coefficient at the interface and the blank size (*i.e.*, the initial size of the work-piece) used in the experiments of [15] are as follows:

Material Properties

Material : JIS S25C Steel,

Young's modulus (E) : 208 GPa,

Poisson's ratio (v): 0.3,

Yield stress (σ_Y): 380.14 MPa,

Hardening coefficient (K) = 467.44 MPa,

Hardening exponent (n) =0.6;

Friction Coefficient

Friction coefficient (f) = 0.25;

Blank Size

Height (H) = 20 mm,

Diameter (D) = 20 mm.

Figure 6.2 shows the comparison of forging load variation with percentage reduction obtained from the code with the experimental results of [15]. The maximum error between the experimental and FE results is observed to be 12.5%. The possible reason for the error could be as follows. The FE code uses the power law model of the hardening curve. However, for the material of [15], it was not possible to model its hardening curve accurately by a power law. Thus, the values of the hardening parameters K and n represent only an approximate response of the material. For this reason, it was decided to change the material for obtaining other results.

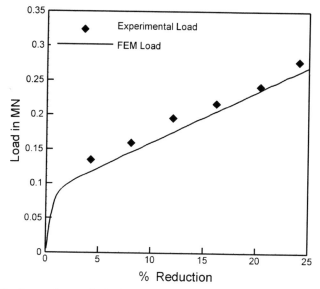

Figure 6.2. Comparison of forging load variation with percentage reduction with experimental result [15]. With permission from Mungi *et al.* [28]. Copyright 2003 Elsevier

6.3.6 Typical Results

Results of this section are obtained by using AISI 1015 steel as the material whose properties are given in [20]. The values of the hardening parameters K and n mentioned below have been obtained by fitting Equation 3.103 through the hardening curve of [20]. In this section, some results are obtained by assuming that the friction at the interface is of sticking type all along the interface. In this case, the second boundary condition of the set at Equation 6.112 is applied at all the interface nodes. This condition is referred to as *complete sticking condition* in the remaining discussion whereas the other condition is referred to as sticking-slipping condition. The material properties of the work-piece, the friction condition at the interface, the blank size and the percentage reduction used in this section are as follows:

Material Properties [20]
Material : AISI 1015 Steel,
Young's modulus (E) : 208 GPa,
Poisson's ratio (v) : 0.3,
Yield stress (σ_Y) : 275 MPa,
Hardening coefficient (K) = 515.23 MPa,
Hardening exponent (n) =0.6;
Friction Condition
Friction coefficient (f) = 0, 0.1, 0.2, 0.3,
Complete sticking condition;
Blank Size
Height (H) = 20 mm,
Diameter (D) = 20 mm;
Percentage Reduction
Percentage reduction (%r) : 20, 35.

Figure 6.3 shows the contact pressure distribution at various values of f as well as for the complete sticking condition. The contact pressure for the complete sticking condition is minimum at the center of the interface (*i.e.*, at point A of Figure 6.1) and attains the maximum value at the edge of the interface (*i.e.*, at point B of Figure 6.1) which is due to the complete sticking. For the sticking-slipping case, the pressure starts developing a peak at the edge of the interface (*i.e.*, at point B of Figure 6.1) with an increase in f. This trend is not predicted by the slab method.

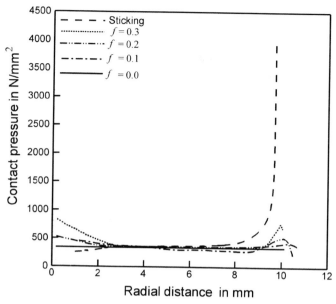

Figure 6.3. Contact pressure distribution at 20% reduction at various values of f and for complete sticking condition

The deformed configurations for $f = 0.2$ and the complete sticking condition are shown in Figure 6.4a,b respectively. It is observed that the deformation is homogeneous at low values of f. As f is increased, the deformation becomes more inhomogeneous, resulting in a larger bulge. Further, the phenomenon of fold over appears at the edge of the interface (*i.e.*, at point B of Figure 6.1) at high values of f and for the complete sticking condition. Figure 6.5a,b shows the equivalent strain distributions respectively for $f = 0.2$ and the complete sticking condition. The distributions are presented without showing the bulge of the domain. It is seen from these figures that the distribution pattern does not change with f. However, the equivalent strain values increase with f everywhere except near the center of the interface (*i.e.*, near point A of Figure 6.1) where they decrease with f. With an increase in f, the equivalent strain levels approach to that of the complete sticking condition. Figure 6.5a indicates that the deformation is more or less homogeneous in the middle of the domain. Further, the figure indicates that it is severe near the edge of the interface (*i.e.*, near point B of Figure 6.1) but much less at the center of the interface (*i.e.*, at point A of Figure 6.1)

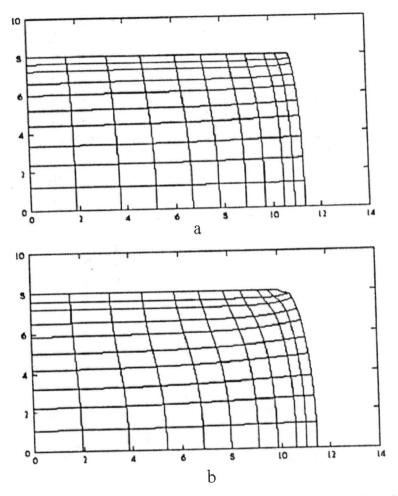

Figure 6.4. Deformed configuration at 20% reduction a. $f = 0.2$; b. Complete sticking condition

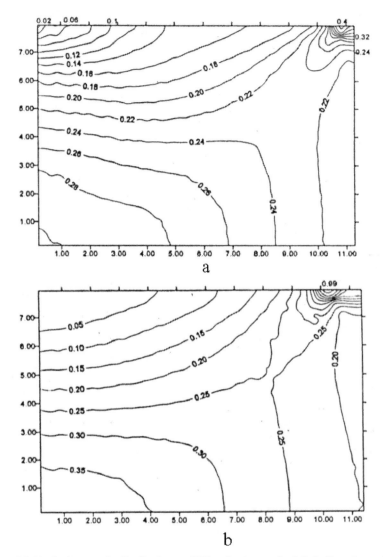

Figure 6.5. Equivalent strain distribution at 20% reduction. **a** $f = 0.2$. **b** Complete sticking condition

The equivalent stress distribution at $f = 0.1$ and $\%r = 35$ is shown in Figure 6.6. Like the equivalent strain distribution, this distribution is also presented without showing the bulge of the domain. Since the equivalent stress and equivalent strain are related by the hardening relation (Equation 3.103), the equivalent stress distribution is similar to that of the equivalent strain. Further, since the deformation is more or less homogeneous in the middle of the domain, the equivalent stress also does not vary much in the middle of the domain. Additionally, since the deformation is severe near the edge of the interface but much less at the center of the interface, the equivalent stress possesses high gradients near the edge of the interface (near point B of Figure 6.1) and much lower values at the center of the interface (at point A of Figure 6.1). In fact, it is observed that, at lesser reduction, the equivalent stress level at the center of the top surface does not reach the yield stress of the material. The maximum values of equivalent strain and equivalent stress occur near point B of Figure 6.1 while their minimum values occur at point A of Figure 6.1.

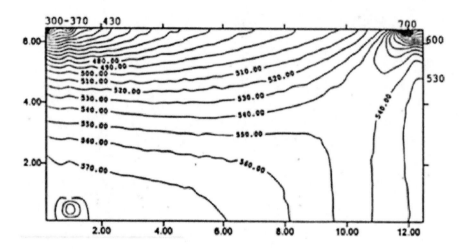

Figure 6.6. Equivalent stress distribution at 35% reduction and $f = 0.1$. With permission from Mungi *et al.* [28]. Copyright 2003 Elsevier

Since the process has been analyzed incrementally, one can observe where the plastic zone originates and how it spreads. As the platen moves down (or as the reduction increases), yielding (equivalent stress = 275 MPa) first occurs at the edge of the interface (*i.e.*, at point B of Figure 6.1). With further movement of the platen, yielding occurs at the center of the work-piece (*i.e.*, at point D of Figure 6.1). With more increase in reduction, the plastic zone spreads both in the radial as well as in the axial direction from these two points.

6.3.7 Residual Stress Distribution

Figure 6.7 shows the equivalent residual stress distribution for AISI 1015 steel, up to 35% reduction, for $H/D = 1$ (*i.e.*, height = diameter = 20 mm) and for $f = 0.1$. Like the equivalent strain distribution, this distribution is also presented without showing the bulge of the domain. It is observed that the values of equivalent residual stress are quite small (in fact, much smaller than the yield stress value of 275 MPa) almost everywhere except in a small region near the edge of the interface (*i.e.*, near point B of Figure 6.1). This is because the deformation is almost homogeneous everywhere except near point B of Figure 6.1.

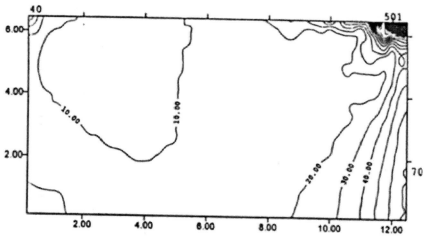

Figure 6.7. Equivalent residual stress distribution at 35% reduction, $H/D = 1$ and $f = 0.1$
With permission from Mungi *et al*. [28]. Copyright 2003 Elsevier

Parametric study of the residual stress distribution is carried out by varying the following four input variables: (i) height to diameter ratio (*i.e., H/D* ratio), (ii) percentage reduction (%*r*), (iii) friction coefficient (*f*) and (iv) the material properties (σ_Y, *K, n; E, v*). The materials chosen are AISI 1015 steel (whose properties are given in Subsection 6.3.6) and AI 1100 aluminum (whose properties are given later). The blank size is chosen such that its volume is equal to that of the blank of height (*H*) and diameter (*D*) equal to 20 mm each. To avoid confusion, the equivalent stress during loading is called the equivalent loading stress.

Effect of height to diameter ratio (H/D):
The pattern of equivalent loading stress distribution does not change much with *H/D* ratio, but the values change differently in different regions. As the *H/D* ratio is increased, the stress values at the center of the interface (*i.e.*, at point A of Figure 6.1) decrease leading to the formation of an elastic zone there. But, the stress values at the center of the work-piece (*i.e.*, near point D of Figure 6.1) increase. Thus, the equivalent loading stress distribution becomes more inhomogeneous in the *axial* direction but only at the core of the work piece. Therefore, the pattern of the equivalent residual stress distribution changes in this region. But, since the equivalent residual stress values are so low in this region, this observation is not of much significance. However, neither the pattern of the equivalent loading stress distribution nor the stress values change at the edge of the interface (*i.e.*, near point B of Figure 6.1). As a result, the maximum value of the equivalent loading stress does not change with the *H/D* ratio. But the maximum value of the equivalent residual stress decreases with the *H/D* ratio.

Effect of percentage reduction (%r):
At low reduction, during loading, an elastic zone is observed near the center of the interface (*i.e.*, near point A of Figure 6.1). With increase in reduction, this elastic zone gets contracted and eventually vanishes at higher reduction. Because of this, the residual stress pattern gets changed at the core of the work-piece only. Again, this result is not of any significance as the residual stress values in this region are

quite low. Further, as expected, both the equivalent stress values (loading as well as residual) increase with reduction everywhere in the domain.

Effect of Friction Coefficient (f):

To study the effect of f, the analysis is carried out for AISI steel and for $H/D = 1$ (*i.e.*, height = diameter = 20 mm). Further, the analysis is carried out only up to 20% reduction, as beyond 20% reduction the reactions at the edge of interface become positive at higher values of f. The equivalent loading stress distributions for $f = 0.2$ and the complete sticking condition are shown in Figure 6.8a,b respectively. The corresponding equivalent residual stress distributions are shown in Figure 6.9a,b.

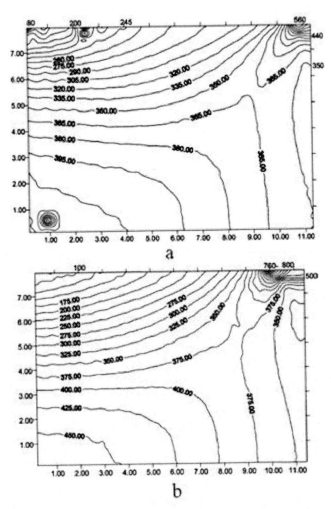

Figure 6.8. Equivalent (loading) stress distribution at 20% reduction and $H/D = 1$. **a** $f = 0.2$; **b** Complete sticking condition. With permission from Mungi *et al.* [28]. Copyright 2003 Elsevier

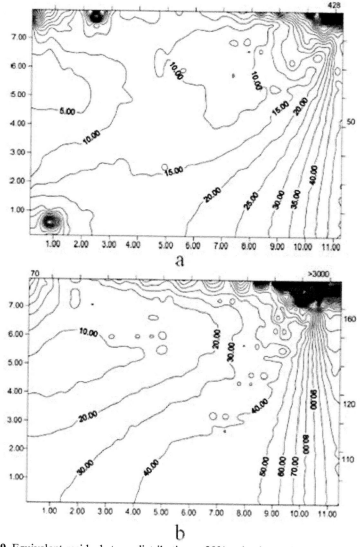

Figure 6.9. Equivalent residual stress distribution at 20% reduction and $H/D = 1$. **a** $f = 0.2$.
b Complete sticking condition. With permission from Mungi *et al.* [28]. Copyright [2003]
Elsevier

As f is increased, the equivalent loading stress values increase everywhere
except near the center of the interface (*i.e.*, near point A of Figure 6.1) where the
values actually decrease. As a result, an elastic zone is formed there. Further, the
location of the maximum equivalent loading stress starts shifting away from the
edge of the interface (*i.e.*, from point B of Figure 6.1). Because of this and the
increase in the value of maximum equivalent loading stress, the high stress
gradient region near the edge of the interface becomes more intense. Thus, the
equivalent loading stress distribution becomes more inhomogeneous and

approaches to the pattern corresponding to the complete sticking condition as f is increased.

Since the equivalent loading stress distribution becomes more inhomogeneous with f, the equivalent residual stress values increase everywhere in the domain. This increase is more severe in the high stress gradient region near the edge of the interface (*i.e.*, near point B of Figure 6.1). In fact, the maximum equivalent residual stress becomes even larger than the maximum equivalent loading stress. This could be the consequence of the reactions at the edge of interface becoming tensile at this level of friction. To make a realistic prediction of the residual stresses at this level of friction, the analysis needs to be repeated after deleting such nodes from the possible contact zone. This analysis will involve an additional set of iterations to determine the possible contact zone. Finally, at higher f, the equivalent residual stress distribution approaches to the pattern corresponding to the complete sticking condition.

Effect of Material Parameters:
The effect of material properties is studied by carrying out the analysis for one more material, namely, AI 1100 aluminum. The analysis is carried out up to 35% reduction, for $H/D = 1$ (*i.e.*, height = diameter = 20 mm) and for $f = 0.1$. The material properties of AI 1100 aluminum are taken from [31]. The values of the hardening parameters K and n mentioned below have been obtained by fitting Equation 3.103 through the hardening curve of [31].

Material Properties
Material : AI 1100 Aluminum,
Young's modulus (E) : 69 GPa,
Poisson's ratio (v) : 0.3,
Yield stress (σ_Y) : 62.74 MPa,
Hardening coefficient (K) = 110.1 MPa,
Hardening exponent (n) =0.68.

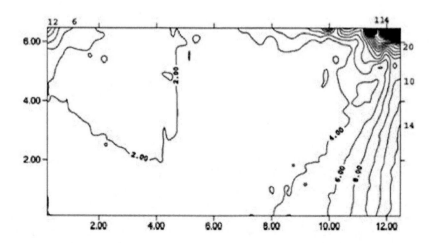

Figure 6.10. Equivalent residual stress distribution in AI 1100 aluminum at 35% reduction, $H/D = 1$ and $f = 0.1$. With permission from Mungi *et al.* [28]. Copyright 2003 Elsevier

Figure 6.10 shows the equivalent residual stress distribution for Al1100 aluminum. Comparison of this figure with Figure 6.7 indicates that the distribution pattern is the same for both materials. Thus, the material properties have no effect on the pattern of equivalent residual stress distribution. However, as expected, the equivalent residual stress values are less for aluminum than for steel. Similar trend is observed for the equivalent loading stress as well.

6.3.8 Damage Distribution, Hydrostatic Stress Distribution and Fracture

To study fracture in the forging problem, two approaches are followed. In the first approach, the continuum damage mechanics model of Lemaitre and Chaboche [32] is used. Since for most metals, the value of damage D up to micro-crack initiation is quite small, we assume that the damage does not affect the constitutive equation of the material. Hence, the deformation and stress fields are determined using the constitutive equation (Equation 6.2) of the undamaged material. The damage increment is calculated using the damage evolution law (Equation 4.219) proposed by Dhar *et al.* [33]:

$$_t\Delta D = c(_t\Delta\varepsilon_{eq}^p) + (a_1 + a_2 \,^t D)(-{}^t Y)(_t\Delta\varepsilon_{eq}^p). \tag{6.126}$$

Here, (c, a_1, a_2) are the material parameters, $^t D$ is the damage at time t, $_t\Delta D$ is the damage increment at time t, $_t\Delta\varepsilon_{eq}^p$ is the equivalent plastic strain increment (Equation 3.93) at time t and $(-{}^t Y)$ is the dissipative part of the thermodynamic force corresponding to the time rate of change of damage (at time t) given by (Equation 4.208):

$$-{}^t Y = \frac{{}^t\sigma_{eq}^2}{2E(1-{}^t D)^2}\left[\frac{2}{3}(1+v) + 3(1-2v)\left(\frac{{}^t\sigma_h}{{}^t\sigma_{eq}}\right)^2\right]. \tag{6.127}$$

Here, E and v are the elastic constants of the material, $^t\sigma_h$ is the hydrostatic part of the stress tensor at time t and $^t\sigma_{eq}$ is the equivalent stress at time t (Equation 3.23). The fracture criterion used is that whenever the damage reaches the critical value D_c at a point, a micro-crack initiates at that point.

In the second approach, the hydrostatic stress criterion proposed by Reddy *et al.* [34] is used. As per this criterion, whenever the hydrostatic stress at a point reaches the value zero, fracture initiates at that point. Note that the critical damage criterion predicts the initiation of a micro-crack while the hydrostatic stress criterion predicts fracture at a macro scale.

Results of this section are obtained for AISI 1090 steel as the material constants needed in the damage evolution law are available for this material [33]. The other material properties are given in [35]. The material properties of the work piece, the friction condition at the interface, the blank size and the percentage reduction used in this section are as follows:

Material Properties [35, 33]
Material: AISI 1090 Steel,
Young's modulus (E): 210 GPa,
Poisson's ratio (ν): 0.3,
Yield stress (σ_Y): 464 MPa,

Hardening relation: $\sigma_{eq} = 1115(\varepsilon_{eq}^p)^{0.19}$,

Parameters in damage evolution law and micro-crack initiation criterion:

$c = 1.898 \times 10^{-2}, a_1 = 9.8 \times 10^{-4} (\text{MPa})^{-1}, a_2 = 1.84 (\text{MPa})^{-1}, D_c = 0.05$;

Friction Condition
Friction coefficient (f) = 0.05;
Blank Size
Height to diameter ratio (H/D) = 0.5, 1, 2;
Percentage Reduction
Percentage reduction (%r): 25–35.
The results reported in this subsection are mostly from [27].

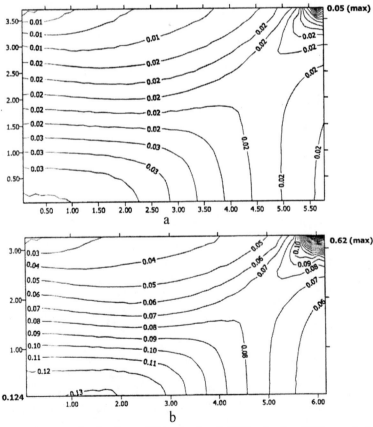

Figure 6.11. Damage distribution. **a** 25% reduction. **b** 35% reduction. With permission from Gupta *et al.* [27]. Copyright 2003 Elsevier

The damage distributions at 25% and 35% reductions are shown in Figure 6.11a,b respectively. This figure shows that the maximum damage occurs at the edge of the interface (*i.e.*, at point B of Figure 6.1). This is because of the severe deformation and high stress gradients present there. The micro-crack initiation first occurs at the edge of the interface at 25% reduction, then at the center of the work-piece (*i.e.*, at point D of Figure 6.1) at 29% reduction and finally at the meridian surface (*i.e.*, at point C of Figure 6.1). At 35% reduction, the damage is above D_c everywhere except near the center of the interface (*i.e.*, near point A of Figure 6.1). The above observation shows that the micro-crack first occurs at the edge of the interface, then at the center of the work-piece and finally at the meridian surface. This observation is in agreement with that of Predeleanu *et al.* [36]. But experimental results [25, 37–39] show that the fracture occurs at the meridian surface where the damage reaches the critical value much later. To understand this phenomenonon, the distribution of the hydrostatic (σ_h) stress is considered.

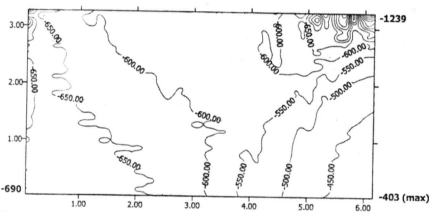

Figure 6.12. Hydrostatic stress distribution at 35% reduction.With permission from Gupta *et al.* [27]. Copyright 2003 Elsevier

Figure 6.12 shows the distribution of hydrostatic stress (σ_h) at 35% reduction. It is observed, from this figure, that the hydrostatic stress is negative everywhere. However, amongst the three locations, namely, the edge of the interface, the center of the work-piece and the meridian surface, it is less negative at the meridian surface. Further, as the reduction is increased, the hydrostatic stress becomes less negative at the meridian surface and more negative at the other two locations. Thus, the possibility of a fracture is higher at the meridian surface compared to the other two locations. Note that the fracture at the meridian surface can be caused either by the axial stress (σ_{zz}) or by the circumferential stress ($\sigma_{\theta\theta}$). However, it is observed that the axial stress is compressive everywhere in the domain [27] and continues to become more compressive with the increase in reduction. Thus, it cannot cause any fracture at the meridian surface.

The circumferential stress distribution ($\sigma_{\theta\theta}$) at 35% reduction is shown in Figure 6.13. It is observed that $\sigma_{\theta\theta}$ is tensile at the meridian surface but compressive at the other two locations (namely, the center of the block and the edge of the interface). Further, $\sigma_{\theta\theta}$ becomes more tensile at the meridian surface and more compressive at the other two locations as the reduction is increased. Thus, the micro-cracks originated at the meridian surface grow with the reduction leading to a fracture.

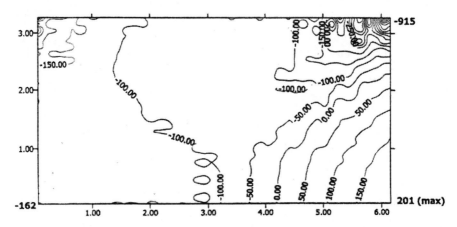

Figure 6.13. Circumferential stress distribution at 35% reduction. With permission from Gupta *et al.*[27]. Copyright [2003] Elsevier

It is observed that, at higher values of f (friction coefficient), the damage D reaches the critical value (D_c) at lesser reduction everywhere in the domain. This happens, because the deformation becomes more inhomogeneous with an increase in f. Further, at higher values of f, the hydrostatic stress (σ_h) becomes less compressive (reaching almost the zero value) and the circumferential stress ($\sigma_{\theta\theta}$) becomes more tensile at the meridian surface. Thus, the fracture occurs at lesser reduction at higher friction.

When the height to diameter ratio (*i.e.*, *H/D* ratio) is either increased or decreased from the value 1, the damage D reaches the critical value at lesser reduction everywhere. Thus, the fracture (at the meridian surface) occurs at lesser reduction when the *H/D* ratio is either increased or decreased from the value 1. However, at higher values of *H/D* ratio, the hydrostatic stress (σ_h) becomes less compressive and the circumferential stress ($\sigma_{\theta\theta}$) becomes more tensile at the center of the work piece, thereby creating a possibility of fracture at that location also (*i.e.*, a central cavity).

6.4 Modeling of Deep Drawing of Cylindrical Cups by Updated Lagrangian Finite Element Method

In deep drawing, the work-piece in the form of a flat sheet (called blank) is forced into a die by means of a punch to form a hollow component (Figure 6.14). Usually, a blank-holder is used to prevent any wrinkling taking place in the flange. The process is carried out without the blank-holder only if there is no possibility of wrinkling, *i.e.*, either the sheet is thick or the deformation is small. In this process, the material bends over the punch and die radii, stretches between the punch contact to die contact and is drawn over the die surface. In the flange region (*i.e.*, over the die surface), the material gets compressed in the tangential direction and stretched in the radial direction. Thus, the deformation and stress distributions are quite complex. In this book, only cylindrical cup drawing will be discussed as it is the simplest deep drawing process.

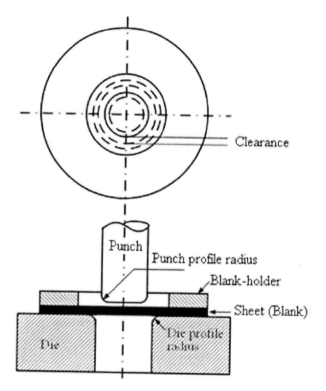

Figure 6.14. Deep drawing process

Since the deep drawing process involves complex deformation and stress distributions, the complete process has not been analysed by simple techniques like the slab method, the upper bound method or the slip line method. Only the analysis of the flange region has been carried out using the simpler techniques [40, 41].

In the simulation of sheet metal forming processes, several finite element models such as membrane, shell and solid (3-D) models have been proposed. The membrane model is the simplest one. Toh and Kobayashi [42] used the membrane elements to determine the optimal blank shape in the square cup drawing process. They used a finite strain formulation with zero blank-holder force. Saran and Samuelsson [43] modeled the behavior of sheet materials by triangular constant strain membrane elements. They used hypoelastic viscoplastic material model with Hill's anisotropic criterion, power law hardening and strain rate sensitivity. Majlessi and Lee [44] also used the membrane elements for the analysis of non-axisymmetric deep drawing problem. The membrane model neglects the stress variation in the thickness direction. As a result, it is not appropriate in problems like deep drawing where the bending deformation is significant.

Onate and Saracibar [45] used viscous voided shell elements to propose a finite element formulation for sheet metal forming. Chou et al. [46] simulated sheet metal forming using plane strain shell elements. Their formulation uses a stress resultant constitutive law based on a quadratic yield function, a hardening rule and the associated flow rule. Shi et al. [47] used the DKT (discrete Kirchoff triangular) shell elements to predict the desired blank shape based on one step simulation algorithm. Since their formulation involves integration in the thickness direction, it takes almost as much computational time and memory as the formulation with solid elements. Chou et al. [48] employed shell elements for the analysis of forming of sheets with planar isotropy. A stress resultant constitutive law based on Hill's anisotropic yield function is used in their formulation. In the shell model, only the middle surface of the sheet is considered as the domain. In that case, it becomes difficult to consider the contact conditions at the sheet-punch or sheet-die interfaces as these conditions apply at the outer surfaces of the sheet. The contact conditions can be easily handled in the solid model.

Menezes and Teodosiu [49] used eight-noded solid elements to simulate deep drawing. Their formulation uses augmented Lagrangian method, Jaumann stress rate tensor and Green-Lagarnge strain tensor. They have reported that the deformation obtained using solid elements is more realistic. Further, the number of elements required to obtain a realistic result is lesser with the 3-D model than the shell model. Colgan and Monaghan [50] used the FEA program AutoForm (which employs solid elements) to model a cup formation. They tried to determine the most important factors influencing the drawing process utilizing the design of experiments and statistical analysis.

The predominant failure modes in the deep drawing process are wrinkling of flanges and thinning and fracture of walls. In many cases, it is possible to eliminate these defects by an appropriate choice of the blank-holder force as this force controls the material flow in the die cavity. Osakada et al. [51] proposed a control algorithm in the FEM program to determine the optimum blank-holder force so as to avoid wrinkling and thinning. Lorenzo et al. [52] suggested a closed loop control sustem based on fuzzy logic which is interfaced with an FEM code. The control system continuously monitors some relevant process parameters and suggests the most effective adjustment of the blank-holder force so as to obtain maximum height without wrinkling or tearing.

When the material of the sheet (also called the blank) is anisotropic, the final cup does not possess the uniform height. This phenomenon is called as *earing*. The initial shape of the blank which minimizes the earing is called the optimum blank shape. Determination of the optimum blank shape is an important part of the blank design. Various techniques exist to obtain the optimum blank shape. Among them, the inverse approach is a name given to the method which locates the positions of the material points on the intial blank from their corresponding positions on the final product. Guo *et al.* [53] proposed this method to evaluate the (large) plastic strains in the deep drawing process. They used triangular membrane elements and the deformation theory of plasticity. Kim and Huh [54] applied the inverse approach method to the multi-stage deep drawing process to determine the optimum blank shape from the desired final shape. The backward tracing algorithm is another method for obtaining the optimum blank shape. In this method, the final desired configuration is traced backward either to an intermediate pre-form or to the initial blank. Ku *et al.* [55] applied the backward tracing algorithm to blank design in three-dimensional forming. Pegada *et al.* [56] used the LS-DYNA package to apply the backward tracing algorithm to obtain the optimum blank shape. They used Barlat's planar anisotropic yield criterion and Belytschko-Lin-Tsay elements.

6.4.1 Domain and Boundary Conditions

For anisotropic materials, the problem of circular cup drawing is not an axisymmetric problem. Therefore, it is analyzed as a three-dimensional problem. However, because of the orthotropy of the sheet and the symmetry of the geometry and boundary conditions about the two axes, only a quarter of the sheet is selected as the domain. The domain at time $t = 0$, along with the coordinate system, is shown in Figure 6.15. The punch, die and blank-holder are assumed rigid. Sticking friction condition is assumed at the punch-sheet interface while the friction at the die-sheet interface is modeled by the Coulomb's law. Again, it is assumed that the friction coefficient is such that all the nodes at the die-sheet interface slip. Since a loss of contact at the punch-sheet interface is possible, it is incorporated in the boundary conditions. It is assumed that enough blank-holder force is applied to maintain the contact at the die-sheet interface. Further, the blank-holder force is assumed to be unifomly distributed and applied in incremental fashion.

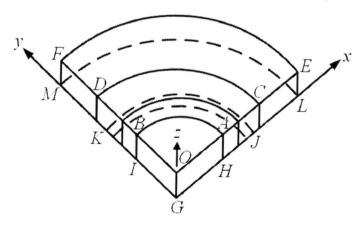

Figure 6.15. Domain of the problem. It is the initial configuration of a quarter sheet.

The boundary conditions *at the start of the analysis* are as follows. Whether the condition is essential or natural is also mentioned against each condition.

Boundary conditions at the start of analysis:
Sheet-punch interface (surface OAB):
As stated earlier, a node at the interface may be in contact with the punch or may lose the contact. It depends on the nature (compressive or tensile) of the z-component of the punch reaction vector at the node. The algorithm to determine the contact status of the node is described in the next subsection. When a node is in contact, the sticking friction condition is assumed as stated earlier, *i.e.*, x and y components of the incremental displacement vector $_t\Delta u$ at the node become zero. Further, the z-component of $_t\Delta u$ must be equal to the incremental punch displacement Δu^*. If a node loses contact, the free surface boundary condition applies, *i.e.*, all the three components of the incremental stress vector $_t\Delta t$ become zero at the node. Thus, the boundary conditions on the surface OAB become:
At contacting node:

$$_t\Delta u_x = 0, \; _t\Delta u_y = 0, \quad _t\Delta u_z = \Delta u^*, \quad \text{(essential)}. \tag{6.128a}$$

At non-contacing node:

$$_t\Delta t_x = 0, \; _t\Delta t_y = 0, \; _t\Delta t_z = 0, \text{(natural)}. \tag{6.128b}$$

Free surfaces ABDCA, EFMLE and GJKG:
At a free surface, all three components of the incremental stress vector $_t\Delta t$ are zero at the node. Thus, the boundary conditions on the surfaces ABDCA, EFMLE and GJKG are given by

$$_t\Delta t_x = 0, \; _t\Delta t_y = 0, \; _t\Delta t_z = 0, \text{(natural)}. \tag{6.129}$$

Plane of symmetry OACELJHGO (in x-z plane):
At a plane of symmetry, the normal component of the incremental displacement vector $_t\Delta u$ and both the shear components of the incremental stress vector $_t\Delta t$ are zero. Since the plane OACELJHGO lies in x-z plane, the boundary conditions on this surface become

$$_t\Delta t_x = 0, \; _t\Delta t_z = 0, \quad \text{(natural)}, \tag{6.130a}$$

$$_t\Delta u_y = 0, \text{(essential)}. \tag{6.130b}$$

Plane of symmetry OBDFMKIGO (in y-z plane):
As stated above, at a plane of symmetry, the normal component of the incremental displacement vector $_t\Delta u$ and both the shear components of the incremental stress

vector $_t\Delta t$ are zero. Since the plane OBDFMKIGO lies in y-z plane, the boundary conditions on this surface are given by

$$_t\Delta u_x = 0, \text{ (essential)}, \tag{6.131a}$$

$$_t\Delta t_y = 0, \quad _t\Delta t_z = 0, \text{ (natural)}. \tag{6.131b}$$

Sheet-die interface (surface JKMLJ):
As stated earlier, the die-sheet interface friction is modeled by the Coulomb's law. Again, it is assumed that the friction coefficient f is such that all the nodes at the interface slip. Then, the magnitude of the frictional stress (*i.e.*, the resultant of $_t\Delta t_x$ and $_t\Delta t_y$) at the interface becomes $\left| f(_t\Delta t_z) \right|$, where $_t\Delta t_z$ is the normal component of the incremental stress vector $_t\Delta t$. Let θ be the angle made by the frictional stress with positive x-axis, then the magnitudes of $_t\Delta t_x$ and $_t\Delta t_y$ become

$$\left| _t\Delta t_x \right| = f \left| _t\Delta t_z \right| \cos\theta, \qquad \left| _t\Delta t_y \right| = f \left| _t\Delta t_z \right| \sin\theta. \tag{6.132}$$

Note that $_t\Delta t_z$ is positive at the interface. Further, since the relative movement of the contact node is in the negative x and y directions, the x and y components of the frictional stress would be in the positive x and y directions. Thus, $_t\Delta t_x$ and $_t\Delta t_y$ also would be positive. Therefore, the boundary conditions in x and y directions, on the surface JKMLJ, become

$$_t\Delta t_x = f(_t\Delta t_z)\cos\theta, \quad _t\Delta t_y = f(_t\Delta t_z)\sin\theta, \quad \text{(natural)}. \tag{6.133a}$$

The third boundary condition is provided by the assumption that enough blank-holder force is applied to maintain the contact of the sheet with the die. Then the normal component (*i.e.*, the z-component) of the incremental displacement vector $_t\Delta u$ must be zero at the interface. Thus, the boundary condition in z-direction, on the surface JKMLJ, is given by

$$_t\Delta u_z = 0, \text{ (essential)}. \tag{6.133b}$$

Surface with blank-holder force (surface CDFEC):
The blank-holder force is applied on the portion CDFEC of the top surface of the sheet. As stated earlier, it is assumed that the blank-holder force is unifomly distributed and applied in incremental fashion. Therefore, the boundary conditions on the surface CDFEC become

$$_t \Delta t_x = 0, \qquad _t \Delta t_y = 0, \qquad _t \Delta t_z = \Delta t^* \quad \text{(natural)}, \qquad (6.134)$$

where Δt^* is the specified value of the incremental blank holding force per unit area.

Change in boundary conditions with punch movement:

As the punch keeps moving down, some nodes under the blank-holder surface move out of the blank-holder and become free. Also, some nodes on the (flat) die surface move on to the die profile radius region. Further, some nodes on the free surface ABDCA move on to the punch profile radius region. Therefore, the boundary conditions of such nodes need to be changed at the end of the increment, whenever necessary. To facilitate the incremental updation of these boundary conditions, the top and bottom surfaces of the sheet are divided into the regions having the same boundary condition as shown in Figure 6.16. The nodes on the top and bottom surfaces are assigned a code depending on the region in which they lie. If, at the end of a particular increment, a node crosses from one region to the neighboring region, its code is changed appropriately.

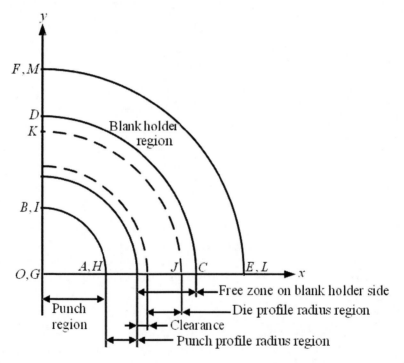

Figure 6.16. Division of the top and bottom surfaces of the sheet as per the boundary conditions

Node penetration into punch profile radius region, punch wall, die profile radius region or die wall

As the punch keeps moving down, some nodes on the free surfaces ABDCA penetrate either the punch profile radius region or the punch wall region. Similarly, some nodes on the free surface GJKG penetrate either the die wall or the die profile radius region. Therefore, the penetration of these nodes is checked at the end of each increment. If a node on these surfaces is found to penetrate, the whole increment is repeated after assigning the required incremental displacement to the node so as to bring it back to the punch or the die profile as the case may be.

The procedure to calculate the required displacement for the case of penetration into the punch profile radius region is explained below. Suppose a node P, at the end of some increment, penetrates to the position P_1 as shown in Figure 6.17. To avoid the penetration, the actual location of the node should be at P_1^* (which is obtained by extending the position vector r of the point P_1 to the punch profile). So, at the end of the increment, first the penetration is checked by comparing $|r|$ (*i.e.*, the magnitude of the position vector r) with the punch profile radius. If $|r|$ is found to be less than the punch profile radius, it means the node has penetrated the punch profile radius region. Then the increment is repeated by specifying the vector PP_1^* as the incremental displacement vector of the node. Similar procedure is followed if a node penetrates either the punch wall or the die profile radius region or the die wall.

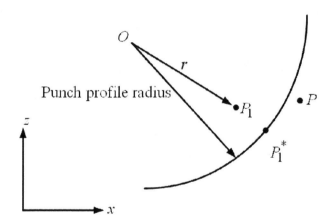

Figure 6.17. Penetration of a node into punch profile radius region

Boundary conditions at die and punch profile radius regions:
When a node on the (flat) die surface moves to the die profile radius region, the boundary conditions (Equations 6.133a,b) get modified because of the change in the normal and tangential directions. Of course, we still assume that the die-sheet interface friction is modeled by the Coulomb's law and the friction coefficient f is such that all the nodes at the die-sheet interface slip. Let $\hat{n}$ be the direction normal to the die profile and $\hat{s}$ and $\hat{t}$ be the two directions tangential to the die profile (Figure 6.18). Further, let $_t\Delta t_n$, $_t\Delta t_s$ and $_t\Delta t_t$ be the components of the incremental stress vector $_t\Delta t$ along $\hat{n}$, $\hat{s}$ and $\hat{t}$ directions. Since the relative

movement of a contact node is along negative $\hat{s}$ direction, the friction force acts along the positive $\hat{s}$ direction. Therefore, the component of the incremental stress vector $_t\Delta t$ along $\hat{t}$ must be zero. Further, the components $_t\Delta t_s$ and $_t\Delta t_n$ should be related by the Coulomb's law. Thus, the friction boundary condition at a node on the die profile radius leads to the following two conditions:

$$_t\Delta t_s - f(_t\Delta t_n) = 0, \quad _t\Delta t_t = 0, \quad \text{(natural)} . \tag{6.135a}$$

The sign in the first part of Equation 6.135a is based on the fact that both the frictional stress $_t\Delta t_s$ and the normal stress $_t\Delta t_n$ are positive. The third boundary condition is provided by the assumption that the node remains in contact with the die. Then, the component of the incremental displacement vector $_t\Delta u$ along the normal direction $\hat{n}$ must be zero. Then, the third boundary condition becomes

$$_t\Delta u_n = 0, \quad \text{(essential)}. \tag{6.135b}$$

In (x, y, z) coordinate system, Equations 6.135a,b can be expressed as

$$_t\Delta t_x(\cos\theta\cos\phi + f\cos\theta\sin\phi) + _t\Delta t_y(\sin\theta\cos\phi + f\sin\theta\sin\phi)$$
$$+ _t\Delta t_z(\sin\phi - f\cos\phi) = 0$$
$$- _t\Delta t_x\sin\theta + _t\Delta t_y\cos\theta = 0$$
$$\text{(natural)}$$

$$, \tag{6.136a}$$

$$- _t\Delta u_x(\cos\theta\sin\phi) - _t\Delta u_y(\sin\theta\sin\phi) + _t\Delta u_z(\cos\phi) = 0, \quad \text{(essential)} , \tag{6.136b}$$

where the angles θ and ϕ are defined in Figure 6.18.

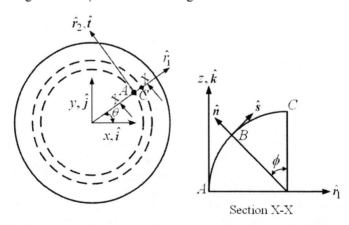

Section X-X

Figure 6.18. Normal and two tangential directions at the die profile radius region

At the interface between the sheet and punch profile radius region, a node may be in contact with the punch or may lose the contact. It depends on the nature (compressive or tensile) of the normal component of the punch reaction vector at the node. The algorithm to determine the contact status of the node is described in the next subsection. When a node is in contact, the sticking friction condition is assumed as stated earlier. Therefore, the boundary conditions at the punch profile radius region remain the same as given by Equations 6.128a, b.

6.4.2 Contact Algorithm

The contact boundary conditions (Equation 6.128) can be applied only in an iterative fashion. Thus, besides the Newton-Raphson iterations, there is an additional set of iterations in the solution of the deep drawing problem. First, for the specified incremental punch displacement and incremental blank-holder force, the contact iterations are carried out to determine the status (contact or non-contact) of the sheet-punch interface nodes. Then the Newton-Raphson iterations are carried out to minimize the unbalanced force vector. The stepwise algorithm for the contact iterations can be described as follows.

(i) *First Contact Iteration*:

- In the first iteration, it is assumed that all the punch-sheet interface nodes are in contact. Thus, the contact boundary condition given by Equation 6.128a is applied to all the interface nodes.
- Next, the finite element equation (Equation 6.88) is solved to find $_t\{\Delta U\}^{(1)}$, *i.e.*, the initial guess to the incremental displacement vector.
- Next, the incremental nodal reaction vector is found by multiplying the original coefficient matrix (*i.e.*, $^t[K]$ without the modifications which are done while applying the essential boundary conditions) by $_t\{\Delta U\}^{(1)}$. Then, the nodal reaction vector is updated. This vector is nothing but the global external force vector $^{t+\Delta t}\{F\}_{ex}$.
- At the end of first contact iteration, the non-contact nodes are identified using the following condition. A node at the interface goes out of contact if the normal component of the nodal reaction vector at that node becomes tensile. For a node *l* on the flat portion of the punch bottom, the normal component of the nodal reaction vector becomes tensile if $(^{t+\Delta t}F_{ex})_{lz} > 0$ where $(^{t+\Delta t}F_{ex})_{lz}$ is the z component of the vector $^{t+\Delta t}\{F\}_{ex}$ at node *l*. For nodes on the punch profile radius region, the normal component would be a linear combination of $(^{t+\Delta t}F_{ex})_{lz}$ and $(^{t+\Delta t}F_{ex})_{lr}$, the r component of the vector $^{t+\Delta t}\{F\}_{ex}$ at node *l*.

(ii) *Second Contact Iteration*:

- Now, the non-contact boundary condition (Equation 6.128b) is applied to all the non-contact nodes.
- Then, the finite element equation (Equation 6.88) is solved again to find the new initial guess to the incremental displacement vector.
- Next, the new $^{t+\Delta t}\{F\}_{ex}$ is found by the method described above.
- Finally, a check is made to find whether any additional nodes are losing contact using the condition mentioned above.

(iii) *Further Contact Iterations*:
- The second contact iteration is repeated until there is no change in the contact status between the two successive contact iterations.

6.4.3 Typical Results

The results presented in this and next subsections are from [57–58] for the aluminum alloy AA2090-T3, which is actually an anisotropic material. However, in this section, only the isotropic analysis is carried out. The material properties of the sheet, the geometric properties of the sheet and set-up and the other process parameters (friction coefficient and blank-holder force) of the problem are as follows [59].

Material Properties
Material : AA2090-T3,
Young's modulus (E): 69 GPa,
Poisson's ratio (v): 0.33,
Yield stress (σ_Y): 280 MPa,

Hardening relation: $\sigma_{eq} = 646(0.025 + \varepsilon_{eq}^p)^{0.227}$;

Geometric Properties of Sheet and Set-up
Sheet diameter (D) = 158.76 mm,
Sheet thickness (t) = 1.6 mm,
Punch diameter (D_p) = 97.46 mm,

Punch profile radius (r_p) = 12.7 mm,

Die opening diameter (D_d) = 101.48 mm,

Die profile radius (r_d) = 12.7 mm;

Other Process Parameters
Friction coefficient (f) = 0.1;
Blank-holder force = 5500 N.

Figure 6.19 shows the variation of punch force with punch displacement. The punch force is simply obtained by adding the z-components of the nodal reactions at the punch contact nodes. (Since the domain is only a quarter of the sheet, this sum has to be multiplied by four to make it the punch force for the whole sheet). It is observed that the punch force increases steadily even after the blank completely comes out of the blank-holder. This may be due to the plastic deformation of the sheet (due to bending) at the die profile radius region and the frictional dissipation at the sheet-die interface.

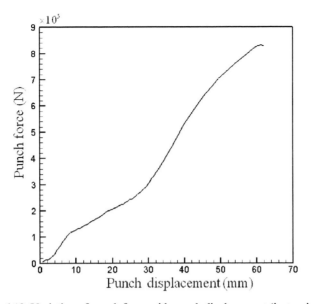

Figure 6.19. Variation of punch force with punch displacement (isotropic case)

Figure 6.20 shows the deformed configuration. Since, only the isotropic analysis is carried out in this section, there is no ear formation. The deformed configuration of the next section, which is based on anisotropic analysis, shows the formation of ears.

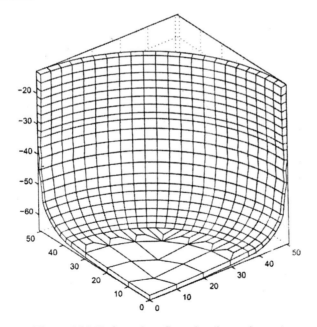

Figure 6.20. Deformed configuration (isotropic case)

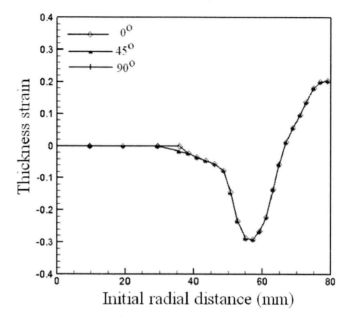

Figure 6.21. Thickness strain distribution (isotropic case)

Figure 6.21 shows the thickness strain distribution along a radial direction. The thickness strain is calculated as the ratio of the change in the distance between the corresponding nodes along the thickness direction to the original thickness. Since the nodes also move along the surface of the sheet, the distance between them does not exactly represents the final thickness of the sheet.

The figure shows that the final thickness of the cup varies considerably from the initial sheet thickness. The thickness of the sheet under the punch is found to be more or less equal to the initial sheet thickness. However, the thickness of the sheet, which was above the flat portion of the die, is observed to be slightly larger than the initial thickness. The minimum thickness (*i.e.*, the maximum thickness strain) usually occurs in the portion of the sheet which is closer to the punch profile radius region. In the final configuration (*i.e.*, in the cup), it occurs in the portion of the wall which is closer to the cup bottom. The minimum thickness is almost 70% of the initial thickness. Because of this thinning, the fracture usually occurs in this region of the cup.

Because of the assumption of isotropy, the thickness strain variation is found to be the same along every radial direction. However, in anisotropic materials, the thickness strain variation is different in different directions. This is shown in next subsection.

6.4.4 Anisotropic Analysis, Ear Formation and Parametric Studies

In this section, the analysis is carried out by treating the sheet material AA2090-T3 as an orthotropic material. The anisotropy is modeled by Barlat's Yld2004-18p yield criterion (Equation 4.118). The components of the matrices $[C']$ and $[C'']$

(Equations 4.119 and 4.120) which appear in the linear transformation (Equation 4.117) are as follows [59, 60]

$$[C'] = \begin{bmatrix} 0 & 0.0698 & -0.9364 & 0 & 0 & 0 \\ -0.0791 & 0 & -1.0030 & 0 & 0 & 0 \\ -0.5247 & -1.3631 & 0 & 0 & 0 & 0 \\ 0 & 0 & 0 & 0.9543 & 0 & 0 \\ 0 & 0 & 0 & 0 & 1.0237 & 0 \\ 0 & 0 & 0 & 0 & 0 & 1.0690 \end{bmatrix}, (6.137a)$$

$$[C''] = \begin{bmatrix} 0 & -0.9811 & -0.4767 & 0 & 0 & 0 \\ -0.5753 & 0 & -0.8668 & 0 & 0 & 0 \\ -1.1450 & 0.0792 & 0 & 0 & 0 & 0 \\ 0 & 0 & 0 & 1.4046 & 0 & 0 \\ 0 & 0 & 0 & 0 & 1.0516 & 0 \\ 0 & 0 & 0 & 0 & 0 & 1.1471 \end{bmatrix}, (6.137b)$$

The other parameters (i.e., other material properties of the sheet, the geometric properties of the sheet and set-up, the friction coefficient and the blank-holder force) are the same as in Subsection 6.4.3.

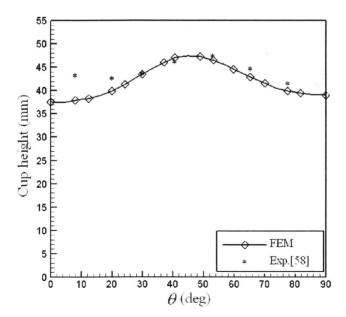

Figure 6.22. Comparison of cup height variation with circumferential angle with experimental result [58]

The finite element code is validated by comparing the cup height variation (with circumferential angle θ) with experimental results of [58]. Figure 6.22 shows the comparison of the predicted cup height variation with the experimental one. It shows a good agreement between the two results except near $\theta = 0°$. It is expected that, with a refinement of the mesh, the agreement would be better near this point also.

Figure 6.23 shows the variation of punch force with punch displacement for the anisotropic case. Here, the variation is similar to the isotropic case up to 40 mm of punch displacement, beyond which the punch force remains almost constant. Further, the punch force is smaller for the given punch displacement for the anisotropic case. The reason for this is as follows. Actually, the yield stress of the sheet material is different in different directions with the maximum value being in the rolling direction. However, in the isotropic analysis, it is assumed to be the same in every direction with its value being equal to the yield stress in the rolling direction. Thus, in the isotropic analysis, the sheet is assumed to be stiffer and therefore, a larger value of the punch force is predicted.

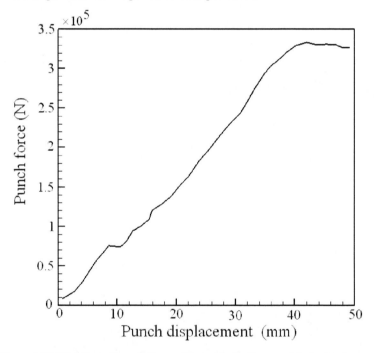

Figure 6.23. Variation of punch force with punch displacement (anisotropic case)

Figure 6.24 shows the deformed configuration for the isotropic case. The ear formation can be clearly observed from this figure. The ear is formed at 45°direction to the rolling direction. This earing can be reduced by optimizing the initial shape of the sheet (*i.e.*, blank shape) as explained in the next subsection.

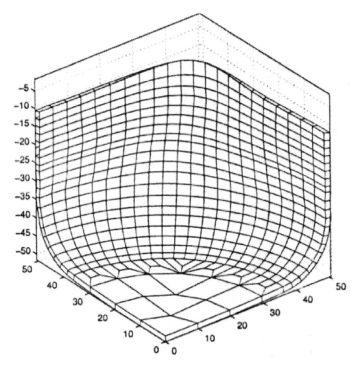

Figure 6.24. Deformed configuration (anisotropic case)

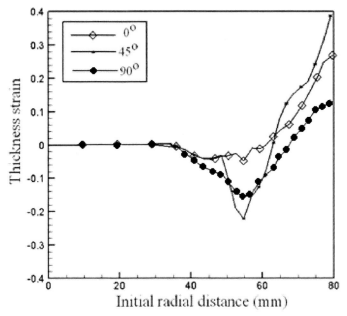

Figure 6.25. Thickness strain distribution (anisotropic case)

Figure 6.25 shows the thickness strain distributions along 0°, 45° and 90° directions (to the rolling direction). As expected, the thickness strain distributions are different along these directions. The maximum magnitude of the thickness strain (*i.e.*, the minimum thickness) is observed along 45°. Further, it is less than the corresponding value for the isotropic case.

Various process parameters affect the final states of the deformation and stress in the drawn cup. These parameters include the geometric parameters (sheet thickness, punch profile radius, die profile radius), material properties, blank-holder force, lubrication at the interface *etc.* The effects of some of these parameters are analyzed in this subsection.

6.4.4.1 Effect of Die Profile Radius

The material properties, the other process parameters and the remaining geometric properties are kept the same as before. Three cases (r_d = 10, 12.7, 16 mm) are analyzed to study the effect of the die profile radius r_d. Figure 6.26 shows the variation of punch force with punch displacement for these three cases. It is observed that the punch force decreases with an increase in the die profile radius. The explanation for this could be as follows. As the die profile radius increases, the material flows easily at the die corner, which in turn leads to a decrease in the punch force. However, the die profile radius should not be too large as it increases the punch travel.

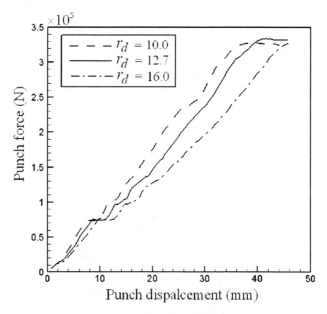

Figure 6.26. Variation of punch force with punch displacement for different die profile radii

Figure 6.27 shows the thickness strain distributions along 45° direction for different die radii. The maximum magnitude of the thickness strain decreases with the die radius.

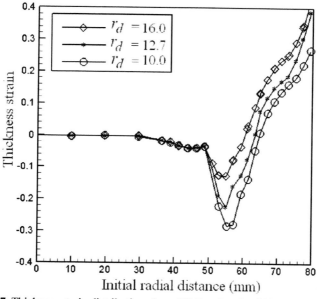

Figure 6.27. Thickness strain distribution along 45°direction for different die profile radii

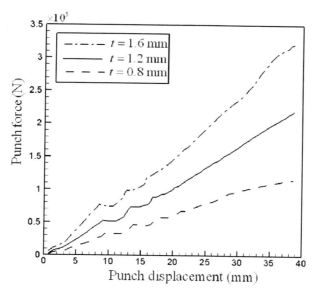

Figure 6.28. Variation of punch force with punch displacement for different sheet thicknesses

6.4.4.2 Effect of Sheet Thickness

The material properties, the other process parameters and the remaining geometric properties are kept the same as before. Three cases (t = 0.8, 1.2, 1.6 mm) are analyzed to study the effect of the sheet thickness t. Figure 6.28 shows the variation of punch force with punch displacement for these three cases. Increase in sheet thickness implies an increase in the sheet volume. Therefore, a larger force is required to achieve the same punch displacement. Thus, the punch force increases with the punch displacement as shown in Figure 6.28.

Figure 6.29 shows the thickness strain distributions along $0°$ direction for different sheet thicknesses. The maximum magnitude of the thickness strain increases with the sheet thickness.

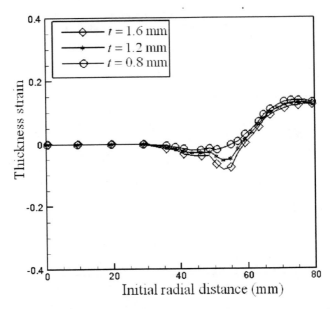

Figure 6.29. Thickness strain distribution along $0°$ direction for different sheet thicknesses

6.4.4.3 Effect of Material Properties

The punch force variation (with the punch displacement) and the thickness strain distribution (along the rolling direction) are studied for one more aluminum alloy (AA6022-T4), the material properties of which are given below [61].

Material Properties

Material: AA6022-T4,

Young's modulus (E): 70 GPa,

Poisson's ratio (ν): 0.33,

Yield stress (σ_Y): 162 MPa,

Hardening relation: $\sigma_{eq} = 396 - 234\exp(-6.745\varepsilon_{eq}^p)$,

The matrices $[C']$ and $[C'']$ appearing in the linear transformation (Equation 4.117) are as follows [61]:

$$[C'] = \begin{bmatrix} 0 & -0.755194 & -0.799378 & 0 & 0 & 0 \\ -0.773630 & 0 & -0.865580 & 0 & 0 & 0 \\ -1.04756 & -1.088160 & 0 & 0 & 0 & 0 \\ 0 & 0 & 0 & 1.016290 & 0 & 0 \\ 0 & 0 & 0 & 0 & 0.993625 & 0 \\ 0 & 0 & 0 & 0 & 0 & 0.624258 \end{bmatrix},$$

(6.138a)

$$[C''] = \begin{bmatrix} 0 & -1.120720 & -1.056340 & 0 & 0 & 0 \\ -1.146560 & 0 & -1.132990 & 0 & 0 & 0 \\ -0.763656 & -0.954688 & 0 & 0 & 0 & 0 \\ 0 & 0 & 0 & 1.009770 & 0 & 0 \\ 0 & 0 & 0 & 0 & 0.994796 & 0 \\ 0 & 0 & 0 & 0 & 0 & 1.208880 \end{bmatrix},$$

(6.138b)

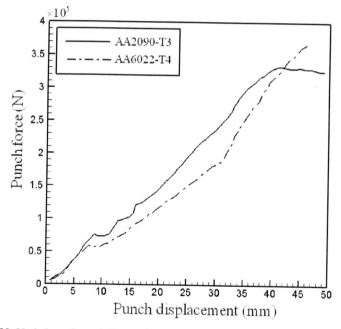

Figure 6.30. Variation of punch force with punch displacement for two different aluminum alloys

Figure 6.30 shows the variation of punch force with punch displacement for two different aluminum alloys: AA2090-T3 and AA6022-T4. Since the yield stress of AA6022-T4 is lower than that of AA2090-T3, the required punch force is less for AA6022-T4. However, the punch force for AA6022-T4 does not seem to remain constant after 40 mm of punch travel.

Figure 6.31 shows the thickness strain distributions along 45° direction for two different aluminum alloys: AA2090-T3 and AA6022-T4. It is observed that, for the material AA6022-T4, the magnitude of the negative thickness strain is slightly more but that of the positive thickness strain is less.

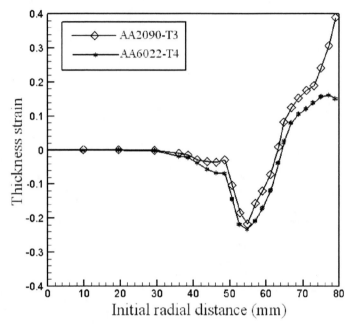

Figure 6.31. Thickness strain distribution along 45° direction for two different aluminum alloys

6.4.5 Optimum Blank Shape

The method proposed by Pegada *et al.* [56] is used to optimize the blank shape so as to minimize the earing. This method can be briefly explained as follows. Figure 6.32a shows a typical cup height variation with the circumferential angle θ, when a circular blank is used. In this method, the circular shape is modified by adding some material where the cup height is less and removing some material where the cup height is more while maintaining the constant volume. So, the stepwise optimization algorithm can be described as follows.

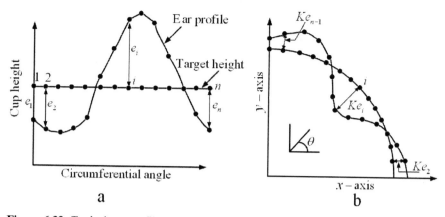

Figure 6.32. Typical ear profile and modification of blank shape. **a** Typical ear profile. **b** Shape modification

First Optimization Iteration:

- First, the cup height variation is obtained by choosing the circular blank shape.
- Next, a *target height* is established by making the areas above and below the target height to be equal (Figure 6.32a).
- Next, the *shape deviation* e_i at circumferential node i is determined as the difference between the cup height at the node and the target height (Figure 6.32a). This is done for all the circumferential nodes: $i = 1, 2,.., n$ where n is the number of circumferential nodes.
- Finally, the maximum shape deviation $(e_{max} - e_{min})$ is calculated where e_{max} is the maximum value and e_{min} is the minimum value of the shape deviations over all the circumferential nodes.

Second Optimization Iteration:

- The circular shape is modified by changing the radius at circumferential node i by the amount Ke_i where K, the shape correction factor, is chosen suitably (Figure 6.32b). This is done for all the circumferential nodes: $i = 1,2,...,n$.
- Again, the cup height variation is obtained corresponding to the modified blank shape and the difference $(e_{max} - e_{min})$ is calculated as explained above.

Further Optimization Iterations:

- The second optimization iteration is repeated until the difference $(e_{max} - e_{min})$ is less than some specified percentage (say $x\%$) of the maximum height.

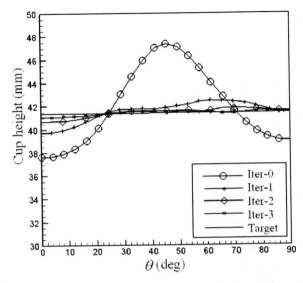

Figure 6.33. Cup height variation with circumferential angle θ in different iterations

Figure 6.33 shows the cup height variations with the circumferential angle θ in different iterations. As the number of iterations increases, the maximum shape deviation decreases from 9.8 to 0.4 mm. Sometimes, the value of the shape correction factor K needs to be changed during the iterations, otherwise the maximum shape deviation does not decrease with the iterations. Figure 6.34 shows the initial blank shape and the optimum blank shape obtained after the third iteration.

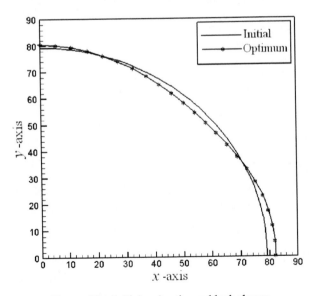

Figure 6.34. Initial and optimum blank shapes

6.5 Summary

In this chapter, first the finite element modeling of metal forming processes using the updated Lagrangian formulation is presented. The formulation is based on the incremental strain-displacement and stress-strain relations developed in Chapter 4. Incremental logarithmic strain tensor is used as a measure of finite incremental deformation. Further, a special updating scheme is employed to make the incremental stress tensor objective. Equilibrium equation at time $t + \Delta t$ is converted first into an integral form and then into a set of non-linear algebraic equations using the Galerkin version of the finite element method. Like that of the Eulerian formulation, the finite element equations of the updated Lagrangian formulation also are non-linear and need an iterative scheme to obtain the solution. The Newton-Raphson iterative scheme, employed to solve the non-linear incremental equations, is explained in detail. Since, the constitutive relation is in the integral form, the Euler forward integration scheme for its integration is presented. Some divergence handling techniques like the under-relaxation method, line search method and increment cutting method are presented.

Next, the finite element model developed for the updated Lagrangian formulation is applied to an axisymmetric problem of open die cold forging of a cylindrical block of an isotropic material. The friction at the platen-block interface is modeled by sticking friction and Coulomb's law. Since this is a displacement control problem, an arc length technique is used to accelerate the convergence. Because of axisymmetry, only a typical r-z plane of the upper half of the cylindrical block is selected as the domain. The finite element model is verified by comparing the predicted forging load variation (with reduction) with experimental results available in the literature. The contact pressure distribution, deformed configuration, equivalent strain field, equivalent stress field and equivalent residual stress field are presented for a typical set of process variables. A parametric study of the residual stresses is carried out with respect to the four process variables, namely height-to-diameter ratio, reduction, friction coefficient and material properties. Finally, some studies on fracture prediction are presented using the critical damage and hydrostatic stress criteria.

In the end, the finite element model of a three-dimensional problem of deep drawing of a cylindrical cup of an anisotropic material is developed. Friction at the punch is assumed to be of sticking type and that at the die is modeled by Coulomb's law. The blank-holder force is assumed to be equally distributed over the area and is applied incrementally. Because of the orthotropy of the sheet and the symmetry of the geometry and boundary conditions about the two axes, only a quarter of the sheet is selected as the domain. The finite element model is verified by comparing the predicted cup height variation with experimental results available in the literature. Then, for a typical set of process variables, the punch force variation, deformed configuration and thickness strain variation are presented. A parametric study is carried out with respect to the three process variables, namely, the die profile radius, sheet thickness and material properties. Finally, the optimization of the initial shape of the sheet is carried out to minimize the earing.

6.6 References

[1]   Malvern, L.E. (1969), Introduction to the Mechanics of a Continuous Medium, Prentice-Hall Inc., Englewood Cliffs

[2]   Crisfield, M.A. (1991), Non-linear Finite Element Analysis of Solids and Structures, Vol. 1: Essentials, John Wiley and Sons, Chichester.

[3]   Bathe, K.J. (1996), Finite Element Procedures, Prentice-Hall of India, New Delhi.

[4]   Altan, T. (1971), Computer simulation to predict load, stress and metal flow in an axisymmetric closed die forging, in A.L. Hoffmanner (Ed.), Metal Forming, Plenum Press, New York, pp. 249–274.

[5]   Hill, R., Lee, E.H. and Tupper, S.J. (1923), A method of numerical analysis of plastic flow in plane strain and its application to the compression of a ductile material between rough platens, Transaction of ASME, Journal of Applied Mechanics, Vol. 73, pp. 46–52.

[6]   Green, A.P. (1951), A theoretical investigation of the compression of a ductile material between smooth flat dies, Philosphical Magazine, Vol. 42, pp. 900–918.

[7]   Kudo, H. (1960), Some analytical and experimental studies of axisymmetric cold forging and extrusion, I and II, International Journal of Mechanical Sciences, Vol. 2, pp. 102–127.

[8]   Avitzur, B. (1968), Metal Forming: Processes and Analysis, McGraw-Hill, New York.

[9]   Lee, C.H. and Kobayashi, (1971), Analysis of axisymmetric upsetting and plane strain side-pressing of solid cylinders by finite element method, Transaction of ASME Journal of Engineering for Industry, Vol. 93, pp. 445–454.

[10]   Shah, S.N., Lee, C.H. and Kobayashi, S. (1974), Compression of tall, circular solid cylinders between parallel flat dies, in: Proceedings of the International Conference on Production Engineering, Tokyo, pp. 295–300.

[11]   Chen, C.C. and Kobayashi, S. (1978), Rigid-plastic finite-element analysis of ring compression, in : H. Armen and R.F. Jones, Jr. (Eds.), Application of Numerical Methods to Forming Processes, Proceedings of the Winter Annual Meeting of ASME, AMD, Vol. 28, pp. 163–174.

[12]   Oh, S.I. and Kobayashi, S. (1976), Workability of aluminum alloy 7075-t6 in upsetting and rolling, Transaction of ASME, Journal of Engineering for Industry, Vol. 98, pp. 800–806.

[13]   Hartley, P., Sturgess, C.E.N. and Rowe, G.W. (1979), Friction in finite element analysis of metal forming process, International Journal of Mechanical Sciences, Vol. 21, pp. 301–311.

[14]   Hartley, P., Sturgess, C.E.N. and Rowe, G.W. (1980), Influence of friction on the prediction of forces, pressure distributions and properties in upset forging, International Journal of Mechanical Sciences, Vol. 22, pp. 743–753.

[15]   Shima, S., Mori, K. and Osakada, K. (1978), Analysis of metal forming by the rigid-plastic finite element method based on plasticity theory for porous metals, in H. Lippmann (Ed.), Metal Forming Plasticity, Springer, Berlin, pp. 305–317.

[16]   Kobayashi, S. Oh, S.I. and Altan, T. (1989), Metal Forming and the Finite Element Method, Oxford University Press, New York.

[17]   Rowe, G.W., Sturgess, C.E.N., Hartley, P. and Pillinger, I. (1991), Finite Element Plasticity and Metal Forming Analysis, Cambridge University Press, Cambridge.

[18]   Hartley, P. and Pillinger, I. and Sturgess, C.E.N. (Eds.) (1992), Numerical Modeling of Material Deformation Processes: Research, Development and Application, Springer, London.

[19]   Bathe, K.J., Ramm, E. and Wilson, E.L. (1975), Finite element formulations for large deformation dynamic analysis, International Journal of Numerical Methods for Engineering, Vol. 9, pp. 353–386.

[20]   Dadras, P. and Thomas, J.F. (1983), Analysis of axisymmetric upsetting based on flow pattern observation, International Journal of Mechanical Sciences, Vol. 25, pp. 421–427.

[21]   Carter, Jr., W.T. and Lee, D. (1986), Further analysis of axisymmetric upsetting, Transaction of ASME, Journal of Engineering for Industry, Vol. 108, pp. 198–204.

[22]   Michel, B. and Boyer, J.C. (1995), Elasto-visco-plastic finite-element analysis of a cold upsetting test and stress-state validation by residual-stress measurements, Journal of Materials Processing Technology, Vol. 54, pp. 120–128.

[23]   Zhao, G., Wright, E. and Grandhi, R.V. (1996), Computer aided perform design in forging using the inverse die contact tracking method, International Journal of Machine Tools & Manufacture, Vol. 36, pp. 755–769.

[24]   Joun, M.S., Lee, S.W. and Chung, J.H. (1998), Finite element analysis of a multi-stage axisymmetric forging process, International Journal of Machine Tools & Manufacture, Vol. 38, pp. 843–854.

[25]   Kim, H.S., Im, Y.T. and Geiger, M. (1999), Prediction of ductile fracture in cold forging of aluminum alloy, Transaction of ASME, Journal of Manufacturing Science and Engineering, Vol. 121(3), pp. 336–344.

[26]   Yang, D.Y., Im, Y.T., Yoo, Y.C., Park, J.J., Kim, J.H., Chun, M.S., Lee, C.H., Lee, Y.K., Park, C.H., Song, J.H., Kim, D.Y., Hong, K.K., Lee, M.C. and Kim, S.I. (2000), Development of integrated and intelligent design and analysis system for forging processes, Annals of CIRP, Vol. 49, pp. 177–180.

[27]   Gupta, S., Reddy, N.V. and Dixit, P.M. (2003), Ductile fracture prediction in axisymmetric upsetting using continuum damage mechanics, Journal of Materials Processing Technology, Vol. 141, pp. 256–265.

[28]   Mungi, M.P., Rasane, S.D. and Dixit, P.M. (2003), Residual stresses in cold axisymmetric forging, Journal of Materials Processing Technology, Vol. 142, pp. 256–266.

[29]   Riks, E. (1972), The application of Newton's method to the problem of elastic stability, Transaction of ASME, Journal of Applied Mechanics, Vol. 39, pp. 1060–1066.

[30]   Batoz, J.L. and Dhatt, G. (1979), Incremental displacement algorithms for non-linear problems, International Journal of Numerical Methods for Engineering, Vol. 14, pp. 1262–1267.

[31]   Park, J.J. and Kobayashi, S. (1984), Three-dimensional finite element analysis of block compression, International Journal of Mechanical Sciences, Vol. 26(3), pp. 165–176.

[32]   Lemaitre, J. and Chaboche, J.L. (1990), Mechanics of Solid Materials, Cambridge University Press, Cambridge.

[33]   Dhar, S., Sethuraman, R. and Dixit, P.M. (1996), A continuum damage mechanics model for void growth and micro-crack initiation, Engineering Fracture Mechanics, Vol. 53, pp. 917–928.

[34]   Reddy, N.V., Dixit, P.M. and Lal, G.K. (1996), Central bursting and optimal die profile for axisymmetric extrusion, Transaction of ASME, Journal of Manufacturing Science and Engineering, Vol. 118, pp. 579–584.

[35]   Le Roy, G., Embury, J.D., Edward, G. and Ashby, M.F. (1981), A model of ductile fracture based on the nucleation and growth of voids, Acta Metallurgica, Vol. 29, pp. 1509–1522.

[36]   Predeleanu, M., Cordebois, J.P. and Belkhiri, L. (1986), Failure analysis of cold upsetting by computer and experimental simulation, in Proceedings of the NUMIFORM'86 Conference, pp. 277–282.

[37]   Kuhn, H.A. and Lee, P.W. (1971), strain instability and fracture at the surface of upset cylinders, Metallurgical Transactions, Vol. 2, pp. 3197–3202.

[38]   Kobayashi, S. (1970), Deformation characteristics and ductile fracture of 1040 steel in simple upsetting of solid cylinders and rings, Transaction of ASME, Journal of Engineering for Industry, Vol. 92, pp. 391–399.

[39]   Semiatin, S.L., Goetz, T.L., Shell, E.B., Seetharaman, V. and Ghosh, A.K. (1999), Cavitation and failure during hot forging of Ti-6Al-4V, Metallurgical and Materials Transaction A, Vol. 30, pp. 1411–1424.

[40]   Johnson, W. and Mellor, P.B. (1972), Engineering Plasticity, von Nostrand Co. Ltd.

[41]   Chakrabarty, J. (1987), Theory of Plasticity, McGraw-Hill Book Company, New York.

[42]   Toh, C.H. and Kobayashi, S. (1985), Deformation analysis and blank design in square cup drawing, International Journal of Machine Tool Design and Research, Vol. 25, pp. 15–32.

[43]   Saran, M.J. and Samuelsson, A. (1990), Elastic-viscoplastic implicit formulation for finite element simulation of complex sheet forming processes, International Journal for Numerical Methods in Engineering, Vol. 30, pp. 1675–1697.

[44]   Majlessi, S.A. and Lee, D. (1993), Deep drawing of square-shaped sheet metal parts, part 1: Finite element analysis, Transaction of ASME, Journal of Engineering for Industry, Vol. 115, pp. 102–109.

[45]   Onate, E. and Saracibar, C.A.D. (1990), Finite element analysis of sheet metal forming problems using a selective viscous bending/membrane formulation, International Journal for Numerical Methods in Engineering, Vol. 30, pp. 1577–1593.

[46]   Chou, C.H., Pan, J. and Tang, S.C. (1994), Analysis of sheet metal forming operations by a stress resultant constitutive law, International Journal for Numerical Methods in Engineering, Vol. 37, pp. 717–735.

[47]   Shi, X., Wei, Y. and Ruan, X. (2001), Simulation of sheet metal forming by a one-step approach: Choice of element, Journal of Materials Processing Technology, Vol. 108, pp. 300–306.

[48]   Chou, C.H., Pan, J. and Tang, S.C. (1994), An anisotropic stress resultant constitutive law for sheet metal forming, International Journal for Numerical Methods in Engineering, Vol. 39, pp. 435–449.

[49]   Menezes, L.F. and Teodosiu, C. (2000), Three dimensional numerical simulation of deep drawing process using solid finite elements, Journal of Materials Processing Technology, Vol. 97, pp. 100–106.

[50]   Colgan, M. and Monaghan, J. (2003), Deep drawing process: Analysis and experiment, Journal of Materials Processing Technology, Vol. 132, pp. 35–41.

[51]   Osakada, K., Wang, C.C. and Mori, K.I. (1995), Controlled FEM simulation for determining history of blank holding force in deep drawing, Annals of CIRP, Vol. 44, pp. 243–246.

[52]   Lorenzo, R.D., Fratini, L. and Micari, F. (1999), Optimal blank holder force path in sheet metal forming processes: an AI based procedure, Annals of CIRP, Vol. 48, pp. 231–234.

[53]   Guo, Y.Q., Batoz, J.L., Detraux, J.M. and Duroux, P. (1990), Finite element procedures for strain estimation of sheet metal forming parts, Annals of CIRP, Vol. 30, pp. 1385–1401.

[54] Kim, S.H. and Huh, H. (2001), Finite element inverse analysis for the design of intermediate dies in multi-stage deep drawing with large aspect ratio, Journal of Materials Processing Technology, Vol. 113, pp. 779–785.

[55] Ku, T.W., Lim, H.J., Choi, H.H., Hwang, S.M. and Kang, B.S. (2001), Implementation of backward tracing scheme of the FEM to blank design in sheet metal forming, Journal of Materials Processing Technology, Vol. 111, pp. 90–97.

[56] Pegada, V.P., Chun, Y. and Santhanam, S. (2002), An algorithm for determining the optimum blank shape for deep drawing of aluminum cups, Journal of Materials Processing Technology, Vol. 125/126, pp. 743–750.

[57] Chakka, V.M. (2006), Optimum Blank Shape Design for Cylindrical Cup Drawing Using Finite Element Method, M.Tech. Thesis, Department of Mechanical Engineering, Indian Institute of Technology Kanpur.

[58] Raja, S. (2007), Optimum Blank Shape Design for Cylindrical Cup Drawing Using Various Anisotropic Criteria, M.Tech. Thesis, Department of Mechanical Engineering, Indian Institute of Technology Kanpur.

[59] Yoon, J.W., Barlat, F., Dick, R.E. and Karabin, M.E. (2006), Prediction of six or eight ears in a drawn cup based on a new anisotropic yield function, International Journal of Plasticity, Vol. 22, pp. 174–193.

[60] Barlat, F., Aretz, H., Yoon, J.W., Karabin, M.E., Brem, J.C. and Dick, R.E. (2005), Linear transformation-based anisotropic yield functions, International Journal of Plasticity, Vol. 21, pp. 1009–1039.

[61] Yoon, J.W., Barlat, F., Gracio, J.J. and Rauch, E. (2005), Anisotropic strain hardening behavior in simple shear for cube textured aluminum alloy sheets, International Journal of Plasticity, Vol. 21, pp. 2426–2447.

Finite Element Modeling of Orthogonal Machining Process

7.1 Introduction

Machining processes are difficult to model for various reasons. Unlike metal forming processes, where almost the whole work-piece gets plastically deformed, in machining processes, the plastic deformation is localized near the cutting edge. Therefore, we need to analyze only a small region of the work-piece around the cutting edge (called the cutting zone). As a result, the selection of the domain dimensions and the appropriate boundary conditions becomes a difficult task. Further, even at a moderate cutting speed, the strain rates are quite high, almost of the order of 10^4 per second. Further, the temperature rise is also quite large. As a result, the viscoplasticity and temperature-sofening effects become more important compared to strain-hardening. Therefore, the material properties associated with these two effects should be known for a range of strain rates and temperatures occurring in typical machining processes. These properties are not readily available. Additionally, to incorporate the temperature rise in the analysis, one needs to solve the heat transfer equation governing the temperature field in conjunction with the usual three equations governing the deformation field. For plastic deformation, these equations are coupled, and hence difficult to solve. We can decouple this problem as follows. We first estimate the average temperature in the cutting zone either experimentally or by simple analytical methods. Then we solve the governing equations of the deformation field by evaluating the material properties at the estimated average temperature of the cutting zone.

Two methods exist for analyzing the machining process. In the first method, it is assumed that the chip formation is continuous and the shape of the chip is known in advance. Thus, the process is analyzed as a steady-state process. This method is called the Eulerian method. In this method, a chip separation criterion is not required. In the second method, the process is analyzed from the beginning to the steady state chip formation. This is called the updated Lagrangian formulation. In this method, a chip separation criterion is required to predict the chip geometry.

Early applications of finite element method [1–4] to the machining process were mainly Eulerian. The main objective of many of these studies was to predict the temperature distribution and therefore, the determination of deformation and stress fields was only an intermediate step. These studies considered the machined material as rigid-plastic. But, later applications of Eulerian formulation to machining process [5, 6] also included viscoplastic effects. All of these applications have considered only orthogonal machining. The first finite element study of the machining process using an updated Lagrangian formulation was made by Strenkowski and Carrol [7]. This was for orthogonal machining. A critical value of the equivalent plastic strain was used to model the separation of a chip. Later on, several researchers [8–10] used the updated Lagrangian formulation for analyzing two- and three-dimensional machining processes. Most of these studies have used an FEM package: ABAQUS [8], MARC [9] or DEFORM [10]. The criterion used for chip separation has been based on controlled crack propagation [8] or some geometrical considerations [9]. Remeshing technique has been used to simulate the chip formation in [10].

In this chapter, we consider only the Eulerian formulation of orthogonal steady-state machining process.

7.2 Domain, Governing Equations and Boundary Conditions for Eulerian Formulation

7.2.1 Domain

In the present formulation, it is assumed that the problem is *decoupled*. We further assume that the elastic deformation is small. As stated above, the visco-plasticity and temperature effects are more dominant compared to the strain-hardening effects. To keep things simple, we assume that the material exhibites no hardening but only visco-plasticity. Thus, we assume the material to be *rigid-viscoplastic*. Further, the temperature softening is accounted for by evaluating the material properties at an average temperature occuring in the cutting zone. Additionally, we analyse the process when it has reached a steady state. Then the transient term in the equation of motion vanishes. We further assume that the body forces are negligible.

We choose a small region of the work-piece around the cutting edge (called the cutting zone) as the control volume, *i.e.*, the domain of the problem. To make the problem two-dimensional, we assume that the width of cut is large compared to the dimensions of the cutting zone. The domain, along with the coordinate system, is shown in Figure 7.1. It is a region in the cross-sectional plane of the work-piece perpendicular to the cutting edge. Point E is the projection of the cutting edge. The z-axis is along the cutting edge or the width of cut. The boundaries AB and EF are actually circular. But, since the cutting zone dimensions are small compared to the work-piece radius, they are taken to be straight. The angle α is equal to the rake angle of the cutting tool. The distance h is called the *tool-chip contact length*. It is given by Equation 3.162 [2]:

$$h = \frac{f \sin \theta}{\sin \phi \cos(\theta + \alpha - \phi)} ,$$
(7.1)

where ϕ is the shear angle (*i.e.*, the inclination of the shear plane with the direction of the cutting velocity), θ is the angle between the shear force and the resultant force and f is the feed. In orthogonal machining, the shear angle can be estimated by measuring the cutting ratio r and using the following relation:

$$r = \frac{\sin \phi}{\cos(\phi - \alpha)} .$$
(7.2)

The angle θ can be computed by measuring the cutting force F_c and thrust force F_t by a dynamometer and using the following relation:

$$\cos \theta = \frac{F_c \cos \phi - F_t \sin \phi}{\sqrt{F_c^2 + F_t^2}} .$$
(7.3)

The boundaries AH, HG, FG and CD are placed sufficiently away from the cutting edge projection E so as to simplify the boundary conditions on these boundaries by taking the advantage of the uniform velocity fields existing there. Further, the boundaries AH, FG and CD are chosen parallel to the shear plane so as to facilitate the mesh generation. So, the domain shape is different to that of Figure 3.22.

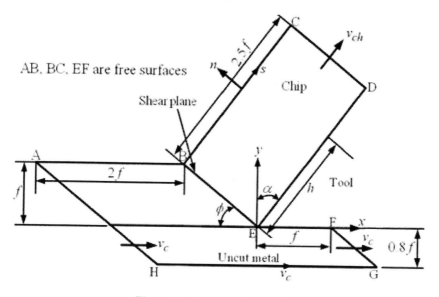

Figure 7.1. Domain of the problem

7.2.2 Governing Equations

For the *decoupled problem of the rigid-viscoplastic material* with zero body force, the velocity field v, the strain rate field $\dot{\varepsilon}$, the hydrostatic stress (or pressure) field p and the deviatoric stress field σ' in the control volume are governed by the following equations: strain rate-velocity relations, stress-strain rate relations, equations of motion and incompressibility constraint. For the purpose of finite element formulation, these equations need to be expressed in the component form. Since the problem is two-dimensional, the velocity vector has two non-zero components and the strain rate and the deviatoric stress tensors have three non-zero independent components each. In terms of the components with respect to the coordinate system of Figure 7.1, The governing equations are as follows:

(i)   *Strain rate - velocity relations*:

Let (v_x, v_y) be the non-zero components of the velocity vector v and $(\dot{\varepsilon}_{xx}, \dot{\varepsilon}_{yy}, \dot{\varepsilon}_{xy})$ be the non-zero independent components of the strain rate tensor $\dot{\varepsilon}$. Then, the strain rate-velocity relations (Equation 3.66) become

$$\dot{\varepsilon}_{xx} = \frac{\partial v_x}{\partial x}, \quad \dot{\varepsilon}_{yy} = \frac{\partial v_y}{\partial y},$$
$$\dot{\varepsilon}_{xy} = \frac{1}{2}\left(\frac{\partial v_x}{\partial y} + \frac{\partial v_y}{\partial x}\right). \tag{7.4}$$

(ii)   *Rigid – viscoplastic deviatoric stress - strain rate relations*:

As stated above, we neglect the hardening and consider only the visco-plasticity. For, non-hardening visco-plastic materials, the relation (Equation 3.165) between the deviatoric stress and the strain rate tensors gets modified as [5]

$$\sigma'_{xx} = 2\eta_2\dot{\varepsilon}_{xx},$$
$$\sigma'_{yy} = 2\eta_2\dot{\varepsilon}_{yy}, \tag{7.5}$$
$$\sigma'_{xy} = 2\eta_2\dot{\varepsilon}_{xy},$$

where $(\sigma'_{xx}, \sigma'_{yy}, \sigma'_{xy})$ are the non-zero independent components of the deviatoric stress tensor σ'. Here, the proportionality factor η_2, for non-linear visco-plastic behavior, is given by

$$\eta_2 = a(\dot{\varepsilon}_{eq})^m + \frac{\sigma_Y}{3\dot{\varepsilon}_{eq}}, \tag{7.6}$$

where σ_Y is the yield stress of the material, $\dot{\varepsilon}_{eq}$ is the equivalent strain rate (defined by a relation similar to Equation 3.156) and a and m are the material constants representing the material visco-plasticity. To account for the temperature softening, the material constants σ_Y, a and m are evaluated at the estimated average temperature of the cutting zone.

(iii) *Equations of motion*:

As stated earlier, we neglect the body forces. Further, we analyse the process when it has reached a steady-state. Therefore, the transient term ($\rho \partial v_i / \partial t$) vanishes. Then, the equations of motion (Equation 3.220), in the component form, take the form

$$\rho \left(v_x \frac{\partial v_x}{\partial x} + v_y \frac{\partial v_x}{\partial y} \right) = -\frac{\partial p}{\partial x} + \left(\frac{\partial \sigma'_{xx}}{\partial x} + \frac{\partial \sigma'_{xy}}{\partial y} \right),$$

$$\rho \left(v_x \frac{\partial v_y}{\partial x} + v_y \frac{\partial v_y}{\partial y} \right) = -\frac{\partial p}{\partial y} + \left(\frac{\partial \sigma'_{xy}}{\partial x} + \frac{\partial \sigma'_{yy}}{\partial y} \right).$$

$$(7.7)$$

In the rolling problem (Section 5.3 of Chapter 5), we neglected the first term of the equation of motion as the acceleration was small. But, in the machining process, the acceleration is not negligible. Hence, we retain this term.

(iv) *Incompressibility constraint*:

The incompressibility constraint (Equation 3.166), in the component form, becomes

$$\dot{\varepsilon}_{xx} + \dot{\varepsilon}_{yy} = 0. \tag{7.8}$$

7.2.3 Boundary Conditions

The boundary conditions of the problem are the same as mentioned in Section 3.10. Note that, since the problem is two-dimensional, only two boundary conditions are needed on each boundary instead of three. Whether a boundary condition is essential or natural is also indicated against each boundary condition. *Boundaries AH, HG and FG*:

As stated earlier, the boundaries AH, HG and FG are chosen sufficiently away from the cutting edge projection E. Therefore, we can assume that the velocity vector has only x-component at these boundaries. Further, the velocity actually varies linearly from point H to point A and from point G to point F. But, since the distances AH and FG are very small compared to the work-piece radius, we assume the velocity to be uniform over these boundaries. Let v_c be the specified *cutting velocity*. Then, the boundary conditions at the boundaries AH, HG and FG become (Equation 3.267)

$$v_x = v_c, \quad v_y = 0, \qquad \text{(essential)}. \tag{7.9}$$

Boundary CD:
The boundary CD is also chosen sufficiently away from the cutting edge projection E. Therefore, we assume that, at this boundary also, the velocity vector is uniform over the whole boundary. Let v_{ch} be the *chip velocity*. The chip velocity can be calculated from the cutting velocity using the conservation of mass equation over the uncut depth and the chip thickness (Equation 3.268):

$$v_{ch} = v_c r, \tag{7.10}$$

where r is the cutting ratio given by Equation 7.2. Note that the chip velocity makes an angle α with y-axis. Then, the boundary conditions at the boundary CD become (Equation 3.269)

$$v_x = v_{ch} \sin \alpha, \quad v_y = v_{ch} \cos \alpha, \quad \text{(essential).} \tag{7.11}$$

Stress free boundaries AB, BC and EF:
The boundaries AB, BC and EF are stress-free surfaces. On the stress-free surfaces, the stress vector is zero at every point. Therefore, the boundary conditions at the boundaries AB, BC and EF can be expressed as (Equation 3.270)

$$t_x = 0, \quad t_y = 0, \quad \text{(natural),} \tag{7.12}$$

where t_x and t_y are the Cartesian components of the stress vector $\mathbf{t_n}$. Sometimes an alternate set of boundary conditions is used on these boundaries. This set is as follows. Since the direction of the velocity vector at the boundaries AB and EF is always along x-axis, the boundary condition (Equation 7.12) may be modified to specify v_y to be zero instead of t_y being zero. Thus, on the boundaries AB and EF, we can use the following boundary conditions:

$$v_y = 0, \quad \text{(essential),} \tag{7.13a}$$
$$t_x = 0, \quad \text{(natural).} \tag{7.13b}$$

Unlike the boundaries AB and EF, the boundary BC is inclined to x-axis. Therefore, on this boundary, we can specify the normal velocity component v_n and the shear stress component t_s to be zero instead of t_x and t_y being zero. Since, the inclination of the boundary BC with y-axis is α, these boundary conditions become

$$v_n \equiv -v_x \cos \alpha + v_y \sin \alpha = 0, \quad \text{(essential),} \tag{7.13c}$$
$$t_s \equiv t_x \sin \alpha + t_y \cos \alpha = 0, \quad \text{(natural).} \tag{7.13d}$$

The modified boundary conditions are expected to give a more accurate velocity field.

Tool-chip interface ED:
Velocity along the tool-chip interface ED can be approximated by the following relation (Equation 3.272):

$$v_\xi = \frac{v_{ch}}{3}\left(1+8\frac{\xi}{h}\right)^{1/2}, \qquad \text{for } \xi \le h,$$

$$= v_{ch}, \qquad\qquad \text{for } \xi > h,$$

(7.14)

where ξ is the distance measured along the boundary from point E. Thus, the value of v_ξ varies from $v_{ch}/3$ at point E to v_{ch} when ξ is equal to h. Note that, the boundary ED makes an angle α with y-axis. Then, the boundary conditions at boundary ED become (Equation 3.273)

$$v_x = v_\xi \sin\alpha, \quad v_y = v_\xi \cos\alpha, \quad \text{(essential)}. \tag{7.15}$$

7.3 Finite Element Formulation

As a first step towards the finite element formulation, we develop the integral form of the equations of motion (Equation 7.7) and the incompressibility constraint (Equation 7.8). This is called the mixed pressure-velocity formulation. We use the weighted residual method, described in Chapter 5, for this purpose. Normally, these equations are first made non-dimensional and then they are converted into an integral form. The procedure for making them non-dimensional is similar to the one described in Section 5.3.2 of Chapter 5. Here, we skip this step.

7.3.1 Integral Form

Let v_x, v_y and p be the functions of (x,y) which satisfy the essential boundary conditions exactly. Then, as stated in Chapter 5, these functions constitute an approximate solution to the problem consisting of the governing equations (Equations 7.4, 7.5, 7.7 and 7.8) and the boundary conditions (Equations 7.9, 7.11, 7.12 or 7.13, 7.15) if the following integral of the weighted residue is made zero:

$$\int_\Omega \left\{ (\dot{\varepsilon}_{xx}+\dot{\varepsilon}_{yy})w_p + \left[-\rho\left(v_x\frac{\partial v_x}{\partial x}+v_y\frac{\partial v_x}{\partial y} \right) - \frac{\partial p}{\partial x} + \left(\frac{\partial \sigma'_{xx}}{\partial x}+\frac{\partial \sigma'_{xy}}{\partial y} \right) \right]w_x \right.$$

$$\left. + \left[-\rho\left(v_x\frac{\partial v_y}{\partial x}+v_y\frac{\partial v_y}{\partial y} \right) - \frac{\partial p}{\partial y} + \left(\frac{\partial \sigma'_{xy}}{\partial x}+\frac{\partial \sigma'_{yy}}{\partial y} \right) \right]w_y \right\} \mathrm{d}x\mathrm{d}y = 0.$$

(7.16)

Here, w_p, w_x and w_y are functions of (x, y), called the weight functions, which are arbitrary except that they satisfy the homogeneous version of the essential boundary conditions. The functions v_x, v_y and p are called the approximation to the solution. In order to weaken the continuity requirements on the approximation, we simplify the second and third terms of the integrand of Equation 7.16 using a procedure similar to that of Section 5.3.3 of Chapter 5. Then, we get

$$\int_\Omega \left\{ -\left(\dot\varepsilon_{xx} + \dot\varepsilon_{yy}\right) w_p + \rho \left[\left(v_x \frac{\partial v_x}{\partial x} + v_y \frac{\partial v_x}{\partial y} \right) w_x + \left(v_x \frac{\partial v_y}{\partial x} + v_y \frac{\partial v_y}{\partial y} \right) w_y \right] \right.$$
$$\left. - p\left(\dot\varepsilon_{xx}(w) + \dot\varepsilon_{yy}(w)\right) + 2\eta_2 \left[\dot\varepsilon_{xx}\dot\varepsilon_{xx}(w) + \dot\varepsilon_{yy}\dot\varepsilon_{yy}(w) + 2\dot\varepsilon_{xy}\dot\varepsilon_{xy}(w) \right] \right\} dxdy$$
$$- \int_{\Gamma_x} w_x t_x \, dl - \int_{\Gamma_y} w_y t_y \, dl = 0.$$

$$(7.17)$$

Here, the quantities $\dot\varepsilon_{xx}(w)$, $\dot\varepsilon_{yy}(w)$ and $\dot\varepsilon_{xy}(w)$ are given by relations similar to Equation 7.4:

$$\dot\varepsilon_{xx}(w) = \frac{\partial w_x}{\partial x}, \quad \dot\varepsilon_{yy}(w) = \frac{\partial w_y}{\partial y},$$
$$\dot\varepsilon_{xy}(w) = \frac{1}{2}\left(\frac{\partial w_x}{\partial y} + \frac{\partial w_y}{\partial x} \right),$$

$$(7.18)$$

and Γ_x and Γ_y are respectively the boundaries on which the components of the stress vectors t_x and t_y are specified. Further, the deviatoric stress–strain rate relations (Equation 7.5) have been used to eliminate σ'_{ij} from the area integral and the decomposition of the stress tensor (Equation 2.100) and the Cauchy's relation (Equation 2.64) have been used to express the boundary integrals in terms of the components t_x and t_y of the stress vector:

$$(-p + \sigma'_{xx})n_x + \sigma'_{xy}n_y = \sigma_{xx}n_x + \sigma_{xy}n_y = t_x,$$
$$\sigma'_{xy}n_x + (-p + \sigma'_{yy})n_y = \sigma_{xy}n_x + \sigma_{yy}n_y = t_y.$$

$$(7.19)$$

Here, n_x and n_y are the components of a unit vector normal to the parts of the boundares on which t_x and t_y are specified.

For the convenience of finite element formulation, it is desirable to express the integral form (Equation 7.17) in an array notation. For this purpose, we define the following arrays:

$$\{\dot{\varepsilon}\} = \left\{ \begin{array}{c} \dot{\varepsilon}_{xx} \\ \dot{\varepsilon}_{yy} \\ (\sqrt{2})\dot{\varepsilon}_{xy} \end{array} \right\}, \quad \{\dot{\varepsilon}(w)\} = \left\{ \begin{array}{c} \dot{\varepsilon}_{xx}(w) \\ \dot{\varepsilon}_{yy}(w) \\ (\sqrt{2})\dot{\varepsilon}_{xy}(w) \end{array} \right\}, \quad \{m\} = \left\{ \begin{array}{c} 1 \\ 1 \\ 0 \end{array} \right\},$$

$$\{\nabla_x v\} = \left\{ \begin{array}{c} \dfrac{\partial v_x}{\partial x} \\ \dfrac{\partial v_y}{\partial x} \end{array} \right\}, \quad \{\nabla_y v\} = \left\{ \begin{array}{c} \dfrac{\partial v_x}{\partial y} \\ \dfrac{\partial v_y}{\partial y} \end{array} \right\}, \quad \{w\} = \left\{ \begin{array}{c} w_x \\ w_y \end{array} \right\}.$$

(7.20)

Then, the integral form (Equation 7.17) becomes

$$\int_\Omega \left\{ -\left(w_p \{m\}^T \{\dot{\varepsilon}\}\right) + \rho\left(v_x \{w\}^T \{\nabla_x v\} + v_y \{w\}^T \{\nabla_y v\}\right) - \left(\{\dot{\varepsilon}^T(w)\} \{m\} p\right) \right.$$
$$\left. + 2\eta_2 \{\dot{\varepsilon}^T(w)\} \{\dot{\varepsilon}\} \right\} dxdy - \int_{\Gamma_x} w_x t_x dl - \int_{\Gamma_y} w_y t_y dl = 0.$$

(7.21)

7.3.2 Approximations for Velocity Components and Pressure

The integral form is similar to that for the rolling problem (Equation 5.108) except for the second term which involves the acceration. Equation 7.21 contains the first derivatives of the velocity components but no derivatives of the pressure. Thus, a C^0-continuity approximation for the velocity components and a discontinuous approximation for the pressure can be chosen. However, for the reasons mentioned in Section 5.3.4 of Chapter 5, we choose a bi-linear approximation (*i.e.*, a C^0 continuity approximation) for the pressure and a higher order C^0 continuity approximation (*i.e.*, a second order serendipity approximation) for the velocity components. The corresponding element is shown in Figure 5.4b where the four corner nodes have three degrees of freedom (v_x, v_y, p) and the remaining four nodes have only two degrees of freedom (v_x, v_y). The domain is discretized into n_e number of such elements.

The approximation for the velocity components over a typical element e is given by

$$\{v\} = [N^v]\{v\}^e,$$

(7.22)

where

$$\{v\} = \begin{Bmatrix} v_x \\ v_y \end{Bmatrix}, \tag{7.23}$$

$$[N^v] = \begin{bmatrix} N_1 & 0 & N_2 & 0 & - & - & - & - & N_8 & 0 \\ 0 & N_1 & 0 & N_2 & - & - & - & - & 0 & N_8 \end{bmatrix} \tag{7.24}$$

is the velocity shape function matrix containing the two-dimensional second order serendipity shape functions N_i $(i = 1, 8)$ and

$$\{v\}^{eT} = \left\{ (v_x)_1^e \quad (v_y)_1^e \quad - \quad - \quad - \quad - \quad (v_x)_8^e \quad (v_y)_8^e \right\} \tag{7.25}$$

is the elemental velocity vector containing the nodal values of the velocity components at all eight nodes of the element e. The approximation for the pressure is given by

$$p = \{N^p\}^T \{p\}^e, \tag{7.26}$$

where the pressure shape function vector

$$\{N^p\}^T = \left\{ N_1^p \quad N_2^p \quad N_3^p \quad N_4^p \right\} \tag{7.27}$$

contains the two-dimensional bi-linear Lagrangian shape functions N_i^p $(i = 1, 4)$ and the elemental pressure vector

$$\{p\}^{eT} = \left\{ p_1^e \quad p_2^e \quad p_3^e \quad p_4^e \right\} \tag{7.28}$$

contains the nodal values of the pressure at the four corner nodes of the element e.

As in the case of finite element formulations of Chapters 5 and 6, here also we use the Galerkin version of the finite element formulation. Then, the weight functions (w_x, w_y) for the velocity components and the weight function w_p for the pressure are constructed by choosing the same shape functions as those used in approximating the velocity components and the pressure respectively. Thus,

$$\{w\} = [N^v]\{w_v\}^e, \tag{7.29}$$

$$w_p = \{w_p\}^{eT} \{N^p\}, \tag{7.30}$$

where the elemental weight vectors

$$\{w_v\}^{eT} = \left\{(w_x)_1^e \quad (w_y)_1^e \quad - \quad - \quad - \quad - \quad (w_x)_8^e \quad (w_y)_8^e\right\}, \tag{7.31}$$

$$\{w_p\}^{eT} = \left\{(w_p)_1^e \quad (w_p)_2^e \quad (w_p)_3^e \quad (w_p)_4^e\right\}, \tag{7.32}$$

contain respectively the nodal values of w_x and w_y at all eight nodes and the nodal values of w_p at the four corner nodes of the element e. These nodal values are known but arbitrary.

As the domain is discretized into area elements, the boundaries Γ_x and Γ_y get automatically discretized into three-noded line elements. Let n_{bx} and n_{by} be the number of line elements on Γ_x and Γ_y respectively. For the evaluation of the boundary integrals of Equation 7.21, we need expressions for the weight functions w_x and w_y over a typical line element b which would be consistent with their expressions (Equation 7.29) over the area elements. Such expressions are

$$w_x = \{w_x\}^{bT}\{N\}^b, \tag{7.33}$$

$$w_y = \{w_y\}^{bT}\{N\}^b, \tag{7.34}$$

where the boundary shape function vector

$$\{N\}^{bT} = \left\{N_1^b \quad N_2^b \quad N_3^b\right\}, \tag{7.35}$$

contains the one-dimensional quadratic Lagrangian shape functions and the boundary elemental weight vectors

$$\{w_x\}^{bT} = \left\{(w_x)_1^b \quad (w_x)_2^b \quad (w_x)_3^b\right\}, \tag{7.36}$$

$$\{w_y\}^{bT} = \left\{(w_y)_1^b \quad (w_y)_2^b \quad (w_y)_3^b\right\}, \tag{7.37}$$

contain respectively the nodal values of w_x and w_y at all three nodes of the line element b. These nodal values are known but arbitrary. We also approximate the variations of t_x and t_y along a typical line element b of the boundaries Γ_x and Γ_y by the following expressions:

$$t_x = \{N\}^{bT}\{t_x\}^b, \tag{7.38}$$

$$t_y = \{N\}^{bT}\{t_y\}^b, \tag{7.39}$$

where the vectors

$$\{t_x\}^{bT} = \left\{(t_x)_1^b \quad (t_x)_2^b \quad (t_x)_3^b\right\}, \tag{7.40}$$

$$\{t_y\}^{bT} = \left\{(t_y)_1^b \quad (t_y)_2^b \quad (t_y)_3^b\right\} \tag{7.41}$$

contain respectively the nodal values of t_x and t_y at all three nodes of the line element b.

7.3.3 Finite Element Equations

Differentiating the approximations for the velocity components (Equation 7.22) and the expressions for the weight functions (Equation 7.29), we obtain the following expressions for various arrays appearing in the integral form:

$$\{\dot{\varepsilon}\} = [B]\{v\}^e, \tag{7.42}$$

$$\{\dot{\varepsilon}(w)\} = [B]\{w_v\}^e, \tag{7.43}$$

$$\{\nabla_x v\} = [B_x]\{v\}^e, \tag{7.44}$$

$$\{\nabla_y v\} = [B_y]\{v\}^e, \tag{7.45}$$

where

$$[B] = \begin{bmatrix} \dfrac{\partial N_1}{\partial x} & 0 & \dfrac{\partial N_2}{\partial x} & 0 & - & - & \dfrac{\partial N_8}{\partial x} & 0 \\[2mm] 0 & \dfrac{\partial N_1}{\partial y} & 0 & \dfrac{\partial N_2}{\partial y} & - & - & 0 & \dfrac{\partial N_8}{\partial y} \\[2mm] \dfrac{1}{\sqrt{2}}\dfrac{\partial N_1}{\partial y} & \dfrac{1}{\sqrt{2}}\dfrac{\partial N_1}{\partial x} & \dfrac{1}{\sqrt{2}}\dfrac{\partial N_2}{\partial y} & \dfrac{1}{\sqrt{2}}\dfrac{\partial N_2}{\partial x} & - & - & \dfrac{1}{\sqrt{2}}\dfrac{\partial N_8}{\partial y} & \dfrac{1}{\sqrt{2}}\dfrac{\partial N_8}{\partial x} \end{bmatrix}, \tag{7.46}$$

$$[B_x] = \begin{bmatrix} \dfrac{\partial N_1}{\partial x} & 0 & \dfrac{\partial N_2}{\partial x} & 0 & - & - & \dfrac{\partial N_8}{\partial x} & 0 \\[2mm] 0 & \dfrac{\partial N_1}{\partial x} & 0 & \dfrac{\partial N_2}{\partial x} & - & - & 0 & \dfrac{\partial N_8}{\partial x} \end{bmatrix}, \tag{7.47}$$

$$[B_y] = \begin{bmatrix} \dfrac{\partial N_1}{\partial y} & 0 & \dfrac{\partial N_2}{\partial y} & 0 & - & - & \dfrac{\partial N_8}{\partial y} & 0 \\[2mm] 0 & \dfrac{\partial N_1}{\partial y} & 0 & \dfrac{\partial N_2}{\partial y} & - & - & 0 & \dfrac{\partial N_8}{\partial y} \end{bmatrix}. \tag{7.48}$$

Substituting Equations 7.26, 7.29, 7.30 and 7.42–7.45 for p, $\{w\}$, w_p, $\{\dot{\varepsilon}\}$, $\{\dot{\varepsilon}(w)\}$, $\{\nabla_x v\}$, $\{\nabla_y v\}$ over a typical area element e and Equations 7.33, 7.34, 7.38 and 7.39 for w_x, w_y, t_x and t_y over a typical line element b into the integral form (Equation 7.21), we get

$$\sum_{e=1}^{n_e}\left(\{w_p\}^{e\mathrm{T}}[k_{pv}]^e\{v\}^e + \{w_v\}^{e\mathrm{T}}[k_{vp}]^e\{p\}^e + \{w_v\}^{e\mathrm{T}}[k_{vv}]^e\{v\}^e\right),$$

$$= \sum_{b=1}^{n_{bx}}\left(\{w_x\}^{b\mathrm{T}}\{f_x\}^b\right) + \sum_{b=1}^{n_{by}}\left(\{w_y\}^{b\mathrm{T}}\{f_y\}^b\right). \tag{7.49}$$

Here, the elemental coefficient matrices $[k_{pv}]^e$, $[k_{vp}]^e$ and $[k_{vv}]^e$ are given by

$$[k_{pv}]^e = -\int_{A^e}\{N^P\}\{m\}^\mathrm{T}[B]\mathrm{d}x\mathrm{d}y,$$

$$\tag{7.50}$$

$$[k_{vp}]^e = [k_{pv}]^{e\mathrm{T}}, \tag{7.51}$$

$$[k_{vv}]^e = \int_{A^e}\left(\rho v_x[N^v]^\mathrm{T}[B_x] + \rho v_y[N^v]^\mathrm{T}[B_y] + 2\eta_2[B]^\mathrm{T}[B]\right)\mathrm{d}x\mathrm{d}y, \tag{7.52}$$

where A^e is the domain of a typical area element e and the elemental right side vectors are given by

$$\{f_x\}^b = \left(\int_{\Gamma_x^b}\{N^b\}\{N^b\}^\mathrm{T}\,\mathrm{d}l\right)\{t_x\}^b, \tag{7.53}$$

$$\{f_y\}^b = \left(\int_{\Gamma_y^b}\{N^b\}\{N^b\}^\mathrm{T}\,\mathrm{d}l\right)\{t_y\}^b, \tag{7.54}$$

where Γ_x^b and Γ_x^b are the domains of a typical line element b on the boundaries Γ_x and Γ_y respectively. Note that, we have not substituted the approximations for v_x and v_y in the expression for $[k_{vv}]^e$. The reason for this is as follows. In the present case, the resulting finite element equations become non-linear and, hence, need to be solved iteratively. In an iterative scheme which we adopt, the v_x and v_y in the expression for $[k_{vv}]^e$ are evaluated from the previous iteration.

Now, define the following elemental matrix and the elemental vectors:

$$[k]^e = \begin{bmatrix} [0] & [k_{pv}]^e \\ [k_{vp}]^e & [k_{vv}]^e \end{bmatrix}, \{\delta\}^e = \begin{Bmatrix} \{p\}^e \\ \{v\}^e \end{Bmatrix}, \{w\}^e = \begin{Bmatrix} \{w_p\}^e \\ \{w_v\}^e \end{Bmatrix}. \tag{7.55}$$

Then, Equation 7.49 becomes

$$\sum_{e=1}^{n_e} \left(\{w\}^{e\mathrm{T}} [k]^e \{\delta\}^e \right) = \sum_{b=1}^{n_{bx}} \left(\{w_x\}^{b\mathrm{T}} \{f_x\}^b \right) + \sum_{b=1}^{n_{by}} \left(\{w_y\}^{b\mathrm{T}} \{f_y\}^b \right). \tag{7.56}$$

The area integrals in Equations 7.50 and 7.52 are evaluated numerically by 3×3 Gauss quadrature, described in Section 5.2.2 of Chapter 5. For this purpose, the integrals are transformed to the natural coordinates (ξ,η) using the following transformation:

$$\int_{A^e} (.....)dxdy = \int_{-1}^{1}\int_{-1}^{1} (........)|J|d\xi d\eta, \tag{7.57}$$

where $|J|$ is the determinant of the Jacobian matrix:

$$[J] = \begin{bmatrix} \dfrac{\partial x}{\partial \xi} & \dfrac{\partial x}{\partial \eta} \\ \dfrac{\partial y}{\partial \xi} & \dfrac{\partial y}{\partial \eta} \end{bmatrix}. \tag{7.58}$$

The Jacobian is evaluated from the geometric approximation of the area element. In order to model the curved boundaries of the elements properly, we use the second order serendipity approximation for the element geometry, the same as that used for the velocity components. Thus,

$$x = \{N\}^{\mathrm{T}}\{x\}^e, y = \{N\}^{\mathrm{T}}\{y\}^e, \tag{7.59}$$

where

$$\{N\}^{\mathrm{T}} = \{N_1 \quad N_2 \quad - \quad - \quad - \quad N_8\} \tag{7.60}$$

contains the two-dimensional second order serendipity shape functions and

$$\{x\}^{e\mathrm{T}} = \{x_1^e \quad x_2^e \quad - \quad - \quad - \quad x_8^e\},$$
$$\{y\}^{e\mathrm{T}} = \{y_1^e \quad y_2^e \quad - \quad - \quad - \quad y_8^e\} \tag{7.61}$$

contain the nodal values of the coordinates (x,y) at all eight nodes of the element e. The line integrals in Equations 7.53 and 7.54 are evaluated numerically using 3 point Gauss quadrature. For this purpose, the integrals are transformed to the natural coordinate ς by the following transformation:

$$\int_{\Gamma^b} (.....)dl = \int_{-1}^{1}(.....)\left| J_b \right| d\varsigma, \tag{7.62}$$

where $\left| J_b \right|$ is the Jacobian for the line element:

$$\left| J_b \right| = \left(\left(\frac{dx}{d\varsigma} \right)^2 + \left(\frac{dy}{d\varsigma} \right)^2 \right)^{1/2}. \tag{7.63}$$

To evaluate the Jacobian, we need the geometric approximation of the line element. The approximation, which would be consistent with the geometric approxomination of the area element (Equation 7.59), is given by

$$x = \{N\}^{bT}\{x\}^b, \quad y = \{N\}^{bT}\{y\}^b, \tag{7.64}$$

where the vector $\{N\}^{bT}$ (defined by Equation 7.35) contains the one-dimensional quadratic Lagrangian shape functions and

$$\{x\}^{bT} = \left\{ x_1^b \quad x_2^b \quad x_3^b \right\}, \quad \{y\}^{bT} = \left\{ y_1^b \quad y_2^b \quad y_3^b \right\}, \tag{7.65}$$

contain the nodal values of the coordinates (x, y) at all the three nodes of the line element b.

After using the assembly procedure of Section 5.2.3 of Chapter 5, the finite element equations (Equation 7.56) become

$$\{W\}^T[K]\{\Delta\} = \{W\}^T\{F\} \tag{7.66}$$

where $\{W\}$, $[K]$ and $\{F\}$ are respectively the global weight vector, the global coefficient matrix and the global right side vector. The vector $\{\Delta\}$ contains the unknowns of the problem, namely, the nodal values of the velocity components and the pressure. It may be called as the global unkown vector. Since the nodal values of weight functions are arbitrary, the finite element equations (Equation 7.66) take the form

$$[K]\{\Delta\} = \{F\}. \tag{7.67}$$

7.3.4 Application of Boundary Conditions, Solution Procedure and Evaluation of Secondary Quantities

Boundary conditions need to be applied to Equation 7.67 before it is solved. The essential boundary conditions (Equations 7.9, 7.11, 7.13a and 7.15) are applied using the procedure described in Section 5.2.4 of Chapter 5. The natural boundary condition (Equation 7.13b) is applied simply by making the corresponding elements of the right side vector zero. The boundary conditions (Equations 7.13c, d) are special boundary conditions. They are applied using the procedure similar to that of the friction boundary condition of Section 5.3.5 of Chapter 5.

Equation 7.52 indicates that the coefficient matrix $[K]$ is not constant but depends on the unknown nodal values of the velocity components. Therefore, the finite element equations (Equation 7.67) are non-linear algebraic equations and need to be solved by an iterative scheme. In the present case, we use the following iterative scheme. In this scheme, we linearize the expression for $[K]$ by evaluating v_x, v_y and η_2 in Equation 7.52 from the previous iteration. To start the iterations, suitable guess values are used for v_x, v_y and η_2. In the first few iterations, the value of η_2 tends to become very large. This leads to some computations difficulties. These difficulties are overcome by prescribing a suitable cut-off for η_2. In the present work, this cut-off is chosen to be 1400. The iterations are continued until the nodal variables converge within 1% between the two successive iterations.

Once the nodal values of velocity components are obtained by solving Equation 7.67, one can proceed to determine the secondary quantities like the boundaries of the primary shear deformation zone (PSDZ), the average shear strain rate $(\dot{\gamma})$ and the average shear stress (τ).

First, the components of the strain rate tensor $(\dot{\varepsilon}_{ij})$ are calculated from the nodal values of the velocity components using Equation 7.42. These components are obtained at the Gauss points as they are observed to be accurate at these points. Then, the equivalent strain rate $(\dot{\varepsilon}_{eq})$ is calculated at the Gauss points from the following relation (relation similar to Equation 3.156):

$$\dot{\varepsilon}_{eq} = \left(\frac{2}{3}\dot{\varepsilon}_{ij}\dot{\varepsilon}_{ij}\right)^{1/2}. \tag{7.68}$$

A certain fraction of the maximum value of $\dot{\varepsilon}_{eq}$ over the whole domain is used as a cut-off value in determining the boundaries of the PSDZ. All the Gauss points at which $\dot{\varepsilon}_{eq}$ is higher than the cut-off value are considered to lie inside the PSDZ.

The boundaries of the PSDZ so determined are not, in general, either straight or parallel to the shear plane. However, in machining literature, the PSDZ is assumed to be a rectangle with sides parallel to the shear plane. The width of this rectangle is considered as an important secondary variable. For our PSDZ, we evaluate its *mean width ds* by dividing the area of the PSDZ by the dimension of the shear plane (*i.e.*, length BE of Figure 7.1).

If we choose a coordinate system in which one axis (*i.e.*, axis 1) is along the shear plane (*i.e.*, line BE of Figure 7.1) and the other axis (*i.e.*, axis 2) perpendicular to it, then the state of deformation in the PSDZ can be approximated as that of a pure shear. It means, out of the three independent components of the strain rate tensor $(\dot{\varepsilon}_{ij})$ and the deviatoric stress tensor (σ'_{ij}), only the shear components become non-zero. Thus,

$$\dot{\varepsilon}_{11} = 0, \quad \dot{\varepsilon}_{22} = 0, \quad \dot{\varepsilon}_{12} \neq 0,$$
$$\sigma'_{11} = 0, \quad \sigma'_{22} = 0, \quad \sigma'_{12} \neq 0. \tag{7.69}$$

In general, these components vary from point to point in the PSDZ. However, in machining literature, usually only one value of the shear strain rate and only one value of the shear stress is associated with the deformation in the PSDZ. We can interpret these values as the values of $\dot{\varepsilon}_{12}$ and σ'_{12} averaged over the PSDZ. The value of the shear strain rate reported in the machining literature is normally denoted by $\dot{\gamma}$ and represents the rate of change of angle between small line elements that were originally along the axes 1 and 2. On the other hand, $\dot{\varepsilon}_{12}$ represents only half the rate of change of this angle. Thus,

$$\dot{\gamma} = 2(\dot{\varepsilon}_{12})_{av} \Big|_{\text{PSDZ}} . \tag{7.70}$$

Although, $\dot{\gamma}$ is simply called the shear strain rate in the machining literature, we shall refere to it as the *average* shear strain rate. The symbol τ is usually used in the machining literature to denote the shear stress. Thus,

$$\tau = (\sigma'_{12})_{av} \Big|_{\text{PSDZ}} . \tag{7.71}$$

In the machining literature, τ is called the *shear flow stress*. However, we shall refer to it as the *average* shear stress. We now express $\dot{\varepsilon}_{12}$ and σ'_{12} in terms of the equivalent strain rate $\dot{\varepsilon}_{eq}$. Substituting $\dot{\varepsilon}_{11} = \dot{\varepsilon}_{22} = 0$ (Equation 7.69) in the expression for $\dot{\varepsilon}_{eq}$ (Equation 7.68), we get

$$\dot{\varepsilon}_{12} = \frac{\sqrt{3}}{2}\dot{\varepsilon}_{eq}. \tag{7.72}$$

Similarly, since $\sigma'_{11} = \sigma'_{22} = 0$ (Equation 7.69), the expression (Equation 3.23)

$$\sigma_{eq} = \left(\frac{3}{2}\sigma'_{ij}\sigma'_{ij}\right)^{1/2} \tag{7.73}$$

for the equivalent stress becomes

$$\sigma_{eq} = \sqrt{3}\sigma'_{12}. \tag{7.74}$$

When the deviatoric stress–strain rate relations (Equation 7.5) are combined with the definitions of the equivalent strain rate (Equation 7.68) and equivalent stress (Equation 7.73), we get the following relationship between σ_{eq} and $\dot{\varepsilon}_{eq}$:

$$\sigma_{eq} = (3\eta_2)\dot{\varepsilon}_{eq}. \tag{7.75}$$

Eliminating σ_{eq} from Equations 7.74 and 7.75, we get

$$\sigma'_{12} = (\sqrt{3}\eta_2)\dot{\varepsilon}_{eq}. \tag{7.76}$$

Substituting Equations 7.72 and 7.76 for $\dot{\varepsilon}_{12}$ and σ'_{12}, Equations 7.70 and 7.71 for the *average* shear strain rate $\dot{\gamma}$ and the *average* shear stress τ become

$$\dot{\gamma} = \sqrt{3}(\dot{\varepsilon}_{eq})_{av} \Big|_{\text{PSDZ}} ,$$
$$\tau = \sqrt{3}\eta_2(\dot{\varepsilon}_{eq})_{av} \Big|_{\text{PSDZ}} . \tag{7.77}$$

The equivalent strain rate ($\dot{\varepsilon}_{eq}$) values at the Gauss points of the PSDZ are used to calculate the *average* shear strain rate $\dot{\gamma}$ and the *average* shear stress τ. The equivalent strain rate $\dot{\varepsilon}_{eq}$ is averaged over the PSDZ to obtain $\dot{\gamma}$ whereas τ is determined by taking the average of the product of η_2 and $\dot{\varepsilon}_{eq}$ where η_2 is calculated from Equation 7.6.

7.4 Results and Discussion

The finite element equations developed in the previous section have been solved for mild steel material for various sets of machining conditions. First, the formulation is validated by comparing the mean width (*ds*) of the PSDZ , the average shear strain rate ($\dot{\gamma}$) and the average shear stress (τ) with the experimental results of Kececioglu [11]. Next, the effects of two important machining parameters, namely the cutting velocity (v_c) and feed (*f*), on the mean width (*ds*) of the PSDZ, the average shear strain rate ($\dot{\gamma}$) and the average shear stress (τ) are studied. Finally, for a typical set of machining conditions, the primary shear deformation zone (PSDZ), the contours of equivalent strain rate ($\dot{\varepsilon}_{eq}$) and the contours of equivalent stress (σ_{eq}) are presented.

The values of material constants (for mild steel) used in the present work are as follows:

- Density (ρ): 7860 kg/m^3;
- Viscoplastic material constants:

$$\sigma_Y = 350\text{MPa}, \quad a = 0.6246 \times 10^8, \quad m = -0.884 \quad \dot{\varepsilon}_{eq} < \left(1/\sqrt{3}\right)10^4,$$

$$\sigma_Y = 450\text{MPa}, \quad a = 0.8845 \times 10^8, \quad m = -0.867 \quad \dot{\varepsilon}_{eq} \geq \left(1/\sqrt{3}\right)10^4.$$

The values of *a* and *m* have been obtained by fitting Equation 7.6 through the experimental results of Jain and Gupta [12]. The above values of *a* are different from those of Joshi *et. al.* [5] because, their definition of equivalent strain rate $\dot{\varepsilon}_{eq}$ differs from Equation 7.68 by a factor of $\sqrt{3}$.

The finite element mesh with 44 eight-noded elements and 165 nodes, used in the present study, is shown in Figure 7.2.

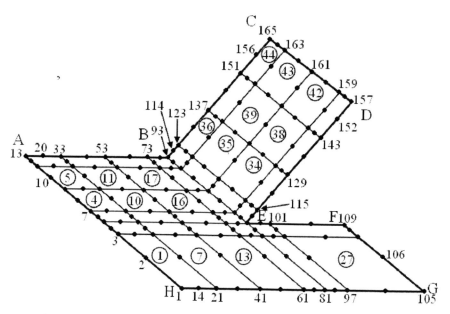

Figure 7.2. Finite element mesh with 44 eight-noded elements and 165 nodes

7.4.1 Validation of the Formulation

The validation is carried out for the following two sets of machining conditions (the same as used in [11])

- Set 1: Cutting velocity (v_c) = 3.73 m/s, Feed (*f*) = 0.3 mm/rev, Rake angle (α) in degrees = 19.5;
- Set 2: Cutting velocity (v_c) = 3.73 m/s, Feed (*f*) = 0.1 mm/rev, Rake angle (α) in degrees = 33.

First, the boundaries of the PSDZ are determined for the two sets of machining conditions using 3% cut-off. However, only the width of the PSDZ is reported in the machining literature by assuming it to be a rectangle with sides parallel to the shear plane. Therefore, for the comparison purpose, the *mean width ds* of our PSDZ is calculated by dividing its area by the dimension of the shear plane (*i.e.*, length BE of Figure 7.1). Further, the *average* shear strain rate $\dot{\gamma}$ and the *average* shear stress τ are calculated for the two sets of machining conditions by averaging respectively $\dot{\varepsilon}_{eq}$ and $\eta_2 \dot{\varepsilon}_{eq}$ over the PSDZ. The values of *ds*, $\dot{\gamma}$ and τ obtained from the present formulation for the two sets of machining conditions are compared with the corresponding experimental values from [11] in Table 7.1. There seems to be good agreement with the experimental results.

Table 7.1. Comparison of *ds*, $\dot{\gamma}$ and τ with experimental values of Kececioglu [11]

Quantity	Set 1			Set 2		
	Present	**Expt. [11]**	**% Error**	**Present**	**Expt. [11]**	**% Error**
ds (μm)	84.52	90.50	- 6.6	18.59	18.00	3.2
$\dot{\gamma}(\times 10^{-4} / s)$	3.83	3.90	- 1.7	15.14	17.60	- 13.9
τ (MPa)	474.90	494.70	- 4.1	540.50	562.40	- 3.2

7.4.2 Parametric Studies

In this section, the effects of two important machining parameters, namely the cutting velocity (v_c) and feed (*f*), on the mean width (*ds*) of the PSDZ, the average shear strain rate ($\dot{\gamma}$) and the average shear stress (τ) are studied.

- Mean width of the PSDZ (ds)

The mean width of the PSDZ (*ds*) decreases slightly with the cutting velocity (v_c) when the feed is kept constant. On the other hand, *ds* increases with the feed when the cutting velocity is kept constant. This result is in agreement with experimental results of [11].

- Average shear strain rate ($\dot{\gamma}$):

With an increase in cutting velocity, the average shear strain rate increases. However, it decreases with the feed.

- Average shear stress (τ):

When the feed is kept constant, the average shear stress increases with the cutting velocity, which is in agreement with the experimental results of [12]. However, when the cutting velocity is kept constant, the average shear stress decreases almost linearly, with the feed. Kececioglu's [11] experimental results support the decreasing trend of τ with feed.

7.4.3 Primary Shear Deformation Zone, Contours of Equivalent Strain Rate and Contours of Equivalent Stress

Figures 7.3–7.5 show respectively the primary shear deformation zone (PSDZ), the contours of equivalent strain rate ($\dot{\varepsilon}_{eq}$) and the contours of equivalent stress (σ_{eq}) for a typical set of machining conditions. A cut-off value of 3% has been used in obtaining the boundaries of the PSDZ.

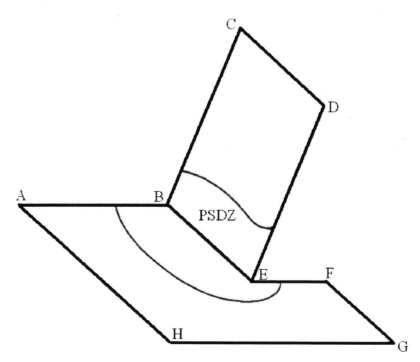

Figure 7.3. Primary shear deformation zone. From Joshi *et al.* [5]. Copyright [1994] Elsevier

Figures 7.4 and 7.5 are the same as Figure 10a, b of Joshi *et al.* [5] except for the following modification. As stated earlier, the definition of $\dot{\varepsilon}_{eq}$ of Joshi *et al.* [5] differs from Equation 7.68 by a factor of $\sqrt{3}$. Further, the definition of σ_{eq} of Joshi *et al.* [5] differs from Equation 7.73 by a factor of $1/\sqrt{3}$. Therefore, the numerical values of $\dot{\varepsilon}_{eq}$ and σ_{eq} in Figures 7.4 and 7.5 are obtained by multiplying the corresponding values of Figure 10a, b of Joshi *et al.* [5] by the factors of $1/\sqrt{3}$ and $\sqrt{3}$ respectively.

It is observed that the maximum values of $\dot{\varepsilon}_{eq}$ and σ_{eq} occur near the cutting edge and are of the order of 2.31×10^4 s^{-1} and 866 MPa.

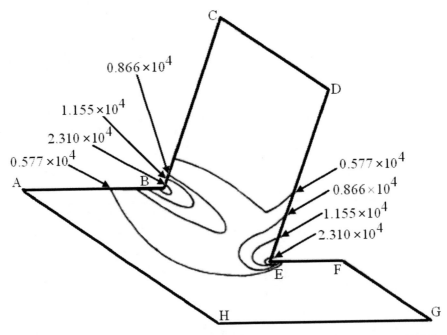

Figure 7.4. Contours of equivalent strain rate (s^{-1}). From Joshi *et al.* [5]. Copyright [1994] Elsevier

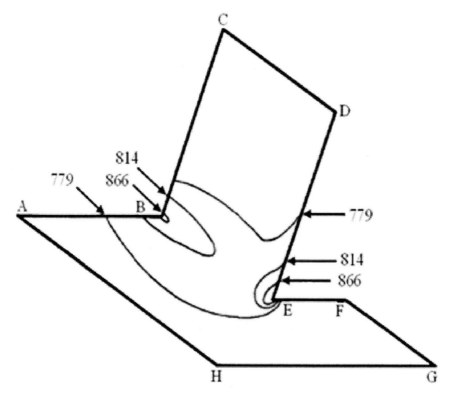

Figure 7.5. Contours of equivalent stress (MPa). From Joshi *et al.* [5]. Copyright [1994] Elsevier

7.5 Summary

In this chapter, first the Eulerian formulation of an orthogonal steady-state machining process is described. Unlike the updated Lagrangian formulation, where the chip formation is predicted using a chip seperation criterion, this formulation assumes the existence of the chip. Further, this formulation is different to that of the Eulerian formultion of the plane-strain rolling problem of Chapter 5 in several aspects. The first difference is that the plastic deformation is not spread over a large part of the work-piece, but is localized near the cutting edge. Therefore, the domain selected is a small region around the shear plane. While selecting the domain, it is assumed that the shear angle and the angle between the shear force and the resultant force have been determined experimentally. Further, the width of cut is assumed large enough to make the problem two-dimensional. The second difference is that, unlike in the rolling pocess, the strain rates and the temperature rise are quite significant. Therefore, one needs to determine the temperature rise and incorporate visco-plasticity and temperature softening in the constitutive equation. To estimate the temperature rise, one needs to solve the heat transfer equation in conjuction with the equations govering the plastic deformation. In the present work, the problem is decoupled by assuming that the average temperature rise in the cutting zone for typical machining conditions can be estimated experimentally or analytically. Then, the temperature softening is incorporated by evaluating the viscoelastic properties of the material at the average temperature of the cutting zone. The friction at the tool-chip contact is very complex and is not fully understood. Therefore, it is taken care of by using simple estimates of the tool-chip contact length and the contact velocity based on empirical knowledge.

The development of the finite element equations is similar to that of the plane-strain rolling problem of Chapter 5. The Eulerian finite element model is verified by comparing the predictions of the mean width of the primary shear deformation zone (PSDZ), the average shear strain rate and the average shear stress with experimental results. Then the parametric study of these three quantities is carried out with respect to two machining variables, namely the cutting velocity and the feed. Finally, for a typical set of machining conditions, the shape of the PSDZ and the contours of the equivalent strain rate and the equivalent stress are presented.

7.6 References

[1]   Tay, A.O., Stevenson, M.G. and de Vahl Davis, G. (1974), Using the finite element method to determine temperature distributions in orthogonal machining, Proceedings of the Institution of Mechanical Enginners, Vol. 188, pp. 627–638.

[2]   Tay, A.O., Stevenson, M.G., de Vahl Davis, G. and Oxley, P.L.B. (1976), A numerical method for calculating temperature distributions in machining, from force and shear angle measurements, International Journal of Machine Tool Design and Reearch., Vol. 16, pp. 335–349.

[3]   Murarka, P.D., Barrow, G. and Hinduja, S. (1979), Influence of the process variables on the temperature distribution in orthogonal machining using the finite element method, International Journal of Mechanical Sciences, Vol. 21, pp. 445–456.

[4]   Iwata, K., Osakada, K. and Terasaka, Y. (1984), Process modeling of orthogonal cutting by the rigid-plastic finite element method, Transactions of ASME, Journal of Engineering Materials and Technology, Vol. 106, pp. 132–138.

[5]   Joshi, V.S., Dixit, P.M. and Jain, V.K. (1994), Viscoplastic analysis of metal cutting by finite element method, International Journal of Machine Tools and Manufacture, Vol. 34, pp. 553–571.

[6]   Kim, K.W. and Sin, H.C. (1996), Development of a thermo-viscoplastic cutting model using finite element method, International Journal of Machine Tools and Manufacture, Vol. 36, pp. 379–397.

[7]   Strenkowski, J.S. and Carrol, J.T. (1985), A finite element model of orthogonal metal cutting, Transactions of ASME, Journal of Engineering for Industry, Vol. 107, pp. 349–354.

[8]   Lei, S., Shin, Y.C., and Incropera, F.P. (1999), Thermo-mechanical modeling of orthogonal machining process by finite element analysis, International Journal of Machine Tools and Manufacture, Vol. 39, pp. 731–750.

[9]   Mamalis, A.G., Horvath, M., Branis, A.S. and Manolakos, D.E. (2001), Finite element simulation of chip formation in orthogonal metal cutting, Journal of Materials Processsing Technology , Vol. 110, pp. 19–27.

[10]   Ceretti, E., Lazzaroni, C., Menegardo, L. and Altan, T. (2000), Turning simulations using a three-dimensional FEM code, Journal of Materials Processsing Technology, Vol. 98, pp. 99–103.

[11] Kececioglu, D. (1958), Shear strain rate in metal cutting and its effect on shear flow stress, Transactions of ASME, Vol. 80, pp. 158–168.
[12] Jain, V.K. and Gupta, B.K. (1987), Effects of accelerated tests on shear flow stress in machining, Transactions of ASME, Journal of Engineering for Industry, Vol. 109, pp. 206–212.

8

Background on Soft Computing

8.1 Introduction

There are many situations where imprecise or incomplete information is available about a problem until an approximate solution of the problem is obtained. To give a simple example, in most of the mechanics problem, one has to use a coefficient of friction whose value is not known precisely. While one can measure the mass of the body quite easily and values of applied forces may be known, the coefficient of friction may not be known precisely. For example Merium and Kraige [1] give a table of friction for various contacting surfaces, but mention that a variation of 25–100% or more from those values could be expected in an actual application, depending on prevailing conditions of cleanliness, surface finish, pressure, lubrication, and velocity. Notwithstanding the prevalent imprecision in the values of coefficient of friction, most of the time one does carry out precise (hard) computations with the most likely value of the coefficient of friction. However, such hard computations without a mention of imprecision in the solution have limited practical utility.

Coming to the areas of machining and metal forming, we encounter many imprecise (and often uncertain) parameters. It should not be surprising to see a large variation in the estimated and experimental cutting forces in orthogonal machining when Merchant's single shear plane model is used for estimating the forces. This variation may be due to imprecise knowledge of coefficient of friction at the tool-chip interface and shear flow stress of the work-material and approximation in the Merchant's model. Similarly, in the metal forming problems, the material properties and coefficient of friction may be known imprecisely. At the same time, there are approximations in the mathematical models.

Soft computing-based methods acknowledge the presence of imprecision and uncertainty, while attempting to find reasonably useful solutions. Soft computing became recognized through the efforts of Lotfi Zadeh, the father of "fuzzy sets". In 1991, he established the Berkely Initiative in Soft Computing (BISC). Soft computing exploits the tolerance for uncertainty and imprecision to achieve greater tractability and robustness, and lower cost of solution [2]. The three popular constituents of soft computing are fuzzy sets, neural networks and genetic algorithms.

Fuzzy set theory helps to carry out computations with imprecisely defined parameters. Sometimes the parameters may be defined in a language form. For example, someone may just say a very good surface finish. In his language a surface roughness of 0.8 μ or less may mean definitely a very good surface finish and as the surface roughness increases beyond 0.8 μ, it gradually (not abruptly) deviates from the meaning of a very good surface finish. With the help of fuzzy set theory, one may predict somewhat less precise but usable values of surface roughness. Fuzzy set theory has been used in traditional hard computing by taking the variables in fuzzy form as well as modeling in its own right. The latter use of fuzzy set theory has been both admired and condemned [3].

Neural networks are motivated by the working of the human brain and are good in learning the behavior of a system. For example, when one observes that with increasing percentage reduction of strip thickness, the roll force increases, one can immediately conclude that the roll force increases with the percentage reduction and an approximate relation between them can be found. This type of model estimation has been carried out with techniques like multiple regressions. However, neural networks can understand the behavior with a reasonable accuracy in the presence of a number of input and output parameters.

The genetic algorithms are used for finding out the maxima and minima of a function in a heuristic manner. The obtained global maxima and minima may not be exact but are expected to be very close to exact ones. Moreover, the objective and constraint functions need not be continuous and may even be expressed in the form of language. This makes it very suitable to be used with fuzzy set theory.

These are three very popular techniques, though there are many similar techniques. Usually the various techniques of soft computing do not compete with each other, but are complementary. In this chapter, we will provide a background on neural networks, fuzzy sets and genetic algorithms.

8.2 Neural Networks

There are a number of textbooks available on neural networks [4–7]. The description of this section is intended just to provide a background for understanding the chapters ahead. Neural networks are motivated by the functioning of brain, which consists of a number of neurons. The network in the brain is called biological neural network, whereas we build artificial neural networks for solving physical problems. The artificial neural network (ANN) may be very different from a biological neural network. From now onwards we will use the terms ANN or neural networks to mean artificial neural networks. Neural networks are systems which can acquire, store and utilize knowledge gained from experience. Neural network techniques have been found capable of learning from a dataset to describe the non-linear and interaction effects with great success.

As a very simple example of how a neural network can be used, consider a dependent variable z related to independent variables x and y in the following manner:

$$z = x^2 + y^2, \tag{8.1}$$

If just provided a few datasets in the form of triplet (x, y, z), the neural network must be able to understand that the function is of the form given in Equation 8.1. The important point is that too many exemplars should not be required. The total data by which the neural network understands the relation between the variables is called training data. After the network has been trained based on the training data, it has to be tested with a few data called testing data. We will discuss ANNs in detail in subsequent subsections. In the following subsection, we provide a very brief description of biological neural networks.

8.2.1 Biological Neural Networks

The brain consists of a densely interconnected set of nerve cells, or basic information processing units, called neurons. The human brain incorporates nearly 10^{11} neurons and 10^{14} connections through synapses between them. Although each neuron has a very simple structure, a combination of such elements constitutes a tremendous processing power. As shown in Figure 8.1, a neuron consists of a cell body, soma, a number of fibers called dendrites, and a single long fiber called the axon. Dendrites form a very fine bush of thin fibers around the neuron's body. Dendrites receive information from neurons through axons (long fibers) that serve as transmission lines. An axon is a long cylindrical connection that carries impulses from the neuron. At the end part of an axon, various branches terminate at the surface of other neurons or on the dendrites. The axon-dendrite contact organ is called a synapse, through which the neuron introduces its signal to the neighboring neuron. Signals are propagated from one neuron to another by complex electrochemical reactions. Signals travel in the axon in the form of electrical impulses. Synapses convert the electrical signals into chemical ones. Chemical substances released from the synapses cause a change in the electrical potential of the cell body. When the potential of the cell body reaches its threshold, an electrical pulse, action potential, is generated. This pulse is transmitted through the axon to reach the other synapses, causing them to increase or decrease the potential of cell bodies. Usually, each neuron has one axon to transmit the signal and thousands of synapses to receive the signals from the other neurons. Generation of electrical impulse by the cell body is called firing of the neuron. If the incoming impulses help in firing of a neuron, they are called excitatory impulses. If they hinder the process of firing, they are called inhibitory.

In response to the stimulation pattern, neurons demonstrate long-term changes in the strength of their connections. Neurons can also form new connections with other neurons. Even entire collections of neurons may sometimes migrate from one place to another. These mechanisms form the basis for learning in the brain. This phenomenon is called plasticity. The plasticity diminishes with age. It has also been found that a typical human brain loses about 2–5% of its total neurons by the time it reaches 50 years of age.

A human brain can be considered as a highly complex, non-linear and parallel information-processing system. Information is stored and processed in a neural network simultaneously throughout the whole network, rather than at specific locations. In other words, in neural networks, both data and its processing are global rather than local. Owing to the plasticity of the network, connections between neurons leading to the 'right judgment' are strengthened while those leading to the 'wrong judgment' become weak. As a result, neural networks have the ability to learn through the experience. Learning is a fundamental and essential characteristic of biological neural networks. The ease and naturalness with which they can learn motivated us to emulate a biological neural network in a computer. However, the types of artificial neural networks, which are described in this chapter, are highly simplified versions of actual biological networks.

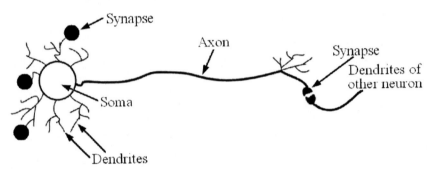

Figure 8.1. A typical biological neuron

8.2.2 Artificial Neurons

The first model of an artificial neuron was proposed by McCulloch and Pitts in their 1943 paper [8] and is called *Threshold Logic Unit* (TLU) or a *Linear Threshold Gate*. Figure 8.2 provides a graphical representation of such a unit with n real-valued input x_i, each input being associated with a parameter w_i. Parameter w_i is also known as a "synaptic weight" or simply "weight", in analogy with biological synapses. A TLU performs a weighted sum operation followed by a threshold operation such that if the value of the sum is greater or equal than a threshold θ, then the output y of the unit is 1, otherwise it is –1. Stated mathematically,

$$y(x) = \begin{cases} 1 & \text{if } \sum_{i=1}^{n} w_i x_i - \theta \geq 0, \\ -1 & \text{if } \sum_{i=1}^{n} w_i x_i - \theta < 0. \end{cases} \tag{8.2}$$

Thus, the neuron will "fire", that is, it will emit an instantaneous "1" signal if the threshold is exceeded; otherwise, it will emit –1. (One can also have a neuron, which will emit 0, when it is not fired.) The weighted sum in Equation 8.2 is called the neuron activation. The function $y(x)$ is called activation function. In Equation 8.2, activation function is called a *sign function*, because it provides +1 or −1 depending on the sign of $\sum_{i=1}^{n} w_i x_i - \theta$. If the activation function emits 0 in the unfired state, it is called a *step function*.

If the inputs are binary then a TLU becomes a Boolean function. However, all Boolean functions cannot be realized by it. For example, AND and OR can be realized by it, but a single unit cannot realize XOR (exclusive or) gate. It is interesting to mention here that NAND and NOR gates can be realized by a single TLU. Since NAND and NOR are universal logic gates, *i.e.*, any logic gate can be obtained by using some NAND (or NOR) gates, TLU is also universal. Any Boolean function can be realized by a suitable network of TLUs.

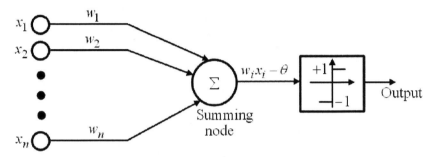

Figure 8.2. An artificial neuron model proposed by McCulloch and Pitts

Example 8.1: How can you realize AND and OR gate using a TLU?

Solution: Truth table of an AND gate is shown in Table 8.1. Observing this truth table, we notice that if the sum of inputs is 2, the output is 1, otherwise it is zero. Hence, we can realize an AND gate with a TLU with the following activation function:

$$y(x) = \begin{cases} 1 & \text{if } \sum_{i=1}^{2} x_i - 2 \geq 0, \\ 0 & \text{if } \sum_{i=1}^{2} x_i - 2 < 0. \end{cases} \tag{8.3}$$

Note that all weights are 1.

Table 8.1. Truth table of an AND gate

Inputs		Output
x_1	x_2	y
0	0	0
0	1	0
1	0	0
1	1	1

Similarly, the truth table for an OR gate is shown in Table 8.2. Here, we observe that if the sum of the inputs is less than 0, the output is 0, otherwise it is 1. Hence, we can realize an OR gate with a TLU with the following activation function:

$$y(x) = \begin{cases} 1 & \text{if } \sum_{i=1}^{2} x_i - 1 \geq 0, \\ 0 & \text{if } \sum_{i=1}^{2} x_i - 1 < 0. \end{cases} \tag{8.4}$$

Table 8.2. Truth table of an OR gate

Inputs		Output
x_1	x_2	y
0	0	0
0	1	1
1	0	1
1	1	1

Example 8.2: Show that a TLU can realize a NAND gate. Also show that the combination of TLUs can realize an XOR gate.

Solution: NAND gate is basically an AND gate followed by a NOT gate. A NOT gate changes 1 to 0 and *vice versa*. The truth table for NAND gate is shown in Table 8.3. This can be realized by a TLU having the following processing function:

$$y(x) = \begin{cases} 1 & \text{if } \sum_{i=1}^{2} -x_i + 1 \geq 0, \\ 0 & \text{if } \sum_{i=1}^{2} -x_i + 1 < 0. \end{cases} \tag{8.5}$$

It can be easily shown that NAND gate is a universal gate, which means that AND, OR and NOT gates can easily be realized using this gate [9].

Table 8.3. Truth table of a NAND gate

Inputs		Output
x_1	x_2	y
0	0	1
0	1	1
1	0	1
1	1	0

Table 8.4 is the truth table of XOR or exclusive OR. The output can be written as a function of input using Boolean algebra. For this purpose, inputs which provide output as 1 are combined by AND operator and these combinations are added using OR operation. Thus, in the present case,

$$y = \bar{x}_1 . x_2 + \bar{x}_2 . x_1 , \tag{8.6}$$

where dot, plus and bar over a variable denote, respectively, AND, OR and NOT operators. Thus, using AND, OR and NOT operators, we can realize XOR. Using NAND gates alone AND, OR and NOT can be obtained as shown in Figure 8.3.

In the previous examples, the activation function provided discrete output. A number of different activation functions can be used with neurons. One common function has the form

$$f(net) = \frac{2}{1 + \exp(-\lambda \, net)} - 1 \; , \tag{8.7}$$

where $net = \sum_{i=1}^{n} w_i x_i + b$. In this b is called bias. Equation 8.7 is called *sigmoidal function*. This function is also called hyperbolic tangent activation function [5]. For any input, this function is bounded between -1 and $+1$. In this, the parameter $\lambda > 0$ determines the steepness of the continuous function $f(net)$ near $net=0$. This can be seen by differentiating Equation 8.7 and putting $net = 0$. Rate of change of the function with respect to net is given by

$$f'(net) = \frac{\mathrm{d}\, f(net)}{\mathrm{d}\, net} = \frac{2\lambda \exp(-\lambda \, net)}{(1 + \exp(-\lambda \, net))^2} \; . \tag{8.8}$$

Its value at $net = 0$ is $\lambda/2$. Thus, the rate of change of the function at the origin is proportional to λ. When λ becomes very large, $f'(net)$ becomes very large. In the limit $\lambda \to \infty$, the continuous function becomes *sign* function defined by Equation 8.2. At $\lambda = 0$, the function becomes constant function. Similar type of function is

$$f(net) = \frac{1}{1 + \exp(-\lambda \, net)} \; , \tag{8.9}$$

which is bounded between 0 and 1. This function is also called log sigmoidal function. A unipolar ramp activation function is given by

$$f(net) = net \; . \tag{8.10}$$

Usually the activation functions having continuous derivatives are preferred, so that they can be conveniently used in network training algorithms involving gradient-based optimization methods. Figure 8.4 shows the model of a neuron with a continuous activation function.

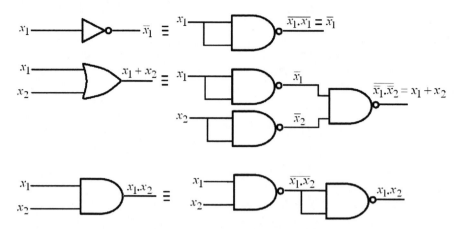

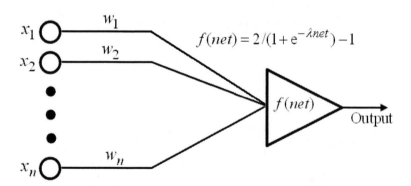

Figure 8.3. Obtaining NOT, OR and AND gates using combinations of NAND gates

Table 8.4. Truth table of a XOR gate

Inputs		Output
x_1	x_2	y
0	0	0
0	1	1
1	0	1
1	1	0

$$f(net) = 2/(1 + e^{-\lambda net}) - 1$$

Figure 8.4. Model of a neuron with continuous activation function

8.2.3   Perceptron: The Learning Machine

In the previous subsection, a model of a single neuron has been presented. The neuron can behave in a particular way depending on its weights and bias. However, it must have the ability to learn through exemplar in order to emulate the behavior of a biological system. In 1958, Rosenblatt introduced the first learning machine, discrete (binary) perceptron [10]. The perceptron will be as shown in Figure 8.2 with the ability to adjust its weights and bias with supplied training data. The learning method in which the data in the form of input and output is supplied and the network is trained to minimize the error between predicted and desired (target) output is called supervised learning. The error is reduced by making small adjustments in the weights to reduce the difference between the predicted and the desired output of the perceptron. The initial weights are randomly assigned and then updated to obtain the output consistent with the training examples. For a perceptron, the process of weight updating is simple. If at iteration p, the predicted output is $o(p)$ and the desired output (target) is $d(p)$, then the error is given by

$$e(p) = d(p) - o(p) \quad \text{where } p = 1, 2, 3 \dots \dots \tag{8.11}$$

At each iteration a fresh training data is presented to the perceptron. If the error $e(p)$, is positive, we need to increase the perceptron's output $o(p)$, but if it is negative, we need to decrease $o(p)$. Taking into account that each perceptron input contributes $x_i(p) \times w_i(p)$ to the total input $X(p)$, we find that if input value $x_i(p)$ is positive, an increase in its weight $w_i(p)$ tends to increase perceptron output $o(p)$. On the other hand, if $x_i(p)$ is negative, an increase in $w_i(p)$ tends to decrease $o(p)$. Thus, the following perceptron learning rule can be established:

$$w_i(p+1) = w_i(p) + \alpha\, x_i(p) \times e(p), \tag{8.12}$$

where α is the learning rate.

Example 8.3: Taking $\theta = 1$ and guess values of the weights as $[0 \quad 0]^T$, train a perceptron to learn the operation of an OR gate. Take learning rate as 1.

Solution: Truth table of an OR gate is shown in Table 8.2. The perceptron's predicted output at any iteration is given by

$$o(p) = \begin{cases} 1 & \text{if} \quad w_1 x_1 + w_2 x_2 - 1 \geq 0 \\ 0 & \text{if} \quad w_1 x_1 + w_2 x_2 - 1 < 0 \end{cases}, \tag{8.13}$$

and weights are updated in the following manner:

$$\begin{Bmatrix} w_1(p+1) \\ w_2(p+1) \end{Bmatrix} = \begin{Bmatrix} w_1(p) \\ w_2(p) \end{Bmatrix} + \begin{Bmatrix} x_1 \\ x_2 \end{Bmatrix} (d(p) - o(p)). \tag{8.14}$$

Here p refers to iteration number. Now, let us start the training process.

Iteration 1 (p=1)

$$\begin{Bmatrix} w_1(1) \\ w_2(1) \end{Bmatrix} = \begin{Bmatrix} 0 \\ 0 \end{Bmatrix}, \quad \begin{Bmatrix} x_1 \\ x_2 \end{Bmatrix} = \begin{Bmatrix} 0 \\ 0 \end{Bmatrix}, \quad d(1) = 0 \text{ and } o(1)=0$$

Thus, $\quad \begin{Bmatrix} w_1(2) \\ w_2(2) \end{Bmatrix} = \begin{Bmatrix} 0 \\ 0 \end{Bmatrix}$

Iteration 2 (p=2)

$$\begin{Bmatrix} x_1 \\ x_2 \end{Bmatrix} = \begin{Bmatrix} 0 \\ 1 \end{Bmatrix}, \quad d(2) = 1 \text{ and } o(2) = 0$$

Hence,

$$\begin{Bmatrix} w_1(3) \\ w_2(3) \end{Bmatrix} = \begin{Bmatrix} 0 \\ 0 \end{Bmatrix} + \begin{Bmatrix} 0 \\ 1 \end{Bmatrix} = \begin{Bmatrix} 0 \\ 1 \end{Bmatrix}$$

Iteration 3 (p = 3)

$$\begin{Bmatrix} x_1 \\ x_2 \end{Bmatrix} = \begin{Bmatrix} 1 \\ 0 \end{Bmatrix}, \quad d(3) = 1 \text{ and } o(3) =0$$

Hence,

$$\begin{Bmatrix} w_1(4) \\ w_2(4) \end{Bmatrix} = \begin{Bmatrix} 0 \\ 1 \end{Bmatrix} + \begin{Bmatrix} 1 \\ 0 \end{Bmatrix} = \begin{Bmatrix} 1 \\ 1 \end{Bmatrix}$$

Iteration 4 (p=4)

$$\begin{Bmatrix} x_1 \\ x_2 \end{Bmatrix} = \begin{Bmatrix} 1 \\ 1 \end{Bmatrix}, \quad d(4) = 1 \text{ and } o(4) =1$$

Hence, no change in the weights will be observed. Further iterations will show that now errors are zero and all the four conditions of OR gate are satisfied. Hence, the weights are $w_1 = w_2 = 1$.

In the same way, we can train to make an AND gate. However XOR gate can be realized by this perceptron. The decision function of this perceptron is $(w_1 x_1 + w_2 x_2 - \theta)$. Neuron emits output 0 or 1 depending on the sign of this function. This means that line $w_1 x_1 + w_2 x_2 - \theta = 0$ separates the points giving output 0 and 1; points falling on one side of the decision function correspond to output 1 and points falling on the other side correspond to output 0. The dataset which can be separated like this is called linearly separable. AND and OR gates are linearly separable, whereas the XOR is not. This can be seen very easily. For $x_1 = 0$ and $x_2 = 0$, XOR gate provides output 0, suggesting $-\theta < 0$. Therefore, threshold has to be positive. For $x_1 = 1$ and $x_2 = 0$, XOR gate provides output 1. This gives

$$w_1 - \theta \geq 0 \quad \text{or} \quad w_1 \geq \theta.$$

Similarly, for $x_1 = 0$ and $x_2 = 1$, the output is 1. This gives

$$w_2 - \theta \geq 0 \quad \text{or} \quad w_2 \geq \theta,$$

For $x_1 = 1$ and $x_2 = 1$, the values of decision function is $w_1 + w_2 - \theta$. Since both w_1 and w_2 are greater or equal to θ, this value becomes greater or equal to θ, a positive quantity. Thus, the output has to be 1. However, for this combination of input values, XOR gate should give 0, which cannot be achieved by this type of perceptron.

For neurons with continuous activation function, a popular learning rule is delta (δ) learning rule introduced by McClelland and Rumelhart in 1986 [11]. This is based on the steepest descent method. In this method, for minimizing the function, we move towards the direction of steepest descent, which is the negative of the gradient vector. For a particular pattern, the error is given by

$$e = \frac{1}{2}(d - o)^2 . \tag{8.15}$$

For input vector with n variables, the predicted output o is given by

$$o = f(w_1 x_1 + w_2 x_2 + \ldots\ldots + w_n x_n) = f(\sum_{i=1}^{n} w_i x_i) . \tag{8.16}$$

Note that here we have not included the threshold or bias. If we keep x_n always equal to -1, then w_n can play the role of threshold. Hence, threshold or bias is a special type of weight associated with a fixed input. With this the error becomes a function of weights as given below:

$$e = \frac{1}{2}\left(d - f(\sum_{i=1}^{n} w_i x_i)\right)^2 = \frac{1}{2}(d - f(net))^2 . \tag{8.17}$$

The i-th component of the gradient vector of the above error function with respect to weights is given by

$$(\nabla e)_i = -(d - f(net)) f'(net) x_i , \tag{8.18}$$

where $f'(net)$ is the first derivative of the activation function with respect to the net input to the neuron. For minimizing the error, we move towards negative direction of the gradient vector. Hence, the weights are updated in the following manner:

$$(w_i)_{new} = (w_i)_{old} + \eta (d - f(net)) f'(net) x_i , \tag{8.19}$$

where η is a constant called learning rate. For closeness to true steepest descent the learning rate should be very small; however, one may get stuck in local minima. Large value of learning rate makes the training process faster and may help in reaching global minima, but the algorithm may sometimes become unstable. Hence, the selection of proper value of learning rate is crucial.

If the weights are updated after every presentation of an input pattern, it is called incremental learning. If the weight updating is performed after all the patterns have been presented, it is called batch learning. The incremental learning is preferred due to requirement of less storage and suitability for online application

Single-layer perceptrons make decisions in the same way, regardless of the activation function used by the perceptron [12]. It means that a single-layer perceptron can classify only linearly separable patterns, regardless of whether we use a hard-limit (discrete) or soft-limit (continuous) activation function. Advanced forms of neural networks, for example, multi-layer perceptrons trained with *back-propagation algorithm*, can overcome the limitation of Rosenblatt's perceptron.

8.2.4 Multi-Layer Perceptron Neural Networks

A multi-layer perceptron (MLP) is a feedforward neural network with one or more hidden layers. A feedforward network has a sequence of layers consisting of a number of neurons in each layer. The output of neurons of one layer becomes input to neurons in the succeeding layer. Typically a network consists of an input layer consisting of neurons corresponding to input variables, at least one middle or hidden layer of computational neurons, and an output layer of computational neurons. The input signals are propagated in a forward direction on a layer-by-layer basis. A multi-layer perceptron with one hidden layer is shown in Figure 8.5.

The first layer, called an input layer, receives data from the outside world. The last layer is the output layer, which sends information out to users. Layers that lie between the input and output layers are called hidden layers and have no direct contact with the environment. Their presence is needed in order to provide complexity to network architecture for modeling non-linear functional relationship. After choosing the network architecture, the network is trained by providing data in the form of several input-output pairs. During the training process, the network adjusts its weights to minimize the error between the predicted and desired outputs.

The most common algorithm for adjusting the weights is the back propagation algorithm. Here, the training process involves two passes. In the forward pass, the input signals propagate from the network input to output. In the reverse pass, the calculated error signals propagate backwards through the network where they are used to adjust the weights. The error signal is the mean squared error given by

$$E = \frac{1}{2} \sum_{k=1}^{K} (d_k - o_k)^2 , \tag{8.20}$$

where d_k is the desired k-th output and o_k is the predicted k-th output of the network. K is the number of neurons in the output layer.

Any efficient optimization method can be used for minimizing the error through weight adjustment. The calculation of the output is carried out layer by layer in the forward direction. The output of one layer is the input to the next layer. In the reverse pass, the weights of the output neurons are adjusted first, since the target value of each output neuron is available to guide the adjustment of the associated weights. Next, the weights of the middle layers are adjusted. Since the middle layers have no target values, errors of the succeeding layers, after proper transformations, are propagated back through the network, layer by layer. Hence, this algorithm is termed as back propagation algorithm. The trained neural network has to be tested by supplying testing data. If the testing error is much more compared to the training error, the network is said to over-fit the data. A properly fitted network will give nearly equal training and testing error.

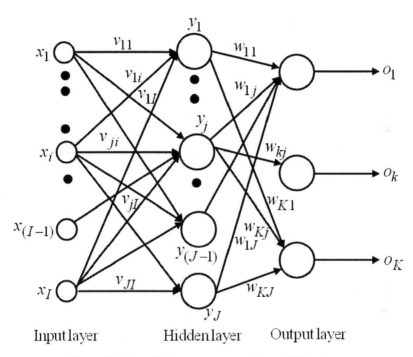

Figure 8.5. A multi-layer perceptron with one hidden layer

Let the input vector be $[x_1 \; x_2 \cdots x_I]^T$ where $x_I = 1$ to account for bias. Let the output vector for hidden layer be $[y_1 \; y_2 \ldots \ldots y_J]^T$. Here also $y_J = 1$ to account for bias. Lastly, the desired (target) output vector is $[d_1 \; d_2 \ldots \ldots d_K]^T$. Weight matrix corresponding to the hidden and output layer is

$$W = \begin{bmatrix} w_{11} & w_{12} & w_{1J} \\ w_{21} & w_{22} & w_{2J} \\ \vdots & \vdots & \vdots \\ \vdots & \vdots & \vdots \\ w_{K1} & w_{K2} & w_{KJ} \end{bmatrix},$$

where w_{ij} is the weight associated between i-th neuron of the output layer and j-th neuron of the hidden layer. Activation vector $\textbf{\textit{net}}$ of the output layer is

$$\textbf{\textit{net}} = \textbf{\textit{Wy}}. \tag{8.21}$$

For a specific pattern p:

$$E_p = \frac{1}{2}\sum_{k=1}^{K}(d_{pk} - o_{pk})^2. \tag{8.22}$$

This error can be minimized using steepest descent method. Thus, incremental change in the weight will be

$$\Delta w_{kj} = -\eta\frac{\partial E_p}{\partial w_{kj}}, \quad \text{for } k = 1, 2,......, K, \quad j = 1, 2,....., J \tag{8.23}$$

The net input signal going to k-th output neuron is

$$net_{pk} = \sum_{j=1}^{J} w_{kj}y_j, \tag{8.24}$$

and the output of that neuron is

$$o_{pk} = f_k(net_{pk}), \tag{8.25}$$

where f is the activation function of the neuron. Using chain rule, we can write

$$\frac{\partial E_p}{\partial w_{kj}} = \frac{\partial E_p}{\partial net_{pk}}\frac{\partial net_{pk}}{\partial w_{kj}} = y_j\frac{\partial E_p}{\partial net_{pk}} = y_j\frac{\partial E_p}{\partial o_{pk}}\frac{\partial o_{pk}}{\partial net_{pk}}. \tag{8.26}$$

Now from Equation 8.22,

$$\frac{\partial E_p}{\partial o_{pk}} = -\left(d_{pk} - o_{pk}\right), \tag{8.27}$$

and from Equation 8.25,

$$\frac{\partial o_{pk}}{\partial net_{pk}} = f_k'(net_{pk}). \tag{8.28}$$

Thus, from Equations 8.23 and 8.26–8.28,

$$\Delta w_{kj} = \eta (d_{pk} - o_{pk}) f_k'(net_k) y_j . \tag{8.29}$$

This way, the weights can be updated until the error becomes less than the prescribed value.

For updating the weights v_{ij} of the previous layer, we apply the steepest descent procedure again. Thus,

$$\Delta v_{ji} = -\eta \frac{\partial E_p}{\partial v_{ji}} \quad \text{for } j = 1, 2, ..., J \text{ and } i = 1, 2, ...I. \tag{8.30}$$

Proceeding in the similar way,

$$\frac{\partial E_p}{\partial v_{ji}} = \frac{\partial E_p}{\partial (net_{pj})} \frac{\partial (net_{pj})}{\partial v_{ji}} = \frac{\partial E_p}{\partial (net_{pj})} \frac{\partial (\sum\limits_{i=1}^{I} v_{ji} x_i)}{\partial v_{ji}} = x_i \frac{\partial E_p}{\partial (net_{pj})} . \tag{8.31}$$

Further, since $y_j = f_j(net_{pj})$,

$$\frac{\partial E_p}{\partial (net_{pj})} = \frac{\partial E_p}{\partial y_j} \frac{\partial y_j}{\partial net_{pj}} = \frac{\partial E_p}{\partial y_j} f_j'(net_{pj}) . \tag{8.32}$$

Now,

$$\frac{\partial E_p}{\partial y_j} = \frac{\partial}{\partial y_j} \left\{ \frac{1}{2} \sum_{k=1}^{K} \left(d_{pk} - f_k(net_{pk}) \right)^2 \right\}$$

$$= -\sum_{k=1}^{K} (d_{pk} - o_{pk}) \frac{\partial}{\partial y_j} \left\{ f(net_{pk}) \right\}$$

$$= -\sum_{k=1}^{K} (d_{pk} - o_{pk}) f_k'(net_{pk}) \frac{\partial (net_{pk})}{\partial y_j}$$

$$= -\sum_{k=1}^{K} (d_{pk} - o_{pk}) f_k'(net_{pk}) w_{kj} .$$

Hence, Equation 8.30 becomes

$$\Delta v_{ji} = \eta f_j'(net_j) x_i \sum_{k=1}^{K} (d_{pk} - o_{pk}) f_k'(net_{pk}) w_{kj} . \tag{8.33}$$

This is called the generalized delta rule. We observe that error has been propagated back by means of the weights between the last and hidden layer.

The algorithm of back-propagation training is given below for a network having (I-1) input neuron and K output neurons (See Figure 8.5).

Step1: First a learning rate $\eta > 0$ and the maximum value of error E_{max} is chosen. Weights W (connection strengths between hidden and output layers) and V (connection strengths between input and hidden layers) are initialized at small random values, W being of size ($K \times J$) and V of size ($J \times I$). Biases are included in the weights. Begin with iteration count ITR = 1, pattern number $p = 1$ and error $E_{(p-1)} = 0$.

Step 2: Training steps start here. Input is presented and the outputs of all the neurons in the hidden layer are computed, which is fed as input to the output layer. Then, the outputs of the neurons of the output layer are computed.

Step 3: Error value is computed using the formula

$$E_p = E_{(p-1)} + \frac{1}{2} \sum_{k=1}^{K} (d_k - o_k)^2 . \qquad (8.34)$$

Step 4: Weights of the output layer are adjusted using

$$(w_{kj})_{new} = w_{kj} + \Delta w_{kj} \quad \text{for } k = 1,2,3,\ldots K \text{ and } j = 1,2,3\ldots J$$

where Δw_{kj} is computed from Equation 8.29.

Step 5: Hidden layer weights are adjusted using

$$\left(v_{ji}\right)_{new} = v_{ji} + \Delta v_{ji} \quad \text{for } j = 1,2,3\ldots J \text{ and } i = 1,2,3\ldots I$$

where Δv_{ji} is computed from Equation 8.33.

Step 6: If $p < P$ then p and ITR are incremented by 1 each and we go to Step 2. Else we go to Step 7.

Step 7: The training cycle is completed. The program is terminated if $E_p < E_{max}$ and weights are stored. If $E_p > E_{max}$, then we set $E_{(p-1)} = 0$ and $p = 1$ and initiate the new training cycle by going to Step 2.

The effectiveness and convergence of back propagation algorithm depend significantly on the value of the learning rate η. In general, however, the optimum value of η depends on the problem being solved, and there is no single learning constant value suitable for different training cases. While gradient descent can be an efficient method for obtaining the weight values that minimize an error, error surfaces frequently possess properties that make the procedure slow to converge. When broad minima yield small gradient values, then a larger value of η will result in a more rapid convergence. However, for problems with steep and narrow minima, a small value of η must be chosen to avoid overshooting the solution.

Thus, it is better to choose η experimentally for each problem. It is to be noted that only small learning constants guarantee a true gradient descent.

Some methods, like the *momentum method*, can be used for increasing the rate of convergence. The method involves supplementing the current weight adjustments with a fraction of the most recent weight adjustment. This is usually done according to the formula

$$\Delta w(t) = -\eta \nabla E(t) + \alpha \Delta w(t-1),$$
(8.35)

where the arguments t and $t-1$ are used to indicate the current and the most recent training step, respectively, and α is a positive momentum constant. The second term, indicating a scaled most recent adjustment of weights, is called the momentum term.

The back propagation algorithm discussed above is based on the steepest descent method. Other optimization methods may also be used to develop the training algorithm. The Levenberg-Marquardt algorithm has been used with considerable success [6]. This method is a modified version of Newton's method. In the classical Newton's method, the error signal E is approximated by a quadratic form:

$$E(z) \simeq E(z_0) + g^T (z - z_0) + \frac{1}{2}(z - z_0)^T H (z - z_0)$$
(8.36)

In the above expression, z denotes the vector containing all the weights and biases of the network, z_0 denotes the vector containing current weights and biases, g is the gradient vector computed as

$$g = \left[\frac{\partial E(z)}{\partial z_1} \quad \frac{\partial E(z)}{\partial z_2} \quad \frac{\partial E(z)}{\partial z_3} \cdots\cdots\cdots \frac{\partial E(z)}{\partial z_N} \right]^T,$$
(8.37)

evaluated at the current value of the weights and H is the Hessian matrix given by

$$H = \begin{bmatrix} \dfrac{\partial^2 E(z)}{\partial z_1^2} & \dfrac{\partial^2 E(z)}{\partial z_1 \partial z_2} & \cdots\cdots\cdots & \dfrac{\partial^2 E(z)}{\partial z_1 \partial z_N} \\[2ex] \dfrac{\partial^2 E(z)}{\partial z_2 \partial z_1} & \dfrac{\partial^2 E(z)}{\partial z_2^2} & \cdots\cdots\cdots & \dfrac{\partial^2 E(z)}{\partial z_2 \partial z_N} \\[2ex] \cdots\cdots\cdots\cdots\cdots\cdots\cdots\cdots\cdots\cdots\cdots \\ \cdots\cdots\cdots\cdots\cdots\cdots\cdots\cdots\cdots\cdots\cdots \\ \dfrac{\partial^2 E(z)}{\partial z_N \partial z_1} & \dfrac{\partial^2 E(z)}{\partial z_N \partial z_2} & \cdots\cdots\cdots & \dfrac{\partial^2 E(z)}{\partial z_N^2} \end{bmatrix},$$
(8.38)

evaluated at the current values of the weights. The minimum of Equation 8.36 is found by differentiating it with respect to z and equating it to $\mathbf{0}$. Thus, the new vector can be found from

$$\mathbf{g} + \mathbf{H}(z_{new} - z_0) = \mathbf{0}. \tag{8.39}$$

Hence,

$$z_{new} = z_0 - \mathbf{H}^{-1}\mathbf{g}. \tag{8.40}$$

The direction of $\mathbf{H}^{-1}\mathbf{g}$ is called the Newton's direction. If the Hessian matrix is not positive definite, the Newton's direction may point toward a local maximum or a saddle point.

In the Levenberg-Marquardt algorithm, the term $\lambda \mathbf{I}$ is added to the Hessian matrix to make it positive definite, where $\mathbf{I}$ is the identity matrix and λ is positive constant. Thus, using Levenberg-Marquardt method, the weights and biases are updated as

$$z_{new} = z_0 - (\mathbf{H} + \lambda \mathbf{I})^{-1}\mathbf{g}. \tag{8.41}$$

As $\lambda \to 0$, the above expression reduces to that used in Newton's method. Similarly, if $\lambda \to \infty$, the expression corresponds to steepest descent method.

If we take, the error function given by Equation 8.22 and denote $(d_{pk} - o_{pk})$ by e_k, then

$$E(z) = \frac{1}{2} \sum_{k=1}^{K} e_k^2. \tag{8.42}$$

The (i, j) component of the Hessian matrix can be written as

$$\left[\nabla^2 E(z)\right]_{ij} = \frac{\partial^2 E}{\partial z_i \partial z_j} = \sum_{k=1}^{K} \left(\frac{\partial e_k}{\partial z_i} \frac{\partial e_k}{\partial z_j} + e_k \frac{\partial^2 e_k}{\partial z_i \partial z_j} \right). \tag{8.43}$$

In the above expression, the second term will be very small near the minimum. Hence,

$$\left[\nabla^2 E(z)\right]_{ij} \simeq \sum_{k=1}^{K} \frac{\partial e_k}{\partial z_i} \frac{\partial e_k}{\partial z_j}. \tag{8.44}$$

This approximation avoids the need for carrying out double differentiation. The parameter λ can be updated during the iterations. To begin with, a high value of the parameter may be taken, which will keep on reducing as the iterations proceed.

8.2.5 Radial Basis Function Neural Network

The supervised training of the neural networks can be viewed as a curve fitting process. The network is presented with training pairs, each consisting of a vector from an input space and a vector from the output space. Through a defined learning algorithm, the network performs the adjustments of its weights so that the error between the actual and desired outputs is minimized relative to some optimization criterion. The trained network performs the interpolation in the output vector space, which is referred to as the generalization property. In this subsection, we describe a radial basis function neural network as an alternative to multi-layer perceptron neural network to carry out this task.

The radial basis function (RBF) network consists of three layers: an input layer, a single layer of non-linear processing neurons, and an output layer. Figure 8.6 shows a typical network. For a network having K neurons in the output layers and J neurons in the hidden layer, the output of RBF is calculated according to

$$o_i = f_i(x) = \sum_{j=1}^{J} w_{ij}\phi_j(x, x_j^c) = \sum_{j=1}^{J} w_{ij}\,\phi_j\left(\left\|x - x_j^c\right\|_2\right) \text{ where } i = 1, 2 \dots K \ , (8.45)$$

where x is the input vector, $\phi_j(.)$ is function from set of all positive real number to set of real numbers, $\left\|.\right\|_2$ denotes the Euclidean norm, w_{ij} are the weights in the output layer, and x_j^c are the RBF centers in the input vector space. For brevity, we will use $\left\|.\right\|$ to mean Euclidean norm, omitting subscript 2. For each neuron in the hidden layer, Euclidean distances between its associated center and the input to the network are computed. The output of the neuron in a hidden layer is a non-linear function of the distance. The most common function is Gaussian function given by

$$\phi_j(x) = \exp\left\{-\frac{\left\|x - x_j^c\right\|^2}{2\sigma_j^2}\right\}, \tag{8.46}$$

where σ_j^2 is called the variance, which controls the spread of the distribution about the center. Some other functions are:

$$\text{Multiquadrics: } \phi_j(x) = \left(\left\|x - x_j^c\right\|^2 + c_j^2\right)^{1/2}, \tag{8.47}$$

$$\text{Inverse multiquadrics: } \phi_j(x) = \left(\left\|x - x_j^c\right\|^2 + c_j^2\right)^{-1/2}, \tag{8.48}$$

$$\text{Thin plate splines}: \phi_j(x) = \left\| x - x_j^c \right\|^2 \log \left\| x - x_j^c \right\|, \tag{8.49}$$

where c_j is called the spread parameter. In the Gaussian function, σ_j can be called the spread parameter. The output of the network is computed as a weighted sum of hidden layer outputs. Once the centers are chosen, the output of the i-th neuron in the output layer for q-th training data can be computed as

$$o_i(q) = \sum_{j=1}^{J} w_{ij} \phi_j(x(q), c_j) \quad \text{where} \quad q = 1, 2, \dots Q, \tag{8.50}$$

where Q is the total number of the training pairs. Arranging in the vector and matrix form, we have

$$\begin{Bmatrix} o_i(1) \\ o_i(2) \\ \cdot \\ \cdot \\ o_i(Q) \end{Bmatrix} = \begin{bmatrix} \phi_1(x(1), c_1) & \phi_2(x(1), c_2) & \dots & \phi_J(x(1), c_J) \\ \phi_1(x(2), c_1) & \phi_2(x(2), c_2) & \dots & \phi_J(x(2), c_J) \\ \dots\dots\dots\dots\dots\dots\dots\dots\dots\dots\dots\dots\dots\dots\dots\dots \\ \dots\dots\dots\dots\dots\dots\dots\dots\dots\dots\dots\dots\dots\dots\dots\dots \\ \phi_1(x(Q), c_1) & \phi_2(x(Q), c_2) & \dots & \phi_J(x(Q), c_J) \end{bmatrix} \begin{bmatrix} w_1 \\ w_2 \\ \cdot \\ \cdot \\ w_J \end{bmatrix}$$

or

$$o_i = \phi w, \tag{8.51}$$

where o_i is the predicted network output vector for the i-th neuron of the output layer, w is the vector of the weights associating the hidden layer neurons to the i-th neuron of the output layers and ϕ is the matrix of RBF non-linear mapping performed by the hidden layer. Because the centers are fixed, the mapping performed by the hidden layer is fixed as well. Therefore, the network-training task is to determine the appropriate settings of the weights in the network output layer so that the performance of the network mapping is optimized in some sense. A common optimization criterion to use is the mean-squared error between the actual and desired network outputs.

Let d_i be a vector of size Q containing the desired outputs for the training pattern. Then, the error as a function of network weights is given by

$$E(w) = \frac{1}{2}(d_i - \phi w)^{\mathrm{T}}(d_i - \phi w) \tag{8.52}$$

In order to minimize the error, the weights are found by differentiating the above expression with respect to w and setting it to zero. Thus,

$$-\boldsymbol{\phi}^{\mathrm{T}}d_i + \boldsymbol{\phi}^{\mathrm{T}}\boldsymbol{\phi}w = 0.$$ (8.53)

Solving for w, we get

$$w = (\boldsymbol{\phi}^{\mathrm{T}}\boldsymbol{\phi})^{-1}\boldsymbol{\phi}^{\mathrm{T}}d_i.$$ (8.54)

Thus, the problem of network training is reduced to matrix inversion and product and the training process becomes faster.

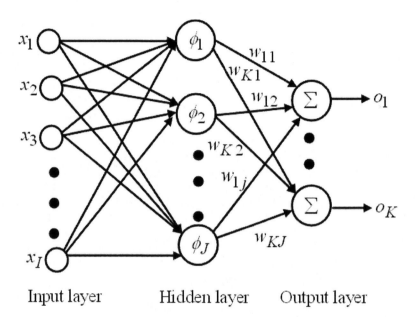

Figure 8.6. A radial basis function neural network

8.2.6 Unsupervised Learning

The neural networks discussed in the previous sections used supervised learning algorithms, which are based on error corrections rules. In these algorithms, an error value is generated from the actual response of the network and the desired response. After that, the weights are modified such that the error is gradually reduced. In unsupervised learning, there is no feedback from the environment for assessing the correctness of the mapping. In other words, there is no "teacher". Instead the network must be able to discover by itself any categories, patterns, or features possibly present in the data. Networks that are able to infer pattern relationship without being supervised are also called self-organizing.

There are many unsupervised learning rules. One rule was proposed by Hebb in his seminal work, "The organization of Behavior" [13]. This is called the Hebbian learning rule. It makes the weight strength proportional to the product of the firing rates of the two interconnected neurons. That is, when two connected neurons fire at the same time and repeatedly, the synapse's strength is increased.

Competitive learning [14] is an unsupervised learning procedure in which the neurons of a network learn to recognize clusters of similar input vectors. The network detects regularities and corrections among the input vectors and adapts the future response of the units to similar inputs. In competitive networks, output units compete among themselves for activation. The simplest competitive learning network consists of a single layer of output neurons to which all inputs are connected. All the units are presented with given input vectors but only one output neuron is activated at any given time: the so-called winner neuron.

In this book, we will mostly discuss the application of supervised neural networks. Hence, we are not discussing unsupervised networks in any further detail. Interested readers may refer to the books on neural networks for this topic.

8.3 Fuzzy Sets

Most students become familiar with the set theory in their pre-engineering classes. A set is a collection of elements. The elements of the universe are either a member or non-member of a set. Such types of sets are called classical or crisp sets. On the other hand, there are sets whose boundaries are imprecisely or vaguely defined. These sets were named Fuzzy Sets by Zadeh in his classic paper [15]. A certain element of the universe may be the member of a fuzzy set to varying degrees. If it is certain that the element is the member of the set, its membership grade in the set will be 1. Similarly, if an element surely does not belong to the set, its membership grade will be zero. There may also be elements whose membership grades will be between 0 and 1. Examples of fuzzy sets are quite common in nature. For example, the set of tall persons is a fuzzy set. One will be able to see a number of tall persons, whom he will surely consider a member of this set. Similarly, short persons will also be found, who will surely not form the part of the set. However, one will come across persons who can neither be called tall nor short. They can form part of the fuzzy set of tall persons with varying degrees.

An example where a classical or crisp set models the real word poorly is as follows. A production manger wants to purchase a high-speed machine. Suppose, his preference is for a machine that can give him a production rate of 50 components per hour. Now if he gets a machine producing 49 components per hour, his satisfaction level will not drop abruptly. He may even be willing to procure a machine producing only 48 components per hour, provided the price is less. Of course, he may not accept a machine producing only 25 components per hour. Thus, it is clear that his satisfaction level will gradually decrease from 1 to 0. This type of behavior can be captured by fuzzy sets only. In this section, we will review the basic concepts of fuzzy set theory. There are a number of books available on this subject [16–18].

8.3.1 Mathematical Definition of Fuzzy Set

In order to define a fuzzy set, it is useful to introduce the concept of 'characteristic' or 'discrimination' function. Let X denote the universal set (a set containing all the possible elements of concern in some particular context). The process by which individuals from X are determined to be either members or nonmembers of a set can be defined by a characteristic (discrimination) function. For a given set A, this function assigns a value $\mu_A(x)$ to every $x \in X$ such that

$$\mu_A(x) = \begin{cases} 1 & \text{if and only if } x \in A, \\ 0 & \text{if and only if } x \notin A, \end{cases} \tag{8.55}$$

Thus, the function maps elements of the universal set to the set containing 0 and 1. This can be indicated by

$$\mu_A : \quad X \rightarrow \{0, 1\} \ . \tag{8.56}$$

The characteristic function of a crisp set assigns a value of either 1 or 0 to each individual in the universal set. This function can be generalized such that the values assigned to the elements of the universal set fall within a specified range and indicate the membership grade of these elements in the set in question. Thus, larger values denote higher degrees of membership and *viceversa*. Such a function is called a membership function and the set defined by it a fuzzy set. The membership function μ_A, by which a fuzzy set A is usually defined has the form

$$\mu_A : X \rightarrow [0, 1], \tag{8.57}$$

where [0, 1] denotes the interval of real numbers from 0 to 1 inclusive of the end points.

If x_i is an element of the universal set and μ_i is its grade of membership in A, then A is written as

$$A = \mu_1 / x_1 + \mu_2 / x_2 + \dots\dots\dots\dots + \mu_n / x_n , \tag{8.58}$$

where the slash is employed to link the elements of A with their grades of membership in A and the '+' sign indicates that the listed pair of elements and membership grades collectively form the definition of the set A. For the case in which a fuzzy set A is defined on a universal set that is finite and countable, we may write

$$A = \sum_{i=1}^{n} \mu_i / x_i \ . \tag{8.59}$$

Here, 'Σ' does not mean summation but a collection only. Similarly, when X is an interval of real numbers, a fuzzy set A is often written in the form

$$A = \int_X \mu_A(x)/x. \tag{8.60}$$

In the above expression, ' $\int_X$ ' does not mean integration, but means collection. Here, $\mu_A(x)$ is the membership function, which assigns a membership grade to every real number.

8.3.2 Some Basic Definitions and Operations

In this subsection, we review some basic definitions and operations defined on fuzzy sets. These may be considered as the generalization of the definitions and operations on crisp set.

Empty fuzzy set: A fuzzy set is empty if and only if its membership function is zero on X.

Equality of fuzzy sets: Two fuzzy sets A and B are equal, written as $A=B$, if and only if $\mu_A(x) = \mu_B(x)$ for all x in X. In other words, two fuzzy sets are equal if and only their membership functions are equal.

Complement: The complement of a fuzzy set A is defined by the membership function

$$\mu_{A'}(x) = 1 - \mu_A(x). \tag{8.61}$$

To give an example, if a temperature of 40 °C has a membership of 0.8 in a set of 'high temperature', it will have a membership of 0.2 in the complementary set of 'cold temperature'.

Containment: A is said to be contained in B if and only if $\mu_A \leq \mu_B$.

Fuzzy union: The union of fuzzy sets A and B, denoted by $A \cup B$, results in the maximum membership grades

$$A \cup B = \int \max[\mu_A(x), \mu_B(x)]/x. \tag{8.62}$$

Fuzzy intersection: The intersection of fuzzy sets A and B denoted by $A \cap B$ results in the minimum membership grades

$$A \cap B = \int \min[\mu_A(x), \mu_B(x)]/x. \tag{8.63}$$

Fuzzy product: The product of A and B, denoted by AB, results in the ordinary product of two membership grades

$$AB = \int \mu_A(x) \mu_B(x)/x. \tag{8.64}$$

Fuzzy sum: The algebraic sum of A and B is denoted by $A+B$ and is defined by

$$A+B = \int (\mu_A(x) + \mu_B(x))/x .\tag{8.65}$$

One can put the limit on $\mu_A(x) + \mu_B(x)$ such that, if this quantity is more than one, the membership grade of algebraic sum is 1.

Absolute difference: The absolute difference of A and B is denoted by $|A - B|$ and is defined by

$$|A - B| = \int |\mu_A(x) - \mu_B(x)|/x .\tag{8.66}$$

Height of a fuzzy set: The height of a fuzzy set is the largest membership grade obtained by any element in that set.

Normality: The fuzzy set A is called normal if the upper bound of its membership grades over X is unity, *i.e.*,

$$Sup_x \mu_A(x) = 1 .\tag{8.67}$$

The height of a normalized fuzzy set is 1. Figure 8.7 shows the membership function of a normal set.

The α-cut of a fuzzy set: An α-cut of a fuzzy set A is a crisp set A_α that contains all the elements of the universal set X that have the membership grade in A greater than or equal to the specific value of α. This definition can be written as

$$A_\alpha = \{x \in X | \ \mu_A(x) \geq \alpha\} .\tag{8.68}$$

In Figure 8.7 all x values contained within A and B constitute α-cut.

Convexity: A fuzzy set is convex if and only if each of its α-cuts is a convex set. Equivalently, we may say that a fuzzy set is convex if and only if

$$\mu_A(\lambda r + (1-\lambda)s) \geq \min[\mu_A(r), \mu_A(s)] \quad \text{for all } r, s \in X \text{ and all } \lambda \in [0,1]. \tag{8.69}$$

Figure 8.7 depicts the membership of a convex set whereas Figure 8.8 shows the membership function of a non-convex set.

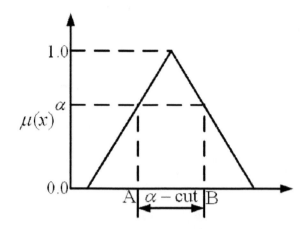

Figure 8.7. Membership function of a normal set

Figure 8.8. Membership function of a non-convex set

8.3.3 Determination of Membership Function

Membership functions are subjective but not arbitrary. For example, in the set of tall men, a 170 cm high person may get a membership grade of 0.8 or 0.9, but whatever may be the choice, a man of 172 cm height will always get a higher membership grade than a man of 170 cm height. Compared to the literature on mathematical aspects of fuzzy set theory, there are only a few papers devoted to the determination of membership function.

As fuzzy sets are usually intended to model people's cognitive states, the membership functions can be determined from either simple or sophisticated elicitation procedures. Various persons simply draw or otherwise specify different membership curves appropriate to a given problem. These persons are typically experts in the problem area. Sometimes they are given a more constrained set of possible curves from which they choose. Under more complex methods, persons can be tested using psychological methods. For example, if we show two machined surfaces to a customer and ask if the surface finish is good, his response may be quicker for the smoother machined surface. Thus, the response time may form a basis for deciding the proper membership grades.

While there is a vast (infinite) array of possible membership function forms, most actual fuzzy control operations draw from a very small set of different curves, for example triangular form of fuzzy members. This simplifies the problem to choosing just the central value and the extreme values on either side. If an expert provides low (l), most likely (m) and high (h) estimate of a parameter, the triangular membership function may be constructed by taking the membership grade as 1.0 at m and 0.5 at l and h. The triangular membership constructed in such a manner is shown in Figure 8.9. The triangle in this figure has vertices at l' and m', which may be called extreme low and extreme high estimates of the parameter. It is interesting to observe at this stage that the low and high estimates of the expert are assigned a membership grade of 0.5 and not zero, because in the scale of 0–1, the membership grade of 0.5 and above indicate the possibility of the element belonging to the set. An element having a membership grade lower than 0.5 is more likely to be a non-member. Sometimes construction of membership function in such a manner may make the extreme low value negative, which may be prohibited by physics. In such a case, the extreme low value may be 0 and the left side of the triangle may be constructed by joining the origin with vertex representing the membership grade at the most likely estimate, as shown in Figure 8.10. The triangular membership function is expressed mathematically as

$$\mu(x) = \begin{cases} 0 & x < l' \\ \dfrac{x - l'}{m - l'} & l' \le x \le m \\ \dfrac{h' - x}{h' - m} & m \le x \le h' \\ 0 & x > h' \end{cases}, \tag{8.70}$$

where

$$l' = \begin{cases} m - 2(m - l) & \text{for } m \ge 2(m - l), \\ 0 & \text{for } m < 2(m - l), \end{cases} \tag{8.71}$$

and

$$h' = m + 2(h - m) \tag{8.72}$$

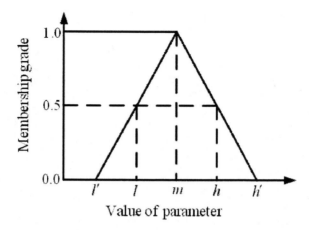

Figure 8.9. A triangular membership function ($m \geq 2(m-l)$)

Similar to triangular membership function, a trapezoidal membership may also be constructed if the most likely estimate lies between m_1 and m_2 instead of at single value m. Such a membership function is shown in Figure 8.11 and can be expressed mathematically as

$$
\mu(x) = \begin{cases}
0 & x < l' \\
\dfrac{x-l'}{m_1-l'} & l' \leq x \leq m_1 \\
1 & m_1 \leq x \leq m_2 \; , \\
\dfrac{h'-x}{h'-m_2} & m_2 \leq x \leq h' \\
0 & x > h'
\end{cases}
\tag{8.73}
$$

where the expressions for l' and h' are given by Equations 8.71–8.72.

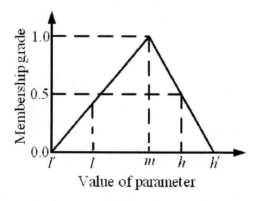

Figure 8.10. A triangular membership function ($m \leq 2(m-l)$)

Because of ease of computation, linear membership functions are preferred; however non-linear membership functions are also used in order to mimic real-life situations closely. Zadeh's π function [19] allows a more gradual drop from $\mu = 1$ down to $\mu = 0.5$ (the crossover point) and the more rapid drop from the crossover point to $\mu = 0$. Some processes in nature follow normal distribution, for example surface roughness produced in a machining process or the yield strength value of a material. In such a case, taking a Gaussian membership function may be more appropriate. A Gaussian membership function may be expressed as

$$\mu(x) = \exp\left(\frac{-(x-c)^2}{2\sigma^2}\right),$$
(8.74)

where c represents the most likely estimate and σ controls the spread. If the probability distribution is Gaussian, then approximately 99.73% values lie within 3σ on either side of the c. Using this fact, σ may be calculated from the low and high estimates of the parameter as follows:

$$\sigma = \frac{h-l}{6},$$
(8.75)

Various other membership functions have been proposed by the researchers.

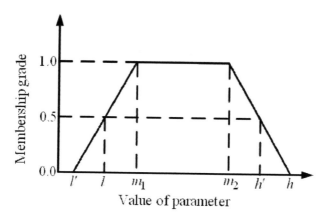

Figure 8.11. A trapezoidal membership function

Sometimes information taken in the form of frequency histograms or other probability curves is used as the basis to construct a membership function. There are a variety of possible conversion methods, each with its own mathematical and methodological strengths and weaknesses. For example, Civanlar and Trussell [20] base the membership function on probability density function $p(x)$ in the following way:

$$\mu(x) = \begin{cases} \lambda\, p(x) & \text{if } \lambda\, p(x) \le 1, \\ 1 & \text{if } \lambda\, p(x) > 1, \end{cases} \tag{8.76}$$

where λ is a constant. However, it should always be remembered that membership functions are not (necessarily) probabilities.

Optimal membership function can be estimated by means of a machine learning or optimization method. In particular, neural networks and evolutionary algorithms have been employed with success in this aim. First application of genetic algorithm for determination of membership function seems to be made by Karr [21]. He applied genetic algorithm to the design of fuzzy logic controller of the cart-pole problem. He chose Gaussian membership functions and obtained the parameters of the function by minimizing the squared difference between the cart and center of the track keeping the pole balanced. For other works on the determination of membership function, one can refer to [22–25].

8.3.4 Fuzzy Relations

Fuzzy relations relate elements of a number of universes to one another through the Cartesian product of the universes. The "strength" of the relation among the elements of an ordered n-tuple is a matter of degree and can vary with continuity between 0 and 1. An n-ary fuzzy relation R over universes $U_1, U_2, \ldots\ldots\ldots\ldots$ U_n is a fuzzy set over their product space, $R \subseteq U_1 \times U_2 \times U_3 \times \ldots\ldots \times U_n$. The membership function R is of the form $\mu_R(u_1, u_2, \ldots\ldots, u_n)$, with $u_i \in U_i$ for $i = 1, 2, \ldots\ldots, n$.

One simple example of a binary fuzzy relation may be 'x is close to y' where $x \in (X = \{1, 2, 3\})$ and $y \in (Y = \{1, 2, 10\})$. The binary relation can be expressed in the form of a relation matrix as follows:

$$R = \begin{bmatrix} 1 & 0.5 & 0 \\ 0.5 & 1 & 0 \\ 0 & 0 & 0 \end{bmatrix}. \tag{8.77}$$

In the above matrix, the rows represent the elements of X, columns the elements of Y and the elements represent the membership grades of the relation. For example, pairs (1, 1) and (2, 2) have the membership grade of 1, the pair (1, 2) has a membership grade 0.5 and pair (1, 10) has a membership grade of 0. The decision of how close is 'close' might have been taken by an expert.

A common relation used in the fuzzy logic systems is "If x is A, then y is B". For example, in turning "If feed is high, surface roughness is high" may have a membership grade of 1. The statement "If depth of cut is low, surface roughness is low" may have a membership grade of 0.5 or 0.6. These types of relations will be discussed later.

8.3.5 Extension Principle

One of the most basic concepts of fuzzy set theory, which can be used to generalize crisp mathematical concepts to fuzzy sets, is the extension principle. Suppose the function $f(x)$ maps the elements of the crisp set X to the crisp set Y; then using extension principle the definition of function f may be extended to a fuzzy set A. Just like we have the function $y = f(x)$, we will have the function $B = f(A)$, where A and B are the fuzzy sets.

Let X be a Cartesian product of universes $X = X_1 \times \ldots \ldots \times X_r$, and $A_1, \ldots, A_r$ be r fuzzy sets in $X_1, \ldots \ldots, X_r$ respectively. If f is a mapping from X to a universe Y i.e., $y = f(x_1 x_2, \ldots \ldots x_r)$, then the extension principle allows us to define a fuzzy set B in Y by

$$B = \{(y, \mu_B(y)) \mid y = f(x_1, \ldots \ldots, x_r), \quad (x_1, \ldots \ldots, x_r) \in X\}, \tag{8.78}$$

where

$$\mu_B(y) = \begin{cases} \max\limits_{(x_1, \ldots \ldots x_r) \in f^{-1}(y)} \min\{\mu_{A_1}(x_1), \ldots \ldots, \mu_{A_r}(x_r)\} & \text{if } f^{-1}(y) \neq \varnothing \\ 0 & \text{otherwise} \end{cases}, \tag{8.79}$$

where f^{-1} is the inverse of f. For $r = 1$, the extension principle reduces to

$$B = f(A) = \{(y, \mu_B(y)) \mid y = f(x), \quad x \in X\}$$

where

$$\mu_B(y) = \begin{cases} \max\limits_{x \in f^{-1}(y)} \mu_A(x) & \text{if } f^{-1}(y) \neq \varnothing \\ 0 & \text{otherwise} \end{cases}. \tag{8.80}$$

An example illustrates this. Let A be a set of fuzzy numbers close to 3 and is defined as follows:

$$A = 0/-1 + 0.1/0 + 0.2/1 + 0.5/2 + 1/3 + 0.5/4 + 0.1/6.$$

Let the function be $f(x) = x^2$. Then by applying the extension principle, B becomes

$$B = f(A) = 0.2/1 + 0.1/0 + 0.5/4 + 1/9 + 0.5/16 + 0.1/36.$$

This has been obtained as follows. Corresponding to an element 1 in B, there are two images in A i.e. -1 and 1. Among them, the membership grade corresponding to -1 is 0 and that corresponding to 1 is 0.2. Hence, the maximum membership grade of 0.2 corresponds to 1 in B. Corresponding to elements 0, 4, 9, 16, 36 there is one image for each in A. Hence, the membership grade will correspond to their images in A.

8.3.6 Fuzzy Arithmetic

A convex and normalized fuzzy set defined on R (set of real numbers) whose membership function is piecewise continuous is called a fuzzy number. A fuzzy number can be thought of as a generalization of an interval number. It can be represented by an infinite number of interval numbers with varying degrees of membership grades. Fuzzy arithmetic deals with fuzzy numbers. Mathematical operations of fuzzy set are defined at an α-cut. Thus, the mathematical operations are similar to interval arithmetic operations. However, with each interval arithmetic operation, a membership grade is associated. Thus, a typical fuzzy number A may be represented by

$$A_\alpha = [a_1^\alpha, a_2^\alpha],$$
(8.81)

where A_α is the interval corresponding to the membership grade of α, a_1^α and a_2^α are the lower and upper limits of the interval.

Some of the operations of fuzzy arithmetic are as follows:

(a) **Fuzzy addition:** Addition of two fuzzy numbers at an α-cut in R is defined by

$$A_\alpha(+)B_\alpha = [a_1^\alpha, a_2^\alpha](+)[b_1^\alpha, b_2^\alpha] = [a_1^\alpha + b_1^\alpha, a_2^\alpha + b_2^\alpha].$$
(8.82)

Addition of two triangular fuzzy numbers is a triangular fuzzy number. Figure 8.12 depicts the addition of two triangular fuzzy numbers. In this case, the addition can be performed by adding the intervals of two numbers corresponding to membership grade of 0 and 1. Thus, the x-coordinates of three vertices of resultant number are the sum of the x-coordinates of the corresponding vertices of the two numbers.

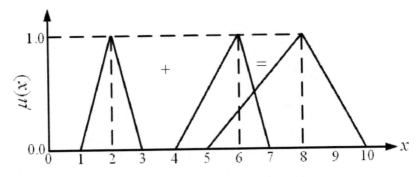

Figure 8.12. Addition of two triangular fuzzy numbers

(b) **Fuzzy subtraction:** Subtraction of two fuzzy numbers at an α-cut in R is defined by

$$A_\alpha(-)B_\alpha =[a_1^\alpha ,a_2^\alpha](-)[b_1^\alpha ,b_2^\alpha]=[a_1^\alpha - b_2^\alpha ,a_2^\alpha - b_1^\alpha].$$ (8.83)

Subtraction of two triangular fuzzy numbers is also a triangular fuzzy number.

(c) **Fuzzy multiplication:** Multiplication of two fuzzy numbers at an α-cut in R is defined by

$$A_\alpha(\times)B_\alpha =[a_1^\alpha ,a_2^\alpha](\times)[b_1^\alpha ,b_2^\alpha]=[a_1^\alpha \times b_1^\alpha ,a_2^\alpha \times b_2^\alpha].$$ (8.84)

Multiplication of two triangular fuzzy numbers is not a triangular fuzzy number. However, as an approximation, we can obtain only three vertices by performing the multiplication of the interval numbers at the membership grade of 0 and 1.

(d) **Fuzzy division-** Division of two fuzzy numbers at an α - cut in R is defined by

$$A_\alpha(\div)B_\alpha =[a_1^\alpha ,a_2^\alpha](\div)[b_1^\alpha ,b_2^\alpha]=[a_1^\alpha \div b_2^\alpha ,a_2^\alpha \div b_1^\alpha].$$ (8.85)

Like multiplication, division of two triangular fuzzy numbers is not a triangular fuzzy number.

8.3.7 Fuzzy Sets *vs* Probability

Fuzzy sets are often incorrectly assumed to indicate some form of probability. Following differences exist between fuzzy set and probability theory:

1. Membership grades are not probabilities. The summation of probabilities on a finite universal set must be equal to 1, while there is no such requirement for membership grades.
2. Probability is an objective characteristic, whilst the membership grade is subjective, although it is natural to assign a lower membership grade to that event, which considered from the aspect of probability has a lower probability of occurrence.
3. When imprecision and indeterminacy have statistical and random characteristics, probability theory is employed, whilst fuzzy set theory is used to deal with vague information and subjectivity of the judgment.
4. Probability theory generally requires that an event must be precisely defined and adequate statistical data should be available, which is not the case with fuzzy set theory.

An example is often provided for differentiating fuzziness from probability. Suppose two types of water bottle marked A and B are on the market. On the sticker of bottle A is written "Probability that this water contains poison is 0.1". On the other bottle is written "The membership grade of this water in the set of poisonous substances is 0.1". The first statement means that there is a good chance that out of 10 bottles purchased, 1 bottle will contain poison leading to death. Hence, a sensible person will not drink the water from the bottle A, no matter how thirsty he is. The second statement means that in general, there is very little harm caused by drinking the water. The degree of similarly between a poisonous substance and water is very less, *i.e.,* 0.1. Hence, this water can be consumed if no better water is available.

8.3.8 Fuzzy Logic

The basic objects of logic are propositions, which have a truth-value, like, for instance, "Maximum temperature of New Delhi in summer is 48 $^{\circ}$C". In classical logic, this statement can be either true or false; the truth-value often being indicated by 1 and the false value by 0. Classical logic fails in many situations. For example, consider the following classical paradoxes of logic:

1. A barber of a village makes a statement: "I give a haircut to those persons who do not cut their hair themselves." The question is that who gives a haircut to the barber if this statement is true. If the barber does not cut his own hair, then by this statement he should cut his own hair. Thus there is a contradiction.

2. On a piece of paper is written "Statement on the other side of the paper is false". The other side contains "Statement on the other side is true". Now, what truth-value should be assigned to the first statement? If we consider it to be true, in order to make the second statement false, it should be false. It can also not be considered false, because in that case, the other statement will be true making the first statement true. Thus, we can neither assign a value '1' nor '0' to this statement.

Fuzzy logic solves this problem by accepting values between 0 and 1. For example, in both paradoxes, the truth-value of 0.5 may be assigned to the statement. Fuzzy logic provides the theoretical foundation for reasoning about imprecise propositions. Engineering applications of fuzzy logic are mostly in controls and decision support systems.

8.3.9 Linguistic Variables and Hedges

Fuzzy set theory is often called a method of computing with language. Here, we often deal with linguistic variables. A linguistic variable is a fuzzy variable. For example, the statement 'Product is beautiful' implies that the linguistic variable *product* takes the linguistic value *beautiful*. A linguistic variable is often associated with fuzzy set quantifiers called hedges. Hedges are terms that modify the meaning of fuzzy variables. Hence, some people call them linguistic modifiers. They include terms such as *very, usually, somewhat, likely, quite, more or less, slightly* etc.

Hedges act as unary operators on fuzzy sets. If we have a fuzzy set for a linguistic value, then the fuzzy set for the modified linguistic value can be obtained by operating the original fuzzy set by the operation corresponding to the hedge that modified the original meaning. For example, *very* performs concentration by reducing the membership values of all members and creates a new subset. From the set of 'beautiful product', it derives the subset of 'very beautiful product'. Suppose a product is rated as 0.9 in the set of 'beautiful product'. Then, in the set of 'very beautiful product', its membership grade will be 0.81, which is obtained by the following operation:

$$\mu_A^{very}(x) = [\mu_A(x)]^2 . \tag{8.86}$$

Extremely serves the same purpose as very, but does it to a greater extent. This operation can be performed by raising the power of $\mu_A(x)$ to 3:

$$\mu_A^{extremly}(x) = [\mu_A(x)]^3 . \tag{8.87}$$

Thus, the product will have a membership grade of 0.729 in the set of extremely beautiful products. *Very very* is just applying hedge 'very' twice. Thus, It is given as a square of the operation of 'very':

$$\mu_A^{very\,very}(x) = [\mu_A^{very}(x)]^2 = [\mu_A(x)]^4 . \tag{8.88}$$

Thus, the product will have a membership grade of 0.6561 in the set of very very beautiful products.

More or less is the operation of dilation that expands a set and thus increases the degree of membership of fuzzy elements. This operation is presented as

$$\mu_A^{more\ or\ less} = \sqrt{\mu_A(x)} . \tag{8.89}$$

Thus, the product which has a membership grade of 0.9 in the set of beautiful products will have membership grade of 0.9487 in the set of more or less beautiful products. To provide more dilation, one can take the cube root of the membership values instead of square roots.

Indeed is the operation of intensification that intensifies the meaning of the whole sentence. It can be done by increasing the degree of membership grades above 0.5 and decreasing those below 0.5. The hedge *indeed* may be given by

$$\mu_A^{indeed}(x) = 2\,[\mu_A(x)]^2 \qquad \text{if } 0 \le \mu_A(x) \le 0.5,$$
$$\mu_A^{indeed}(x) = 1 - 2\,[1 - \mu_A(x)]^2 \qquad \text{if } 0.5 \le \mu_A(x) \le 1. \tag{8.90}$$

It may be noted that the procedure described in this subsection is just an attempt to convert language into mathematics. The meaning of 'language' differs with place and culture. The expressions for hedges should be decided after careful consideration.

8.3.10 Fuzzy Rules

In 1973, Zadeh published his revolutionary paper [26]. This paper outlined a new approach to the analysis of complex systems, in which he suggested capturing human knowledge in fuzzy rules. A fuzzy rule can be defined as a conditional statement in the form:

$$\text{IF } x \text{ is } A$$
$$\text{THEN } y \text{ is } B$$

where x and y are linguistic variables and A and B are linguistic values determined by fuzzy sets on the universes of discourse X and Y respectively. For example, consider these two fuzzy rules in the context of machining:

Rule 1: IF material is hard
    THEN *cutting speed* should be *low*.
Rule 2: IF material is *soft*
    THEN *cutting speed* should be *high*.

Here, 'hard', 'soft', 'low' and 'high' are fuzzy sets. These sets have to be designed carefully using the available knowledge.

An expert's knowledge or the knowledge obtained from the analysis can be put in the form of various fuzzy rules. These rules are used in fuzzy reasoning. Fuzzy reasoning includes two distinct parts: evaluating the rule antecedent (the IF part of the rule) and applying the result to the consequent (the THEN part of the rule). In a classical rule base system, if the rule antecedent is true, then the consequent is also true. In fuzzy systems, where the antecedent is a fuzzy statement, rules may be applicable to some extent, or in other words partial firing of the rules is allowed. If the antecedent is true to some degree of membership, then the consequent is also true to the same degree.

8.3.11 Fuzzy Inference

Fuzzy inference can be defined as a process of mapping from a given input to an output using the theory of fuzzy sets. The most commonly used fuzzy inference technique is the so-called Mamdani method. In 1975, Professor Ebrahim Mamdani of London University built one of the first fuzzy systems to control a steam engine and boiler combination [27]. He applied a set of fuzzy rules supplied by experienced human operators. The Mamdani-style fuzzy inference process is performed in four steps: fuzzification of the input variables, rule evaluation, aggregation of the rule outputs and finally defuzzification. To see how things fit together, we examine a simple two-input one-output problem that includes three rules. The problem is of deciding the cutting force in machining based on the work-material and depth of cut, assuming that other parameters including the feed remain fixed. This example is just for illustrating the concepts. In practice, there will be more rules.

Rule 1:
    IF work-material is hard
    AND depth of cut is high.
    THEN cutting speed should be low.

Rule 2:
   IF work-material is hard
   AND depth of cut is low.
   THEN cutting speed should be medium.
Rule 3:
   IF work-material is soft
   AND depth of cut is high.
   THEN cutting speed should be medium.
Rule 4:
   IF work-material is soft
   AND depth of cut is low.
   THEN cutting speed should be high.
The basic steps for Mamdani-style fuzzy inference for our problem are described below.

Step 1: Fuzzification
The first step is to take the crisp inputs of material hardness and depth of cut and determine the degree to which these inputs belong to each of the appropriate fuzzy sets. Let us assume that the range of material hardness is 100 to 200 BHN and the range of depth of cut is 1 to 5 mm. For material hardness, two fuzzy sets are defined—'hard' and 'soft'. For depth of cut, two fuzzy sets are defined—'high' and 'low'. The membership functions for these sets are assumed to be linear and are shown in Figure 8.13. Assume that the work-piece material has a hardness of 125 BHN and a membership grade of 0.75 and 0.25 in the sets of 'soft' and 'hard' materials respectively. Also assume a depth of cut of 3 mm, which has a membership grade of 0.5 in both 'low' and 'high' depth of cut.

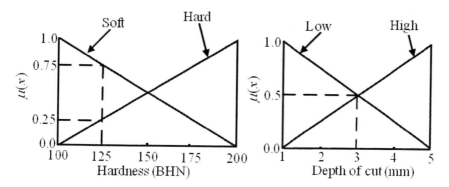

Figure. 8.13. Fuzzification of hardness and depth of cut

Step 2: Rule Evaluation
The second step is to take the fuzzified inputs and apply them to the antecedents of the fuzzy rules. If a given fuzzy rule has multiple antecedents, the fuzzy operator (AND or OR) is used to obtain a single number that represents the result of antecedent evaluation. This number (the truth-value) is then applied to the consequent membership functions. If either one of the two rule antecedents is

applicable, the OR operator is used. The classical fuzzy operation *union* can be used for this purpose. If both of the two rule antecedents are compulsory, the AND operator is applicable and the fuzzy operation *intersection* is used. The same method is extended to the case of multiple antecedents.

In our problem, the strength of the four rules is 0.25, 0.25, 0.5 and 0.5 respectively. This has been obtained by applying intersection operation on the AND parts of the rule. In the first rule, the material is hard with a membership grade of 0.25 and depth of cut is high with a membership grade of 0.5. Hence, the strength of the rule is minimum (because of AND operator) of two *i.e.* 0.25. Thus, the truth-value of the statement "Cutting speed is low." is 0.25. Similarly, the truth-value of the other three rules is 0.25, 0.5 and 0.5 respectively.

Let us assume that, based on the expert's information, the cutting speed is represented as shown in Figure 8.14. Here, the range of cutting speed has been taken between 60 m/min and 100 m/min. Three sets– slow speed, medium speed and high speed– have been created. The most common method of correlating the rule consequent with the truth-value of the rule antecedent is simply to cut the consequent membership function at the level of antecedent truth. The method is called clipping or correlation minimum. Since the top of the membership function is sliced, the clipped fuzzy set loses some information. However, clipping is preferred because it involves less complex and faster mathematics, and generates an aggregate output surface that is easier to defuzzify. The other method may be to scale down the membership function, so that its highest membership function is equal to rule strength. Figure 8.15 shows the clipped membership function (by thick lines) as result of applying various rules.

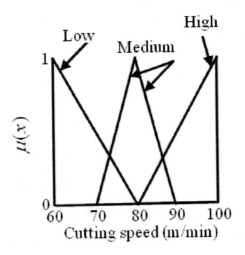

Figure 8.14. Representation of cutting speed in the form of fuzzy sets

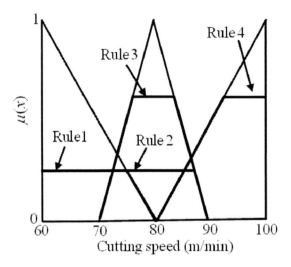

Figure 8.15. Clipping of membership functions due to application of rules

Step 3: Aggregation of the Rule Outputs

Aggregation is the process of unification of the outputs of all rules. In other words, we take the membership function of all rule consequents previously clipped and combine them into a single fuzzy set by applying fuzzy union operation. Thus, the input of the aggregation process is the list of clipped consequent membership functions, and output is one fuzzy set. Figure 8.16 shows the aggregated output. Note that the strengths of the four rules are 0.25, 0.25, 0.5 and 0.5 respectively. As a result of applying fuzzy union operation, one will get the output as shown in Figure 8.16.

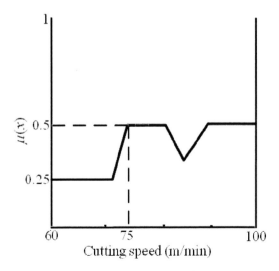

Figure 8.16. Aggregated output

Step 4: Defuzzification
As the decision-maker requires a crisp value of cutting speed, the output obtained in the third step has to be defuzzified. Defuzzification is the conversion of a fuzzy quantity into a precise output. There are various popular defuzzification methods. They are described below:

Height Method
This method is also known as the *maximum membership principle*. In the height method, the number with maximum degree of membership is chosen. However, this can be applied only if output contains a maximum peak. In the problem chosen, there is no peak. Hence, the method cannot be applied for this problem.

Centroid Method
This method is the most prominent and physically appealing of all the defuzzification methods. The method is also known as *center of area* or *center of gravity method*. The crisp value is obtained by taking the position of center of gravity in x-axis. Thus, the crisp value is given by

$$x^* = \frac{\int x\,\mu(x)dx}{\int \mu(x)\,dx}. \tag{8.91}$$

Weighted Average Method
In this method the output is obtained by the weighted average of each maximum output of the set of rules stored in the knowledge base of the system. The weighted average defuzzification technique can be expressed as

$$x^* = \frac{\sum\limits_{i=1}^{n} (x_{max})_i\, \mu_i}{\sum\limits_{i=1}^{n} \mu_i}, \tag{8.92}$$

where x^* is the defuzzified output, $(x_{max})_i$ is the maximum output of each rule, and μ_i is the strength associated with each rule. This method is computationally faster and easier and gives fairly accurate result.

Mean-max Method
This method, also called *middle of maxima,* is a slight generalization of the height method to the case where there is more than one value of maximum degree of membership. It takes as x^* the midpoint between the smallest and the largest number having maximum degree of membership.

Center of Largest Area
If the output fuzzy set has at least two convex sub-regions, then the center of gravity of the convex sub-region with the largest area is used as crisp value.

First (or Last) of Maxima

This method is generalization of the height method for the case in which the output membership function has more than one maximum. Then either the first or the last of the maxima is used as defuzzified value.

In our problem, applying the Mean-max method, the crisp value of cutting speed comes out at 87.5 m/min. Hence, for given values of hardness and depth of cut, the cutting speed should be 87.5 m/min.

In the Mamdani inference model, described above, fuzzy sets are used in rule antecedents and consequents. The Sugeno model proposed by Sugeno and co-workers [28–29] uses crisp value in consequent. The format of Sugeno type fuzzy rule is,

If x is A and y is B, then z is $f(x, y)$,

where A and B are fuzzy sets and $f(x, y)$ is a crisp function. When $f(x, y)$ is a first-degree polynomial, the resulting fuzzy inference system is called a first-degree Sugeno fuzzy model. When f is a constant, we get a zero-degree Sugeno fuzzy model. The Sugeno-type models are computationally more efficient compared to Mamdani-type models.

8.4 Genetic Algorithms

There are a number of evolutionary algorithms, which mimic natural evolutionary principles for optimizing. Among them, genetic algorithms are very powerful evolutionary optimization techniques, which do not require the derivatives of the objective and constraint functions. These are so named because they follow the principles of natural genetics. Professor John Holland of the University of Michigan, Ann Arbor, first envisaged the concept of these algorithms [30]. Now, there are many variants of these algorithms. We will briefly describe two of them *i.e.*, binary-coded and real-coded genetic algorithms. For details, one can refer standard textbooks [31, 32].

There are a number of advantages of using genetic algorithms (GAs):

- GAs are parallel-search procedures that can be implemented on parallel-processing machines for very fast computations.
- GAs are applicable to both continuous and discrete design variable optimization problems.
- They are suitable for combinatorial optimization problems, where the solution space contains finite set of points.
- GAs are stochastic and are less likely to get trapped in local minima, which inevitably are present in most of the practical applications.
- GAs are very suitable for solving multi-objective problems.

8.4.1 Binary Coded Genetic Algorithms

These are the original genetic algorithms. In this book, we will refer a binary coded genetic algorithm as BGA and a real coded genetic algorithm as RGA. The term GA will be used as a general term to mean both types of genetic algorithms. In BGA, the design variables of the optimization problem are coded in binary form. Thus, instead of operating on real values of design variables, we operate on the binary values. Thus, a solution point is represented by a string (chromosome) consisting of '0's and '1's. Each '0' and '1' value is called a bit and is analogous to a gene. The mapping between the binary and real form can be easily established. Suppose the i-th variable is represented by a sub-string S_i, then its real value is given by

$$x_i = x_i^L + \frac{x_i^U - x_i^L}{2^{l_i} - 1}(\text{decoded (decimal) value of } S_i), \qquad (8.93)$$

where x_i^L and x_i^U are the lower and upper bounds of the variable and l_i is the length of the string. The higher is the length of the string, the higher is the precision.

Example 8.4: The optimized thickness and width of a beam of rectangular cross-section are represented by 10101100, the sub-string lengths of both variables being 4. Lower and upper bounds of thickness are 4 cm and 8 cm respectively, whereas the corresponding values for the width are 8 cm to 16 cm. Find out the solution values in real form. What is the precision of the design variables?

Solution: There are two design variables- thickness and width. Let us denote their optimized values as x_1 and x_2. Using Equation 8.93,

$$x_1 = 4 + \frac{8-4}{2^4-1}(2^3 + 2^1) = 6\frac{2}{3}\text{cm and } x_2 = 8 + \frac{16-8}{2^4-1}(2^3 + 2^2) = 14\frac{2}{5}\text{cm}.$$

In this problem, each variable is represented by 4 bits, hence it can take total 16 values. In other words, 15 divisions of the given ranges are possible. Range (difference between upper and lower bounds) of thickness is 4 cm and that of width is 8 cm. As the precision of a variable is equal to the length of one division of the range, the precision of thickness is 4/15 cm and that of width is 8/15 cm.

In GA, we start from a population of solution points and find out the fitness value for each point. The closer the function value at a point to the desired objective, the higher will be the fitness function. Thus, in a maximization problem, the fitness function may be taken as equal to the objective function. In order to carry out GA operations, often a positive value of fitness is desired. In that case, fitness function may be taken as the addition of a positive constant and objective function. Positive constant should be taken just more than the maximum possible negative value of the function, based on our rough estimate. Too high a positive constant will make the fitness values of all points almost equal and may create difficulty in some cases. For the minimization problems, the fitness function should be such that it gives higher value for lower objective function. For non-negative objective function $f(x)$, one such function can be

$$F(x) = \frac{1}{1 + f(x)} ,$$
(8.94)

where a value of 1 has been added in the denominator to avoid division by 0, in case the function becomes negative.

The optimization process using BGA is as follows. Initially a population consisting of various combinations of design (or decision) variables is chosen. Each combination is coded in the form of a binary string, each string being called a chromosome. This population is operated by three genetic operators *viz.*, reproduction, crossover and mutation and a new generation consisting of chromosomes with better fitness is formed. The procedure is repeated until convergence is obtained. In the following subsections, we describe the three genetic operators.

8.4.1.1 Reproduction

This is the first operator applied on a population. In reproduction, good strings in a population are assigned a large number of copies. The reproduction can be carried out in a number of ways [31]. In the tournament selection, tournaments are played between two solutions and the winning solution is taken. By tournament playing we mean that two solutions are compared and the solution having the better fitness is chosen. Each solution participates in exactly two tournaments in a random manner. Thus, the best solution gets two copies in the population and the worst having lost both the tournaments gets eliminated. Other solution may get zero, one or two copies in the population. Figure 8.17 illustrates the procedure pictorially. The population consists of four members. It is assumed that the fitness value of each member is proportional to his height. Four tournaments are played and the taller member wins. In this way, we get two copies each of the tallest and second tallest. Note that other possibilities also exist depending on how the teams are formed. Thus, probability plays a role here.

In proportionate selection, copies proportional to fitness values are taken. Supposing the fitness values of a 4-member population are 25, 50, 10 and 40. Sum of the fitness values is 125. Dividing the fitness values by this number, we get the probabilities of survival of different members. In this case, the probabilities are 0.2, 0.4, 0.08 and 0.32. Expected numbers of copies are found by multiplying these probabilities with the size of the population, in this case 4. Thus, the members are expected to have 0.8, 1.6, 0.32 and 1.28 copies. This means that if the reproduction operator is carried out a large number of times, on an average these will be the number of copies. However, in any single operation a particular member may get 0, 1, 2, 3 or 4 copies. For achieving this operation in a computer, the following procedure may be adopted:

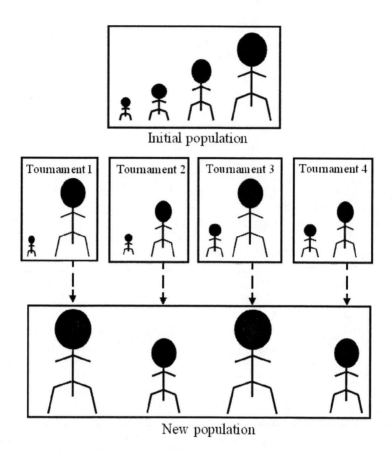

Figure 8.17. Reproduction using tournament selection

- We make ranges proportional to the probabilities between 0 and 1. In our case, the ranges are 0–0.2, 0.2–0.6, 0.6–0.68 and 0.68–1.0.
- Generate random numbers equal to the number of members in the population. In whatever range a particular number falls, the corresponding chromosome is selected.

This method of selection is called roulette wheel selection (RWS), because the same operation can be achieved mechanically by spinning a wheel a number of times. The wheel (shown in Figure 8.18) is divided into divisions equal to population size, where the size of each division is proportional to the fitness of the corresponding member. The wheel is spun and is allowed to stop. Then the member whose division stops before a fixed pointer is selected.

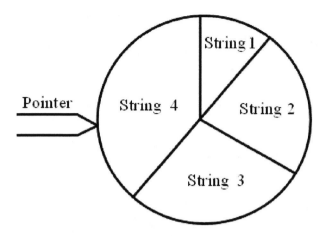

Figure 8.18. A roulette wheel

The proportionate selection operator has scaling problem. If the fitness value of one member is more, the member gets selected quite often. Similarly, if the fitness values of all members are more or less the same, all members have equal probability of getting selected. The tournament selection does not have this ranking problem. The scaling difficulty can be eliminated by using a ranking selection operator. In this method, solutions are sorted according to their fitness values and the ranks are assigned, the worst member getting the rank 1. The proportionate selection is then applied based on these ranks.

8.4.1.2 Crossover
In crossover operation, new chromosomes are created by exchanging the information between two chromosomes.   To accomplish this, the following procedure is adopted. If the population size is N, $N/2$ pairs are formed at random. Two chromosomes (strings) in each pair are called parents. Taking each pair at a time, a random crossover site is selected. Then, two offspring (children) are produced by exchanging all the bits on the right side of the cross-over site. More specifically, this is called a single point cross-over operator. In double cross-over, two different crossover sites are chosen at random. This divides the strings into three substrings. The crossover operation is completed by exchanging the bits lying between the two crossover sites. This can be generalized to n-point crossover. Figure 8.19 illustrates the single and double point crossover operations. The single point crossover preserves the structure of the parent structure to the maximum extent. The extent of string preservation reduces with increase of crossover sites. However, it is difficult to say which type of crossover is better.

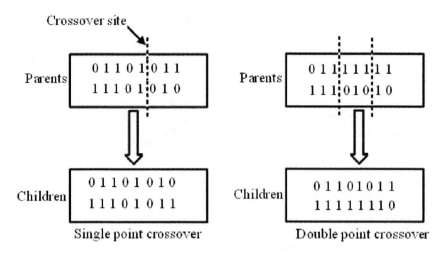

Figure 8.19. Single and double point crossover operations

Crossover operation is carried out with some probability. This is because, some good strings have to be preserved. If a crossover probability of p_c is chosen, then 100 p_c % of the strings are used for cross-over and the remaining strings are copied as they are to the next population. A common practice is to choose about ¾-th strings for the crossover.

8.4.1.3 Mutation

Mutation changes the bits of the chromosomes with some low probability (typically 0.01) of mutation. It is needed to provide some diversity in the population. It serves the crucial role of preventing the system from getting stuck to the local optimum. Only reproduction and mutation operations do not guarantee true optimum points. Mutation can randomly create a very good chromosome. It may also create a very bad chromosome, but it will hopefully not get transferred in the next generation. Figure 8.20 illustrates the operation of mutation. One way to carry out the mutation operation is to generate a random number between 0 and 1 for each bit. If the random number is less than the mutation probability (say 0.01), the bit is changed. For example, in Figure 8.20, the third bit from the right gets changed from 0 to 1.

Sequential application of reproduction, crossover and mutation completes one generation. The population keeps on evolving through a number of generations, until it is observed that average fitness of the population has been steady since past 4–5 generations. After that the GA process is stopped and one run is said to be completed. The best string in that run represents the optimum solution. As GA is probabilistic, several runs may be carried out with different initial population for finding out the global optima.

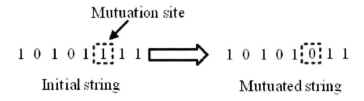

Figure 8.20. A mutation operation

8.4.2 Real Coded Genetic Algorithms

Instead of operating on strings by converting the real variables into binary numbers, one can perform the operations of genetic algorithm on real numbers themselves. This procedure is called real coded genetic algorithm (RGA). RGA is of recent origin. RGA uses real numbers instead of binary and is more suitable for continuous search space. In binary coded genetic algorithms, for achieving sufficient precision, a variable has to be represented by a number of bits. Also, the more bits in the chromosome, the greater the needed population size [33]. Binary coded genetic algorithms (BCA) suffer from the problem of Hamming cliffs, *i.e.*, often in order to make a very small change in real parameter, a number of bits need to be changed. For example, 1000 and 0111 are very close to each other in real space, the first being 8 and the second 7. However, 4 bits have to be changed for going from 7 to 8. This reduces the efficiency of GA. RGAs do not face this problem, though in GAs this problem can also be avoided by using gray codes instead of binary [34]. The procedure of the real coded genetic algorithm is same as that of the binary coded genetic algorithm. Initial population is random and population keeps on evolving towards betterment in successive generations. In each generation, the population is operated by three main operators—*reproduction, crossover* and *mutation*—to create a new population. If no significant improvement in the average fitness value of the population is observed for a few successive generations, convergence is assumed. The reproduction operator is the same as in BGA. However, crossover and mutation have to be carried out in a different way.

In the crossover operation, new members are created by exchanging the information between two parent members. In RGA, the term crossover is really a misnomer. A simulated binary crossover operator (SBX) [31] can be used, which has a similar search power to that in a single-point crossover in BGA. This works as follows. Choose a random number $u_i \in [0, 1]$. Calculate

$$\beta = \begin{cases} (2u_i)^{\frac{1}{\eta_c+1}} & \text{if } u_i \leq 0.5, \\ \left(\dfrac{1}{2(1-u_i)}\right)^{\frac{1}{\eta_c+1}} & \text{otherwise,} \end{cases} \tag{8.95}$$

where η_c is a crossover index which is a non-negative real number. A large value of η_c gives a higher probability for creating a 'near-parent' solution and a small value of η_c allows distant solutions to be selected as offspring. Offspring are given by

$$x_i^{c1} = 0.5[(1+\beta)x_i^{p1} + (1-\beta)x_i^{p2}],$$

$$x_i^{c2} = 0.5[(1-\beta)x_i^{p1} + (1+\beta)x_i^{p2}],$$

(8.96)

where x_i^{c1}, x_i^{c2} denote the i-th variables of the chromosomes of two child-members and x_i^{p1}, x_i^{p2} the i-th variables the chromosomes of two parent-members. The crossover operation is performed with a crossover probability p_c.

Mutation provides a local perturbation in order to provide diversity to population and reduce the possibility of getting trapped in local optimum. In BGA, mutation is carried out by altering one or more bits in the chromosome string. In RGA, a mutation operator based on polynomial mutation [31] can be used. Accordingly, the mutated value y_i of x_i is given by

$$y_i = x_i + (x_i^u - x_i^l)\overline{\delta}_i.$$

(8.97)

Here, x_i^u and x_i^l are the upper and lower bound values of the i-th variable. The parameter $\overline{\delta}_i$ is given by

$$\overline{\delta}_i = \begin{cases} (2r_i)^{1/(\eta_m+1)} - 1, & \text{if } r_i < 0.5, \\ 1 - [2(1-r_i)]^{1/(\eta_m+1)}, & \text{if } r_i \geq 0.5, \end{cases}$$

(8.98)

where η_m is mutation operator and r_i is a random number in [0, 1]. We shall discuss the implementation details of RGA later.

8.5 Soft Computing *vs* FEM

After giving a background of soft computing in this chapter and FEM in Chapter 5 of the book, it is appropriate to compare the applicability of soft computing techniques in comparison to FEM. FEM is a well established technique for solving differential equations. The success of this method depends on how well the physical problem has been represented in the form of differential equations and how accurately the model parameters can be determined. The model parameters have to be obtained from the experiments, and many times their determination is very difficult, time consuming and expensive. In physics, as we model a problem at a more fundamental level, the more insight may be obtained about the problem.

However, the complexity in obtaining the model parameter increases. For example, for determining the cutting forces in the machining of steel, a continuum model will require the data on flow stress of the material and friction. If the same thing is modeled at molecular level, the determination of the model parameters will require expensive experiments.

If we use neural network models to obtain the cutting forces, we can directly obtain the data containing the values of cutting forces for different feed, depth of cut and cutting speed combinations. In this procedure, we even do not need the information of flow stress and friction. However, even this type of modeling can be called physics-based modeling but not at a fundamental level. This does not mean that the neural network approach discourages the modeling at a fundamental level. In fact, if the reliable finite element model is available, the training data may be obtained from simulations instead of real experiments. Finite element simulations may require large computational time, but the trained neural network will provided the results in much less time.

Fuzzy logic can provide the approximate results by means of expert rules. In this form it may be an alternative to FEM or neural network. In the other form, it helps us to carry out computation with fuzzy numbers. Thus, it can be incorporated in FEM for carrying out the computations with fuzzy parameters.

Genetic algorithms and other evolutionary schemes can be used for optimization. The objective and constraint functions required by genetic algorithms may be obtained by FEM. Thus, this constituent of soft computing can also be used along with FEM to extract useful information.

To conclude, soft computing and FEM techniques can be complementary to one another. At the same time, any of these techniques can also be applied independently depending on the nature of the problem. The subsequent chapter will provide various examples of their applications.

8.6 Summary

Soft computing differs from conventional (hard) computing in that it is somewhat tolerant of imprecision, uncertainty, partial truth and approximation. In this chapter, we have introduced three constituents of soft computing, *viz.* neural networks, fuzzy sets and genetic algorithms. In neural networks, multi-layer perceptron networks based on back propagation algorithm and radial basis function networks have been discussed in more detail. These can be used for function approximation. A very brief description of some other types of networks has also been provided. In the section on fuzzy sets, the concepts of fuzzy sets, fuzzy arithmetic and fuzzy logic have been introduced. An example has been provided to show the decision-making process using fuzzy logic. Genetic algorithms fall in the category of evolutionary optimization techniques. Other similar techniques are simulated annealing, ant colony optimization, particle swarm technique *etc.* In view of the success and popularity of genetic algorithms, we have focused our discussion on genetic algorithms. Binary coded and real coded genetic algorithms have been discussed. The subsequent chapters will discuss the application of soft computing to metal forming and metal removing processes.

8.7 References

[1]   Merium, J.L. and Kraig, L.G. (1999), Engineering Mechanics, Vol.2, Dynamics, John Wiley & Sons (Asia), Singapore.

[2]   Zadeh, L.A. (1994), Fuzzy logic, neural networks and soft computing, Communications of the ACM, Vol. 3, No. 3, pp. 77–84.

[3]   Kosko, B. (1994), Fuzzy thinking: the New Science of Fuzzy Logic, Flamingo, London.

[4]   Zurada, J.M. (1997), Introduction to Artificial Neural Systems, Jaico Publishing House, Mumbai.

[5]   Hassoun, M.H. (1999), Fundamentals of Artificial Neural Networks, Prentice-Hall of India, New Delhi.

[6]   Ham, F.M. and Kostanic, I. (2001), Principles of Neurocomputing for Science and Engineering, McGraw-Hill, New York.

[7]   Haykin, S. (2004), Neural Networks: A Comprehensive Foundation, Prentice-Hall of India, New Delhi.

[8]   McCulloch, W.S. and Pitts W.H. (1943), A logical calculus of the ideas immanent in nervous activity, Bulletin of Mathematical Biophysics, Vol. 5, pp. 115–133.

[9]   Tocci, R.J. (2000), Digital System: Principle and Applications, sixth edition, Prentice-Hall of India, New Delhi.

[10]   Rosenblatt, F., (1958), The perceptron: a probabilistic model for information storage in the brain, Psychological Review, Vol. 65, pp. 386–408.

[11]   McClelland, T.L. Rumelhart, D.E. and the PDP Research Group (1986), Parallel Distributed Processing, The MIT Press, Cambridge.

[12]   Shynk, J.J. (1990), Performance surfaces of a single-layer perceptron, IEEE Transactions on Neural Networks, Vol. 1, pp. 268–274.

[13]   Hebb, D. (1949), The Organization of Behavior, Wiley, New York.

[14]   Jang, J.-S.R., Sun, C.-T. and Mizutani, E. (2002), Neuro-Fuzzy and Soft Computing-A Computational Approach to Learning and Machine Intelligence, Prentice-Hall of India, New Delhi.

[15]   Zadeh, L.A. (1965), Fuzzy sets, Information and Control, Vol.8, pp.338–353.

[16]   Klier, G.J. and Folger, T.A. (1993), Fuzzy Sets, Uncertainty and Information, Prentice-Hall of India Private Limited, New Delhi.

[17]   Kaufmann A. and Gupta, M.M. (1985), Introduction to Fuzzy Arithmetic: Theory and Applications, Von Nostran Reinhold Company Inc., New York.

[18]   Bojadziev G. and Bojadziev, M. (1995), Advances in Fuzzy Systems-Application and Theory, Vol. 5, World Scientific, Singapore.

[19]   Zadeh, L.A. (1976), A fuzzy-algorithmic approach to the definition of complex or imprecise concepts, International Journal of Man-Machine Studies, Vol. 8, pp. 249–291.

[20]   Civanlar, M.R. and Trussel, H.J. (1986), Constructing membership functions using statistical data, Fuzzy Sets and Systems, Vol. 18, pp. 1–14.

[21]   Karr, C.L. (1991), Design of an adaptive fuzzy controller using a genetic algorithm, Proceedings of 4th International Conference on Genetic Algorithms, San Diego, CA, pp. 450–456.

[22]   Arslan, A. and Kaya, M. (2001), Determination of fuzzy logic membership function using genetic algorithms, Fuzzy Sets and Systems, Vol. 118, pp. 297–306.

[23]   Garibaldi, J.M. and Ifeachor, E.C. (1999), Application of simulated annealing fuzzy model tuning to umbilical cord acid-base interpretation, IEEE Transactions on Fuzzy Systems, Vol. 7, pp. 72–84.

[24] Homaifar, A. and McCormic, E. (1995), Simultaneous design of membership functions and rule sets for fuzzy controllers using genetic algorithms, IEEE Transactions on Fuzzy Systems, Vol. 3, pp. 129–139.

[25] Bagis, A. (2003), Determination of fuzzy membership function with tabu search-an application to control, Fuzzy Sets and Systems, Vol. 139, pp. 209–225.

[26] Zadeh, L.A. (1973), Outline of a new approach to the analysis of complex systems and decision processes, Information Science, Vol. 9, pp.43–80.

[27] Mamdani, E.H. (1976), Advances in linguistic synthesis of fuzzy controllers, International Journal of Man-Machine Studies, Vol. 8, pp.669–678.

[28] Takagi, T. and Sugeno, M. (1985), Fuzzy identification of systems and its applications to modeling and control, IEEE transactions on System, Man and Cybernetics, Vol. 15, pp. 116–132.

[29] Sugeno, M. and Kang, G.T. (1988), Structure identification of fuzzy model, Fuzzy Sets and Systems, Vol. 28, pp. 15–33.

[30] Holland, J.H. (1975), Adaptation in Natural and Artificial Systems, University of Michigan Press, Ann arbor.

[31] Deb, K. (2003), Multi-Objective Optimization using Evolutionary Algorithms, John Wiley & Sons, Singapore.

[32] Goldberg, D.E. (1989), Genetic Algorithms in Search, Optimization and Machine Learning, Addison-Wesley, Reading, MA.

[33] Goldberg, D.E., Deb K. and Clark, J.H. (1992), Genetic algorithms, noise, and the sizing of populations, Complex Systems, Vol. 6, pp. 333–362.

[34] Grefenstette, J.J. (1986), Optimization of control parameters for genetic algorithms, IEEE Transactions on Systems, Man, and Cybernetics, Vol. 16, 122–128.

9

Predictive Modeling of Metal Forming and Machining Processes Using Soft Computing

9.1 Introduction

The finite element method has been a very effective tool in the modeling of metal forming and machining processes as it provides detailed information regarding the product during and after the processes. Analysis of the process often requires non-linear elasto-plastic finite element formulation. The finite element method can also be used for finding out the stress distribution in the tool and stress/vibration analysis of the machines. Unlike the work material, the tools and machines undergo only elastic deformations. In spite of this, a non-linear analysis is often needed. The major drawback of the finite element method is that it requires a large computational time.

This chapter describes the application of soft computing methods that can predict important variables of the processes. Soft computing methods rely more on the data than the physics of the process, although knowledge of the physics of the process may augment the effectiveness of the soft computing methods. For example, in fuzzy-based systems, the rule base may be prepared from the physical laws of the process. Also, in many cases, data can be obtained from computer simulations based on the physics of the process. Such data may be supplemented with experimental data. Before applying a soft computing technique, it is very useful to carry out a statistical analysis of the data. Such an analysis may provide a valuable insight about the process and help in the design of soft computing tools. Also, in order to extract maximum information from limited datasets, data should preferably be collected in a planned manner. In view of this, in this chapter the design of experiments and some useful statistical techniques have been described first. After that the application of neural networks for the modeling of metal forming and machining processes has been discussed. This is followed by a description of the modeling using fuzzy sets. Section 9.6 describes a neuro-fuzzy inference system for carrying out the same task. Finally, a section has been devoted to describing the method of computing the important process variables considering the fuzziness in input variables.

9.2 Design of Experiments and Preliminary Study of the Data

Assume that a certain parameter depends on several independent variables. In order to study the effect of the variables on the parameter, one has to generate data either from experiments or from numerical simulation. A systematic way of studying this effect is through a proper design of experiments. Proper design of experiments brings out a large amount of information with the limited number of experiments. The traditional method of experimentation is the 'one factor at a time' method. In this, only one out of several variables (also called factors) is changed at a time, keeping all other independent variables constant at some values. Although this approach is simple and one gets inference before all experiments are over, it does not uncover the effect of interaction among variables. Moreover, this approach is inefficient and costly. Most of the time, the effect of one variable on the dependent parameter may be strongly influenced by the value of other independent variables. This is called the interaction effect, which cannot be estimated properly in the 'one factor at a time' method.

In the full factorial method of experimentation, each variable is divided into different levels. In some cases the variables take only discrete values and they need not be numbers *e.g.*, presence or absence of a lubricant in metal forming. Here, the presence of the lubricant may be designated as level 1 and absence of lubricant as level 2. In other cases, the variables take analogue values. For example, the percentage reduction in rolling, which may vary between a certain range, say 8% to 24%. In that case, the whole range can be divided into levels. For example, in a two-level design, one can assign level 1 to percentage reduction of 8–16% and level 2 to 16–24%. One can decide to divide the range into three or more levels if more number of experiments can be conducted. Once all the factors have been divided into a number of levels, all possible combinations of levels are considered. Total number of combinations of factors is dependent on the number of variables (factors) and the levels. For example, if there are 7 factors at 2 levels, total combinations would be 2^7, *i.e.*, 128. Thus, in full factorial method, one would need to do 128 experiments. If the variables are divided into 3 levels, total combinations will be 3^7, *i.e.*, 2187, an enormously high number. Thus, many times, full factorial design is not feasible and the fractional factorial method is to be used.

In the fractional factorial method as suggested by Fisher [1] and Plackett and Burman [2], out of various possible combinations, some are selected for study. The selection of the combinations is done in a systematic way so as to bring out the main and interaction effects of the variables. The orthogonal arrays are constructed with a limited number of experiments as a subset of the full factorial layout. An equal number of each level of each factor is represented in the array. For each pair of factors, every combination of factor levels exists and occurs equally often. The technique of orthogonal arrays reduces the size of the experiments to a practicable level. However, some information is lost in this process. Therefore, while adopting this method, technical knowledge of the persons involved in the experiment is very important to ensure that the loss of information is relatively insignificant. To give an example, first consider the full factorial design for three factors at two levels. The first level is represented by -1 and the second level by $+1$. The 2^3 factorial design is as per Table 9.1. In this table, column A×B indicates the interaction effect of factor A×B. The level $+1$ indicates that both A and B are at the same level and -1 indicates that both are at different levels. Similarly, A×B×C is the column of interaction of three factors. If this interaction is not important, then experiments corresponding to any one level of A×B×C can be chosen. For example, we can make an array consisting of only four experiments corresponding to $+1$ level of A×B×C, thus preparing Table 9.2. Table 9.2 is called the fractional factorial table denoted by $OA_4(2^3)$. The OA indicates orthogonal array. The subscripts 4 indicates the number of experiments in the array. The quantity 2 in the bracket indicates the number of level and superscript 3 indicates the number of factors. Note that one could have made the orthogonal arrays with the rows corresponding to level -1 of A×B×C. Thus, the $OA_4(2^3)$ table may not be unique, but it is a balanced table as each factor is equally represented and it is also orthogonal, as for each pair of factors, every combination of factor level exists and occurs equally often. This will allow one to extract and separate out the effects of different factors and their two-way interaction.

For different types of orthogonal arrays, one can refer to [3]. In orthogonal arrays, the columns correspond to factors and the rows correspond to experiments. One can also choose an array having more columns than the variables and delete extra columns. The resulting table will also be an orthogonal array.

Table 9.1. A 2^3 full factorial array

Experiment	A	B	C	A×B	B×C	C×A	A×B×C
1	−1	−1	−1	+1	+1	+1	−1
2	−1	−1	+1	+1	−1	−1	+1
3	−1	+1	−1	−1	−1	+1	+1
4	−1	+1	+1	−1	+1	−1	−1
5	+1	−1	−1	−1	+1	−1	+1
6	+1	−1	+1	−1	−1	+1	−1
7	+1	+1	−1	+1	−1	−1	−1
8	+1	+1	+1	+1	+1	+1	+1

With the designed experiments, one can find the main and interaction effects of a factor. The main effect indicates the individual contribution of the factors to the total variability inherent in the experimental results. For a two level factor, the main effect is obtained as

$$\text{Effect of a factor} = \frac{\sum \text{responses at level 2 of the factor} - \sum \text{responses at level 1 of the factor}}{\text{half the number of experiments}},$$

(9.1)

where 'Σ' denotes the summation.

If the experiments have been conducted at three levels of each factor, level 1 indicating the low level, level 2 indicating the high level and level 3 indicating the middle level, then one can find the linear as well as quadratic effect for a factor. The linear effect will be given by Equation 9.1. The quadratic effect will be given by,

$$\text{Quadratic effect} = \frac{\sum \text{responses at level } 2 - 2\sum \text{responses at level } 3 + \sum \text{responses at level } 1}{\text{one third the number of experiments}}.$$

(9.2)

If the effect of a factor is purely linear, its quadratic effect will be zero.

Table 9.2. An OA_4 (2^3) fractional factorial array

Experiment	A	B	C	A×B	B×C	C×A	A×B×C
2	-1	-1	+1	+1	-1	-1	+1
3	-1	+1	-1	-1	-1	+1	+1
5	+1	-1	-1	-1	+1	-1	+1
8	+1	+1	+1	+1	+1	+1	+1

Apart from the main effects, one might need to know the effect of interaction. This effect is found in a similar manner. In the orthogonal array, we can make the columns corresponding to interaction of two variables and then treat that column as corresponding to a separate factor. For example, in Table 9.2, A×B is treated like a factor with levels +1 and −1 for finding out its effect.

Example 9.1: For knowing the tool life in the machining of medium carbon steel with a TiN coated carbide tool, an experimental study was carried out. Initially, cutting speed v, feed f and depth of cut d were taken at two levels and full factorial experiments were conducted as *per* Table 9.3. Find out the main effect of the parameters v, f and d and the interaction effect of f and v. Also make an $OA_4(2^3)$ fractional factorial array and find the main effects of the parameter from that array.

Table 9.3. A 2^3 array for studying the dependence of tool life on cutting parameters

S. No.	v (m/min)	f (mm/rev)	d (mm)	T (min)
1	135	0.04	0.3	160
2	135	0.04	1.2	120
3	135	0.32	0.3	110
4	135	0.32	1.2	20
5	270	0.04	0.3	60
6	270	0.04	1.2	50
7	270	0.32	0.3	7
8	270	0.32	1.2	2

Solution: Here, for each variable, one level corresponds to low values and the other level corresponds to high value. Using Equation 9.1, the main effects of v, f and d on tool life are as follows:

$$\text{Effect of } v = \frac{(60+50+7+2)-(160+120+110+20)}{4} = -97.75,$$

$$\text{Effect of } f = \frac{(110+20+7+2)-(160+120+60+50)}{4} = -62.75,$$

$$\text{Effect of } d = \frac{(120+20+50+2)-(160+110+60+7)}{4} = -36.25.$$

Thus, it is seen that in the given ranges of the process parameters, the cutting speed has the maximum effect on the tool life followed by the feed and depth of cut. The negative value of the parameters indicates that increasing these parameters decreases the tool life.

For finding out the interaction effect of f and v, it will be helpful to make Table 9.4. Designating the levels of f and v by '+1' and '−1', we easily obtain two levels '+1' and '−1' for $f \times v$. From this table, we find the interaction effect of f and v as follows:

$$\text{Effect of } f \times v = \frac{\sum \text{tool life at level '+1'} - \sum \text{tool life at level '−1'}}{4},$$

$$= \frac{(160+120+7+2)-(110+20+60+50)}{4} = 12.25.$$

Thus, it is seen that compared to the main effect, the interaction effect is smaller.

Table 9.4. Table for interaction effect of f and v, the values in bracket indicating the level

S. No.	v (m/min)	f (mm/rev)	fxv	T (min)
1	135 (-1)	0.04 (-1)	(+1)	160
2	135 (-1)	0.04 (-1)	(+1)	120
3	135 (-1)	0.32 (+1)	(-1)	110
4	135 (-1)	0.32 (+1)	(-1)	20
5	270 (+1)	0.04 (-1)	(-1)	60
6	270 (+1)	0.04 (-1)	(-1)	50
7	270 (+1)	0.32 (+1)	(+1)	7
8	270 (+1)	0.32 (+1)	(+1)	2

Table 9.5 is an $OA_4(2^3)$ prepared on the basis of Table 9.2. From this table, main effect of v, f and d is calculated as

$$\text{Effect of } v = \frac{(60+2)-(120+110)}{2} = -84,$$

$$\text{Effect of } f = \frac{(110+2)-(120+60)}{2} = -34,$$

$$\text{Effect of } d = \frac{(120+2)-(110+60)}{2} = -24.$$

The effects calculated from the fractional factorial array are different to those calculated from full fractional array, but give the same qualitative picture showing that the tool life is influenced most by the cutting speed followed by the feed and the depth of cut. Thus, the calculation of the effects may provide valuable information about the process. However, this information must be used with caution. It is possible to see a very small main effect of a parameter, when the parameter contributes almost same amount to response at low and high level, but contributes very differently at middle values.

Table 9.5. An $OA_4(2^3)$ fractional factorial array

S.No.	v (m/min)	f (mm/rev)	d (mm)	T (min)
1	135	0.04	1.2	120
2	135	0.32	0.3	110
3	270	0.04	0.3	60
4	270	0.32	1.2	2

9.3 Preliminary Statistical Analysis

Before neural network or fuzzy set modeling, it is advisable to carry out the preliminary statistical analysis for finding the significance of various parameters. Parameters that do not have a significant effect on the response can be eliminated from the model. In this section, we shall review a few useful statistical methods, which are helpful in modeling using soft computing.

9.3.1 Correlation Analysis

We can find out the correlation between a variable and its response or between two variables. Let x_i and y_i denote i-th data of variables X ad Y respectively, then estimate of correlation coefficient is given by

$$r(X,Y) = \frac{\sum_{i=1}^{n}(x_i - \bar{x})(y_i - \bar{y})}{\sqrt{\sum_{i=1}^{n}(x_i - \bar{x})^2 \sum_{i=1}^{n}(y_i - \bar{y})^2}}, \tag{9.3}$$

where n is the number of data, $\bar{x}$ is the average value of X and $\bar{y}$ is the average value of Y. It can be shown that $-1 \le r \le 1$. The value of -1 indicates perfect negative linear relation and the value of $+1$ indicates perfect positive linear relation. If $r=0$, there is no linear association between X and Y. However, this does not mean that there is no association between X and Y. There may be a non-linear association between X and Y. Also, it is to be noted that r value being close to $\pm$ 1 is a necessary condition for strong linear association, but is not sufficient. The value can also be high due to outliers. Moreover, the r value close to ± 1 may not be a cause-effect relation, but rather a coincidence.

In the case of a multi-variable problem, a correlation matrix may be constructed whose elements provide the correlation coefficient between two pairs of variables. For example, if the variables are X, Y and Z, the correlation matrix will be as shown in Table 9.6. Note that this matrix is symmetric as $r(X, Y) = r(Y, X)$ *etc* and the elements in the leading diagonal are always 1 as a variable has a strong positive linear relation with itself. If two independent variables have a strong linear relationship, only one should be used to model the behavior of dependent variable on the independent variable.

Table 9.6. A correlation matrix for three variables

	X	**Y**	**Z**
X	1	$r(X, Y)$	$r(X, Z)$
Y	$r(Y, X)$	1	$r(Y, Z)$
Z	$r(Z,X)$	$r(Y, Z)$	1

To know whether the particular value of r really indicates a correlation is dependent on how many data were used for calculating the correlation coefficient. A high value of correlation coefficient calculated based on a large number data has high reliability for linear association. The statistical way of ascertening this is through hypothesis testing, which is described in the following subsection.

Example 9.2: For the example of Equation 9.1, find out the correlation matrix showing the dependence of the variables.

Solution: We construct the correlation matrix using the software SPSS version 12.0. The reader may verify it by hand calculations. The matrix is given as

	v	f	d	T
v	1	0	0	−0.669
f	0	1	0	−0.577
d	0	0	1	−0.333
T	−0.669	−0.577	−0.333	1

It is seen that the cutting speed has the highest negative correlation with the tool life, followed by the feed and the depth of cut. Correlation of each variable with itself is 1. Hence, the diagonal terms are one. Also, v, f and d are the independent variables whose values have been chosen based on an orthogonal array; hence the correlation coefficient for any pair of these variables is zero.

9.3.2 Hypothesis Testing

In many scientific processes we make assumptions and test them for validity. Suppose after taking a sample, we notice that data of the sample does not support our hypothesis. This difference with our hypothesis may be due to the hypothesis being wrong or due to the random process of the sample being biased. Therefore, the tests are carried out to find out whether or not the difference between sample results and hypothesis is due to chance. If the difference is not due to chance, it is called a significant difference. The techniques that find out whether a difference is significant or not are called the test of significance. The whole procedure is known as the testing of a hypothesis.

A hypothesis is a statement supposed to be true till it is proved false. There are two types of hypotheses. The null hypothesis (H_0) is a statement about a single population characteristic usually having a specific value. For example,

$$H_0: \text{The mean flow stress of the material is 300 MPa,} \qquad (9.4)$$

is a null hypothesis. The null hypothesis can also be a statement concerning two or more population parameters, specifying the difference between them. For example, for the wire drawing of two different metals through similar dies, one hypothesis can be

$$H_0: \text{The friction coefficient in both the cases is the same.} \qquad (9.5)$$

The alternative hypothesis (H_1) is a statement about population characteristics usually being larger (or smaller) than a specific value. This is called a one-sided or one-tailed test. More generally, it may be population characteristics being different to a specific value, which is called a two-sided or two-tailed test. For example,

$$H_1: \text{The mean flow stress of the material is larger than 300 MPa,} \qquad (9.6)$$

and

$$H_1: \text{The mean flow stress of the material is smaller than 300 MPa} \qquad (9.7)$$

are one-sided alternative hypotheses, whereas

$$H_1: \text{The mean flow stress of the material is not 300 MPa} \qquad (9.8)$$

is a two-sided alternative hypothesis. Once the null and alternative hypotheses have been set up, the next job is to test these hypotheses. There are different tests of significance, and some of them are described here.

First, we describe the test of significance for the mean of large samples. This test can be applied to find out whether the difference between sample mean and population mean is significant or not. We illustrate this with an example. Suppose the mean flow stress of a material has been quoted in a handbook as 300 MPa with a standard deviation of 20 MPa. If 500 test pieces were tested and the mean flow stress of these test pieces was 290 MPa, can we say that there is insignificant difference between mean flow stress tested and that quoted in the handbook? To answer this question, the following steps are executed.

- Set the null hypothesis:

$$H_0: \text{The sample of 500 test-pieces has been drawn from a population with mean flow stress of 300 MPa.} \qquad (9.9)$$

- Calculate a test statistic:

$$z = \frac{\bar{x} - M}{\sigma / \sqrt{n}} , \qquad (9.10)$$

where x is the sample mean, M is the population mean, σ is the standard deviation and n is the sample size. Thus, in the present case,

$$z = \frac{290 - 300}{20 / \sqrt{500}} = -11.18 . \qquad (9.11)$$

- Select the appropriate confidence level, say 95%. This means if we keep testing different samples of the population (with mean 300 MPa and standard deviation 20 MPa) for a large number of times, 95% samples should tally with the results of our test. We expect 5% samples to defy the test results. In other words, with 95% confidence level, the probability of incorrectly rejecting the hypothesis is 0.05. This is called type I error. When we fail to reject a null hypothesis in spite of it being wrong, we make type II error. Obviously, an attempt to minimize type I error increases type II error. Thus, a trade-off has to be made between these two types of errors. We denote the probability of making type I error by α, which is called significance level. We compare the test statistics with a statistic z_α, which corresponds to a significance level of α. The table providing the values of statistic z_α is available in the books on statistics. You may remember that in a normal distribution, 95% values fall within ± 1.96 σ. Hence, the statistic z_α with which the test-statistic will be compared for level of significance of 0.05 (or 95% confidence level) is 1.96. Here, we expect that the null hypothesis will be incorrectly rejected with a probability of 0.05. The α value of 0.05 is the most commonly used level of significance for practical problems. In the problems where the cost of making type I error is very high, we choose a lower value of α, typically 0.01. In the problems, where the cost of making type I error is much less compared to the cost of making type II error, we can use the higher value of α, say 0.1.

- In the last step, we check whether $|z| < z_\alpha$. If $|z|$ is less than z_α, we conclude that the difference is not significant for rejecting the null hypothesis. In the present case $|z| > z_\alpha$, therefore, the difference is significant and the null hypothesis is rejected. Thus, we reject the hypothesis at 0.05 significance level that the test pieces have been drawn from a population with a mean flow stress of 300 MPa.

Many times the population standard deviation σ may not be known. In that case, σ can be replaced by the sample standard deviation s.

The test of significance can also be carried out for the difference of the means of two large samples. Let $\bar{x}$ be the mean of a sample of size n_1 from a population having the mean M_1 and the variance σ_1^2. Similarly, let $\bar{y}$ be the mean of a sample of size n_2 drawn from a population having the mean M_2 and the variance σ_2^2. If the null hypothesis to be tested is

$$H_0:\ M_1{=}M_2\ , \tag{9.12}$$

the test-statistic is given by

$$z = \frac{\bar{x} - \bar{y}}{\sqrt{\sigma_1^2 / n_1 + \sigma_2^2 / n_2}}\ . \tag{9.13}$$

If σ_1^2 and σ_2^2 are unknown, then

$$z = \frac{\bar{x} - \bar{y}}{\sqrt{s_1^2 / n_1 + s_2^2 / n_2}}\ , \tag{9.14}$$

where s_1 and s_2 are the sample standard deviations.

When the sample size is small, then the test of significance is based on Student's t-test. Here, the t-statistic for the test of significance for the mean is

$$t = \frac{\bar{x} - \mu}{s / \sqrt{n}}\ . \tag{9.15}$$

where μ is the population mean. Tables of t-distribution are available in books on statistics. For α level of significance (confidence level $(1{-}\alpha){\times}100$), one can obtain t_α, if the degree of freedom is known. For one parameter sample of size n, the degree of freedom is equal to $(n{-}1)$. If $|t| < t_\alpha$, the difference between sample mean and population mean is not significant. If $|t| > t_\alpha$, the difference is significant.

The test of significance for the difference of two means of two small samples is carried out when it is necessary to ascertain that samples come from the populations having the same mean. For this purpose, the following steps are executed:

- State the null hypothesis:

    H_0: $\bar{x}$ and $\bar{y}$ do not differ significantly. (9.16)

- Compute the test statistic:

$$t = \frac{\bar{x} - \bar{y}}{s\sqrt{1/n_1 + 1/n_2}}\ , \tag{9.17}$$

where

$$s^2 = \frac{(n_1 - 1)s_1^2 + (n_2 - 1)s_2^2}{n_1 + n_2 - 2}. \tag{9.18}$$

In this case, the degree of freedom is $n_1 + n_2 - 2$.

- Compare the t value with t_α. If $|t| < t_\alpha$, accept H_0 at α level of significance.

In hypothesis testing, often the p-value is estimated. It is the maximum probability that the observed hypothesis will be rejected on the basis of a given sample even though it is true (type-I error). Thus, the smaller the p-value, the more unlikely is the rejection of the hypothesis. The p-value can also be understood as the value of α at which the test statistic becomes critical. For smaller p-values, the procedure may fail to reject a false null hypothesis, thus making type-II error. One has to make a balance between the two types of errors.

For finding out if a particular value of r is significant or not, the test of significance may be employed. Here, we make the following null hypothesis:

H_0: There is no linear relationship between data. $\qquad$ (9.19)

The alternative hypothesis is

H_1: There is a positive linear relationship (one-tailed) between the two variables. $\qquad$ (9.20)

To test the hypothesis, r is compared with critical values of the correlation coefficient r_α [3]. For a given α, r_α will be dependent on the degrees of freedom, which is 2 less than the number of observations. If $r > r_\alpha$ we reject H_0 at a level of significance of α; otherwise we do not reject the hypothesis.

In many cases, a hypothesized value for the parameter may not be available; instead parameters of a population are estimated based on the sample. Suppose the mean of a sample is 60. This does not mean that population mean will also be 60. One might have obtained sample mean of 60 by chance. In the next sample, one may get the value other than 60. The probability that the population mean is exactly 60 is 0. However, if we say that the population mean is between 59 and 61, then there is some probability of its being true. If we increase the interval size, the probability that the interval encompasses the population mean will increase. This probability is called confidence level and is often expressed as a percentage. Commonly used confidence levels are 95% and 99%. If the confidence level is 95%, it means that a certain assumption (such as the population mean lies in a prescribed interval) will be true in 95% of the cases. It is obvious that for the estimation of mean with higher confidence level, the interval of the estimation has to be increased.

There is a direct relationship between hypothesis and confidence interval estimation. A hypothesis test for H_0: $M=M_0$ against H_1: $M=M_0$ will be rejected at a significance level of α if M_0 is not in $(1-\alpha)\times100\%$ confidence limit for M.

Example 9.3: A material is machined by two different milling cutters. After the machining is over, the surface roughness is measured using a stylus type surface measuring instrument. Surface roughness is measured at 12 places. One cutter provided the surface roughness values: 2.96, 2.31, 3.43, 2.74, 3.21, 3.31, 3.44, 2.14, 2.86, 2.04, 2.88, 3.19 μm, whilst the other cutter provided the surface roughness values: 3.25, 2.21, 1.97, 2.52, 2.43, 3.03, 2.06, 2.12, 1.67, 2.53, 2.86, 2.65μm. Can we say that both cutters produce significantly different surface roughness values?

Solution: The first cutter produces a surface roughness with a mean of 2.88 μm and standard deviation of 0.49 μm, whilst the second cutter produces surface roughness with a mean of 2.44 μm and standard deviation of 0.46μm. Just by comparing the means, it appears that the surface roughness generated by the second cutter is lower than that generated by the first cutter. Now, we state the null hypothesis:

H_0: Surface roughness values generated by the cutters do not differ significantly. We compute the test statistic, using Equation 9.17. Here, $n_1=n_2=12$. Hence, from Equation 9.18,

$$s^2 = \frac{0.49^2 + 0.46^2}{2} = 0.2259 \quad \text{or } s = 0.475.$$

Then, from Equation 9.17,

$$t = \frac{2.88 - 2.44}{0.475\sqrt{(1/12)+(1/12)}} = 2.269.$$

The degree of freedom of the data is $(n_1 + n_2 - 2) = 22$. For 95% confidence level and 22 degrees of freedom, the t_α value read from statistical table is 2.074. Since here $|t|$ is greater than this value, we reject the hypothesis. For 99% confidence level and 22 degrees of freedom, the t_α value is 2.819. Since $|t|$ is less than this value, we cannot reject the hypothesis at 99% confidence level. Thus, the hypothesis can be rejected at 95% confidence level but not at 99% confidence level. At 95% confidence level, the probability that the algorithm is incorrectly rejected is 0.05, whereas it is 0.01 at 99% confidence level. The smaller the probability of incorrectly rejecting the hypothesis, the greater is the reluctance to reject the hypothesis. That's why in the present case, although the hypothesis is rejected at 95% confidence level, it could not be rejected at 99% confidence level.

Example 9.4: Observe the correlation matrix of Example 9.2. Find if there is a significant correlation between the process parameters and tool life.

Solution: For six degrees of freedom and 95% confidence level, the critical value of correlation coefficient is 0.707 [3]. Thus, at 95% confidence level, there is no significant correlation.

9.3.3 Analysis of Variance

Suppose a certain response, say roll torque in a rolling process, is dependent on a number of variables (also called factors). The variation in the response is due to variation in the factors as well as random effects. In analysis of variance (ANOVA), we determine the effect of various factors in a systematic manner. For this purpose, experiments must be properly designed and should have been completed. Then the significance of various components associated with factor effects is assessed by comparison with the residual. For this purpose, an F-test is employed. The F-test is employed for comparing the variances. Suppose we are required to compare the variances of two samples of size n_1 and n_2, then the test statistic will be

$$F = \frac{\text{Larger sample variance}}{\text{Smaller sample variance}}. \tag{9.21}$$

The critical value is found from the F-tables. The critical values of F are given as a function of degrees of freedom of two samples and α.

Consider an experimental design in which m experiments are performed at each level. First, we find the total sum of squares (TSS) as per the following formula:

$$TSS = \sum_{i=1}^{n} y_j^2 - \frac{\left(\sum_{i=1}^{n} y_j\right)^2}{n}, \tag{9.22}$$

where n is the total number of experiments and y_j is the response for the j-th experiment. Total degrees of freedom is equal to $n-1$. Then we find the main effects for each factor. For a k-level factor with m observations corresponding to each level, the sum of squares for a factor A is given by

$$SS_A = \frac{(A_1)^2 + \ldots\ldots\ldots + (A_k)^2}{m} - \frac{\left(\sum_{j=1}^{n} y_j\right)^2}{n}, \tag{9.23}$$

where k is the number of levels for factor A and A_i denotes the sum of responses at level i. The degree of freedom for factor A, df_A is equal to $k-1$. The similar type of formula can be used for finding out the interaction effects. If A is an a-level factor and B is a b-level factor, then

$$SS_{A\times B} = \frac{(A_1B_1)^2 + (A_1B_2)^2 + (A_2B_1)^2 + \dots\dots\dots + (A_aB_b)^2}{m} - SS_A - SS_B - \frac{\left(\sum\limits_{j=1}^{n} y_j\right)^2}{n}, \quad (9.24)$$

where m is the number of observations corresponding to each combination of factor levels. The degree of freedom $df_{A\times B}$ is $(a-1)(b-1)$. For three-way interaction,

$$SS_{A\times B\times C} = \frac{(A_1B_1C_1)^2}{m} + \dots\dots + \frac{(A_aB_bC_c)^2}{m}$$

$$- SS_A - SS_B - SS_C - SS_{A\times B} - SS_{A\times C} - SS_{B\times C} - \frac{\left(\sum\limits_{j=1}^{n} y_j\right)^2}{n}. \quad (9.25)$$

with the degree of freedom $df_{A\times B\times C} = (a-1)(b-1)(c-1)$. If the number of observations are different for different combinations, the formulae for sum of squares for the interaction of two factors become

$$SS_{A\times B} = \frac{(A_1B_1)^2}{m_{11}} + \frac{(A_1B_2)^2}{m_{12}} + \frac{(A_2B_1)^2}{m_{21}} + \dots\dots\dots + \frac{(A_aB_b)^2}{m_{ab}}$$

$$- SS_A - SS_B - \frac{\left(\sum\limits_{j=1}^{n} y_j\right)^2}{n}, \quad (9.26)$$

where m_{11}, m_{12} etc. are the respective numbers of observations. For three factor A, B, C, interaction effect of three factors is given by

$$SS_{A\times B\times C} = \frac{(A_1B_1C_1)^2}{m_{111}} + \dots\dots\dots + \frac{(A_aB_bC_c)^2}{m_{abc}}$$

$$- SS_A - SS_B - SS_{A\times B} - SS_{A\times C} - SS_{B\times C} - \frac{\left(\sum\limits_{j=1}^{n} y_j\right)^2}{n}. \quad (9.27)$$

Residual (error) sum of squares (SS_e) is the TSS minus the total of the SS of all effects. The residual degree of freedom can be obtained as

$$df_e = df_{total} - df_A - df_B - df_C - df_{A\times B} - df_{A\times C} - df_{B\times C} - df_{A\times B\times C}. \quad (9.28)$$

Note that when there is only one observation per experiment, error sum of squares will become 0 and we cannot compare the sum of squares of factors or interactions with the error sum of squares in the form of ratios. In such cases, a higher order interaction is usually considered to be non-existent, and its sum of squares is attributed to the error sum of squares.

Table 9.7. ANOVA table

Source	df	SS	MSS=SS/df	F-ratio	p-value
A	df_A	SS_A	$MSS_A = SS_A/df_A$	MSS_A/MSS_e	....
B	df_B	SS_B	$MSS_B = SS_B/df_B$	MSS_B/MSS_e	....
....	....	....	....	....	....
A×B	$df_{A \times B}$	$SS_{A \times B}$	$MSS_{A \times B} = SS_{A \times B}/df_{A \times B}$	$MSS_{A \times B}/MSS_e$	....
......	....	....	....	....	....
Error	df_e	SS_e	$MSS_e = SS_e/df_e$	1	....
Total	df_{total}	TSS			

After finding the sums of squares for various effects and error, we find the corresponding mean sums of squares (MSS) by dividing by the corresponding degrees of freedom. Then we divide the mean sums of squares of each effect by the error sum of squares and call these ratios F-ratios. The F-ratio compares the variance attributed to a particular (main or interaction) factor effect with the variance attributed to randomness in order to assess the significance of the effect. If the F-ratio is large (more than 4), we say the effect is significant. Normally, the critical values from the one-sided F-tables are utilized for some prescribed level of significance. Table 9.7 shows a typical ANOVA table.

Example 9.5: The feed forces obtained during the machining of gray cast iron with ceramic cutting tool are given in Table 9.8. Prepare an ANOVA table for this.

Table 9.8. Feed force for different cutting conditions

f (mm/rev)	v (m/min)	d (mm)	Feed force (N)		
			Replicate 1	Replicate 2	Replicate 3
0.04	100	1	118	121	130
0.04	100	1.5	225	240	260
0.16	100	1	173	168	160
0.16	100	1.5	100	110	108
0.04	400	1	63	61	67
0.04	400	1.5	85	91	98
0.16	400	1	172	141	141
0.16	400	1.5	226	217	221

Solution: ANOVA table is shown in Table 9.9. It has been obtained through SPSS version 12.0. In this table, the p-values up to three decimals have been shown. As the p-value is less than 0.05 for the factors f, v and d and their interactions, the factors and their interactions are significant at 95% confidence level.

Table 9.9. ANOVA table

Source	df	SS	MSS=SS/df	F-ratio	p-value
f	1	5953.500	5953.500	58.752	0.000
v	1	4537.500	4537.500	44.778	0.000
d	1	9048.167	9048.167	89.291	0.000
$f{\times}v$	1	35882.667	35882.667	354.105	0.000
$f{\times}d$	1	7072.667	7072.667	69.796	0.000
$v{\times}d$	1	600.00	600.00	5.921	0.027
$f{\times}v{\times}d$	1	18481.500	18481.500	182.383	0.000
Residual	16	1621.333	101.333		
Total	23	83197.333			

9.3.4 Multiple Regression

If a dependent variable y is the function of m independent variables, $viz.$, x_1, $x_2, \ldots \ldots x_m$, then the multiple linear regression model may be written as

$$y = c_0 + c_1 x_1 + c_2 x_2 + \ldots \ldots \ldots \ldots + c_m x_m + e , \qquad (9.29)$$

where e is the error term. In the case where, there are n observations, one can write the following n equations:

$$
\begin{aligned}
y_1 &= c_0 + c_1 x_{11} + c_2 x_{12} + \ldots \ldots \ldots + c_m x_{1m} + e_1, \\
y_2 &= c_0 + c_1 x_{21} + c_2 x_{22} + \ldots \ldots \ldots + c_m x_{2m} + e_2, \\
&\ldots \ldots \ldots \ldots \ldots \ldots \ldots \ldots \ldots \ldots \ldots \ldots \ldots \ldots \ldots \\
y_n &= c_0 + c_1 x_{n1} + c_2 x_{n2} + \ldots \ldots \ldots + c_m x_{nm} + e_n,
\end{aligned}
\qquad (9.30)
$$

where y_i indicates the i-th observed value of y and x_{ij} indicates i-th observed value of x_j. In the matrix form, we can write the above set of equations as

$$
\begin{Bmatrix} y_1 \\ y_2 \\ \cdot \\ \cdot \\ \cdot \\ y_n \end{Bmatrix} =
\begin{bmatrix}
1 & x_{11} & x_{12} & \ldots \ldots \ldots x_{1m} \\
1 & x_{21} & x_{22} & \ldots \ldots \ldots x_{2m} \\
\cdot & \cdot & \cdot & \ldots \ldots \ldots \cdot \\
\cdot & \cdot & \cdot & \ldots \ldots \ldots \cdot \\
\cdot & \cdot & \cdot & \ldots \ldots \ldots \cdot \\
1 & x_{n1} & x_{n2} & \ldots \ldots \ldots x_{nm}
\end{bmatrix}
\begin{Bmatrix} c_0 \\ c_1 \\ \cdot \\ \cdot \\ \cdot \\ c_m \end{Bmatrix} +
\begin{Bmatrix} e_1 \\ e_2 \\ \cdot \\ \cdot \\ \cdot \\ e_m \end{Bmatrix} ,
\qquad (9.31)
$$

or

$$Y = XC + E . \qquad (9.32)$$

Our attempt should be to minimize the error vector **E**. If all the terms in the vector **E** are zero, the model is perfect. Usually we minimize **E** in least square sense. The sum squared error is given by

$$\mathbf{E}^T\mathbf{E}=(\mathbf{Y}-\mathbf{XC})^T(\mathbf{Y}-\mathbf{XC})=\mathbf{Y}^T\mathbf{Y}-\mathbf{Y}^T\mathbf{XC}-\mathbf{C}^T\mathbf{X}^T\mathbf{Y}+\mathbf{C}^T\mathbf{X}^T\mathbf{XC}. \tag{9.33}$$

By making use of the property that the transpose of a scalar is equal to the scalar itself, we can write

$$\mathbf{Y}^T\mathbf{XC}=\mathbf{C}^T\mathbf{X}^T\mathbf{Y}. \tag{9.34}$$

Hence,

$$\mathbf{E}^T\mathbf{E}=\mathbf{Y}^T\mathbf{Y}-2\mathbf{C}^T\mathbf{X}^T\mathbf{Y}+\mathbf{C}^T\mathbf{X}^T\mathbf{XC}. \tag{9.35}$$

Minimizing this with respect to **C**, we get

$$\frac{\partial(\mathbf{E}^T\mathbf{E})}{\partial\mathbf{C}}=-2\mathbf{X}^T\mathbf{Y}+2\mathbf{X}^T\mathbf{XC}=\mathbf{0}. \tag{9.36}$$

Thus, the error will be minimized, if

$$\mathbf{X}^T\mathbf{XC}=\mathbf{X}^T\mathbf{Y}, \tag{9.37}$$

or

$$\mathbf{C}=(\mathbf{X}^T\mathbf{X})^{-1}\mathbf{X}^T\mathbf{Y}. \tag{9.38}$$

The coefficient vector can be found either by solving for **C**, from Equation 9.37 using any equation solver routine or from Equation 9.38. The matrix $(\mathbf{X}^T\mathbf{X})^{-1}\mathbf{X}^T$ is called the pseudo-inverse of **X**.

The procedure described above for multiple linear regression can also be applied for non-linear regression, when y can be expressed as a polynomial function of the dependent variables. It should be noted that the fitted multiple regression model should also be tested for some data which have not participated in the fitting of the model. For knowing the performance of the fitting, coefficient of determination (R^2) can be calculated, which is given by

$$R^2=\frac{\sum_{i=1}^{n}(y_i-\bar{y})^2-\sum_{i=1}^{n}(y_i-\hat{y})^2}{\sum_{i=1}^{n}(y_i-\bar{y})^2}, \tag{9.39}$$

where $\bar{y}$ indicates the mean and $\hat{y}$ is the predicted value. The R^2 lies between 0 and 1. The closer the value to 1, the better the fitting.

A regression model can also be developed for predicting the lower and upper bounds of the dependent variable. With this, y can be represented as a triangular fuzzy number by assigning a membership grade of 0.5 to the lower and upper bound estimates and a membership grade of 1 to the most likely estimate obtained by a standard multiple regression procedure. The detailed procedure for fuzzy linear regression has been developed by Tanaka [4]. Based on this reference, a simple method for finding the lower and upper estimate is described here.

The lower and upper estimates of the variable can be expressed as

$$y_l(x) = c_0^l + c_1^l x_1 + \ldots\ldots\ldots + c_m^l x_m,$$
$$y_u(x) = c_0^u + c_1^u x_1 + \ldots\ldots\ldots + c_m^u x_m, \tag{9.40}$$

where $c_0^l, c_1^l, \ldots\ldots\ldots c_m^l$ are the coefficients for predicting the lower estimate and $c_0^u, c_1^u, \ldots\ldots\ldots c_m^u$ are the coefficients for predicting the upper estimate of y. Assume that n observations are available. The coefficients can be obtained by solving the following optimization problem:

$$\text{Minimize } \sum_{i=1}^{n} (c_0^u - c_0^l) + (c_1^u - c_1^l)x_{i1} + \ldots\ldots\ldots + (c_m^u - c_m^l)x_{im}, \tag{9.41}$$

subject to

$$\left. \begin{array}{l} c_0^l + c_1^l x_{i1} + \ldots\ldots\ldots + c_m^l x_{im} \le y_i \\ c_0^u + c_1^u x_{i1} + \ldots\ldots\ldots + c_m^u x_{im} \ge y_i \end{array} \right\} \quad \text{for } i = 1, 2, \ldots\ldots, n. \tag{9.42}$$

This ensures that the sum of the differences between the upper and lower estimates is minimized.

Example 9.6: In a rolling process, the roll radius is 65 mm and the inlet thickness of the strip is 1mm. Material is steel with yield strength 324 MPa and b and n equal to 0.052 and 0.295 respectively. In this rolling process, the variations of roll force and roll torque (for unit width of the strip) with percentage reduction r and friction coefficient f are as *per* Table 9.10. This data has been obtained by running an FEM code. Assuming that the roll force and roll torque vary linearly with r and f, fit the multiple linear regression models.

Table 9.10. Variation of roll force and roll torque with r and f

S. No.	R	f	Roll torque per unit strip width (kN-m/m)	Roll force per unit strip width (MN/m)
1	10	0.06	1.7169	1.5768
2	10	0.09	1.8032	1.6606
3	10	0.1	1.8319	1.6943
4	11	0.06	1.9164	1.6795
5	11	0.1	2.0532	1.8024
6	12	0.06	2.1223	1.7962
7	12	0.08	2.2047	1.8830
8	12	0.1	2.2822	1.9227
9	13	0.06	2.3305	1.8981
10	13	0.1	2.5158	2.0405
11	14	0.06	2.5403	1.9981
12	14	0.1	2.7564	2.1616

Solution: Following the procedure discussed in this section, a multiple linear regression model is fitted. The equations for roll torque and roll force per unit width are obtained as follows:

$$T_r = -0.7478 + 0.2191\,r + 4.052917\,f,$$

$$F_r = 0.2424 + 0.1124\,r + 3.3468\,f.$$

Table 9.11 shows the predicted roll force and roll torque data for a unit strip width. We observed that the predicted values are very close to the FEM values of Table 9.10. Thus, for small ranges of r and f, a linear approximation may be employed. The coefficient of determination for the fitting can be determined using Equation 9.39. It is 0.997 for T_r and 0.996 for F_r, indicating an excellent fitting.

Table 9.11. Predicted roll force and roll torque for a strip width of unity

S. No.	T_r (kN-m/m)	F_r (MN/m)
1	1.6864	1.5672
2	1.8080	1.6676
3	1.8485	1.7011
4	1.9055	1.6796
5	2.0676	1.8135
6	2.1246	1.7920
7	2.2056	1.8589
8	2.2867	1.9259
9	2.3437	1.9044
10	2.5058	2.0383
11	2.5628	2.0168
12	2.7249	2.1507

The regression equations predicting the lower and upper bound of roll torque and roll force are also obtained following the procedure outlined in this subsection. These expressions are given by

$$T_r^l = -0.5931 + 0.2098\,r + 3.27\,f, \qquad T_r^u = -0.7667 + 0.2311r + 2.8767\,f\ ;$$

$$F_r^l = 0.2754 + 0.1089\,r + 3.2814\,f, \qquad F_r^u = 0.1633 + 0.1237r + 2.9375\,f.$$

Table 9.12 shows the upper and lower bound estimates of roll torque and roll force. A comparison of this table with Table 9.10 shows that, in all cases, the FEM data falls in the closed interval of lower and upper bound estimates. Reader should verify these results by writing their own code.

Table 9.12. Upper bound and lower bound estimates of roll torque and roll force

S. No.	Roll torque/width (kN-m/m)		Roll force/width (MN/m)	
	Lower bound	Upper bound	Lower bound	Upper bound
1	1.7011	1.7169	1.5621	1.5768
2	1.7992	1.8032	1.6606	1.6649
3	1.8319	1.8320	1.6934	1.6943
4	1.9109	1.9480	1.6711	1.7005
5	2.0417	2.0630	1.8024	1.8180
6	2.1207	2.1791	1.7801	1.8242
7	2.1861	2.2366	1.8457	1.8830
8	2.2515	2.2942	1.9113	1.9417
9	2.3305	2.4102	1.8891	1.9479
10	2.4613	2.5253	2.0203	2.0652
11	2.5403	2.6413	1.9981	2.0717
12	2.6711	2.7564	2.1293	2.1892

9.4 Neural Network Modeling

Let us consider the problem of predicting roll force and roll torque in plain strain rolling as a function of process variables. The roll force F_r and roll torque, T_r per unit width are dependent on the yield strength of the material $(\sigma_Y)_0$, hardening parameters b and n, coefficient of friction f, roll radius R, initial thickness of the sheet h_1 and percentage reduction r. Then the roll force and roll torque can be expressed as

$$F_r = (\sigma_Y)_0\, h_1\, F\left(R/h_1, r, b, n, f\right), \tag{9.43}$$

$$T = (\sigma_Y)_0\, h_1^2\, G\left(R/h_1, r, b, n, f\right), \tag{9.44}$$

where F and G are functions of five non-dimensional parameters. A block diagram of a neural network is as shown in Figure 9.1, where the black box is dependent on the type of the network used and its design. A background on neural networks is provided in Chapter 8. In this section, we describe the procedure for neural network modeling. The following subsection describes the procedure in the context of modeling a neural network for the roll force and roll torque prediction, but the procedure can be applied to modeling of any dependent variable in metal forming or machining processes.

Figure 9.1. Neural network as a black box

9.4.1 Selection of Training and Testing Data

Selection of training and testing data is very crucial. Training data are supplied to a neural network and the network adjusts its weights and/or any other parameter in order to minimize the error between predicted and actual known values of roll force and roll torque. It is possible to reduce the error in prediction for training data as much as we want by increasing the number of network parameters like weights. However, such a network may cause over-fitting problem. An over-fitted network memorizes the data for which it has been trained, but predicts very poorly for unseen data. This is called a poor generalization capability. In order to have a good generalization capability, the training error is not reduced indefinitely. Instead, for each designed network, apart from training error, testing error is also found. Testing error is the error in the prediction for testing data. In the literature many researchers use the term 'cross-validation' for 'testing'. Design of the network is finalized based on the training as well as the testing error. Once the design has been finalized, the performance of the network can be studied by supplying some validation data to it. Researchers using the term 'cross-validation' for 'testing' use the term 'testing' for 'validation'.

The number of training and testing data and type of data are very crucial to the design of a network. There is no well-established formula for finding out the number of training and testing data. A number of suggestions have been offered. One simple way is to divide the total data as two thirds training and one third testing data. One can also keep about 10% of the available data for validation and divide the remaining data into the training and testing sets. Some researchers have provided empirical expressions relating the total number of neurons and training data. For example, Lawrence and Petterson [5] have suggested having the training data between 2 and 10 times the total (input + hidden + output) neurons. Training data is selected randomly or by using the design of experiments. Full or fractional factorial experiments can be used. Dixit and Chandra [6] have suggested a simple way for choosing the training data. Considering that neural networks are poorer in extrapolation than in interpolation, they suggest that for n inputs, the minimum number of training data should be such that it encompasses the corners of n-dimensional hypercube. For example, in the case of a three input problem, there is a total of eight combinations of high and low values of the input parameter, all of which should be taken in the training set. In addition, the parameter having high influence on the dependent output variable should find more representation in the training set compared to an independent parameter having less influence. Equations 9.1 and 9.2 can form a basis for judging the influence of the parameter.

Kohli and Dixit [7] have used a simple method for assessing the number of testing data. The method is based on the simple theory of probability. We assess the performance of the fitted network based on the testing data. If the error of prediction is below a specified value for all testing data, the network is passed. If the testing data shows more error, the neural network is modified. Let us assume that we are passing a network based on m testing data and also that, for the passed neural network, the ratio of incorrect predictions to total predictions is p. Thus, the probability that the network makes correct prediction for a random input dataset is $(1-p)$. The total probability P of making m correct predictions is given by

$$P = (1 - p)^m .$$
(9.45)

For a highly reliable testing procedure, this probability should be quite low unless there is an ideal neural network model for which $p = 0$. The network designer can specify the values of p as well as P. For a highly reliable network, P and p should be small. However, the smaller is P and/or p, the larger is the value of m. It is to be noted that for practical problems, one should not expect a very low value of p, because there may be inherent errors in the experimental data also.

Example 9.7: A neural network is tested based on 10 testing data and predictions for all the data are found satisfactory. What is the probability that the neural network model that provides correct predictions only 90% of time, will get selected based on the testing?

Solution: Here, $p = 0.1$ and $m = 10$. From Equation 9.41, $P = (0.9)^{10} = 0.35$. Thus, we see that there is a significant probability that the neural network model will provide correct predictions only 90% of the time. If we increase the number of testing data to 20, P will be 0.12. It means that if a network is selected based on the testing by 20 random data, the probability that the neural network provides the correct prediction only 90% of the time will be only 0.12, a significantly lower value.

All the data used in the network has to be normalized to lie between a common range, say between 0.1 and 0.9. A linear mapping can do this. If x_{max} and x_{min} are the maximum and minimum values of a parameter and x_{nmax} and x_{nmin} are the normalized maximum and minimum values, then the normalized value of the parameter is given by

$$x_n = x_{nmin} + \frac{x_{nmax} - x_{nmin}}{x_{max} - x_{min}}(x - x_{min}),$$ (9.46)

where x is the actual value of the parameter.

9.4.2 Deciding the Processing Functions

In Chapter 8 we introduced hyperbolic tangent, log sigmoidal and unipolar ramp functions. In multilayer perceptron networks (MLP), hyperbolic tangent and log sigmoidal functions are usually used in the hidden layer neurons. Output neurons often use pure linear functions, which provide the weighted sum of the inputs obtained from the neurons in the previous layers. One can use other different types of processing functions. However, there is no solid evidence to show that one particular type of processing function is better than the others. We suggest the use of either log sigmoidal or hyperbolic tangent functions for hidden layer neurons and pure linear for the output neurons. In the case of radial basis function (RBF) neural networks, a wide choice of processing functions is available. Among these Gaussian and multiquadrics have been widely used. Some authors have observed multiquadrics to perform better for data interpolation; however again there is no solid evidence for it. One argument in favor of multiquadrics can be that they represent a global response, because their values increase with increasing distance from the center, whereas the Gaussian functions monotonically decrease with increasing distance from the center and hence have local response characteristics. However, spread parameters have a large influence on the characteristics of a radial basis function.

9.4.3 Effect of Number of Hidden Layers

In radial basis function neural networks, only one hidden layer is used. In multi-layer perceptron neural networks, more than one hidden layer may be used. Sometimes, the networks with a large number of layers and fewer units in each layer may generalize better than shallow networks with many units in each layer. However, introducing more hidden layers increases the complexity of the training process. In principle, it is possible to model any continuous function with one hidden layer alone.

9.4.4 Effect of Number of Neurons in the Hidden Layers

The most critical parameter affecting the accuracy of prediction in a neural network model is the number of neurons in the hidden layers. One can keep on reducing the training error by increasing the number of neurons. However, the testing error may increase. Therefore, for the best performance one needs to have the optimum number of neurons in the hidden layer. Although some guidelines are available for deciding the number of neurons in the hidden layers, most researchers prefer to employ a trial and error procedure. One can start with some minimum number of neurons and keep on increasing the neurons till the overall error (maximum of training and testing error) increases. Conversely, one can start with a very high number of neurons in the hidden layer and keep on reducing the neurons till the performance starts deteriorating. However, it is to be noted that often the increase or decrease in the performance may not be monotonic with the number of hidden neurons. Therefore, it may be necessary to test the performance of network for all possible numbers of hidden neurons in a range. Previous experience may serve as a good guideline in deciding the range.

In RBF neural networks, the centers (neurons) in the hidden layer are increased progressively till the error starts increasing. The strategy adopted in the toolbox of the MATLAB® package is as follows. Initially, the hidden layer has no neurons. The network is simulated and the input vector with the greatest error is made the new center. The network is again trained and the input vector with greatest error is included as center. This procedure keeps on repeating till the mean squared error falls below the goal.

9.4.5 Effect of Spread Parameter in Radial Basis Function Neural Network

In radial basis function neural networks, the performance is highly sensitive to spread parameter. To illustrate this, we make an attempt to fit the function $y = x(10 - x)$ $0 \le x \le 10$ with six training data at $x = 0, 2, 4, 6, 8$ and 10. We use the radial basis function given by Equation 8.46, reproduced for one input case as

$$\phi_j(x, \sigma_j) = \exp\left\{-\frac{\left\|x - x_j^c\right\|_2^2}{2\sigma_j^2}\right\}.$$

Here, the spread parameter is the square root of the varaiance σ_j^2. Figure 9.2 shows a function fitted by radial basis function neural networks with two different spread parameters. The actual function has also been plotted. It is observed that the smaller value of spread parameter fits a non-smooth function with many peaks and valleys. The larger value of spread parameter provides a smooth curve; however, increasing the spread parameter beyond a limit makes the resulting simultaneous equations highly ill-conditioned and may pose difficulty in solving. Figure 9.3 fits the following function based on 10 uniformly spaced data:

$$y = \begin{cases} x+1 & 0 \le x \le 5, \\ 11-x & 5 < x \le 10. \end{cases}$$

This function has a slope discontinuity at $x = 5$. It is seen in the figure that the actual curve is well approximated by a Gaussian radial basis function with a variance of 18.1622. This shows the importance of selecting proper spread parameters.

When a Gaussian function is used, the variance σ_j^2 is commonly set according to the following simple heuristic relationship [8]:

$$\sigma_j^2 = \frac{d_{max}^2}{K}, \tag{9.47}$$

where d_{max} is the maximum Euclidean distance between the selected centers and K is the number of the centers. As this is only a heuristic relationship, the optimum spread parameter should be searched in the vicinity of the value obtained by Equation 9.47.

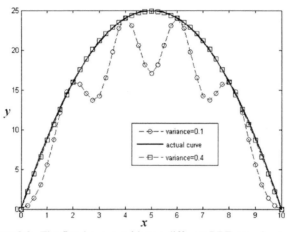

Figure 9.2. The fitted curves with two different RBF neural networks

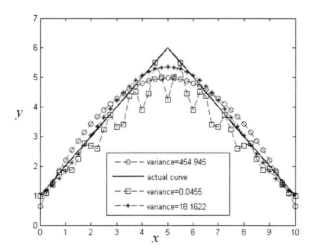

Figure 9.3. The fitted curves with three different RBF neural networks

9.4.6 Data Filtration

Data filtration is the process of removing spurious data that might have come by the error in data collection. When replicates are collected, one can ignore the values out of $\pm 2\sigma$ or $\pm 3\sigma$ limits. Also, once the network has been properly fitted and the error in prediction for some data is more than twice the root mean squared error, those data should be examined. It is possible that some of these are spurious data. Those data that have been confirmed to be correct should be retained in the training set and the network should be modeled again.

9.4.7 Lower and Upper Estimates

Many times, lower and upper estimates of the parameter are required along with the most likely estimate. With lower, upper and most likely estimates known, the parameter can be represented as a fuzzy number. A methodology to predict the lower and upper bounds of any non-linear function has been proposed by Ishibuchi and Tanaka [9]. This requires a simple modification of the back propagation algorithm. It is based on the concept that, during the training of a neural network for the prediction of an upper estimate, if the predicted value is more than the actual (target) value, a reduced value of the error has to be used by the back propagation algorithm. On the other hand, if the predicted value is less than the target value, the full error is to be considered. However, in the early iterations of the algorithm, even if the predicted value is more than the actual value, nearly full error has to be used to ensure that predictions are not too far away from the target values. As the iterations of the back propagation algorithm proceed, the weightage of the error keeps on reducing for the predictions which are more than the actual one. Thus, the error for a particular data (pattern) p is

$$
E_p = \begin{cases} (d_p - o_p)^2 / 2 & \text{for } d_p \ge o_p, \\ w(i)(d_p - o_p)^2 / 2 & \text{for } d_p < o_p, \end{cases} \tag{9.48}
$$

where d_p is the actual (target) output, o_p is the predicted output and $w(i)$ is the weight factor. The weight factor keeps on decreasing with iterations. It should have the following two properties:

Property 1: The weight factor lies between 0 and 1, i.e., $0 < w(i) \le 1$.

Property 2: As $i \to \infty$, $w(i) \to 0$.

Different types of decreasing functions can be used for the weight factor. If the weight factor decreases rapidly as the iterations proceed, the upper bound estimate will be far from the target values, although convergence rate will be faster. If the weight factor decreases slowly, the upper bound estimate will be nearer to the actual values. However, first the convergence will be slow and second for some data, the predictions may not be true upper bound estimates. Thus, there is no unique function $w(i)$, which can be recommended. For the prediction of roll force and roll torque in cold flat rolling process, Dixit and Chandra [6] used

$$w(i) = \frac{1}{1 + (i/4000)^3},$$
(9.49)

whereas for the prediction of surface roughness in a turning process, Kohli and Dixit [7] used

$$w(i) = \frac{1}{1 + (i/500)^3}.$$
(9.50)

Numerical experiments help one to arrive at a proper decision regarding the weight factor. The procedure for finding the lower estimate is the exactly reverse. In this case, the error function is given by

$$E_p = \begin{cases} (d_p - o_p)^2 / 2 & \text{for } d_p < o_p, \\ w(i)(d_p - o_p)^2 / 2 & \text{for } d_p \geq o_p. \end{cases}$$
(9.51)

Here, as the iterations proceed, the weightage of the error is reduced for the cases where the predicted value is lower than the target value.

For predicting the upper and lower bounds using a radial basis function neural network, the weights can be obtained as interval numbers as explained in the section on multiple regression. Then, one set of weights predicts the lower bound estimate, whilst the other set predicts the upper bound. This procedure was used by Sonar et al. [10] for predicting the surface roughness in a turning process.

Example 9.8: We present one example of predicting the surface roughness in the turning process, as reported in [7, 10]. For predicting the surface finish, the cutting speed (v), feed (f), depth of cut (d) and acceleration of radial vibration (a) are used. Table 9.13 shows the training data and Table 9.14 shows the testing data. Initially 16 training and 8 testing data were taken and the best possible network was fitted with this data. The testing data for which the error in prediction was more than a prescribed value were transferred to the training set and new data were added in lieu of that. In this way, finally there were 19 training and 11 testing data.

Table 9.13. Training dataset in wet turning by HSS tool

S. No.	v (m/s)	d (mm)	f (mm/rev)	a (m/s^2)	Ra (μm)
1	107.80	0.3	0.04	0.55	1.74
2	105.12	0.3	0.16	0.97	3.23
3	104.80	0.6	0.04	2.92	2.74
4	106.02	0.6	0.16	2.66	2.91
5	27.71	0.3	0.04	0.95	2.06
6	26.99	0.3	0.16	0.88	5.20
7	27.71	0.6	0.04	0.59	2.87
8	26.99	0.6	0.16	1.42	6.20
9	46.55	0.3	0.08	0.73	3.21
10	64.56	0.4	0.04	0.66	2.13
11	78.10	0.6	0.12	2.10	4.57
12	73.95	0.3	0.05	0.76	2.52
13	38.50	0.3	0.06	0.47	3.37
14	34.71	0.6	0.08	0.63	3.67
15	74.13	0.4	0.10	2.48	4.80
16	36.87	0.5	0.12	1.07	4.55
17	48.14	0.6	0.08	0.58	4.52
18	106.47	0.3	0.08	0.65	2.26
19	23.45	0.3	0.04	0.42	1.99

Table from Kohli and Dixit [7]. Copyright [2005] Springer

Table 9.14. Testing dataset in wet turning by HSS tool

S. No	v	d	f	a	Ra				*(%)
					Exp.	L.E.	M.L.	U.E.	
1	42.98	0.4	0.10	0.67	4.24	3.62	4.05	5.22	4.54
2	35.96	0.3	0.12	0.95	5.21	4.01	4.71	6.08	9.56
3	76.43	0.6	0.05	1.23	2.91	2.45	3.24	3.66	−11.39
4	72.92	0.6	0.04	0.79	2.81	2.40	2.95	3.43	−4.96
5	32.87	0.3	0.12	0.48	4.29	4.00	4.59	5.92	−6.89
6	48.75	0.6	0.04	0.65	3.18	2.56	3.19	3.83	−0.15
7	103.55	0.6	0.08	3.66	3.59	2.45	3.46	3.71	3.52
8	47.52	0.6	0.16	1.73	5.43	5.42	5.80	6.18	−6.72
9	47.17	0.3	0.04	0.90	2.31	2.11	2.49	3.00	−7.93
10	27.35	0.6	0.08	0.72	4.00	3.41	3.82	5.28	4.53
11	54.28	0.6	0.04	0.55	2.78	2.65	3.17	3.67	−14.04

* - Deviation of most likely value from experimental value
Units of variables in Table 9.14 are same as in Table 9.13.
L.E. – Predicted lower estimate of surface roughness
M.L. – Predicted most likely estimate of surface roughness
U.E. – Predicted upper estimate of surface roughness
Exp. – Experimental Values
Table from Kohli and Dixit [7]. Copyright [2005] Springer

The fitted network was tested with 29 validation data (Table 9.15). These data were different from training and testing data. Figure 9.4 shows the lower, upper and most likely estimates using MLP network and compares them with the experimental values. It is seen that in most cases, the most likely estimates are close to the experimental values. Moreover, the experimental values fall in between the lower and upper estimates. Figure 9.5 shows the results using RBF network. Here also a good agreement is found between predictions and experiments.

Table 9.15. Validation dataset in wet turning by HSS tool

S. No.	v (m/s)	d (mm)	f (mm/rev)	a (m/s^2)	Ra (μm)
1	34.71	0.6	0.08	0.63	3.67
2	64.56	0.4	0.04	0.66	2.13
3	29.39	0.6	0.16	1.12	5.22
4	20.88	0.4	0.12	0.67	4.79
5	54.91	0.4	0.08	2.80	3.53
6	38.50	0.3	0.06	0.47	3.37
7	34.71	0.6	0.08	0.57	3.73
8	48.14	0.6	0.08	0.58	4.52
9	64.56	0.4	0.04	0.54	2.84
10	32.87	0.3	0.12	0.55	5.26
11	54.28	0.6	0.04	0.71	3.76
12	29.39	0.6	0.16	0.98	5.44
13	20.88	0.4	0.12	0.73	3.76
14	54.91	0.4	0.08	0.73	3.48
15	38.50	0.3	0.06	0.56	3.77
16	48.70	0.4	0.04	0.54	2.39
17	61.88	0.5	0.08	0.98	4.68
18	60.56	0.5	0.08	0.77	3.14
19	100.85	0.5	0.16	1.49	3.11
20	26.48	0.4	0.04	0.37	2.03
21	49.25	0.3	0.06	0.86	3.13
22	106.34	0.3	0.04	0.94	2.65
23	45.95	0.6	0.06	1.16	3.70
24	101.80	0.6	0.06	2.70	2.51
25	45.99	0.3	0.16	1.15	5.87
26	104.20	0.3	0.16	2.23	3.36
27	45.95	0.6	0.16	2.28	5.06
28	46.97	0.6	0.16	2.43	6.35
29	103.10	0.6	0.16	3.32	4.72

Table from Kohli and Dixit.[7]. Copyright [2005] Springer

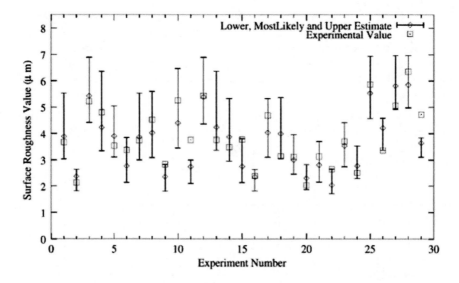

Figure 9.4. Predicted values *vs* experimental values of surface roughness in wet turning of steel with HSS tool using MPL network. From Kohli and Dixit [7]. Copyright [2005] Springer

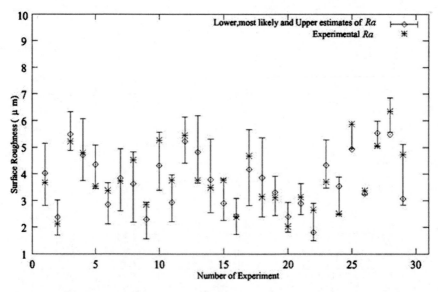

Figure 9.5. Predicted values *vs* experimental values of surface roughness in wet turning of steel with HSS tool using RBF network. From Sonar *et al.* [10]. Copyright [2006] Springer

9.5 Prediction of Dependent Variables Using Fuzzy Sets

The fuzzy set theory can be used to predict the dependent variables in a process. In Chapter 8, it was discussed how a fuzzy inference system works. Based on a rule base and the values of input variables, output variables can be predicted. Chen and Black [11] proposed a procedure for generating the rule base as well as predicting the dependent variable. They employed the procedure for tool breakage monitoring and named it the Fuzzy-Nets In-Process (FNIP) system. Chen and Savage [12] employed it for the prediction of surface roughness in milling operations. The procedure basically consists of five steps, which are briefly described for a general problem of predictive modeling of a dependent variable.

Step 1: Fuzzification of the Data
In this step, the input-output datasets are fuzzified. This means that the data is divided into a number of fuzzy sets. One particular element of the dataset can be a member of more than one fuzzy sets with different membership grades. Figure 9.6 shows the fuzzified input and output variables for a machining process. It is seen that the cutting speed of 90 m/min has a membership grade of 0.37 in H2 and 0.63 in H1. Similarly, the depth of cut of 0.48 mm has a membership grade of 0.4 in the fuzzy set L1 and 0.6 in H1.

Step 2: Rule Generation
In this step, we generate the rules of the form: "If u is U_1 and v is V_1, then o is O_2," where u and v are the two independent variables and o is a dependent variable. The U_1 and the V_1 are the fuzzy sets. The total number of possible fuzzy rules is equal to the product of fuzzy divisions of each input variable. Thus, for a two input problem, if u is divided into 4 fuzzy sets and v is divided into 7 fuzzy sets, the total rules will be 28. Rules are generated from available datasets, which may be obtained from experiments. Each set consisting of the input variables and the dependent variable can contribute to a rule. For generating the rules, each variable is assigned to a fuzzy set in which it has the maximum membership grade. For each input-output dataset, one rule will be formed. In the rule generation, basically we convert the numerical values of the variable into fuzzy sets. As an example, consider that it is known that for a comfortable condition in the room, a fan has to run at 2500 RPM, when the temperature was 35 °C and relative humidity was 70%. In this case, temperature and humidity are the independent variables and fan speed required to maintain comfortable condition is a dependent variable. Now suppose that 35 °C has the highest membership grade of 0.7 in fuzzy set 'high'. We shall then consider this temperature as 'high' for generating a rule, irrespective of the fact that this temperature may also be called a 'medium' temperature with a lower membership grade, say 0.35. Similarly, suppose that 70% relative humidity has the highest membership grade in the fuzzy set 'medium' and the fan speed of 2500 RPM has the highest membership grade in the fuzzy set 'high'. In that case, this dataset generates the rule: If temperature is high and relative humidity is medium, then the fan speed is high. In this way, each dataset is used for generating the rules. There is a possibility of occurrence of repeated and conflicting rules. Conflicting rules have to be tackled in step 3.

Step 3: Conflict Resolution
Two or more rules conflict when they have the same IF condition, but different THEN consequent. Suppose there are the following two rules:
Rule 1: If u is U_1 and v is V_1, then o is O_2.
Rule 2: If u is U_1 and v is V_1, then o is O_3.
These rules conflict, because one cannot decide whether the output has to be O_2 or O_3. The following strategy may be adopted to resolve the conflict:

- The rule occurring more frequently is chosen over the less frequent rule.
- If each rule occurs as frequently as the other, the degree of each rule is calculated. The degree of a rule is defined as the product of membership grades of input and output variables. In Chapter 8 we defined the strength of a rule as the product of membership grades of input variable. Thus, the degree of a rule is strength of a rule multiplied by the membership grade of output variable. For example, based on data, if u has the highest membership grade of 0.7 in U_1, v has the highest membership grade of 0.8 in V_1 and o has the highest membership grade of 0.6 in O_2, then we generate the following rule:
  If u is U_1 and v is V_1, then o is O_2.
  The degree of this rule is $0.7 \times 0.8 \times 0.6 = 0.336$. In the case of two conflicting rules, the rule with the higher degree is chosen. A criterion can be made that, if the difference in the degrees of the two rules is more than a prescribed value δ, the rule with the higher degree is chosen.
- If the degrees of the two rules are nearly equal, *i.e.*, the difference in the degrees of the two rules is less than δ, then the resolution of input parameters need to be increased by increasing the fuzzy set divisions that represent the variable.

Step 4: Combination of Rules
All the rules are combined. Usually, a large number of rules are generated. These rules can be reduced by using Boolean operations. The reduced set of rules are easier to interpret. As a simple example, consider the following two rules:
    If temperature is high and humidity is medium, then the fan speed is high.
    If temperature is high and humidity is high, then the fan speed is high.
    Then, one can combine these into one rule:
    If temperature is high and humidity is (medium or high), then the fan speed is high.
Further, if the humidity has been divided into three fuzzy sets, low, medium and high, then the above rule may also be written as
    If temperature is high and humidity is (not low), then the fan speed is high.

Step 5: Making Out the Inference
Once the rule base is ready, for a particular combination of input variables, output can be computed following the procedure described in Section 8.3.11. One can also solve the inverse problems by use of rule base. Some extrapolation is also possible.
    Abburi and Dixit [13] have used this method for predicting the surface roughness in a turning process. They first fitted an MLP neural network by training it with the experimental data. The fitted network was used to generate a large

database. This database was used to generate the rule base. The Fuzzification of the data was done as *per* Figure 9.6. With the 29 validation data of Table 9.15, the root mean squared fractional error was 16.79% and 76% data had error less than 20%.

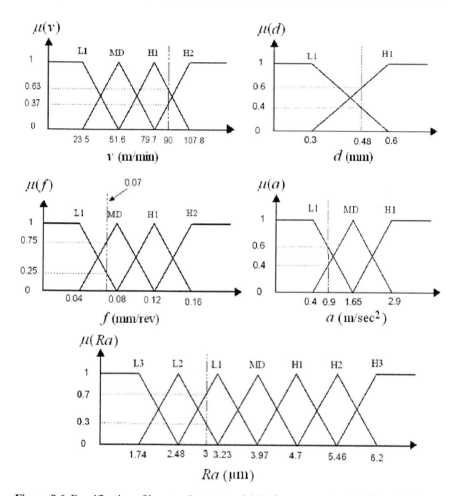

Figure 9.6. Fuzzification of input and output variables for wet turning by HSS tools. From Abburi and Dixit[13]. Copyright [2006] Elsevier

9.6 Prediction Using ANFIS

In this section, we describe a procedure which is a combination of artificial neural network and fuzzy set theory. It is called the adaptive-network-based fuzzy inference system (ANFIS) [14]. It consists of five layers; a typical architecture for two input and one output problem is shown in Figure 9.7. For simplicity, each input x and y is divided into two fuzzy sets only: A_1, A_2 and B_1, B_2.

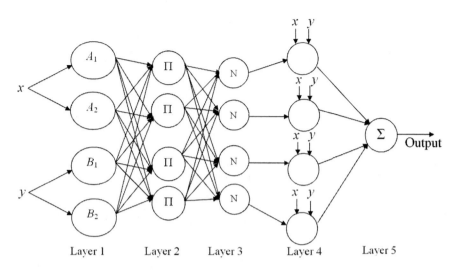

Figure 9.7. An ANFIS architecture

Layer 1: Every neuron i in this layer is an adaptive neuron, with a known output function:

$$O_{1,i} = \mu_{A_i}(x) \qquad \text{for } i = 1, 2,$$
$$O_{1,i} = \mu_{B_{i-2}}(x) \qquad \text{for } i = 3, 4.$$

(9.52)

where 'μ' indicates membership function. It can be any suitable function, for example a generalized-bell-shaped function, given by

$$\mu_{A_i}(x) = \frac{1}{1 + \left| (x - c_i)/a_i \right|^{2b_i}},$$

(9.53)

or a Gaussian function given by

$$\mu_{A_i}(x) = \exp\left(-\frac{(x - c_i)^2}{2\sigma_i^2} \right).$$

(9.54)

Parameters in this (a_i, b_i, c_i and σ_i) are known as premise parameters and are determined during the training of the network.

Layer 2: In this layer each neuron corresponds to one particular rule. The membership grades corresponding to each IF part of the rule reach the neuron and are multiplied. Thus, the output of the neuron corresponds to the firing strength of the rule. Thus,

$$O_{2,ij} = w_{ij} = \mu_{A_i}(x)\mu_{B_j}(y) \qquad \text{for } i = 1, 2 \text{ and } j = 1, 2. \tag{9.55}$$

Layer 3: The outputs from layer 2 reach layer 3. Each neuron in this layer, finds the normalized firing strength. Thus,

$$O_{3,ij} = \overline{w}_{ij} = \frac{w_{ij}}{w_{11} + w_{12} + w_{21} + w_{22}} \qquad \text{for } i = 1, 2 \text{ and } j = 1, 2. \tag{9.56}$$

Layer 4: Each neuron i in this layer is an adaptive neuron with a neuron function:

$$O_{4,ij} = \overline{w}_{ij} f_{ij} = \overline{w}_{ij}(p_{ij}x + q_{ij}y + r_{ij}). \tag{9.57}$$

Parameters p_{ij}, q_{ij} and r_{ij} in this layer are called consequent parameters.

Layer 5: The single neuron in this layer is a fixed neuron computing the overall outputs as the summation of all incoming signals. Thus,

$$O_{5,ij} = \sum_j \sum_i \overline{w}_{ij} f_{ij}. \tag{9.58}$$

Consequent parameters are found in forward pass by least square estimator. The premise parameters in the backwards pass are found by gradient descent method.

Example 9.9: It is desired to develop a model for the prediction of roll torque in the rolling operation. The range of parameters is

$$\%r = 4 - 24, \quad R/h_1 = 50 - 100, \quad f = 0.06 - 0.14.$$

The material properties are fixed at $(\sigma_Y)_0 = 324$ MPa, $b=0.052$, $n=0.295$. A total of 16 training data are generated by an FEM code. Out of these eight are generated according to full factorial design and the remaining eight in a random manner. A total eight data are generated for testing of the network. The training and testing datasets are shown in Tables 9.16 and 9.17. The MATLAB® package is used for modeling and the ANFIS architecture (screen print out of MATLAB®) for this problem is shown in Figure 9.8. Each input variable is divided into 'high' and 'low' and the generalized bell-shaped membership functions are used. Table 9.18 compares ANFIS results with the FEM results. In all the cases the error is less than 10%.

Table 9.16. Training dataset for roll torque prediction

S. No.	R/h_1	f	$\% r$	T (kN-m/m)
1	50	0.06	4	0.4795
2	100	0.06	4	0.9615
3	50	0.14	4	0.5012
4	100	0.06	24	7.7913
5	100	0.14	24	10.8048
6	71	0.09	7	1.3097
7	73	0.1	9	1.8393
8	87	0.12	20	6.4451
9	78	0.11	14	3.4730
10	100	0.14	4	1.0610
11	50	0.06	24	3.6475
12	50	0.14	24	4.4904
13	63	0.08	13	2.3525
14	60	0.08	15	2.6496
15	61	0.07	11	1.8310
16	88	0.065	19	5.1582

Table 9.17. Testing dataset for roll torque prediction

S. No.	R/h_1	f	$\% r$	T (kN-m/m)
1	93	0.07	23	7.0693
2	79	0.12	18	4.9906
3	67	0.13	21	5.1809
4	79	0.11	15	3.8437
5	74	0.09	8	1.5996
6	65	0.11	20	3.9972
7	60	0.08	10	1.6358
8	52	0.07	20	3.1482

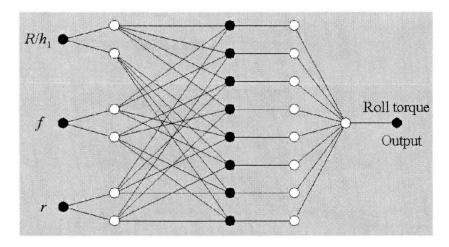

Figure 9.8. ANFIS model for the roll torque prediction

Table 9.18. Result of roll torque prediction for testing dataset

S. No.	Roll torque (kN-m/m)		%
	FEM Results	**ANFIS Results**	**Deviation**
1	7.0693	6.8744	2.7575
2	4.9906	5.0694	1.5786
3	5.1809	5.1551	0.4979
4	3.8437	3.6946	3.8796
5	1.5996	1.4919	6.7326
6	3.9972	3.7641	5.8309
7	1.6358	1.7000	3.9230
8	3.1480	3.1736	0.8129

9.7 Computation with Fuzzy Variables

In the previous sections, the application of neural networks, fuzzy sets and the neuro-fuzzy systems for the prediction of process parameters has been described. It is assumed that a precise value of independent process variables is known. In practice some process variables may be known imprecisely, for example friction in a metal forming process. Imprecise variables can be considered as fuzzy variables. Then fuzzy arithmetic may be employed to compute the output as a fuzzy parameter. The fuzzy arithmetic has been described in Section 8.3.5.

An interesting observation about fuzzy arithmetic is that it provides unnecessary wider intervals at each α-cut when a variable occurs more than once in an expression. For example, consider the expression

$$Y = X + \frac{1}{X}, \tag{9.59}$$

where X and Y are fuzzy numbers. Assume that at a certain α-cut, $X_a = [2, 4]$. The fuzzy arithmetic computations will provide

$$Y_\alpha = [2, 4] + [1, 1] \div [2, 4] = [2, 4] + [1/4, 1/2] = [2.25, 4.5]. \tag{9.60}$$

However, if we consider the fact that the lower limit of X_α is 2, providing corresponding Y as 2.5 and the upper limit of X_α is 4, providing Y as 4.5, the reasonable value of Y_α seems to be [2.5, 4.5]. This is because, in the same experiment, the actual value of X is one quantity and taking different values of X for different terms of the expression is not justified. The number [2.5, 4.5] has a narrower range compared to [2.25, 4.5] computed in Equation 9.60. Similarly, consider the expression

$$Y = X - X. \tag{9.61}$$

For $X_a = [2, 4]$, the arithmetic computation provides

$$Y_\alpha = [2, 4] - [2, 4] \div [2, 4] = [-2, 2], \tag{9.62}$$

which is clearly wrong because the answer should be [0, 0]. To alleviate this problem, Dong and Shah [5] have provided the vertex method for computing the functions of the fuzzy variables.

In the vertex method, at each α-cut, the function is evaluated at all possible combinations of lower and upper limits of the variables. Each variable has one lower and one upper limit. Thus, the n variables form 2^n such combinations. These can be thought as the vertices of the n-dimensional hypercube. The lower limit of the function, at the particular α-cut is the lowest among the values calculated from all possible combinations. Similarly, the upper limit of the function is the highest among the values calculated from all possible combinations. This is valid if there are no maxima and minima in the domain of input variables. If there are maxima and minima, maximum and minimum limits of the function should be chosen amongst the possible combinations of lower and upper limits as well as the maxima and minima points. One difficulty with the vertex method is the curse of dimensionality. As the number of independent variables increases, the number of vertices becomes too high. For a function of 5 independent fuzzy variables, there are a total of 32 vertices. Thus, at each α-cut, one needs to carry out 32 function evaluations. Fortunately, in most of the cases of metal forming and metal cutting, behavior of the dependent variable as a function of independent variable is known qualitatively. One can easily get the idea of which two out of all possible vertices will provide maximum and minimum value of the function respectively. Thus, the computations can be carried out only at two vertices. For example, in Equation

9.43, if material properties and coefficient of friction are considered as fuzzy variables, it can be seen that at an α-cut, the maximum value of roll force will be obtained for a combination, in which $(\sigma_Y)_0$, f and n are kept at the maximum and b at the minimum. Similarly, the minimum value of roll force will be obtained for the lowest values of $(\sigma_Y)_0$, f and n and the highest value of b at the particular α-cut.

For finding the fuzzy dependent variable as a function of fuzzy independent variables, the computations are needed at different α-cuts. If the fuzzy independent variables are triangular or trapezoidal numbers and the function involves only addition and subtraction operations, the computations at two α-cuts are good enough, because the dependent variable will be a triangular/trapezoidal number for which the information at two α-cuts is enough to construct the fuzzy number. However, if the function involves the multiplication and/or divisions, in general, the resulting independent variable will not be a triangular/trapezoidal number and computations at many α-cuts are required in order to construct smooth curves representing fuzzy dependent variable. However, in many cases, the fuzzy dependent variable may be approximated by a triangular/trapezoidal membership function to a reasonable degree of accuracy. We can estimate the order of error in making such an approximation. Let us say the lower limit of a fuzzy independent variable is $\left(a_1 + (a_2 - a_1)\alpha\right)$ where a_1 and a_2 are the lower limits at α-cuts of 0 and 1 respectively. In the same way, the lower limit of the other fuzzy independent variable is $\left(b_1 + (b_2 - b_1)\alpha\right)$. The product of these two variables results in a fuzzy variable with a lower limit of

$$\left(a_1 + \Delta a\, \alpha\right)\left(b_1 + \Delta b\, \alpha\right) = a_1 b_1 + a_1 \Delta b\, \alpha + b_1 \Delta a\, \alpha + \Delta a\, \Delta b\, \alpha^2 , \qquad (9.63)$$

where Δa and Δb denote $(a_2 - a_1)$ and $(b_2 - b_1)$ respectively. This expression is a quadratic function of α, requiring computations at three α-cuts. However, considering that Δa and Δb are small quantities, the product $\Delta a\, \Delta b\, \alpha^2$ may be neglected. In that case, the expression given by Equation 9.63 becomes a linear expression. For $\alpha = 1$, the percentage error considering this approximation is given by

$$\text{percentage error} = \frac{\Delta a\, \Delta b}{a_2 b_2} \times 100 . \qquad (9.64)$$

Thus, assuming that there is a total 10% uncertainty in the estimation of the two variables, i.e., $\Delta a / a_2 = \Delta b / b_2 = 0.1$, the percentage error in making a linear approximation turns out to be only 1%. Here, the maximum error is at an α-cut of 1. If we construct the fuzzy dependent variable based on the computations at α-cuts of 0 and 1, the maximum error will be at some intermediate α-cut, the order of error being the same. Similar estimates of error can be made for fuzzy division and other derived operations. One of the reasons for choosing triangular/trapezoidal fuzzy numbers is the reduced computational requirement with these numbers.

Example 9.10: A solid cylinder of radius a and height h is axially compressed by a pair of rough platens. The approximate load P required to bring the cylinder to the yield point is given by [16]

$$\frac{P}{\pi a^2} = \sigma_Y \left(1 + \frac{2fa}{3h}\right), \tag{9.65}$$

where f is the coefficient of friction and σ_Y is the yield strength. Assume that σ_Y is 300 MPa within ±5% accuracy and, most likely, lower and upper estimates of friction are 0.08, 0.06 and 0.12. For a = 5 mm and h = 15 mm, estimate the load required for yielding.

Solution: With the given data, σ_Y and f are constructed as linear triangular fuzzy numbers. For this purpose, most likely values are assigned membership values 1 and lower and upper estimates are assigned membership grades of 0.5. The fuzzy numbers Y and f are shown in Figure 9.9. Since P is dependent on two fuzzy numbers, vertex method requires four computations at each α-cut. However, it is clear that the high f and high Y combination will give the highest P and low f and low Y will give the lowest P. Hence, at each α-cut, two computations are needed to find the lower and upper limits of P. The dependent variable plotted as a fuzzy number starting from the membership grade of 0.5 is called the possibility distribution of the number. The values having less than 0.5 membership are rarely possible and are therefore not included in the possibility distribution. Figure 9.10 shows the possibility distribution of the forging load. It can be seen that it is almost linear.   Thus, it is possible to make a linear approximation for   possibility distribution in many situations.

We can obtain the fuzzy dependent variable as a fuzzy number, but how do we use this information? One use is that it provides a good qualitative feel about the dependent variable. We can observe the spread in the variable, *i.e.*, how fuzzy that variable is. We can also use this information for a reliability-based design. A fuzzy reliability measure has been introduced [17]. The method is based on the concept of entropy. Therefore, before describing the method, it is essential to describe the concept of entropy.

The term 'entropy' is normally used to describe the degree of uncertainty about an event. For an event consisting of the discrete random variable S_i (i = 1, 2,...., q) with P_i as the associated probabilities, the Shannon entropy is defined as [18]

$$H(S) = \sum_{i=1}^{q} P_i \ln(1/P_i). \tag{9.66}$$

As an example, consider that in a bag there are four white, four black and four red balls. If a ball is drawn randomly, the probability that a ball of a particular color will come is 1/3. Now, the three events are: (1) The ball is red. (2) The ball is black. (3) The ball is white. The entropy of the system is

$$H(S) = \frac{1}{3}\ln 3 + \frac{1}{3}\ln 3 + \frac{1}{3}\ln 3 = 1.0986.$$

Now consider a bag in which there are 1 white, 1 black and 10 red balls. The probabilities of the three events will be 1/12, 1/12 and 5/6. The Shannon entropy for this system will be

$$H(S) = \frac{1}{12}\ln 12 + \frac{1}{12}\ln 12 + \frac{5}{6}\ln(6/5) = 0.5661.$$

We observe that the second system has lesser entropy than the first one, because the second system is less random. (We know that it is dominated by the red balls.) Thus, entropy is a measure of randomness. It is natural that the entropy of a fuzzy set should be a measure of the uncertainty rather than randomness. Analogous to the definition of Shannon entropy, the entropy with a particular membership grade may be defined as [19]

$$d(\mu) = \begin{cases} -[\mu\log_2\mu + (1-\mu)\log_2(1-\mu)] & \text{for } 0 < \mu < 1, \\ 0 & \text{for } \mu = 0,1. \end{cases} \qquad (9.67)$$

In this form, the value of entropy is maximum (and equal to 1) at $\mu = 0.5$, i.e., when the uncertainty is maximum. It is to be noted that De Luca and Termini [19] used the natural logarithm in the entropy expression. Here, the base of the logarithm has been taken as 2 to bound the entropy value between 0 and 1. The entropy given by Equation 9.67 satisfies the following property:

$$d(\mu^*) \leq d(\mu) \text{ given } (\mu^* \geq \mu \text{ for } \mu \geq 0.5 \text{ and } \mu^* \leq \mu \text{ for } \mu \geq 0.5). \qquad (9.68)$$

This is because μ^* is less (or equal) uncertain than μ .

After having defined the entropy of a fuzzy set, we are now in a position to propose a measure of the reliability of a design involving fuzzy parameters. Conventionally, the definition of reliability is based on the probability theory. But when a process is controlled by fuzzy parameters instead of random parameters, the definition of reliability should be based on the uncertainty associated with the subjective information, i.e., on the membership grade of fuzzy parameters. In order to avoid any conflict of terminology, we call this reliability "fuzzy reliability".

For deriving an expression of fuzzy reliability, we introduce one more measure called possibility index, PI. The possibility index quantifies the possibility of the success of a design. Considering Example 9.10, let $P_R(\mu)$ and $P_L(\mu)$ denote the right (upper) and left (lower) limits of the forging power (to cause yielding) at a membership grade of μ. Note that for the particular membership grade μ, the yielding is surely possible when the power P is greater than $P_R(\mu)$. Thus, PI should be chosen to be 1 when P is greater than $P_R(\mu)$. Further, yielding is impossible when P is less than $P_L(\mu)$, which means PI should be zero for this case. For the intermediate values ($P_L(\mu) \le P \le P_R(\mu)$), PI varies from 0 to 1. If it is assumed that it varies linearly, then PI can be defined as

$$PI(\mu, P) = \begin{cases} 0 & \text{if } P \le P_L(\mu), \\ \dfrac{P - P_L(\mu)}{P_R(\mu) - P_L(\mu)} & \text{if } P_L(\mu) < P < P_R(\mu), \\ 1 & \text{if } P \ge P_R(\mu). \end{cases} \qquad (9.69)$$

Note that for a power P, PI is a function of μ. Thus, there is a degree of uncertainty associated with the possibility index. This is the other uncertainty, which affects the reliability. A measure of this uncertainty is the non-probabilistic entropy $d(\mu)$ defined by Equation 9.67. Then, the quantity $(1 - d(\mu))$ can be considered as a measure of certainty. Therefore, a reliability index can be defined as

$$\beta(\mu, P) = PI(\mu, P)\left[1 - d(\mu)\right]. \qquad (9.70)$$

Now, the area under the $\beta - \mu$ curve is taken as the measure of reliability. The maximum value of area corresponds to the case when $PI=1$ for all μ. Therefore, it is taken as 100% reliability. Thus, the reliability can be defined as

$$R_e(\%) = \frac{\displaystyle\int_{0.5}^{1} \beta(\mu, P^*)\,d\mu}{\displaystyle\int_{0.5}^{1} \left(1 - d(\mu)\right)d\mu} \times 100 . \qquad (9.71)$$

For Example 9.10, the variation of fuzzy reliability with load is shown in Figure 9.11.

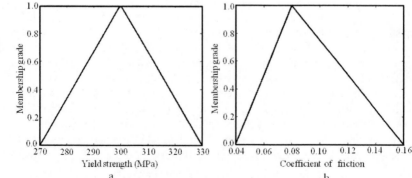

Figure 9.9. Membership functions. **a** Of yield strength. **b** Of coefficient of friction

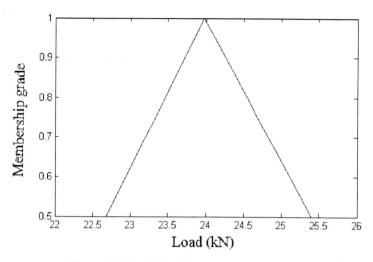

Figure 9.10. Possibilty distribution of forging load

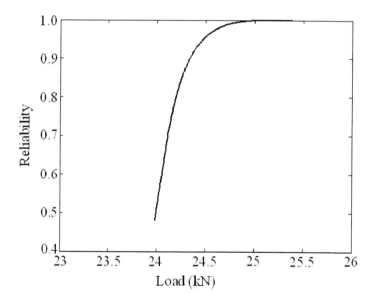

Figure 9.11. Fuzzy reliability of design for different forging loads

9.8 Summary

In this chapter, we have described the application of neural networks and fuzzy set theory in the estimation of the output variables of metal forming or machining processes. This approach becomes highly useful when physics-based modeling is difficult and/or necessary physical parameters needed in the physical model are difficult to determine. The chapter starts with the discussion of the statistical tools. Before applying the soft computing-based methods, it is advisable to carry out statistical analysis including modeling using multiple-regression.

Two most common neural networks are MLP and RBF. Among these, the MLP network requires less data, but needs more time in training, whereas RBF network can be trained in a faster manner, but will need more data. The procedure of training of these networks has a profound effect on the performance of the networks. The discussion pertaining to this aspect has been included, but this area is still open to research. The fuzzy set-based system can also be used for prediction. The main advantage is that it provides better physical feel of the problem and is easy to implement. The combination of neural network and fuzzy set theory has also been discussed. Finally, there is a section on computation using fuzzy parameters.

9.9 References

[1]   Fisher, R.A. (1926), The arrangement of field experiments, Journal of the Ministry of Agriculture of Great Britain, Vol. 33, pp. 503–513.

[2]   Plackett, R.L. and Burman, J.P. (1946), The design of optimum multifactorial experiments, Biometrica, Vol. 33, pp. 305–325.

[3]   Logothetis, N. (1992), Managing for Total Quality: From Deming to Taguchi and SPC, Prentice-Hall of India Pvt. Ltd., New Delhi.

[4]   Tanaka, H. (1987), Fuzzy data analysis by possibilistic linear models, Fuzzy Sets Sys, Vol. 24, pp. 363–375.

[5]   Lawrence, M. and Petterson, A. (1998), BrainMaker User's Guide and Reference Manual, 7-th Edition. California Scientific Software, Nevada City, California

[6]   Dixit, U.S. and Chandra, S. (2003), A neural network based methodology for the prediction of roll force and roll torque in fuzzy form for cold rolling process, Int. J. Adv. Manuf. Technol., Vol. 22, pp. 883–889.

[7]   Kohli, A. and Dixit, U.S. (2005), A neural-network-based methodology for the prediction of surface roughness in a turning process, Int. J. Adv. Mnuf. Technol., Vol. 25, pp. 118–129.

[8]   Haykin S., (1996), Adaptive Filter Theory, 3-rd edn., Prentice-Hall, New York.

[9]   Ishibuchi, H. and Tanaka, H. (1991), Regression analysis with interval model by neural networks. Proceedings of the IEEE Int. Joint Conf. on Neural Networks, Singapore, pp. 1594–1599.

[10] Sonar, D.K., Dixit, U.S. and Ojha, D.K. (2006), The application of a radial basis function neural network for predicting the surface roughness in a turning process, Int. J. Adv. Mnuf. Technol., Vol. 27, pp. 661–666.

[11] Chen, J.C. and Black J.T. (1997), A fuzzy-nets in-process (FNIP) system for tool-breakage monitoring in end-milling operations, Int. J. Mach. Tools and Manuf., Vol. 37, pp. 783–800.

[12] Chen, J.C. and Savage, M. (2001), A fuzzy-net-based multilevel in-process surface roughness recognition system in milling operations, Int J Adv Manuf Technol Vol. 17, pp. 670–676.

[13] Abburi, N.R., and Dixit, U.S. (2006), A knowledge-based system for the prediction of surface roughness in turning process, Robotics and Computer-Integrated Manufacturing, Vol. 22, 4, pp. 363–372.

[14] Jang, J.S.R. (1993), ANFIS: Adaptive-Network-Based Fuzzy Inference System, IEEE Transaction on System, Man and Cybernetics, Vol. 23, pp. 665–685.

[15] Dong, W. and Shah, H.C. (1987), Vertex method for computing functions of fuzzy variables, Fuzzy Sets and Systems, Vol. 24, pp. 65–78.

[16] Johnson W, Mellor, P.B. (1972), Engineering Plasticity, Von Nostrand Reinhold Company, London.

[17] Dixit, U.S. (1997), Cold Flat Rolling: Modeling with Fuzzy Parameters, Anisotropic Effects and Residual Stresses, Ph.D. thesis, IIT Kanpur.

[18] Shannon, C.E., (1948), The mathematical theory of communication, The Bell System Technical Journal, Vol. 27, 379–423, 623–656.

[19] De Luca, A., and Termini, A., (1972), A Definition of Nonprobablistic Entropy in the Setting of Fuzzy Set Theory, Information and Control, Vol. 20, pp. 301–312.

10

Optimization of Metal Forming and Machining Processes

10.1 Introduction

The aim of every engineer is to carry out optimization. We keep on optimizing many things even without using any optimization techniques. One can find endless number of problems in metal forming and machining where optimization can play a major role. The task of the optimization can be divided into three main subtasks:

(1) Formulation of the statement of the optimization problem in terms of the objective function and constraints.
(2) Developing the mathematical model for obtaining the objective function and constraints as a function of the design (decision) variables whose value one needs to determine in the process of obtaining the optimal solution.
(3) Solving the optimization problem using a suitable optimization algorithm.

The first subtask is very important and requires a lot of experience, intuitive knowledge and understanding of the physics. No mathematical technique can be a substitute for the knowledge of the expert. The expert's knowledge has to be formulated in the form of the optimization problem of the following form:

Minimize $f_1(x)$, $f_2(x)$,............, $f_l(x)$,
subject to

$$g_i(x) \le 0 \quad i = 1, 2,..........., m,$$
$$h_i(x) = 0 \quad i = 1, 2,..........., n,$$
$$x_i^l \le x_i \le x_i^u \quad i = 1, 2,..........., k.$$

(10.1)

where $f_i(x)$ is the i-th objective function out of the total l objective functions, $g_i(x)$ is the i-th inequality constraint out of the total m inequality constraints, $h_i(x)$ is the i-th equality constraint out of the total n equality constraints, and x_i^l and x_i^u are the lower and upper bounds of the i-th design variable x_i, total design variables being k. This problem is a multi-objective optimization problem. Note that if some objective function needs to be maximized, it can be converted into a minimization form by multiplying with -1. Similarly, an inequality constraint of the form greater than or equal to can be converted into an inequality constraint of the form less than or equal to by multiplying it with -1.

In most of the optimization problems of metal forming and machining, it is not possible to obtain the closed form expression for the objective function and constraints. Sometimes the value of the objective function and/or the constraint functions can be obtained by an FEM routine. In that case, the FEM routine acts as a black box function. However, it is often better to train a neural network based on FEM data or prepare a rule base for fuzzy inference system, and use a trained neural network or fuzzy inference system as a black box function in an optimization code. This will be a computationally efficient procedure as a well-trained neural network and/or fuzzy inference system will provide much quicker prediction than an FEM routine. Moreover, in some cases, it may not be possible to predict a parameter using the FEM and one has to rely on the neural network model or fuzzy-inference system based on the experimental/experiential knowledge.

The aim of this chapter is to show how finite element method and/or soft computing techniques can help in optimizing metal forming and machining processes. There is a plethora of optimization problems. Only a few representative problems have been discussed. Also, sometimes the traditional optimization algorithms can also be quite effective in solving some optimization problems. However, the discussion about the traditional optimization algorithm is beyond the scope of this book. The readers interested in learning various traditional optimization algorithms may refer to [1–3].

10.2 Optimization Problems in Metal Forming

In metal forming optimization problems, the interest is to produce quality products with minimum energy. In almost all metal forming processes, the optimization can play a major role. Some examples are as follows. In the steady state problem of wire drawing and rolling, the optimal scheduling of passes is an important task. In the forging processes, the optimal design of dies and pre-form shape is important. In deep drawing processes, the decision about the proper blank-holder force in order to avoid tearing and wrinkling forms an optimization problem. In Section 6.4.5, a procedure to obtain optimal blank shape to avoid earing in cylindrical cup deep drawing has been described. In the following subsections, we choose some other typical processes to discuss the formulation of optimization problems and procedures for solving it.

10.2.1 Optimization of Roll Pass Scheduling

In Chapter 1 we introduced the tandem rolling mill. In the tandem rolling mill, a strip is successively reduced in thickness at each stand as it passes through the mill. It can also be employed to change the cross-sectional shape of the product, such as in rod rolling. However, here we will discuss the optimization of cold plane-strain rolling. Designing of an optimum reduction schedule that will give a correct output gauge and satisfactory shape and surface finish with minimum energy is of paramount importance. Rolling mill industries have been carrying out scheduling based on past experience and using rules of thumb. It is a good idea to combine analytical information gained using FEM and expert knowledge in the form of neural network and/or fuzzy set-based models.

In an early paper Avitzur [4] developed a procedure to optimize tandem mill operation by maximizing the production rate subject to certain constraints. He illustrated the procedure by a numerical example for a hypothetical six-stand tandem hot rolling mill. The effects of changes in some of the process variables on productivity and other factors were studied. Brayant and Spooner [5] described a simple approach to schedule design and discussed the on-line correction of shape and related aspects of mill design. Brayant *et al.* [6] presented a methodology for optimum schedule design, which considers the flatness of the strip and the general mill operating conditions. Dixit and Dixit [7] minimized the total specific power considering the constraint of avoiding central burst and alligatoring. They have also tackled the uncertainty in the process and material parameters using fuzzy set theory. Wang *et al.* [8] have applied genetic algorithms to the tandem cold mill scheduling problem.

Some typical constraints in the rolling problem are as follows:
Constraint on the front and back tensions:
In tandem rolling with intermediate stands, the front tension of one stand becomes the back tension of the other stand and is called interstand strip tension. The upper tension limit is usually fixed by the tearing consideration. It has been estimated that the maximum safe level of strip tension to reduce the possibility of tearing is about one-third of the yield stress [6]. The lower limit of the tension may be to have the tension more than the maximum longitudinal compressive residual stress produced in the strip, in order to avoid the buckling of the strip. Anyhow, it should be well above zero, otherwise the strip may loop between the stands.
Constraint on the residual stresses:
One can also put the constraint on the residual stresses. It is believed that a heavy reduction will cause tensile residual stresses on the surface, which will cause crack propagation on the surface. However, the residual stresses also depend on the other process parameters. For implementing this constraint, a sound mathematical model or huge amount of practical data should be available.
Constraint on roll force:
The roll force at any stand should be limited to some value. This value depends on the considerations of mill modulus and roll supporting bearings. The last stand force is critical due to imposed quality requirement like roughness and flatness.

Constraint on roll power:
The roll power should be limited not to cause excessive torque. The roll torque has to be kept below a limit based on work roll neck stress and drive spindle capacity. If the horsepower of the motors is be specified, the roll power at any stand should be limited so as not to cause excessive overloading of the motors. Often it is better to have the same power for all stands. Sometimes the heat transfer consideration plays a role in the maximum allowable roll power. If it is assumed that the entire rolling power is dissipated as convective heat, a simple equation can be written as follows:

$$P = 2\pi RLh(\theta_R - \theta_0),\tag{10.2}$$

where L is the length of the roll, R is the radius of the roll, h is the convective heat transfer coefficient, and θ_R and θ_0 are the temperatures of the roll surface and surroundings, respectively.

Constraint on rolling speed:
The production rate depends on the exit speed of the roll. It is desirable to have the maximum exit speed. The speeds of the other rolls should be adjusted depending on the need of interstand tension. A theoretical discussion on this aspect is available in [9].

Constraints on reduction:
Based on the upper bound analysis, Avitzur *et al.* [10] has shown that the central burst defect tends to be promoted by the small percentage reduction. The same thing is true about alligatoring or split end defects. At a very low reduction, the material near to the roll surface flows properly, but the central portion of the strip does not. This causes the central burst and/or alligatoring defects. Using the relation provided in [11], it can be shown that for avoiding the central burst and split end defects,

$$r \ge \frac{h_{\text{exit}}}{1.81R},\tag{10.3}$$

where r is the fractional reduction and h_{exit} is the exit thickness.

The simple equation for the maximum possible reduction without causing the skidding of the rolls is [12]

$$r \le \frac{8f^2R}{h_{\text{entry}}},\tag{10.4}$$

where h_{entry} is the inlet thickness and f is the coefficient of friction. This equation has been derived by considering the equilibrium of the forces acting at the strip and taking the roll diameter as double the actual roll diameter to take into account the roll flattening effect. The finite element analysis can also find the maximum possible reduction based on the location of the neutral point. If the neutral point reaches the exit, a limit on the maximum possible reduction has been obtained. For unaided entry of the strip in the roll gap,

$$r \le \frac{f^2 R}{h_{\text{entry}}}.$$
(10.5)

This equation is important during the threading operation in the rolling mill. The threading operation starts from the entry of raw material into the first stand and continues up to the acceleration of the mill.

In the last stand, the maximum possible reduction may be limited by the flatness and surface roughness requirement of the strip. It is expected that with increasing reduction, the quality of the strip including the surface finish will deteriorate. Not much work has been done to assess the surface roughness of the rolled strip using mathematical modeling. Neural network modeling can play a vital role in this.

Having discussed the constraints in the tandem mill scheduling, let us discuss the typical objectives in the schedule optimization problem. The common objectives are the minimization of total power and maximization of production rate. Sometimes the constraints can also be incorporated into the objective function. For example, if it is known that the inter-stand tension t_i should lie between two limits $(t_i)_{\text{min}}$ and $(t_i)_{\text{max}}$, one can convert this constraint into an objective function:

$$\text{Minimize } f = \left(t_i - \frac{(t_i)_{\text{min}} + (t_i)_{\text{max}}}{2} \right)^2,$$
(10.6)

thus trying to keep the inter-stand tension near the mean of the given range. Once there are multiple objective functions, the problem can be solved by multi-objective genetic algorithms. Here, all the solutions which provide a good value at least with respect to one objective function are preserved. Thus, one can generate a Pareto-optimal solution-set. In this set, no solution dominates another solution. It means that in the Pareto-optimal set there is no solution which is worse than any other solution from the viewpoints of all the objectives. For example, if two objective functions f_1 and f_2 are to be maximized and there are three solutions viz., (i) $f_1=10$, $f_2=5$, (ii) $f_1=9$, $f_2=4$ and (iii) $f_1=1$, $f_2=11$, then (i) and (iii) are Pareto-optimal solutions. The solution (ii) is not a Pareto-optimal solution, because the first solution dominates it in all respects. (In the first solution, both the objective functions are more than in the second solution.)

Once the Pareto-optimal solutions are obtained, a higher order decision can be taken to choose the best solution. Fuzzy set theory can be employed here. In this, based on the expert's opinion, the membership functions for each objective are constructed. If μ_i is the membership function for i-th objective f_i, then the overall membership grade of a solution may be written as [13]

$$\mu_0 = \alpha \sqrt[n]{\mu_1 \mu_2 .....\mu_n} + (1-\alpha)\min[\mu_1, \mu_2, ..., \mu_n]. \tag{10.7}$$

For $\alpha = 1$, Equation 10.7 reduces to a compensating trade-off, where the high membership grade in one objective function can compensate for a low membership grade in the other. For $\alpha = 0$, Equation 10.7 reduces to non-compensating trade-off, where the performance of a solution is decided by the most poorly performing objective. Putting $\alpha = 0.5$ provides equal weightage to both compensating and non-compensating trade-off.

10.2.2 Optimization of Rolls

The selection of roll diameter is an important decision for rolling mill design. The advantages associated with small and large roll diameters are listed in Table 10.1. It is seen that there are positive points associated with small as well as large diameter rolls.

Table 10.1. Advantages associated with the small and large rolls

Small Rolls	Large Rolls
Less rolling force, less rolling torque and power, less spread	More rigidity (backup rolls may not be needed in some cases), better cooling, less tendency for split edge and alligatoring defects, better roll life

The finite element modeling, empirical relations and experience can provide the quantitative values of the attributes listed in Table 1.1. The membership function can be assigned to various attributes for a particular roll diameter. The roll diameter that maximizes the overall membership grade is chosen.

The roll profile can also be optimized to provide a good quality product. The strip profile, strip shape and edge cracking are influenced by work roll deflections, backup roll deflection, work roll flattening and work thermal expansion. These are compensated by providing crown on work-rolls, backup rolls, or both the rolls. The elastic deformation of the rolls can be analyzed by using the finite element method. For finding the thermal expansion, the temperature analysis needs to be carried out, and then the roll deformation can be found by thermo-elastic analysis. In general, the required crown on the roll will be dependent on the reduction, type of material and other process conditions. Thus, one particular roll profile is not appropriate for all situations. Therefore, there is a role of the optimization in finding the most suitable roll profile.

10.2.3 Optimization of Wire Drawing and Extrusion

The optimization problem in wire drawing and extrusion concerns deciding the number of passes in multi-stage processes and optimization of die profiles. Joun and Hwang [14] optimized wire drawing and extrusion to minimize the total forming energy. The process modeling is carried out using the finite element method. However, the authors fixed the total number of passes *a priori*. Their design variables are die geometry and process conditions. Celano *et al.* [15] have optimized the multi-stage wire drawing process by using simulated annealing. In this work, the number of passes was not fixed *a priori*, although the die angles were fixed. The objective chosen by these authors was to have more or less uniform stress in all the passes.

For optimizing the multipass wire drawing and extrusion process, the major objective can be the minimization of the total forming power. The constraints can be based on the following considerations:

(1) The product should be defect free. For this purpose, the constraints based on critera for defect prevention have to be applied. According to simple hydrostatic stress criteria for defect prevention, the hydrostatic stress should not become tensile in the deformation zone.

(2) The strain distribution in the product should be as uniform as possible.

(3) Die pressure should be sufficiently small to prevent excessive die wear.

(4) Die pressure should be as uniform as possible in order to avoid the tendency of pitting.

(5) The power needed at any stand should be below the capacity at the stand.

The die shape can be optimized by approximating the inner surface of the die by piecewise continuous polynomials. The coefficients of the polynomial can be obtained by solving the optimization problem. There are some papers using this approach. For example, Balaji *et al.* [16] have optimized die profile of extrusion dies. Reddy *et al.* [17] have also carried out die shape optimization. A third degree polynomial die is used and its coefficients are adjusted for minimizing the power.

Lee *et al.* [18] have optimized the die profile for obtaining uniform microstructure in hot extruded product. The die profile is approximated by the Bezier curve and the parameters of the curve are obtained as a result of the optimization. The thermo-coupled rigid-viscoplastic FEM is employed to find the quantities needed for microstructure evolution model. For microstructure evolution, Yada and Senuma's empirical model [19] is employed. This model requires temperature, strain and strain rate which are found by FEM. The objective function is to determine the optimal die profile which provides the least square deviation between the average grain size and the actual grain size. Thus, the objective is

$$\text{Minimize } f = \frac{\sqrt{\Sigma (d_i - d_{avg})^2 V_i}}{V_{total}}, \tag{10.8}$$

where V_{total} is the control volume of the extruded portion of the product, d_i is the grain size of the i-th finite element and V_i is the volume of the i-th finite element. The average grain size d_{avg} is found as

$$d_{avg} = \frac{\sum d_i v_i}{V_{total}}. \qquad (10.9)$$

Lin *et al.* [20] has optimized the die profile for improving the die life in the hot extrusion process. For improving the die life, an optimization scheme coupled with a rigid-viscoplastic finite element analysis has been employed to obtain a die profile providing more uniform surface load distribution. The die profile is represented by a cubic-spline curve. The objective function of the problem is to provide uniform axial normal and shear stress at the interface. The total extrusion load is kept as a constraint.

Ulysse [21] has carried out extrusion die design for flat faced dies. The concept employed by the author is that an optimized die will extrude a profile with uniform exit velocity. Uniform exit flow through a flat faced die can be achieved through the use of bearings and/or pockets (cavities placed at the entrance of the die orifice). By decreasing the bearing length at a particular location of a die opening, the resistance to flow rate is decreased. Similarly, by increasing the bearing length, the resistance to flow rate can be increased. The resistance to flow rate can also be decreased by increasing the local width of pocket around the die orifice. Thus, by varying bearing lengths and pocket widths, one can obtain uniform die exit velocity. The author has developed a numerical model using FEM and optimized the bearing and pocket sizes to minimize the variation in the exit velocities of different sections.

Yan and Xia [22] have optimized the parameters of the profile extrusion process. The die hole layout for a non-symmetric angle profile was obtained. The objective function is the standard deviation of the velocity field to ensure the uniform flow velocity. First, the FEM simulations were performed using an orthogonal array to minimize the number of simulations. The simulations were used to train a neural network module. The fitted neural network module was used as a black box for optimization of the problem using genetic algorithms.

10.2.4 A Brief Review of Other Optimization Studies in Metal Forming

The literature on the optimization problem in metal forming is quite sparse. In this subsection, a brief review of the optimization studies in metal forming is presented. Kleinermann and Ponthot [23] first used optimization for solving the inverse problem of metal forming for identifying the parameters, particularly the material behavior. The updated model was used for the optimization of initial shape and tool shape. The authors presented application examples of initial shape design for superplastic forming of a cup and deep-drawing of a fastener component.

Zhao *et al.* [24] have presented a methodology of optimizing perform shape in the net-shaped forging process. The method approximates the perform shape by B-splines, the control points of which are optimized so as to give the final shape very close to the desired shape. The final shape is predicted by finite element analysis.

Antonio *et al.* [25] have optimized the shape of the first-stage forming tool and the initial work-piece temperature that minimizes the total forming energy and the gap between the final forged shape and the desired one, bounding the maximum temperature reached in a two-stage hot forging process. In this work a genetic algorithm with elitist strategy has been used. In elitist strategy, a core of best individuals is always preserved and is transferred as it is to the new generation. The new population is generated from the old one using four operations: selection, crossover, elimination/substitution and mutation. In the selection operation, the members of the population are ranked according to their fitness. Some highly ranked members form the elite group leaving another group of ordinary members. One parent is selected from the elite group and another from the ordinary group for carrying out crossover operation. The crossover operator transforms two parent-chromosomes into new chromosomes having genes from both the parents. The new members generated from the crossover operation join the original population. At this stage, the population size gets enlarged due to the addition of new members following the crossover operation. The enlarged population is ranked. The solutions with similar genetic properties are eliminated and new randomly generated members are added in lieu of it. Further ranking is carried out and a few worst individuals are eliminated, simulating the death of the old members. After this operation, the size of the population is less than the original size. This is brought to the original size by generating some individuals by mutation. After the new population is obtained, the steps of the genetic algorithm are repeated till a suitable convergence criterion is obtained.

Khoury et al. [26] have listed a few criteria and constraints on the forging process that industry requires. The first criterion is to obtain a good forged part (with proper die filling and absence of defects). The second criterion is to use the just necessary effort for decreasing the wear. The press load capacity and press speed can be taken as constraints. The challenge in the optimization task is to satisfy various conflicting goals simultaneously and to have accurate models of the process.

In sheet metal forming processes such as deep drawing, the blank-holder force (BHF) is a critical parameter. If the blank-holder force is higher, there are chances of tearing the sheet. With a too low blank-holder force, wrinkling occurs. Thus, the blank-holder force has to be kept at an optimum value. Chengzhi et al. [27] have used an adaptive response surface methodology to obtain the optimum variable blank-holder force. The optimization is carried out to avoid wrinkling and fracture. By manipulating the blank-holder force, it is possible to eliminate wrinkling and fracture. Wrinkling may occur at the start of the stroke if BHF is too low and fracture may occur if the BHF is higher. Therefore the authors have adopted the strategy of applying a variable blank-holder force.

The finite element modeling of sheet metal forming is complicated, especially when a thorough defect analysis is required. Liu et al. [28] have carried out the optimization of blank holding force and draw beads restraining force based on the existing FEM software. Draw beads control the flow of the blank into the die cavity. Beads restrict the flow of sheet metal by bending and unbending it during the drawing. Tang et al. [29] have optimized draw bead design in sheet metal forming. Hu et al. [30] have carried out the optimization of forging-extrusion problems using a fuzzy and rough set-based knowledge base. Jansson and Nilsson [31] have optimized sheet metal forming processes using the response surface methodology. The response surface methodology is an optimization method that approximates the function and constrains by simple polynomials, often linear ones.

If there are n design variables to fit a linear function, only $(n+1)$ functions are required, although for more accuracy some more functions can be evaluated. Initially, the function is fitted on a larger domain and the optimum solution is found. Then the domain is shifted to have the optimum solution at the center. At the same time, the domain is reduced and the linear approximations are employed in this region. The new optimum solution is found. This process is repeated till a convergence is obtained. The authors have used a finite element program and optimized the blank holding force, draw beads, die radius *etc.* Sattari *et al.* [32] optimized the deep drawing process for minimizing the thickness variation.

Shi *et al.* [33] have optimized die shape for sheet metal forming processes. The objective function used by them is

$$f = f_r + f_w .$$
(10.10)

The first part f_r, called the rupture criterion, is given by the ratio of thickness strain to the critical allowable thickness strain calculated from the Forming Limit Diagram (FLD). A higher value of f_r indicates a higher possibility of splitting the material. The second part f_w, called the wrinkling criterion, is the absolute value of the ratio of minor and major principal strains. The larger the value of f_w, the greater the possibility of wrinkling occurring.

There have been some attempts to optimize the superplastic forming process [34]. The superplasticity is characterized by a high sensitivity of stress to strain rate. One measure of superplasticity is the rate sensitivity index m given by

$$m = \left(\frac{\partial \ln \sigma}{\partial \ln \dot{\varepsilon}^p} \right)_{T,\{\Lambda\}} ,$$
(10.11)

where σ is the stress, $\dot{\varepsilon}^p$ is the plastic strain and the partial derivative is defined at constant temperature T and microstructural state $\{\Lambda\}$. During superplasticity, m is usually more than 0.4. In most of the optimization work, the forming load is adjusted to ensure that the maximum strain rate coincides with that giving maximum m. Bate *et al.* [34] have shown that the maximization of m need not be the proper goal. The authors have carried out the optimization for maximising the minimum thickness or minimizing the cavitations level.

Parsa and Pournia [35] have optimized initial blank shape using the inverse finite element method. Sosnowski *et al.* [36] have optimized the sheet metal forming tool based on sensitivity analysis. Based on their study, Ganter *et al.* [37] have concluded that prediction of surface defects, residual stresses and springback is still unreliable using numerical methods.

10.3 Optimization Problems in Machining

The economics of machining has been an important area of research in machining starting with the early work of Gilbert [38]. The main objectives are minimization of cost of machining, maximization of the production rate and maximization of profit rate. The major constraints are the constraint on surface roughness, forces coming on the tool and machine power. In general, optimization of machining is a multi-objective problem. The major difficulty in the optimzation is the knowledge about the metal cutting behavior. There should be a model to predict the tool life, a model to predict the job quality and a model to predict the forces and temperature of the tool. Figure 10.1 shows the block diagram of the optimization procedure. This diagram shows that for obtaining the optimal process parameters, one requires models of machining performance, machining cost and machining time. For determining the machining cost and time, one needs the tool life as a function of process parameters. We shall first review the research work in the area of machining processes, mainly confining to two important processes, turning and milling. Compared to studies on turning, milling has been less investigated. However, the optimization methodology of the turning process can also be extended to the milling process. Therefore, after presenting a brief review on the optimization of machining processes, we shall discuss, in some detail, the formulation of a multipass turning process.

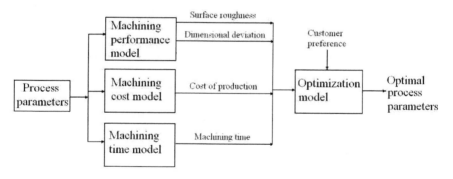

Figure 10.1. Block diagram of the optimization procedure in Machining

10.3.1 A Brief Review of Optimization of Machining Processes

Optimization of process parameters in machining processes has been an area of prime importance for researchers and engineers. In an early work, Gilbert [38] presented an analytical procedure for determining the optimum cutting speed in a single pass turning process. In this work feed and depth of cut were kept fixed. The problem was solved for two objective functions: maximization of the production rate and minimization of cost. This was a preliminary work. It is well known that tool life depends on feed and depth of cut apart from the cutting speed. Also, the consideration of surface roughness has to be an integral part of the machining optimization problem.

Later on most of the researchers have taken either the empirical relations for surface roughness or the ideal surface roughness formula (based on geometric consideration) given by

$$R_a = \frac{f^2}{32R},$$
(10.12)

where R_a is the center line average (CLA) surface roughness value in mm, f is the feed in mm/rev and R is the tool nose radius in mm. It is to be noted that this formula is derived on the basis that only the nose portion of the tool is making contact with the work-piece, a situation typical for turning with low depth of cut. However, a number of researchers have employed this relation even when the depth of cut was more than the nose radius. For large depth of cut other relations based on geometric consideration are available in the literature. For tool life determination, most of the researchers have used extended Taylor's tool life formula:

$$Tv^p f^q d^r = C,$$
(10.13)

where v is the cutting speed in m/min, T is the tool life in minutes, d is the depth of cut in mm and p, q, r and C are the constants for a particular tool and work material combination.

Ermer [39], Petopoulos [40] and Lambert and Walvekar [41] used geometric programming for the optimization of the constrained machining economics. In a geometric programming problem, the constraints and the objective function are expressed as posynomials. A function is called posynomial if it can be expressed as the sum of real power terms each of the form $c_i x_1^{\alpha_{i1}} x_2^{\alpha_{i2}} ....x_n^{\alpha_{in}}$ with $c_i > 0$ and $x_i > 0$ and i range from 1 to the number of terms N. The geometric programming approach could not become popular for two reasons. First, the constraints and objective functions must be expressible in the form of a posynomial and second, as the number of constraints increase, the degree of difficulty in solving a geometric programming problem increases.

Shin and Joo [42] presented a model for the optimization of machining conditions in a multipass turning process consisting of rough and finish passes. They used a dynamic programming method. Gupta *et al.* [43] used integer programming for optimization. They used two steps for the minimization of total production cost. In the first step, minimization of costs for rough and finish passes for various fixed depths of cut is obtained. The values of depths of cut are selected from a series of allowable depths. In the second step, an optimal combination of depths of cut for rough and the finish passes, the optimal number of passes and the minimum total costs are obtained. The second step is solved using integer programming. In this method, as the number of allowable depths of cut called "optimal sub-divisions" increase, the computation time increases. It is to be noted that there are multiple solutions of the problem in many cases, which Gupta *et al.* did not report.

Al-Ahmari [44] has prepared a direct non-linear mathematical model that solves the multipass turning optimization problem. The decision variables are: cutting speed in each pass, feed rate in each pass, depth of cut in each pass and number of passes. The model was solved using LINGO software. Kee [45] has carried out optimization for conventional as well as CNC lathes. In conventional lathes, the speeds and feeds have to be provided in steps, whereas in CNC lathes, speeds and feeds are continuous. This changes the optimization strategy for both types of lathes. Yeo [46] has provided methodology for an effective optimization of machining parameters for multipass turning operation. Using an equal depth of cut strategy, he solved the problem using sequential quadratic programming.

Yang and Tarng [47] have used the Taguchi method to find the optimal cutting parameters for turning operations. Lee and Tarng [48] have obtained the optimal cutting parameters for multipass turning operation using polynomial network and sequential quadratic programming. The polynomial network can learn the relationship between cutting parameters (cutting speed, feed and depth of cut) and cutting performance (surface roughness, cutting force and tool life) through a self-organizing adaptive modeling technique.

An interesting observation is that some researchers have employed an unequal depths of cut strategy for multipass rough turning process, considering it to be superior to an equal depth of cut strategy. However, in most cases, it seems reasonable to take equal cutting depth in the roughing passes and reduced depth of cut in finishing pass in order to have a better surface finish. This is the usual shop floor practice. Attainment of an optimum solution with unequal depths of cut in roughing passes may be due to the presence of multiple optimal solutions, which the traditional methods fail to capture.

Recently, there has been a trend of applying non-traditional optimization techniques for optimizing turning operations. Chen and Tsai [49] applied simulated annealing to the problem of turning process optimization. In their work, an optimization algorithm based on simulated annealing and Hooke-Jeeves pattern search is developed for optimization of multipass turning operations. Baykasoğlu and Dereli [50] also made use of simulated annealing. However, they have not taken into consideration the surface roughness constraint. Genetic algorithm (GA) has also been applied for solving the multipass-turning operation. Alberti and Perrone [51] and Onwubolu and Kumalo [52] have used genetic algorithm for solving multipass turning process. By comparing their results with the results of Chen and Tsai [49], Onwubolu and Kumalo concluded that GA significantly outperforms simulated annealing. However, Chen and Chen [53] have shown that Onwubolu and Kumalo have incorrectly handled the machining model. The solutions obtained by them did not satisfy the total depth constraint. As a result, in their solution the stock removal of turning was less than the requirement. Further, by running a code based on genetic algorithm, Chen and Chen [53] showed that genetic algorithm provides no better solution than simulated annealing solution presented in Chen and Tsai [49].

Wang et al. [54] have presented a methodology using GA for the selection of cutting conditions and tool inserts in multipass turning operations. Their objective function incorporated the surface roughness, cutting force, toollife, material removal rate and chip-breakability. The weightage of these performance measures can be controlled by the user. The performance measures are predicted by a hybrid model based on metal cutting theories and interpolation from an experimental database. Saravanan *et al.* [55] have applied genetic algorithm and simulated annealing for the optimization of turning of cylindrical stock into a continuous finished profile consisting of straight turning, facing, taper-turning and circular machining. They found that the simulated annealing produced marginally better results than the genetic algorithm. Using GA, Wang and Jawahir [56] have optimized multipass turning operations, incorporating the effect of progressive tool wear. Satishkumar *et al.* [57] have optimized multipass turning using three non-traditional optimization techniques—genetic algorithm, simulated annealing and ant colony algorithm. Among these, the performance of the ant colony algorithm was found to the best. Abburi and Dixit [58] applied real coded genetic algorithm for the multi-objective optimization of a multipass turning process. The objectives considered by them are the minimization of the cost of machining and maximization of the production rate.

Compared to turning process optimization, there is a lower number of publications on milling process optimization. A few of the papers are being reviewed here. Armarego *et al.* [59] carried out optimization for single pass peripheral milling. The authors considered the constraint on power, torque and feed force. The variable bounds on feed and cutting speed were also considered. The authors concluded that computer-based optimization can help in increasing production rate and reducing production cost. Also, there is a need to reduce loading time and cost for obtaining the maximum benefit. The same authors [60] have also carried out optimization for peripheral and end-milling operations for maximizing the production rate. Multipass milling is found to be superior to single pass milling from the point of view of machining economics. Tolouei-Rad and Bhidendi [61] have carried out optimization of the milling process for maximum production rate, minimum production cost and maximum profit rate. The following constraints were considered:

(1) Maximum machine power,

(2) Maximum allowed surface roughness,

(3) Maximum cutting force permitted by the rigidity of the machine,

(4) Maximum heat generated by cutting,

(5) Available feed rates and spindle speeds on the machine tool.

Sönmez *et al.* [62] have carried out the optimization of multipass slab milling and face milling for maximum production rate. In their method, the optimum number of passes is found *via* dynamic programming and the optimum variables in each pass are found *via* geometric programming. Tandon *et al.* [63] have carried out optimization of the milling process using an evolutionary computing method called Particle Swarm Optimization (PSO). In their work, artificial neural networks are used to model the cutting forces in pocket milling operation. The optimized values of cutting force and feed could reduce the cutting time up to 35%. Wang *et al.* [64] optimized the multipass milling process using genetic algorithms and a combination of genetic and simulated annealing processes. Reddy and Rao [65] developed a response surface model of surface roughness in end milling considering radial rake angle, nose radius, cutting speed and feed rate. Later on, they optimized for surface roughness using a genetic algorithm. The material was medium carbon steel and the operation was carried out at a constant depth of cut. The same authors [66] have used a genetic algorithm to optimize the dry milling process. Ozcelik et al. [67] have optimized the end milling of Inconel 718 using a genetic algorithm. The process was modeled using neural networks. Palanisamy et al. [68] optimized the end milling process using a genetic algorithm. Wang et al. [69] have carried out the multi-objective optimization of the milling process using a combination of genetic algorithm and simulated annealing. In [70], an expert system, incorporating fuzzy reasoning mechanism, has been presented for the purpose of optimizing parameters and predicting performance measures in high-speed milling of hardened AISI D2. This expert system can optimize the parameters in accordance with the objectives of 'maximizing tool life', 'minimizing surface roughness', and also the attainment of both of these simultaneously.

10.3.2 Optimization of Multipass Turning Process

In this section, we shall describe the optimization of multipass turning process. The same procedure can be extended to the machining of other processes, although mathematical models for the prediction of performance parameters will be different in different processes.

10.3.2.1 Objective Function

Consider the multipass turning of a cylindrical work-piece of length L and initial diameter of D_0, the final diameter being D_f. Maximizing the production rate is equivalent to minimizing the total production time per component T_P, which is expressed as

$$T_P = T_{tR} + \frac{t_c T_{tR}}{T_r} + T_{tF} + \frac{t_c T_{tF}}{T_f} + T_L + t_{ts}, \qquad (10.14)$$

where T_{tR} is the total cutting time of rough machining, t_c the time required for changing a tool, T_r the tool life for rough machining, T_{tF} the total cutting time of finish machining, T_f the tool life for finish machining, T_L the loading and unloading time and t_{ts} the tool setting time. Total cutting time for rough machining is obtained as the summation of cutting times for m roughing passes, i.e.,

$$T_{tR} = \sum_{i=1}^{m} t_{r_i} = \sum_{i=1}^{m} \frac{\pi L D_{i-1}}{v_{R_i} f_{R_i}}, \qquad (10.15)$$

where v_{R_i} and f_{R_i} are the cutting speed and feed respectively at the i-th roughing pass and D_{i-1} the work-piece diameter at the beginning of that pass. The cutting time for finish machining is obtained as

$$T_{tF} = \frac{\pi L D_m}{v_F f_F},$$
(10.16)

where v_F and f_F are the cutting speed and feed respectively at the finish pass and D_m is the diameter after m roughing passes.

A rearrangement of Equation 10.14 provides

$$T_P = T_L + t_{ts} + T_{tR} + T_{tF} + t_c \left(\frac{T_{tR}}{T_r} + \frac{T_{tF}}{T_f} \right).$$
(10.17)

In the above expression, the first two terms are independent of the process parameters and thus may be eliminated from the objective function of the optimization problem for single pass machining. In multipass machining, total tool setting time is given as

$$t_{ts} = (m+1)t_s,$$
(10.18)

where t_s is the setting time for each pass. Therefore, if the number of passes is a design variable, this term should be considered in the optimization problem. The attempt should be made to reduce the job loading, job unloading and the tool setting time. The third and fourth terms in the expression for total production time are dependent on the cutting speed, feed and depths of cut. However, they are known functions and are independent of tool-job combination. No modeling of the machining process is required for the estimation of these two terms. They simply state that cutting speed and feed should be as high as possible and number of passes should be kept at a minimum. The last terms associated with tool change time require the expression for tool life as a function of process parameters, *i.e.*, cutting speed, feed and depth of cut. The tool life is dependent on the tool-job combination, machine rigidity and type of coolant. It also matters how one defines a tool life. From the job-quality point of view, a tool is considered failed when it fails to provide the desired surface roughness and dimensional tolerance. The other criterion of tool failure may be based on the cutting forces encountered. However, the most popular criterion of tool life is the amount of maximum allowed flank wear. Excessive flank wear causes increase in cutting force, needs adjustment in tool setting for obtaining an accurate part, weakens the tool and sometimes spoils the surface finish. There is no unique figure for the amount of permissible maximum flank wear. For finish turning by carbide tools, the maximum flank wear should be limited to 0.4 mm. For rough turning, it may go up to 0.6 mm. Even if the process planner knows the tool failure criterion accurately, it is not easy to obtain the mathematical model for tool life. Recourse may be then made to soft computing methods like neural network and fuzzy set theory as outlined in the previous chapter.

The cost of machining per piece is given as

$$F_c = C_0 T_p + C_t \left(\frac{T_{tR}}{T_r} + \frac{T_{tF}}{T_f} \right),$$
(10.19)

where C_0 is the operating cost per minute and C_t is the tool cost. The operating cost consists of overhead, labor, coolant and electricity costs. Substituting the value of T_p from Equation 10.16, the cost of machining per piece can be written as

$$F_c = C_0 \left[T_L + t_{ts} + T_{tR} + T_{tF} + (t_c + C_t / C_0) \left(\frac{T_{tR}}{T_r} + \frac{T_{tF}}{T_f} \right) \right].$$
(10.20)

From Equation 10.17 and Equation 10.20, it is clear that we can write a single expression for the objective functions of minimization of the production time and minimization of the cost of machining as was realized by the previous researchers too [71]. The objective function to be minimized for both the goals is

$$f = t_{ts} + T_{tR} + T_{tF} + t_c^* \left(\frac{T_{tR}}{T_r} + \frac{T_{tF}}{T_f} \right),$$
(10.21)

where

$$t_c^* = \begin{cases} t_c & \text{for minimum production time,} \\ t_c + C_t / C_0 & \text{for minimum cost of machining.} \end{cases}$$
(10.22)

If the ratio C_t / C_0 is small in comparison to tool change time t_c, one can optimize only for minimum production time. However, if the tool is a expensive tool, $e.g.$, CBN or diamond, both optimization goals will yield different solution. One can obtain a number of solutions by solving the optimization problem for different t_c^* in between t_c and $t_c + C_t / C_0$. All these solutions will be Pareto-optimal solutions. A higher level decision can be taken to choose the best solution among them.

Abburi and Dixit [58] have obtained Pareto-optimal solutions in a different way. The method is described here. The authors observed that, in Equation 10.19, the term associated with C_t is basically the fraction of tool getting consumed in the production of one piece. Denoting it by F_t, the production cost can be expressed as

$$F_c = C_0 T_p + C_t F_t.$$
(10.23)

From the above expression, it is observed that for minimizing F_c both T_p and F_t have to be minimized. Thus, the minimization of cost problem can be converted into the following multi-objective problem:

$$\text{Minimize}\,(T_p \text{ and } F_t),\, \text{subject to the applicable constraints}. \qquad (10.24)$$

The above problem does not require the values of C_0 and C_t. As the minimizations of T_p and F_t are two conflicting objectives, one can obtain various Pareto-optimal solutions. The best amongst the Pareto-optimal solutions can be chosen at a later stage with known cost data.

The theoretical minimum possible value of F_t is zero corresponding to an infinite tool life. Thus, Pareto-optimal solutions will cover a range of F_t from 0 to that corresponding to the minimum possible T_p. However, depending on the value of C_0 and C_t, some of these solutions will increase T_p and F_c simultaneously, which is undesirable. A condition when the increase in T_p would reduce cost can be derived as follows.

Writing Equation 10.23 in differential form:

$$dF_c = C_0\, dT_p + C_t\, dF_t, \qquad (10.25)$$

the requirement $dF_c < 0$ along with the fact that $dF_t < 0$ leads to

$$\frac{dT_p}{dF_t} > -\left(\frac{C_t}{C_0}\right). \qquad (10.26)$$

The left hand side of the above inequality is the Lagrange multiplier λ associated with an equality constraint $(F_t = \text{a prescribed value})$ corresponding to the optimization problem of minimizing T_p with the applicable constraints along with this equality constraint. The prescribed value in the equality constraint is the value of F_t at which the condition is being checked. Note that the Lagrange multiplier λ is a negative quantity. Therefore, the F_t should be lowered only till the magnitude of λ is less than C_t / C_0. With the rough estimate of the costs known at lower level, only those Pareto-optimal solutions need to be generated that do not violate this condition. Equation 10.25 can also be written as

$$dF_c = C_0\left(1 + \frac{C_t}{C_0}\frac{dF_t}{dT_p}\right)dT_p = C_0\left(1 + \frac{C_t}{C_0\,\lambda}\right)dT_p. \qquad (10.27)$$

The expression in the parenthesis should be negative in order to reduce production cost at the expense of some increase in the production time. With this expression, one can assess the difference between the minimum cost and the cost corresponding to minimum production time. If the ratio of tool cost to operating cost is very small, the difference between the two costs will also be small.

10.3.2.2 Machining Constraints

Minimization of total production time per component is carried out by imposing the following constraints:

Tool life constraint: Tool life should not be less than a prescribed length of time. Otherwise frequent tool change will be required. The minimum value of tool life should be at least more than the time required to machine one component. Most likely, when we put this constraint, the constraint will become active. Similarly, one can put the constraint on the maximum value of tool life. If the process parameters are such that they produce very high tool life, there are chances that the tool can be thrown away not fully utilized.

Surface roughness constraint: A constraint that surface roughness should not be more than the prescribed value can be put. Sometimes, excessive better surface finish is not desired. For example, for proper convective or pool boiling heat transfer, a certain minimum amount of surface roughness is desired. Therefore, the surface roughness value may be restricted to lie in a zone. For proper implementation of this constraint, a surface roughness prediction model should be available.

Cutting force constraints: The components of cutting forces should be limited to avoid excessive job and tool deflection and breakage of the tool. The tool and job deflection and the stresses in the job can be found by finite element analysis.

Machine power constraint: The machine power can be calculated using the following formula:

$$\text{Machine power} = \frac{\text{main cutting force} \times \text{cutting speed}}{\text{efficiency of the machine}}. \tag{10.28}$$

The machine power should be limited to avoid excessive overloading of the spindle motor. At the same time, if machine power is much lesser than the power of the spindle motor, the machine is underutilized.

Geometric constraint: The final diameter should be equal to initial diameter of the job minus twice the sum of depths of cut, *i.e.*,

$$D_f = D_0 - 2\sum_{i=1}^{m} d_{ri} - 2d_f, \tag{10.29}$$

where d_{ri} is the depth of cut in the *i*-th roughing pass, d_f is the depth of cut in the finishing pass and *m* is the number of passes. This is a geometric relation and should be satisfied.

Variable bounds: Cutting speed, feed and depth of cut should lie within certain ranges. These ranges are dependent on the type of machine, type of tool and type of material.

In addition to these constraints, there is a constraint that the number of passes *m* is an integer quantity.

10.3.2.3 A Note on the Solution of the Optimization Problem

The machining optimization problem is highly non-linear and possesses multiple solutions. For example, consider the results from the paper of Gupta *et al.* [43] in Table 10.2. For each roughing depth of cut, the optimum cost is shown in Table 10.2. If 12 mm diameter is reduced, one can take two roughing passes of depths of cutting 2.1mm and 2.9mm. The combined optimum cost for two passes is (0.555+0.667)= $1.222. One can also have two roughing passes each of 2.5 mm. Then the combined optimum cost will be equal to 2×0.611= $1.222, *i.e.*, the same as before. In this case, equal depth of cut strategy is appropriate. Clearly there are two solutions providing the same cost. In general, there are a number of such solutions.

For the problems having multiple optimum solutions, the genetic algorithms are highly suitable. When the cutting speed and feed can be varied in a continuous manner like in a Computer Numerically Controlled (CNC) machine, the real-coded genetic algorithm can be used. When the cutting speed and feed can have only discrete variation like in a conventional machine tool having a gearbox for changing the speed and feed, one can employ a binary-coded genetic algorithm.

Table 10.2. A part of the results of Gupta *et al.* [43]

Roughing depth of Cut	Cost/piece in $
1.9	0.549
2.0	0.551
2.1	0.555
2.2	0.569
2.3	0.583
2.4	0.597
2.5	0.611
2.6	0.625
2.7	0.639
2.8	0.653
2.9	0.667
3.0	0.681

Genetic algorithms often provide only near optimum solutions. Also, near the optimum solution, these algorithms have a tendency to get slowed down. Therefore, it is better to combine the traditional optimization method with a genetic algorithm. Genetic algorithms can provide various near optimum solutions. Taking a near optimum solution as an initial guess, the final optimum solution can be obtained using a traditional optimization method. The traditional optimization routine can be run with various near optimum solutions as initial guesses, providing different solutions. Abburi and Dixit [58] have combined a sequential quadratic programming method with a real-coded genetic algorithm. A sequential quadratic programming method uses a quadratic model for objective function and a linear model for the constraint. A non-linear programming problem in which the objective function is quadratic and constraints are linear is called a quadratic programming problem. A sequential quadratic programming method solves a quadratic programming problem at each iteration. At each iteration, the objective function is the quadratic approximation and the constraints are the linear approximation of the constraints. In the objective function, instead of putting the Hessian matrix, a positive definite matrix is put which gradually changes to Hessian matrix as the iterations proceed. The quadratic programming problem may be solved efficiently by a simplex method. The solution of a quadratic programming problem provides a search direction. Then a line search method is used to minimize a merit function that incorporates the objective and constraint functions and the optimum point is found. The whole procedure is repeated from this new point till the convergence is achieved.

Sometimes it is better to solve the optimization problem using an equal depth of cut strategy. This reduces the number of variables to six only, *i.e.*, the cutting speed, feed and depth of cut for roughing passes, number of passes and cutting speed and feed for finishing pass. In order to see the sensitivity of the solution with respect to roughing depth of cut, the following procedure may be employed [58]. We first find the optimum cost or production time in a rough pass corresponding to an equal depth of cut d_r. Let this solution be called x_1. Then we find the optimum solutions for $(1-\delta)d_r$ and $(1+\delta)d_r$ depths of cut, where δ is a small number, say 0.9. Let these solutions be called x_2 and x_3. Now, the following quantity is calculated:

$$\varepsilon = \frac{|x_2 + x_3 - 2x_1|}{0.2 d_r x_3}(d_{max} - d_r).$$
(10.30)

If this quantity is very small, say 0.01, we need not go for unequal depth of cut strategy, as only about 1% change is expected in the solution in that case. If the quantity is large, we can go for unequal depth of cut strategy and solve the optimization problem by taking the solution for equal depth of cut strategy as an initial guess.

Availability of a knowledge base in the form of fuzzy "if…, then…." rules can be very helpful in the optimization task. One can immediately reach into the zone where the optima is expected to lie with the help of rules. Further, optimization can be carried out using traditional or non-traditional optimization methods.

The higher level decision for selecting the process parameters can be taken as follows. For each criterion specified by the manufacturer, we can assign membership grades for each criterion. Then we can find the overall membership grade using Equation 10.7. The solution providing the maximum overall membership grade is taken as the best solution.

10.3.3 Online Determination of Equations for Machining Performance Parameters

Optimization methodology can be employed for online determination of machining performance parameters. With modern CNC machines, it is possible to collect machining data in real time. This data can be used to find the relation between process and performance parameters. Ojha and Dixit [72] have suggested a method for obtaining or updating Taylor's tool life exponent based on shop floor data. For the optimization of process parameters, a reliable tool life equation is needed. On the shop floor, the cutting tools are often used in an inefficient manner due to lack of information about tool life. According to a CIRP working paper [73], "….in USA, the correct cutting tool is selected 50% of the time, the tool is used at the rated cutting speed only 58% of the time and the tools are used up to their full tool life capability ….". Non-availability of reliable tool data on tool life for various tool-job combinations is one of the major causes for this. Obtaining the constants in the extended Taylor's tool life relation given by Equation 10.13 requires a considerable number of tests for obtaining reliable results. This is more so because there is a lot of statistical variation in tool life. Carrying out a large number of tests is not only time consuming, but is also expensive. This becomes a major concern when the job material and tools are expensive. It has been reported that constructing tool life curves at two different cutting speeds using ISO turning tests often requires roughly 40 hrs of machining time [74]. Hegginbotham and Pandey [75] proposed a "variable rate running" method in which 20 kg of material was consumed to estimate the dependency of tool life on cutting speed.

Suppose in its full life a tool operates for time t_1 at a particular operating condition providing a tool life of T_1 and for time t_2 at another operating condition providing the tool life of T_2. Then assuming that there is a linear relationship between the tool wear and time, the following relationship holds good:

$$\frac{t_1}{T_1} + \frac{t_2}{T_2} = 1. \tag{10.31}$$

If the data of m tools is available from beginning till failure, then for the j-th tool, the following relationship is valid:

$$\sum_{i=1}^{n^j} \frac{t_i^j}{T_i^j} = 1, \tag{10.32}$$

where superscript indicates tool number, subscript the operating condition and n^j is the number of cutting conditions at which the j-th tool operates. Using Equation 10.13, we can write

$$T_i^j = \frac{C}{(v_i^j)^p (f_i^j)^q (d_i^j)^r}. \tag{10.33}$$

For finding the exponents p, q and r, the following objective function may be minimized:

$$f = \sum_{j=1}^{m} \left(\sum_{i=1}^{n^j} \frac{t_i^j}{T_i^j} - 1 \right)^2 . \tag{10.34}$$

For minimization of the function f, one may use an evolutionary algorithm like GA, which may provide multiple solutions. The best solution can be chosen based on the physical consideration.

The expressions for the estimation of upper and lower estimates of tool life may also be developed based on the shop floor data. For estimating the lower and upper bounds of the tool life, the following method is proposed in this book. Let the upper bound of the tool life be predicted by

$$T^u = \frac{C_1}{v^{p_1} f^{q_1} d^{r_1}}, \tag{10.35}$$

and the lower bound by

$$T^l = \frac{C_2}{v^{p_2} f^{q_2} d^{r_2}}, \tag{10.36}$$

where C_1, C_2, p_1, p_2, q_1, q_2, r_1 and r_2 are the constants dependent on tool-job combination. If the j-th tool operates at n_j different operating conditions, then

$$\sum_{i=1}^{n^j} \frac{t_i^j}{T_i^{uj}} \le 1, \tag{10.37}$$

where t_i^j is the time j-th tool operates at i-th cutting condition and T_i^{uj} is the upper bound of the tool life at that condition. Similarly, we can write

$$\sum_{i=1}^{n^j} \frac{t_i^j}{T_i^{lj}} \ge 1, \tag{10.38}$$

where T_i^{lj} is the lower bound of the tool life for j-th tool operating at i-th cutting condition. The constants C_1, C_2, p_1, p_2, q_1, q_2, r_1 and r_2 should be such that the inequalities in Equations 10.35 and 10.36 are satisfied and the difference in upper and lower estimates of the tool life is kept at a minimum. Thus, treating C_1, C_2, p_1, p_2, q_1, q_2, r_1 and r_2 as design variables, the following optimization problem can be solved:

$$\text{Minimize } \sum_{j=1}^{m} \sum_{i=1}^{n^j} \left(\frac{t_i^j}{T_i^{lj}} - \frac{t_i^j}{T_i^{uj}} \right), \tag{10.39}$$

subject to constraints of Equations 10.37 and 10.38

For most of the commonly used tools and work-materials C_1, C_2, p_1, p_2, q_1, q_2, r_1 and r_2 are positive. Therefore, this constraint can also be imposed.

Example 10.1: Table 10.3 shows the history of eight tools of the same type, which operated at two cutting conditions, before they failed. Find the expression for the lower, upper and most likely estimates of the tool life with this data.

Solution: For finding the most likely estimate, objective function of Equation 10.34 is minimized using FMINCON routine of MATLAB®. In this case $m=8$ and $n'=2$. The following tool life equation is obtained:

$$vT^{0.2}f^{0.15}d^{0.2} = 273. \tag{10.40}$$

For finding the upper and lower estimates of tool life, Equation 10.39 was taken as the objective function and Equations 10.37 and 10.38 as the constraints. MATLAB® routine FMINCON was used for minimization. The lower estimate is given by

$$vT^{0.17}f^{0.13}d^{0.19} = 260. \tag{10.41}$$

The upper estimate is given by

$$vT^{0.2}f^{0.15}d^{0.16} = 285. \tag{10.42}$$

Table 10.3. Machining history of eight cutting tools

Tool No.	Cutting speed (m/min)	Feed (mm/rev)	Depth of Cut (mm)	Machining time (min)
1	180	0.14	0.6	43.60
	225	0.14	0.9	3.80
2	190	0.20	0.6	18.38
	200	0.16	0.9	9.80
3	185	0.20	0.7	18.33
	195	0.14	0.8	15.19
4	210	0.16	0.6	14.87
	220	0.14	0.9	4.99
5	200	0.16	0.6	20.22
	180	0.20	0.7	14.82
6	190	0.14	0.7	24.55
	185	0.16	0.8	13.61
7	210	0.20	0.7	9.76
	220	0.16	0.6	9.99
8	225	0.20	0.6	10.29
	195	0.14	0.9	6.38

The similar type of strategy can be used for finding the empirical expressions for cutting forces and surface roughness in a given range. The form of the function for cutting force and surface roughness is

$$f = kf^\alpha v^\beta d^\gamma . \tag{10.43}$$

If the experimental data is available, constants k, α, β and γ can be obtained by solving an optimization problem. In this case, we can take the logarithmic of the function and also solve the problem by multiple linear regression.

10.4 Summary

In this chapter we have discussed the optimization problems in metal forming and metal cutting. For solving optimization problems, one can employ the neural network, fuzzy set and genetic algorithm. Some optimization problems of metal forming have been discussed. To give a glimpse of many other optimization problems, a brief review of the literature has been presented. In machining optimization, the most common optimization problem is finding the feed, depth of cut, cutting speed and number of machining passes for maximizing the production time and/or minimizing the cost of machining. A brief review of the optimization of turning and milling processes has been presented. Finally, the optimization of multipass turning has been discussed in detail. The reader can easily extend the optimization procedure to other machining processes.

10.5 References

[1]   Rao, S.S. (1996), Engineering Optimization–Theory and Practice, John Wiley and Sons Inc. and New Age International (P) Ltd. Publishers, New Delhi.
[2]   Arora, J.S. (1989), Introduction to Optimum Design, McGraw-Hill, New York.
[3]   Deb, K. (1998), Optimization for Engineering Design: Algorithms and Examples, Prentice-Hall of India, New Delhi.
[4]   Avitzur, B. (1962), Pass reduction schedule for optimum production of a hot strip mill, Iron Steel Eng., Dec., pp. 104–114.
[5]   Brayant, G.F. and Spooner, P.D. (1973), On-line adoption of tandem mill schedules, Automation of Tandem Mills, Brayant G.F., ed., The Iron and Steel Institute, London.
[6]   Brayant, G.F., Halliday, J.M. and Spooner, P.D. (1973), Optimal scheduling of a tandem cold rolling mill, Brayant G.F., ed., The Iron and Steel Institute, London.
[7]   Dixit, U.S. and Dixit, P.M. (2000), Application of fuzzy set theory in the scheduling of a tandem cold-rolling mill, Transactions of ASME, Journal of Manufacturing Science and Engineering, Vol. 122, pp. 494–500.
[8]   Wang, D.D., Tieu, A.K., de Boer, F.G., Ma, B. and Yuen, W.Y.D. Yuen, (2000), Towards a heuristic optimum design of rolling schedules for tandem cold rolling mills, Engineering Applications of Artificial Intelligence, Vol. 13, pp. 397–406.
[9]   Sekulic, M. R. and Alexander, J. M. (1963), A theoretical discussion of the automatic control of multi-stand tandem cold rolling mills, International Journal of Mechanical Sciences, Vol. 5, pp. 149–163.

[10] Avitzur, B., Van Tyne, C.J. and Turczyn, S. (1988), The prevention of central burst during rolling, Transactions of ASME, Journal of Engineering for Industry, Vol. 110, pp. 173–178.

[11] Zhu, Y.D. and Avitzur, B. (1988), Criterion for the prevention of split ends, Transactions of ASME Journal of Engineering for Industry, Vol. 110, pp. 162–172.

[12] Roberts, W.L. (1978), Cold Rolling of Steel, Marcel Dekker, New York.

[13] Antonsson, E.K. and Otto, K.N. (1995), Imprecision in engineering design, Transactions of ASME Journal of Mechanical Design, Vol. 117(B), pp. 25–32.

[14] Joun, M.S. and Hwang, S.M. (1993), Pass schedule optimal design in multi-pass extrusion and drawing by finite element method, International Journal of Machine Tools & Manufacture, Vol. 33, pp. 713–724.

[15] Celano, G., Fichera, S., Fratini, L. and Micari (2001), The application of AI techniques in the optimal design of multi-pass drawing processes, Journal of Materials Processing Technology, Vol. 113, pp. 680–685.

[16] Balaji, P.A., Sundararajan, T. and Lal, G.K. (1991), Viscoplastic deformation analysis and extrusion die design by FEM, Transactions of ASME, Journal of Applied Mechanics, Vol. 58, pp. 644–650.

[17] Reddy, N.V., Dixit, P.M. and Lal, G.K. (1997), Die design for axisymmetric hot extrusion, International Journal of Machine Tools & Manufacture, Vol. 37, pp. 1635–1650.

[18] Lee, S.K., Ko, D.C. and Kim, B.M. (2000), Optimal die profile design for uniform microstructure in hot extrusion product, International Journal of Machine Tools & Manufacture ., Vol. 40, pp. 1457–1478.

[19] Yada, H. and Senuma, T. (1986), Resistance to hot deformation of steel, Journal of JSTP, Vol. 27, pp. 34–44.

[20] Lin, Z., Juchen, X., Xinyun, W., Guoan, H. (2003), Optimization of die profile for improving die life in the hot extrusion process, Journal of Materials Processing Technology, Vol. 142, pp. 659–664.

[21] Ulysse, P. (2002), Extrusion die design for flow balance using FE and optimization methods, International Journal of Mechanical Sciences, Vol. 44, pp. 319–341.

[22] Yan, H., Xia, J. (2006), An approach to the optimal design of technological parameters in the profile extrusion process, Science and technology of Advanced materials, Vol. 7, pp. 127–131.

[23] Kleinermann, J. and Ponthot, J. (2003), Parameter identification and shape/process optimization in metal forming simulation, Journal of Materials Processing Technology., Vol. 139, pp. 521–526.

[24] Zhao, G., Ma, X., Zhao, X., Grandhi, R.V. (2004), Studies on optimization of metal forming processes using sensitivity analysis methods, Journal of Materials Processing Technology, Vol. 147, pp. 217–228.

[25] Antonio, C.C., Castro, C.F., Sousa, L.C. (2004), Optimization of metal forming processes, Computers and Structures, Vol. 82, pp. 1425–1433.

[26] Khoury, I., Giraud-Moreau, L., Lafon, P., Labergere, C. (2006), Towards an optimization of forging processes using geometric parameters, Journal of Materials Processing Technology, Vol. 177, pp. 224–227.

[27] Chengzhi, S., Guanlong, C., Zhongqin, L. (2005), Determining the optimum variable blank-holder forces using adaptive response surface methodology (ARSM), International Journal of Advanced Manufacturing Technology, Vol. 26, pp. 23–29.

[28] Liu, Q., Liu, W., Ruan, F., Qiu, H. (2007), Parameters' automated optimization in sheet metal forming processes, Journal of Materials Processing Technology, Vol. 187-188, pp. 159–163.

[29] Tang, B., Sun, J., Zhao, Z., Chen, J., Ruan, X. (2006), Optimization of drawbead design in sheet forming using one step finite element method coupled with response

surface methodology, International Journal of Advanced Manufacturing Technology, Vol. 31, pp. 225–234.

[30] Hu, J., Peng, Y., Li, D., Yin, J. (2007), Robust optimization based on knowledge discovery from metal forming simulation, Journal of Materials Processing Technology, Vol. 187–188, pp. 698–701.

[31] Jansson, T., Nilsson, L. (2006), Optimizing sheet metal forming processes- using a design hierarchy and response surface methodology, Journal of Materials Processing Technology, Vol. 178, pp. 218–233.

[32] Sattari, H., Sedaghati, R., Ganeshan, R. (2006), Analysis and design optimization of deep drawing process Part II: Optimization, Journal of Materials Processing Technology, Vol. 184, pp. 84–92.

[33] Shi, X., Chen, J., Peng, Y., Ruan, X. (2004), A new approach of die shape optimization for sheet metal forming processes, Journal of Materials Processing Technolog., Vol. 152, pp. 35–42.

[34] Bate, P. S., Ridley, N., Zhang, B., Dover, S. (2006), Optimization of the superplastic forming of aluminium alloys, Journal of Materials Processing Technology, Vol. 177, pp. 91–94.

[35] Parsa, M.H., Pournia, P. (2007), Optimization of initial blank shape predicted based on inverse finite element method, Finite Elements in Analysis and Design, Vol. 43, pp. 218–233.

[36] Sosnowski, W., Marczewska, I., Marczewski, A. (2002), Sensitivity based optimization of sheet metal forming tools, Journal of Materials Processing Technology, Vol. 124, pp. 319–328.

[37] Gantar, G., Pepelnjak, T., Kuzman, K. (2002), Optimization of sheet metal forming processes by the use of numerical simulations, Journal of Materials Processing Technology, Vol. 130-131, pp. 54–59.

[38] Gilbert, W.W. (1950), Economics of machining. In Machining Theory and Practice. American Society of Metals, 465–485.

[39] Ermer, D.S. (1971), Optimization of constrained machining economics problem by geometric   programming, Transaction of the ASME, Journal of Engineering for Industry, Vol. 93, pp. 1067–1072.

[40] Petropoulos, P.G. (1973), Optimal selection of machining rate variables by geometric programming, International Journal of Production Research, Vol.11, pp. 305–314.

[41] Lambert, P.K. and Walvekar, A.G. (1978), Optimization of multi-pass machining operations, International Journal of Production Research, Vol. 16, pp. 247–259.

[42] Shin, Y.C. and Joo, Y.S. (1992), Optimization of machining condition with practical constraints, International Journal of Production Research, Vol. 30, pp. 2907–2919.

[43] Gupta, R., Batra, J.L. and Lal, G.K. (1995), Determination of optimal subdivision of depth of cut in multipass turning with constraints. International Journal of Production Research, Vol. 33, pp. 2555–2565.

[44] Al-Ahmari, A.M.A. (2001), Mathematical model for determining machining parameters in multipass turning, International Journal of Production Research, Vol. 39, pp. 3367–3376.

[45] Kee, P.K. (1994), Development of computer-aided machining optimization for multi-pass rough turning operations, International Journal of Production Economics, Vol. 37, pp. 215–227.

[46] Yeo, S.H. (1995), A multipass optimization strategy for CNC lathe operations, International Journal of Production Economics, Vol. 40, pp. 209–218.

[47] Yang, W.H. and Tarng, Y.S. (1998), Design optimization of cutting parameters for turning operation based on the Taguchi method, Journal of Material Processing Technology, Vol. 84, pp. 122–129.

[48] Lee, B.Y. and Tarng, Y.S. (2000), Cutting parameter selection for maximizing production rate or minimizing production cost in multistage turning operations, Journal of Material Processing Technology, Vol. 105, pp. 61–66.

[49] Chen, M.C. and Tsai, D.M. (1996), A simulated annealing approach for optimization of multipass turning operation, International Journal of Production Research, Vol. 34, pp. 2803–2825.

[50] Baykasoğlu, A. and Dereli T. (2002), Novel algorithm approach to generate the 'number of passes' and 'depth of cuts' for the optimization routines of multipass machining, International Journal of Production Research, Vol. 40, pp. 1549–1565.

[51] Alberti, N. and Perrone, G. (1999), Multipass machining operations by using fuzzy possibilistic programming and genetic algorithm, Proceedings of the Institution of Mechanical Engineers, Journal of Engineering Manufacture, Vol. 213, pp. 261–273.

[52] Onwubolu, G.C. and Kumalo, T. (2001), Optimization of multipass turning operations with genetic algorithm, International Journal of Production Research, Vol. 39, pp. 3727–3745.

[53] Chen, M.C. and Chen, K.Y. (2003), Optimization of multi-pass turning operations with genetic algorithms: a note, International Journal of Production Research, Vol. 41, pp. 3385–3388.

[54] Wang, X., Da, Z.J., Balaji, A.K. and Jawahir, I.S. (2002), Performance-based optimal selection of cutting conditions and cutting tools in multi-pass turning operations using genetic algorithms, International Journal of Production Research, Vol. 40, pp. 2053–2065.

[55] Saravanan, R., Asokan, P. and Vijayakumar, K. (2003), Machining parameters optimization for turning cylindrical stock into a continuous finished profile using genetic algorithm (GA) and simulated annealing (SA), International Journal of Advanced Manufacturing Technology, Vol. 21, pp. 1–9.

[56] Wang, X. and Jawahir, I.S. (2005), Optimization of multi-pass turning operations using genetic algorithms for the selection of cutting conditions and cutting tools with tool wear effect, International Journal of Production Research, Vol. 43, pp. 3543–3559.

[57] Satishkumar, S., Asokan, P. and Kumanan, S. (2006), Optimization of depth of cut in multi-pass turning using nontraditional optimization techniques. International Journal of Advanced Manufacturing Technology, Vol. 29, Vol. 230–238.

[58] Abburi, N.R. and Dixit, U.S. (2007), Multi-objective optimization of multipass turning processes, International Journal of Advanced Manufacturing Technology, Vol. 32, pp. 902–910.

[59] Armarego, E.J.A, Smith, A.J.R. and Wang, J. (1993), Constrained optimization strategies and CAM software for single pass peripheral milling, International Journal of Production Research, Vol. 31, pp. 2139–2160.

[60] Armarego, E.J.A., Smith, A.J.R. and Wang, J. (1994), Computer-aided constrained optimization analyses and strategies for multi-pass helical tooth milling operation, Annals of the CIRP, Vol. 43, pp. 437–442.

[61] Tolouei-Rad, M. and Bhidendi I.M. (1997), On the optimization of machining parameters for milling operations, International Journal of Machine Tools & Manufacture, Vol. 37, pp. 1–16.

[62] Sönmez, A.İ, Baykasoğlu, A., Dereli, T. and Filiz, İ.H. (1999), Dynamic optimization of multipass milling operation via geometric programming, International Journal of Machine Tools & Manufacture, Vol. 39, pp. 297–320.

[63] Tandon, V., El-Mounayri, H. and Kishawy, H. (2002), NC end milling optimization using evolutionary computation, International Journal of Machine Tools & Manufacture, Vol. 42, pp. 595–605.

[64] Wang, Z. G., Wong, Y. S. and Rahman, M. (2004), Optimization of multi-pass milling using genetic algorithm and genetic simulated annealing, International Journal of Advanced Manufacturing Technology, Vol. 24, pp. 727–732.

[65] Reddy, N. S. K. and Rao, P.V. (2005), Selection of optimum tool geometry and cutting conditions using a surface roughness prediction model for end milling, International Journal of Advanced Manufacturing Technology,, Vol. 26, pp. 1202–1210.

[66] Reddy, N. S. K., Rao, P.V. (2006), Selection of an optimal parametric combination for achieving a better surface finish in dry milling using genetic algorithms, International Journal of Advanced Manufacturing Technology, Vol. 28, pp. 463–473.

[67] Ozcelik, B., Oktem, H. and Kurtaran, H. (2005), Optimum surface roughness in end milling Inconel 718 by coupling neural network model and genetic algorithm, International Journal of Advanced Manufacturing Technology, Vol. 27, pp. 234–241.

[68] Palanisamy, I. Rajendran and S. Shanmugasundaram (2007), Optimization of machining parameters using genetic algorithm and experimental validation for end-milling operations, International Journal of Advanced Manufacturing Technology, Vol. 32, pp. 644–655.

[69] Wang, Z.G., Wong, Y.S., Rahman, M. and Sun, J. (2006), Multi-objective optimization of high-speed milling with parallel genetic simulated annealing, International Journal of Advanced Manufacturing Technology, Vol. 31, pp. 209–218.

[70] Iqbal, A., He, N., Li, L. and Dar, N.U. (2007), A fuzzy expert system for optimizing parameters and predicting performance measures in hard-milling process, Expert Systems with Applications, Volume 32, pp. 1020–1027.

[71] Kiliç, S.E. (1985), Use of one-dimensional search method for the optimization of turning operations, Modeling, Simulation and Control B, Vol. 14, pp. 39–63.

[72] Ojha, D.K., and Dixit, U.S. (2005), Economic and reliable tool life estimation procedure for turning, International Journal of Advanced Manufacturing Technology, Vol. 26, pp. 726–732.

[73] Armarego, E.J.A., Jawahir I.S., Osafaviev, V.A. and Patri, V.K. (1996), Modeling of machining operations. In working paper STC 'C' working group. Hallway Publishers, Paris, France.

[74] Stephenson, D.A. and Agapiou, J.S. (1997), Metal Cutting Theory and Practice, Dekker, New York.

[75] Hegginbotham, W.B. and Pandey, P.C. (1967), A variable rate machining test for tool life evaluation, Proc. 8-th International MTDR Conference, Manchester, September 1967.

11

Epilogue

The objective of this book is to describe basic fundamentals in the modeling of metal forming and machining processes. The techniques described in the book will also be useful for the modeling of other manufacturing processes. In the last 10 chapters we have discussed physics-based finite element modeling as well as soft computing-based modeling.

Chapter 1 provides the introduction of metal forming and machining processes, and issues and challenges involved in modeling these processes. In machining, the major concern is to choose the process variables (like cutting speed, feed, depth of cut, tool geometry *etc.*) so as to obtain the desired shape and surface finish with minimum cost and/or maximum production rate. On the other hand, in metal forming processes, the process variables (die geometry, pass scheduling, initial blank shape *etc.*) need to be chosen so as to obtain a defect-free product at minimum power consumption.

The metal forming and machining processes involve plastic deformation of the work-piece due to stresses applied by some form of tools or dies. In view of this, first, the review of the concept of the stress at a point has been provided in Chapter 2. The stress vector (*i.e.*, traction vector), the stress tensor and the relation between them (*i.e.*, the Cauchy's relation) have been described. Concepts like the principal invariants of the stress tensor, the decomposition of the stress tensor into the hydrostatic and deviatoric parts *etc.*, which are useful for Chapter 3, have also been discussed. Next, the equations of motion have been presented. Even though metal forming and machining processes involve large deformation and are governed by the plastic constitutive equations (*i.e.*, the stress-strain relations), it is instructive to discuss, first, a measure of small deformation (*i.e.*, infinitesimal or linear strain tensor) and linear elastic stress-strain relations for isotropic materials. This material has been provided at the end of the chapter. This chapter also introduces tensors and index notation.

Chapter 3 describes the classical theory of plasticity. Before describing the mathematical relations of plasticity, it discusses experimental observations on plasticity that provide the basis for these relations. Then it covers two yield criteria for isotropic materials: von Mises and Tresca criteria. Next it discusses two measures of plastic deformation: incremental linear strain tensor and the strain rate tensor. The concept of isotropic hardening, *i.e.*, the change in the size of the yield surface due to deformation, has been described next. Two models of isotropic hardening, namely strain hardening and work hardening, have been presented. One way to develop the plastic constitutive equation is to postulate the existence of a plastic potential. Then the flow rule gives the plastic part of the constitutive equation. Assuming that the elastic and plastic parts of the deformation are additive, the incremental stress-strain and the stress-strain rate relations for isotropic materials have been developed. An unloading criterion has been presented. The deformation in metal forming and machining processes is usually accompanied by a large rotation. Therefore, the constitutive equation must be objective, *i.e.*, it must be independent of the rotation. A commonly used measure of objective stress rate, namely the Jaumann stress rate, has been described. Two commonly used formulations of metal forming and machining processes are: updated Lagrangian formulation and Eulerian formulation. Examples of updated Lagrangian formulation and Eulerian formulation in metal forming and Eulerian formulation in machining have been provided. The initial and boundary conditions for these examples are also discussed.

Chapter 4 describes the plasticity of finite deformation. Incremental strain and strain-rate measures for finite deformation have been introduced. The decomposition of these measures into elastic and plastic parts, which are not additive as in the case of small deformations, is described. The modifications in the constitutive equations (both the incremental as well as the rate forms) due to finite deformation have been presented. The chapter also discusses anisotropic yield criteria and the corresponding anisotropic constitutive equations. Some commonly used anisotropic yield criteria like those of Hill and Barlat and his co-workers have been discussed. An anisotropic plane strain criterion has been presented. Then two simple models of kinematic hardening (*i.e.*, the translation of the yield surface without change in shape or size) have also been discussed. It is now well-established that ductile fracture is caused by micro-void nucleation, growth and coalescence. Some ductile fracture models based on this observation, namely the Berg-Gurson model, Goods and Brown void nucleation model, Rice and Tracy void growth model, Thomason void coalescence model and Lemaitre's continuum damage mechanics model have been presented. Some empirical fracture criteria, which are not necessarily based on the above observation but are simple to use, are also described. Finally, some friction models like the friction factor model, the Coulomb's law model and a more general model of Wanheim and Bay have been discussed. Thus the first four chapters provide the background for physics based modeling of the metal forming and machining processes.

Chapter 5 provides examples of finite element modeling of metal forming processes using Eulerian formulation. The background to finite element method is provided in this chapter. As an example of plane strain problems, cold flat rolling is chosen. Similarly, the wire drawing process is chosen as an example of axisymmetric processes. In both of these problems, the constitutive equation of a rigid-plastic isotropic material is used. (Thus, the elastic and anisotropic effects are neglected.) An iterative scheme to solve the non-linear finite element equations is presented. First, the finite element models are validated by comparing the predicted roll force, roll torque and drawing stress with well-known experimental results.

Then the deformation fields and parametric studies with respect to important process variables have been presented for both the examples. The chapter also discusses three-dimensional modeling of metal forming processes. The incorporation of anisotropic effects is also described. Further, some techniques of including the elastic effects and estimating the residual stresses are also discussed. This chapter makes use of the background of Chapter 3.

Chapter 6 provides the examples of finite element modeling using updated Lagrangian procedure. First, using the background material of Chapter 4, the incremental elasto-plastic finite element equations corresponding to a finite increment size are developed. Since these are non-linear equations, the Newton-Raphson iterative scheme for solving them is presented. The forging of a cylindrical block and deep drawing of a cylindrical cup have been taken as an examples of axisymmetric and three-dimensional problems. For the forging problem, the material is assumed to be isotropic. First the finite element model is validated by comparing the predicted forging load variation with experimental results. Then, for typical process conditions, the contact pressure distribution, the deformed shape and the deformation and stress fields are presented. The residual stresses at the end of a forging process are determined and their parametric study with respect to important process variables is carried out. For the deep drawing problem, the material is assumed to be anisotropic. As a result, the process is considered as three-dimensional. First the finite element model is validated by comparing the cup height variation with experimental results. Then, for typical process variables, the deformed shape, the punch load variation and thickness strain variation are presented. Their parametric study with respect to important process variables is carried out. For anisotropic materials, the height of the drawn cup becomes non-uniform, i.e., the cup develops ears. A technique to optimize the initial blank shape, so as to reduce the earing, is described.

Chapter 7 discusses the finite element modeling of machining processes. Only orthogonal machining is considered. The depth of cut is assumed large so as to make the problem two-dimensional. The temperature rise in the machining process is quite significant. The thermo-mechanical problem is decoupled by assuming that the average temperature in the cutting zone is to be estimated either experimentally or by simple analytical techniques. Temperature softening is incorporated by estimating the material properties at this temperature. To account for high strain rates, visco-plasticity effects are included in the constitutive equation. Eulerian formulation is used by assuming that the shear angle, the cutting ratio and the angle between the shear force and the resultant force are determined experimentally. Using the background material of Chapter 3, first the finite element equations are developed. These equations are non-linear. The iterative technique to solve them is similar to that of Chapter 5. Then the finite element model is validated by comparing the predicted average width of the primary shear deformation zone (PSDZ), the average shear strain rate and the average shear stress with experimental results. Next the PSDZ and the deformation and stress fields are presented for typical process conditions. Finally the parametric study of the average shear strain rate and average shear stress is carried out with respect to two important machining variables, namely, the cutting speed and feed.

Chapter 8 provides the background of neural networks, fuzzy set theory and genetic algorithms. Two types of neural networks are described in detail—multi-layer perceptron neural network and radial basis function neural network. The advantage of radial basis function neural network is its faster training time, although in general it requires more training data compared to a multi-layer perceptron neural network. It is also possible to predict the range of the dependent variables using neural networks. Neural networks are useful tools for learning from data. However, they function like a black box and are poor in extrapolation. Fuzzy set theory can be employed to deal with uncertainty and imprecision in data. Fuzzy set theory makes use of linguistic variables. Thus, it is often called "computing with language". Prediction of a variable can be done having a fuzzy rule base. As there are rules showing the dependence of a variable on independent variables, the fuzzy set-based prediction is transparent unlike the black box prediction of neural networks. The prediction by fuzzy set is very convenient when the rule base is available. Some extrapolation may also be tolerated. However, when the rule base is not available, one has to generate rules from the data. Generating the rules requires an enormous amount of data. One strategy is to use neural networks for learning from the data and fuzzy sets for making an expert system for prediction. A number of researchers have used the combination of fuzzy sets and neural networks in a number of different ways.

Genetic algorithms fall into the category of evolutionary optimization methods. The evolutionary optimization methods are heuristic-based techniques, in which optimal solution evolves from iteration to iteration. There are a number of evolutionary optimization techniques such as Ant Colony Optimization, Particle Swarm Optimization, Scatter Search *etc*. It is to be mentioned that some researchers consider only genetic algorithms as the evolutionary optimization procedures, whilst others treat genetic algorithms as a type of evolutionary optimization procedure. Hertz and Kobler [1] define evolutionary optimization procedures as iterative solution techniques that handle a population of individuals and make them evolve according to rules that have to be clearly specified. At each iteration, individuals evolve independently and also due to exchange of information among individuals. This definition includes a gamut of optimization methods in the category of evolutionary optimization methods. However, in this book, we have discussed only genetic algorithms to provide a more focused treatment for application of soft computing techniques to the modeling and optimization of metal forming and machining processes.

Chapter 9 presents examples of application of soft computing methods in machining and metal forming. An introduction to design of experiments and certain statistical techniques has been provided. It is always useful to find statistical properties of data and use this information in further processing of data using soft computing techniques. We have emphasized the need to have a good dataset as well as a well-designed model. Without providing a rigorous mathematical analysis, this chapter mentions the key points to be considered during the process-modeling using soft computing techniques. We have covered neural networks, fuzzy sets and neuro-fuzzy modeling of machining and metal forming processes. A section on computation with fuzzy variables is included with a discussion on a fuzzy reliability measure.

Chapter 10 describes the optimization of metal forming and machining processes. A brief review of research work in this area is presented. The emphasis in this chapter is on formulating the optimization problems for different processes. The issues involved in the solution algorithms have been discussed very briefly. This chapter also discusses multi-objective problems and the associated concept of the Pareto-optimality. The use of fuzzy set theory and neural networks for optimization has been highlighted.

It is expected that this book will serve as a foundation for taking up further computational work in metal forming and machining areas. This could be in the form of finite element and soft computing modeling of other metal forming and machining processes not covered in this book. Apart from this, the book can provide a direction for computational research in other areas of manufacturing processes including non-traditional ones.

11.1 References

[1]   Hertz, A. and Kobler, D. (2000), A framework for the description of evolutionary algorithms, European Journal of Operational Research, Vol. 126, pp. 1–12.

Index

A

aggregation of rules, 489
alternative hypothesis, 510
AND gate, 455
ANFIS, 535
anisotropy, 107
    normal, 223
    planar, 223
    Hill's yield criteria, 226
    Barlat and lion yield criteria (plane
    stress), 227–229
    Barlat and co-workers's yield
    criteria (three dimensional),
        229– 234
    plane-strain (anisotropic yield
    criteria), 236–239
ANOVA, 515
ANOVA table, 517
ant colony optimization, 562
area void fraction, 258
associated flow rule, 143
axon, 453

B

back propagation algorithm, 462
balanced bi-axial test, 224– 225
batch learning, 462
Bauschinger effect, 104
bending, 19–20
biological neural networks, 453

blank-holder force, 557
blanking, 20–22
body force vector, 61
broaching, 30
bulk metal forming, 2
bulk modulus, 92

C

Cauchy's relation, 45
central bursting, 13, 551
characteristic function, 473
chevron cracking, 13
clipping, 488
coefficient of determination, 519–520
compatibility conditions, 82–83
competitive learning, 472
confidence level, 511
conflicting rule, 534
consequent parameter, 537
constitutive equations
    elastic part, 207–208
    plastic part, 208–210
contact algorithm, 405
continuum damage mechanics models,
    258
    critical damage criterion of Dhar
    et al., 262
    Rousselier's model, 262
convexity, 114, 475
correlation analysis, 508